Trends in Control and Decision-Making for Human–Robot Collaboration Systems

Yue Wang · Fumin Zhang
Editors

Trends in Control and Decision-Making for Human–Robot Collaboration Systems

Springer

Editors
Yue Wang
Department of Mechanical Engineering
Clemson University
Clemson, SC
USA

Fumin Zhang
School of Electrical and Computer Engineering
Georgia Institute of Technology
Atlanta, GA
USA

ISBN 978-3-319-40532-2 ISBN 978-3-319-40533-9 (eBook)
DOI 10.1007/978-3-319-40533-9

Library of Congress Control Number: 2016959746

Printed on acid-free paper

This Springer imprint is published by Springer Nature
The registered company is Springer International Publishing AG
The registered company address is: Gewerbestrasse 11, 6330 Cham, Switzerland

To our advisors, students, and family

Preface

This book aims to present new developments in control and decision-making theory in the field of human–robot interaction (HRI). Despite advances in robotics, sensing, and autonomy, human participation is still an essential component in various operations, especially under uncertain and complex environments. In particular, human–robot collaboration (HRC) systems integrate the strengths of human in terms of high-level decision-making, flexibility, dexterity, vision processing, etc., with robot's capabilities of performing repetitive tasks in dangerous situations in order to realize the full potential of autonomous systems. Therefore, a considerable amount of effort has been made in this area, which however still lacks qualitative analysis, performance prediction, and guarantees. This offers little insight into the effective control and decision-making for the collaboration operations, especially in real time.

To fill the gap, the chapters in this book discuss in detail the development of new control and decision-making algorithms for HRC systems for guaranteed joint human–robot system performance. Chapter 1 provides an overview of the extant works in HRI and detailed introduction of main contributions of each chapter. Chapters 2–9 present methods for one human–robot pair collaboration and Chaps. 10–16 develop the control and coordination algorithms for humans to collaborate with multiple robots and swarms. Both physical (e.g., force, vision) and psychological human factors (e.g., trust and regret) are embedded into control and decision-making approaches such as nonlinear control, shared control, switching control, optimal control, and sequential detection. The considered applications include transportation, healthcare, manufacturing, and defense. Robot experiments and simulations with human-in-the-loop are conducted to demonstrate the effectiveness of the proposed algorithms.

Clemson, SC, USA — Yue Wang
Atlanta, GA, USA — Fumin Zhang
April 2016

Acknowledgements

The editors would like to thank all the chapter authors and reviewers that work together on this book. A special acknowledgment goes to all the graduate students in the Interdisciplinary and Intelligent Research (I^2R) Laboratory in the Mechanical Engineering Department at Clemson University, whom assisted the editors to review chapters and provided useful feedbacks to improve the quality of the book. The first editor would like to thank the support from the National Science Foundation under Grant No. CMMI-1454139. The editors would also like to thank Springer Verlag and its staff for the professional support.

Contents

Chapter 1
Introduction

Yue Wang and Fumin Zhang

1.1 Overview

Despite substantive advances in autonomous robots, human participation is still an essential component in sectors such as manufacturing [106], structure health monitoring [54], robotics [29], transportation [42], healthcare [40], and aerospace [1]. Most robots are brittle and need human intervention to handle situations they are not designed/programmed to handle. Meanwhile, humans are superior at high-level decision-making and offer a significant advantage in performing cognitive tasks in complex, dynamic, and uncertain environments [31, 35]. On the other hand, humans are not good at processing large volumes of data, nor of sustaining attention for long periods of time. Extensive manpower involved in repetitive tasks is tedious, burdensome to workforces, inefficient, and affects adversely on worker's health and safety [88]. For example, in the manufacturing sector, the combination of machines' ability to conduct precise and repeatable operations with humans' ability to see, feel, touch, and think will increase efficiency, quality, dexterity, and a system's capability to handle emergencies, thus relieving assemblers and fabricators working in manufacturing plants from unpleasant and burdensome jobs [23, 93, 106, 120]. Lightweight collaborative robots such as Rethink Robotics Baxter [45], Universal Robotics UR5 and UR10 [83], KUKA iiwa [57], Kinova JACO, and MICO [7] have been developed to work side by side with human workers in assembly lines [89]. In the critical infrastructure sector, structural damage can occur at unexpected places and times, causing a public safety threat. The utilization of human supervised mobile sensor

Y. Wang (✉)
Department of Mechanical Engineering, Clemson University,
Clemson, SC 29634, USA
e-mail: yue6@clemson.edu

F. Zhang
School of Electrical and Computer Engineering, Georgia Institute
of Technology, Atlanta, GA 30332, USA
e-mail: fumin@ece.gatech.edu

Y. Wang and F. Zhang (eds.), *Trends in Control and Decision-Making for Human–Robot Collaboration Systems*, DOI 10.1007/978-3-319-40533-9_1

networks enables adaptive structural health monitoring and real-time judgment in places where it is impractical or dangerous to humans [34, 54]. In the transportation systems sector, V2X and machine learning technologies have enabled connected and automated driving. Human becomes the supervisor who can intervene in the operation and is responsible for the action of the joint driver-car system [32, 110]. In the healthcare sector, assistive robotics, rehabilitation robotics, and telesurgery techniques made significant progress to better assist patients and doctors [21, 91]. In the space sector, ground personnel, crew in orbit, and astronauts need to supervise and assist telerobots (e.g., Robonaut 2 (R2), Astrobee [38]) for more effective human–robot interaction (HRI) to improve the efficiency and productivity of human exploration missions [37, 58]. Potential military applications include patrol and surveillance, reconnaissance, motion planning, trajectory tracking, and combat applications, where it is dangerous for human to perform the tasks solely and most beneficial under collaborations between humans and robots [94, 95, 115, 116].

Because all autonomous robots are joint human–machine systems, which are supervised by humans to some degree, the best capabilities result from the coordination and collaboration of humans and robots. Much research has been conducted on the development of HRI [43, 81], and its sister fields, human–computer interaction (HCI) [80, 107], and human–machine interaction (HMI) [18, 90]. Interested readers can refer to books [59, 77, 101, 102], journal special issues [47, 69], theses [24, 25, 31, 39, 111, 114], and conference proceedings [2, 41, 66, 67, 73]. Given the multidisciplinary nature, we give a brief summary below for the state-of-art of relevant works in systems and controls, artificial intelligence (AI), robotics, and human factors engineering. Recent works in supervisory control (human-on-the-loop) [100, 103] consider multi-robot systems [79], swarming networks [28, 29], and multi-operator control [74, 118]. Human-in-the-loop approaches formulate the human factors as external inputs to autonomous agents and embed them into control frameworks [6, 22, 24, 51, 71, 85, 118]. However, these tasks usually require continuous human attention and near constant responses. Some pioneer control theory research has been undertaken to investigate the modeling of human decision-making process [22], workload management based on dynamical queuing [97], and concurrent manual and autonomous control [24]. The concept of affective computing was coined in [86, 87] and refers to the computing that relates to, arises from, and deliberately influences human's emotion [20, 64, 72, 76, 82, 109, 113]. In particular, attention-based control has been utilized in HRI [39, 60, 68]. It considers giving emotional capabilities to computers. Motivation for this research field came from enhancing machine learning algorithms by mimicking human cognitive capabilities. The major areas for affective computing have been monitoring, social assistance, and entertainment [10, 113]. The most nascent research trend in AI includes mixed-initiative interaction which emphasizes human robot mutual understanding and interpretation of each other's initiative where the key is not only the ability of the human to understand and predict robot performance, but also the robot's ability to identify human needs and select intervention points to assume different levels of initiative [5, 14, 18, 36, 49, 52]. In robotics, the combination of autonomous and teleoperations forms the teleautonomous approach where humans and robots cannot work in the same location [3, 4, 27, 44, 78, 112, 117]. However, teleautonomous systems have been employed at a fairly coarse level. Extensive efforts have been put in the research of levels of autonomy (LOAs, a.k.a. autonomy taxonomy) but the results are not satisfactory since they tend to separate autonomy from the joint human–machine team [17, 19, 26, 53, 62, 63, 84, 104, 105, 108]. Existing efforts in adaptive autonomy and sliding autonomy are based on the definition of LOAs [61, 65, 98]. Common analysis approaches used in human subject research, such as NASA TLX subjective workload measure [48, 92], the Likert-type questionnaires for trust, engagement, transparency, naturalness, and cooperation level measurement [12, 15, 50, 55, 56], situational awareness assessment technique (SAGAT) [33], and ANOVA statistical significance test [11, 16], are mostly descriptive with emphasis on the statistical characteristics for design purpose [96, 119, 121].

The exploitation of the full potential of HRC systems will revolutionize everyday life. However, as most extant works in HRI are qualitative and descriptive [8, 46, 70, 75, 99] and remain highly specialized and focused on human–machine interface (HMI) design [13, 43, 99], they may not be compatible with the quantitative and rigorous mathematical languages used in control and decision-making theory. While control theory is able to provide system-level prediction and performance guarantees, existing approaches barely consider the interactions between humans and autonomous systems and are not directly applicable. The model, analysis, and implementation of effective HRC, especially for real-time robotic operations, remain largely an open problem [23, 49]. A fundamental issue that arises in the control of HRC systems is to ensure and improve the robot performance while maintaining human safety and comfort through collaboration. The control and decision-making algorithms must be able to quantify different features of the robotic systems as well as human factors in order to analyze the joint system performance. Therefore, research integrates human factors with the typical regulating mechanisms of control theory and rational reasoning mechanisms of decision-making theory, will open a game-changing realm in systems and controls research. This book seeks to present recent progress on the control and decision-making for human–robot collaboration (HRC) systems, which is an important research area with rapidly growing interest among control practitioners. It is worth noting that although HRC systems could sometimes be considered as the same research subject with HRI, these two terms carry different meanings. HRI is a more general term that includes collaboration while HRC exclusively means humans and robots working together in order to achieve a common goal. The definitions of "collaboration" are multiple: coworking, achieving common goals, and sharing the burden of a mission.

A brief introduction of each chapter and their connections to the book theme is provided below. We categorize the contributed papers into (i) HRC systems with one human–robot pair and (ii) HRC systems with multiple humans, multiple robots, and swarm robotic systems.

1.2 Collaboration Between One Human–Robot Pair

Chapters 2 and 9, "Robust Shared-Control for Rear-Wheel Drive Cars"–"Regret-Based Allocation of Autonomy in Shared Visual Detection for Human–Robot Collaborative Assembly in Manufacturing" present new control and decision-making methods for one human–robot pair collaboration including shared control, cooperative control, switching control, optimal control, impedance control, and sequential decision-making. Physical and psychological human factors (e.g., human arm dynamics, human–robot interaction force, regret in decision-making) as well as learning algorithms are embedded into these control and decision-making algorithms, respectively.

Chapter 2, "Robust Shared-Control for Rear-Wheel Drive Cars" studies the shared control problem between human and automatic controller for the kinematic model of a group of rear-wheel drive cars in different environments. The shared controller established in the chapter is designed to assist the human driver to drive safely, which can override the human steering control in emergency situations. A sharing signal is designed to ensure the two control efforts will not conflict with each other in normal conditions, but will switch to the autonomous controller in emergency situations. Formal properties of the closed-loop system with shared control are established by a Lyapunov-like analysis. This design is based on either absolute positions or correlated positions such as distances to the obstacles and angle differences.

Chapter 3, "Baxter-On-Wheels (BOW): An Assistive Mobile Manipulator for Mobility Impaired Individuals" develops an optimal shared control strategy for a customized 14-DOF dual arm robot manipulator (Baxter) on a wheelchair that can be commanded by patients with severe disability. The objective is to aid the independence of quadriplegic and other mobility impaired individuals through HRC. The human controls the mobile manipulator using a device with less Degrees of Freedom (DoFs) than the mobile manipulator. Detailed descriptions of the integrated system architecture and system kinematics are presented. The manipulator control respects various safety constraints while achieving the desired goal specified by the human. The shared control scheme is then formulated as a constrained optimization problem. The proposed system design and control approaches are experimentally evaluated for two tasks execution: (i) picking and placing bottles and (ii) cleaning board.

Chapter 4, "Switchings Between Trajectory Tracking and Force Minimization in Human–Robot Collaboration" proposes a combined force control and impedance control solution for robot manipulators when they are assisted by human arm movements. A physical HRC framework is developed to switch between trajectory tracking under no human intervention and internal force minimization during the movement toward the human's target. The goal of trajectory tracking is to move the robot's end effector along a given reference trajectory as long as there is no human intervention. The robot movement can also be altered by human intervention such that it moves based on the interaction force which is desired to be minimized. An adaptive observer is designed to estimate the unknown dynamics of the human arm. An optimal control problem is formulated with costs switching between trajectory tracking and equilibrium holding.

Chapter 5, "Estimating Human Intention During a Human–Robot Cooperative Task Based on the Internal Force Model" studies the relation between physical actions and intents of humans in performing dyadic movement tasks. This work proposes a mathematical model for human internal force when performing cooperative manipulation tasks, which is based on the minimum jerk movement of the human and deformation strain of the force. A series of experiments are conducted to validate the proposed model. A control policy is proposed for a robot to emulate the cooperation strategy between humans based on the modeling of internal force interactions for collaborative human–robot physical manipulation. The hypothesis is that human feel more natural to collaborate with robotic manipulators under the proposed cooperation policy.

Chapter 6, "A Learning Algorithm to Select Consistent Reactions to Human Movements" proposes the expanded dual expert (DEA) algorithm to achieve a tradeoff between consistency and adaptiveness for a robot to select control laws to generate reactions to human movements. The learning algorithm is usually employed for the robot to predict human actions and then select appropriate reactions accordingly. Theoretical analysis of the DEA algorithm is provided. Both simulation and experiments are conducted to compare the DEA algorithm with two classic learning algorithms. The results confirm that DEA enables a robot to learn the preferred reaction to pass a human in a hallway environment.

Chapter 7, "Assistive Optimal Control-on-Request with Application in Standing Balance Therapy and Reinforcement" proposes a new control-on-request (COR) approach to design autonomous controllers that assist a physical therapist for patients wearing exoskeleton. An optimal switching control is developed to improve the capability of existing shared control interfaces for using robotic exoskeletons in standing balance therapy. The optimal switching time of the controller and its duration are computed. These COR enhanced interfaces allow users to request on-demand bursts of assistive computer control authority. Simulation results are provided to demonstrate potential benefits.

Chapter 8, "Intelligent Human–Robot Interaction Systems Using Reinforcement Learning and Neural Networks" presents an intelligent control scheme of robotic manipulators to assist human operator to perform a given task with minimum workload demands and optimal performance. The proposed control methodology consists of two feedback loops: a nonlinear robust inner control that makes the robot behave like a prescribed impedance model perceived by the operator, and an optimal outer control that finds the parameters for this model to minimize the tracking error. The inner loop uses a neural network for adaptive tracking of a desired impedance model, and the outer loop uses reinforcement learning to design a desired impedance model to match a simplified human model with unknown parameters.

Chapter 9, "Regret-Based Allocation of Autonomy in Shared Visual Detection for Human–Robot Collaborative Assembly in Manufacturing" integrates the analysis of human decision-making pattern under uncertainty, in particular, regret, into the motion control of industrial robots for human–robot collaborative assembly tasks in manufacturing. A case study on the human–robot shared vision detection of right assembly parts is provided. A Bayesian-based optimal allocation scheme for sensing autonomy is first presented and then modified to integrate a regret function developed based on regret theory in order to provide a suboptimal, however, human-like decision-making strategy. The hypothesis is that an HRC system sharing a same mental model outperforms the system with an optimal robot model different from the human pattern. Baxter robot experiments with human subjects are conducted in a hybrid manufacturing cell in the lab settings. Subjective tests are used to measure human's regret intensity used in the allocation scheme. The experiment results show the effectiveness of the regret-based decision-making scheme in terms of workload and task easiness.

1.3 Collaboration Between Human and Multiple Robots/Swarms

Due to the increased manpower costs, it is desired that a small number of human operators will be able to control flocks of robots, intervening only at key moments to give high-level instructions [9, 30]. Chapters 11 and 16, "Considering Human Behavior Uncertainty and Disagreements in Human-Robot Cooperative Manipulation" "Co-design of Control and Scheduling for Human–Swarm Collaboration Systems Based on Mutual Trust" seek to address this problem by developing the control and coordination algorithms for humans to collaborate with multiple robots and swarms based on robust control, passivity-based control, coverage control, formation control, and real-time scheduling, respectively. Physical and psychological human factors (e.g., human motion, vision, haptic feedbacks, safety in HRI, and human trust in robots) are embedded into these control algorithms, respectively.

Chapter 10, "Considering Human Behavior Uncertainty and Disagreements in Human–Robot Cooperative Manipulation" proposes an optimal control scheme in order to adapt robot to the uncertainty and disagreements in human control actions in human robot load sharing tasks. A general framework is introduced to incorporate human wrench models. The effective wrench for each party has been calculated and the concept of internal wrench is rigorously defined. In the proposed control method, risk-sensitive stochastic optimal control techniques are utilized in order to predict the uncertainty and help the robot to minimize the conflict between the human control forces applied to an object and the autonomous control forces applied by the robots. Experimental data are collected with human subjects to justify the advantage of proactive control strategies.

Chapter 11, "Designing the Robot Behavior for Safe Human–Robot Interactions" addresses the design of autonomous robot motion controller for safe physical HRI from a multiagent system point of view with potential applications in industrial cooperative robots, autonomous cars, and medical robots. This work models both humans and robots as leader/follower agents and defines a safe region (set) for the movement of the follower agent. A safety index is designed for constraining the motion of the follower agent in the safe set based on the current and predicted information of all agents. A nonlinear robust controller is designed to ensure the invariance of the safety set. The ultimate goal is to control the follower agent motion such that it reaches its goal while the constraint due to the safety index is imposed. The dynamics of human motion is unknown and is learned on the fly via two separate parameter estimation algorithms: The safe set algorithm (SSA) uses a constant upper bound based on the average prediction error. The safe exploration algorithm (SEA) considers a time-varying bound for uncertainties and updates this value at every time step. A comparative study is conducted to demonstrate the pros and cons of using SEA over SSA. A combination of the SSA and SEA algorithms is then proposed for system with a time-varying number of agents (topology).

Chapter 12, "When Human Visual Performance Is Imperfect—How to Optimize the Collaboration Between One Human Operator and Multiple Field Robots" develops an optimal decision framework for human and robot combined vision under limited communication and energy budget in a multi-robot field exploration and classification task. This work provides a paradigm that takes advantage of the superior visual performance of humans and the field exploration capabilities of robots, allowing the robot to only ask humans for help when needed, due to a limited chance of communication with the human as well as limited energy/time budget. The robot only needs to understand the extent of humans' visual capabilities and should ask the human within an optimum query range, depending on the energy cost and the noise variance. The simulation results show that the robot can reduce its energy consumption considerably by properly taking advantage of its collaboration and the approach can achieve a good performance with a low bandwidth usage. Several cases are provided to illustrate the underlying methodology and its applicable to collaborative scenarios.

Chapter 13, "Human-Collaborative Schemes in the Motion Control of Single and Multiple Mobile Robots" shows and compares several ways to effectively design human–robot collaborative control of single and multiple mobile robots with applications in aerial robots. First, a taxonomy is proposed for highlighting the main features of such collaborative systems for a single robot. The proposed categories include: (1) physical domain of the robots, (2) level of autonomy, (3) type of force interaction with operator, (4) type of operation whether from close proximity or teleoperation, (5) contact-free versus physical contact, (6) use or lack of onboard sensors, and the length and frequency of operator's interaction with robots. This taxonomy is then extended to the multiple robot case with new categories such as leader–follower or formation control schemes. Several examples of the type of HRC in such schemes are provided throughout the chapter.

Chapter 14, "A Passivity-Based Approach to Human–Swarm Collaboration and Passivity Analysis of Human Operators" presents a passivity-based method for human–swarm interaction. In the system, the operator is assumed to have a device to command a limited number of accessible robots, and to be in front of a monitor which displays certain information fed back from the robots. The intended control objective is to render positions/velocities of the group of kinematic robots synchronized to desired reference by human inputs. A cooperative controller is implemented on every robot and the stability and passivity of human swarm interaction over visual-based feedback is proven by assuming passivity of an appropriately defined human operator decision process. Experimental data are collected to support the use of transfer functions for human decision-making.

Chapter 15, "Human–Swarm Interactions via Coverage of Time-Varying Densities" develops an interface for a human to control a large swarm of robot by specifying density functions, which are translated into a Voronoi distribution that are achieved autonomously by robot swarms. This work uses time-varying density functions to construct suitable abstractions of an entire swarm robot team so that it is amenable to human control. A series of coverage control laws are designed based on the density functions to guarantee optimal coverage by the robot team and externally influence the robot swarm. Robotic experiments are implemented in two different platforms to demonstrate the viability of human–swarm interactions with the proposed schemes.

Chapter 16, "Co-design of Control and Scheduling for Human-Swarm Collaboration Systems Based on Mutual Trust" codesigns the cooperative control and real-time scheduling algorithms for human–swarm collaboration under a mutual trust framework. Distributed formation control laws are developed for the leading agent in each swarm cluster. Cohesive and collision avoidance motion control laws are developed for the following agents in the cluster. The replicator–mutator (RM) dynamics are used to model human networks and dynamically update human's attention allocation to the swarm clusters. A human–swarm mutual trust-based collaboration framework is proposed and humans and swarm clusters are paired based on a "fitness" measurement. A real-time dynamic scheduling algorithm is then codesigned with the agent motion control laws. Simulation results show that the proposed codesign of scheduling and control algorithms can guarantee effective real-time allocation of human resources and ensure acceptable swarm performance.

References

1. Kevin G (2012) Connected and automated vehicle research in the united states. Technical report, US Department of Transportation, Washington, DC, July 2012. Task Force Report
2. Ro-Man IEEE (2012) The 21st IEEE International Symposium on Robot and Human Interactive Communication. France, Paris
3. Ali KS, Arkin R (2000) Multiagent teleautonomous behavioral control. Mach Intell Robot Control 1(1)
4. Ali KS (1999) Multiagent telerobotics: matching systems to tasks. PhD thesis, Georgia Institute of Technology
5. Allen JE, Guinn CI, Horvtz E (1999) Mixed-initiative interaction. IEEE Intell Syst Appl 14(5):14–23
6. Sengstacken A, DeLaurentis D, Akbarzadeh-T M (2010) Optimization of shared autonomy vehicle control architectures for swarm operations. IEEE Trans Syst Man Cybern Part B Cybern 40(4):1145–1157
7. Argall BD (2015) Turning assistive machines into assistive robots. In: SPIE OPTO. International Society for Optics and Photonics, pp 93701Y–93701Y
8. Arkin R, Ali KS (1994) Integration of reactive and telerobotic control in multi-agent robotic systems. In: From animals to animats 3: Proceedings Third International Conference on Simulation of Adaptive Behavior, pp 473–478

9. Bainbridge WA, Hart J, Kim ES, Scassellati B (2008) Fan-out: measuring human control of multiple robots. In: The 17th IEEE international symposium on robot and human interactive communication, RO-MAN, pp 701–706
10. Bates J (1994) The role of emotion in believable agents. Commun ACM 37(7):122–125
11. Bernhardson CS (1975) Type I error rates when multiple comparison procedures follow a significant F test of anova. Biometrics, 229–232
12. Bisantz AM, Seong Y (2001) Assessment of operator trust in and utilization of automated decision-aids under different framing conditions. Int J Ind Ergonomics 28(2):85–97
13. Bolton ML, Bass EJ, Siminiceanu RI (2013) Using formal verification to evaluate human-automation interaction: a review. CSL Tech Rep 43(3)
14. Bradshaw JM, Feltovich PJ, Jung H, Kulkarni S, Taysom W, Uszok A (2004) Dimensions of adjustable autonomy and mixed-initiative interaction. Agents and computational autonomy. Springer, Heidelberg, pp 17–39
15. Cynthia B (2004) Social interactions in HRI: the robot view. IEEE Trans Syst Man Cybern Part C Appl Rev 34(2):181–186
16. Brown MB, Forsythe AB (1974) The ANOVA and multiple comparisons for data with heterogeneous variances. Biometrics, 719–724
17. Bruemmer DJ, Dudenhoeffer DD, Marble JL (2002) Dynamic-autonomy for urban search and rescue. In: AAAI mobile robot competition, pp 33–37
18. Bruemmer DJ, Marble DJ, Dudenhoeffer DD (2002) Mutual initiative in human-machine teams. In: Proceedings of the 2002 IEEE 7th conference on human factors and power plants 2002. IEEE, New Jersey, pp 7–22
19. Burke JL, Murphy RR, Rogers E, Lumelsky VJ, Scholtz J (2004) Final report for the DARPA/NSF interdisciplinary study on human-robot interaction. IEEE Trans Syst Man Cybern Part C Appl Rev 34(2):103–112
20. Calvo RA, D'Mello S (2010) Affect detection: an interdisciplinary review of models, methods, and their applications. IEEE Trans Affective Comput 1(1):18–37
21. Camarillo DB, Krummel TM, Salisbury JK (2004) Robotic technology in surgery: past, present, and future. Am J Surg 188(4):2–15
22. Cao M, Stewart A, Leonard NE (2008) Integrating human and robot decision-making dynamics with feedback: models and convergence analysis. In: 47th IEEE conference on decision and control, pp 1127–1132
23. Charalambous G (2013) Human-automation collaboration in manufacturing: identifying key implementation factors. In: Contemporary Ergonomics and human factors 2013: proceedings of the international conference on Ergonomics and human factors 2013. Cambridge, UK, p 59
24. Chipalkatty R (2012) Human-in-the-loop control for cooperative human-robot tasks. PhD dissertation, Georgia Institute of Technology
25. Clare AS (2013) Modeling real-time human-automation collaborative scheduling of unmanned vehicles. PhD dissertation, Massachusetts Institute of Technology
26. Clough BT (2002) Metrics, schmetrics! How do you track a UAV's autonomy? In: AIAA's 1st technical conference and workshop on unmanned aerospace vehicles
27. Conway L, Volz RA, Walker MW (1990) Teleautonomous systems: projecting and coordinating intelligent action at a distance. IEEE Trans Robot Autom 6(2):146–158
28. Coppin G, Legras F (2012) Autonomy spectrum and performance perception issues in swarm supervisory control. Proc IEEE 100(3):590–603
29. Cummings ML (2004) Human supervisory control of swarming networks. In: 2nd annual swarming: autonomous intelligent networked systems conference
30. Cummings ML, Mitchell PJ (2008) Predicting controller capacity in supervisory control of multiple UAVs. IEEE Syst Man Cybern A Syst Hum 38(2):451–460
31. Desai M (2012) Modeling trust to improve human-robot interaction
32. Dey KC, Yan L, Wang X, Wang Y, Shen H, Chowdhury M, Yu L, Qiu C, Soundararaj V (2015) A review of communication, driver characteristics, and controls aspects of cooperative adaptive cruise control (CACC). IEEE Trans Intell Transp Syst 17(2):491–509

33. Endsley MR (1988) Situation awareness global assessment technique (SAGAT). In: Proceedings of the IEEE 1988 national aerospace and electronics conference, 1988. NAECON 1988, IEEE, New Jersey, pp 789–795
34. Esser B, Pelczarski N, Huston D, Arms S (2000) Wireless inductive robotic inspection of structures. Proc IASTED Int Conf Roboot Appl 8:14–16
35. Feng L, Wiltsche C, Humphrey L, Topcu U (2015) Controller synthesis for autonomous systems interacting with human operators. In: Proceedings of the ACM/IEEE sixth international conference on cyber-physical systems. ACM, New York, pp 70–79
36. Ferguson G, Allen JF, Miller BW (1996) Trains-95: towards a mixed-initiative planning assistant. In: AIPS, pp 70–77
37. Fong T, Provencher C, Micire M, Diftler M, Berka R, Bluethmann B, Mittman D (2012) The human exploration telerobotics project: objectives, approach, and testing. 2012 IEEE aerospace conference. IEEE, New Jersey, pp 1–9
38. Fong T, Zumbado JR, Currie N, Mishkin A, Akin DL (2013) Space telerobotics unique challenges to human-robot collaboration in space. Rev Hum Factors Ergonomics 9(1):6–56
39. Gage A (2004) Multi-robot task allocation using affect. PhD dissertation, University of South Florida
40. Gassert R, Moser R, Burdet E, Bleuler H (2006) MRI/fMRI-compatible robotic system with force feedback for interaction with human motion. IEEE/ASME Trans Mechatron 11(2):216–224
41. Ge S, Khatib O, Cabibihan J, Simmons R, Williams MA (2012) In: Proceedings of the social robotics: 4th international conference, ICSR 2012, (Lecture notes in computer science/Lecture notes in artificial intelligence). Chengdu, China
42. Göhring D, Latotzky D, Wang M, Rojas R (2013) Semi-autonomous car control using brain computer interfaces. In: Intelligent autonomous systems, vol 12. Springer, Heidelberg, pp 393–408
43. Goodrich MA, Schultz AC (2007) Human-robot interaction: a survey. Found Trends Hum Comput Interact 1(3):203–275
44. Graves S, Volz R (1995) Action selection in teleautonomous systems. In: IEEE/RSJ international conference on intelligent robots and systems, 'human robot interaction and cooperative robots', vol 3. IEEE, New Jersey, pp 14–19
45. Guizzo E, Ackerman E (2012) How rethink robotics built its new baxter robot worker. IEEE spectrum
46. Hancock PA, Billings DR, Schaefer KE, Chen JYC, de Visser EJ, Parasuraman R (2011) A meta-analysis of factors affecting trust in human-robot interaction. Hum Factors J Hum Factors Ergonomics Soc 53:517–527
47. Hanheide M, Lohse M, Zender H (2012) Special issue on expectations, intentions and actions in human-robot interaction. Int J Soc Robot 4(2)
48. Hart SG (2006) NASA-task load index (NASA-TLX); 20 years later. In: Proceedings of the human factors and Ergonomics society annual meeting, vol 50. Sage Publications, London, pp 904–908
49. Hayes B, Scassellati B (2013) Challenges in shared-environment human-robot collaboration. Learning 8:9
50. Heerink M, Krose B, Evers V, Wielinga B (2009) Measuring acceptance of an assistive social robot: a suggested toolkit. In: The 18th IEEE international symposium on robot and human interactive communication, RO-MAN, pp 528–533
51. Hirche S, Buss M (2012) Human-oriented control for haptic teleoperation. Proc IEEE 100(3):623–647
52. Horvitz E (1999) Principles of mixed-initiative user interfaces. In: Proceedings of the SIGCHI conference on human factors in computing systems. ACM, New York, pp 159–166
53. Huang H, Messina E, Albus J (2007) Autonomy levels for unmanned systems (ALFUS) framework. NIST Spec Publ 1011-II-1.0
54. Huston D, Esser B, Miller J, Wang X (2003) Robotic and mobile sensor systems for structural health monitoring. In: Proceedings of the 4th international workshop on structural health monitoring, vol 1517. Stanford, CA

55. Serena Ivaldi, Salvatore M Anzalone, Woody Rousseau, Olivier Sigaud, and Mohamed Chetouani. Robot initiative in a team learning task increases the rhythm of interaction but not the perceived engagement. Frontiers in neurorobotics, 8, 2014
56. Jokinen K, Wilcock G (2013) User experience in human-robot interactions. In: Proceedings of the fourth workshop on perceptual quality of systems (PQS 2013), FTW-TECHREPORT-128, FTW, Vienna, Austria
57. Jubien A, Gautier M, Janot A (2014) Dynamic identification of the kuka LWR robot using motor torques and joint torque sensors data. In: World congress of the international federation of automatic control (IFAC), pp 1–6
58. Julia B, Aaron H, Allison T (2014) Advancing safe human-robot interactions with robonaut 2. In: Proceeding of the 12th In-ternational symposium on artificial intelligence, robotics and automation in space, vol 17, p 19
59. Karwowski W, Rahimi M, (eds) Human-robot interaction. Taylor and Francis, Milton Park
60. Kasderidis S, Taylor JG (2003) Attention-based robot control. In: Knowledge-based intelligent information and engineering systems, vol 2774, Lecture notes in computer scienceSpringer, Heidelberg, pp 615–621
61. Kidwell B, Calhoun GL, Ruff HA, Parasuraman R (2012) Adaptable and adaptive automation for supervisory control of multiple autonomous vehicles. In: Proceedings of the human factors and Ergonomics society annual meeting, vol 56. Sage Publications, London, pp 428–432
62. Kirlik A (1993) Modeling strategic behavior in human-automation interaction: why an "aid" can (and should) go unused. Hum Factors 35(2):221–242
63. Klein G (2004) Ten challenges for making automation a "team player" in joint human-agent activity. IEEE Intell Syst 19(6):91–95
64. Koelstra S, Muhl C, Soleymani M, Lee JS, Yazdani A, Ebrahimi T, Pun T, Nijholt A, Patras I (2012) Deap: A database for emotion analysis; using physiological signals. IEEE Trans Affective Comput 3(1):18–31
65. Kortenkamp D, Burridge R, Bonasso P, Schrenkenghost D, Hudson M (1999) An intelligent software architecture for semiautonomous robot control. In: Autonomy control software workshop, autonomous agents
66. Kuzuoka H, Evers V, Imai M, Forlizzi J (2013) In: Proceedings of the 8th ACM/IEEE international conference on human-robot interaction. Tokyo, Japan
67. Lamers MH, Verbeek FJ (2010) In: Human-robot personal relationships third international conference, HRPR 2010, Revised selected papers series: Lecture notes of the Institute for computer sciences, social-informatics and telecommunications engineering, vol 59. Leiden, The Netherlands
68. Lang S, Kleinehagenbrock M, Hohenner S, Fritsch J, Fink GA, Sagerer G (2003) Providing the basis for human-robot-interaction: a multi-modal attention system for a mobile robot. Proceedings of the 5th international conference on multimodal interfaces, pp 28–35
69. Laschi C, Breazeal C, Nakauchi Y (2007) Guest editorial special issue on human-robot interaction. IEEE Trans Robot 23(5)
70. Lee JD, See KA (2004) Trust in automation: designing for appropriate reliance. Hum Factors 46:50–80
71. Lewis B, Sukthankar G (2011) Two hands are better than one: Assisting users with multi-robot manipulation tasks. In: IEEE/RSJ international conference on intelligent robots and systems (IROS), pp 2590–2595
72. Loewenstein G, Lerner J (2003) The role of emotion in decision making. In: Davidson RJ, Goldsmith HH, Scherer KR (eds) The handbook of affective science. Oxford University Press, Oxford, England, pp 619–642
73. Mackay WE, Brewster S, Bødker S (2013) In: Proceedings of the SIGCHI conference on human factors in computing systems. Paris, France
74. Majji M, Rai R (2013) Autonomous task assignment of multiple operators for human robot interaction. In: American control conference, pp 6454–6459
75. McLachlan S, Arblaster J, Liu O, Miro J, Chenoweth L (2005) A multistage shared control method for an intelligent mobility assistant. In: 9th international conference on rehabilitation robotics, pp 426–429

76. Michaud F, Prijanian P, Audet J, Letourneau D (2000) Artificial emotions and social robotics. In: Proceedings of the distributed autonomous robotic systems
77. Michelucci P (ed) The Springer handbook of human computation. Springer, Heidelberg
78. Murphy RR, Sprouse J (1996) Strategies for searching an area with semi-autonomous mobile robots. In: Proceedings of robotics for challenging environments. Citeseer, pp 15–21
79. Nehme CE (2009) Modeling human supervisory control in heterogeneous unmanned vehicle systems. PhD dissertation, Massachusetts Institute of Technology
80. Norman, DA Draper SW (1986) User centered system design; new perspectives on human-computer interaction. L. Erlbaum Associates Inc.,
81. Oleson KE, Billings DR, Kocsis V, Chen JYC, Hancock PA (2011) Antecedents of trust in human-robot collaborations. In: IEEE first international multi-disciplinary conference on cognitive methods in situation awareness and decision support (CogSIMA), pp 175–178
82. Ortony A (2002) On making believable emotional agents believable. In: Trappl R, Petta P, Payr S (eds) Emotions in humans and artifacts, Chap 6. The MIT Press, Cambridge, pp 189–211
83. Ostergaard EH (2012) Lightweight robot for everybody [industrial activities]. IEEE Robot Autom Mag, 19(4):17–18, 2012
84. Parasuraman R, Sheridan TB, Wickens CD (2000) A model for types and levels of human interaction with automation. IEEE Trans Syst Man Cybern Part A Syst Hum 30(3):286–297
85. Parker L, Howard A (2009) Assistive formation maintenance for human-led multirobot systems. EEE Int Conf Syst Man Cybern, 2350–2355
86. Picard RW Affective computing. MIT Press, Cambridge
87. Picard RW (2010) Affective computing: from laughter to IEEE. IEEE Trans Affective Comput 1(1):11–17
88. Rahman SM, Sadrfaridpour B, Wang Y (2015) Trust-based optimal subtask allocation and model predictive control for human-robot collaborative assembly in manufacturing. In: ASME 2015 dynamic systems and control conference. American Society of Mechanical Engineers, pp. V002T32A004–V002T32A004
89. Regalado A (2013) Small factories give baxter the robot a cautious once-over. MIT Technology Review, January 2013. http://www.technologyreview.com/news/509296/small-factories-give-baxter-the-robot-a-cautious-once-over/
90. Riley V (1989) A general model of mixed-initiative human-machine systems. Proceedings of the human factors and Ergonomics society annual meeting, vol 33. Sage Publications, London, pp 124–128
91. Hayley R, Bruce M, Elizabeth B (2014) The role of healthcare robots for older people at home: a review. Int J Soc Robot 6(4):575–591
92. Rubio S, Díaz E, Martín J, Puente J (2004) Evaluation of subjective mental workload: a comparison of SWAT, NASA-TLX, and workload profile methods. Appl Psychol 53(1):61–86
93. Sadrfaridpour B, Saeidi H, Burke J, Madathil K, Wang Y (2015) Modeling and control of trust in human-robot collaborative manufacturing. In: The intersection of robust intelligence (RI) and trust in autonomous systems. Springer, Heidelberg
94. Saeidi H, McLane F, Sadrfaidpour B, Sand E, Fu S, Rodriguez J, Wagner J, Wang Y (2016) Trust-based mixed-initiative bilateral teleoperation control of mobile robotic systems. In: American control conference (ACC)
95. Saeidi H, Wang Y (2015) Trust and self-confidence based autonomy allocation of robotic systems. In: Conference on decision and control
96. Sanders T, Oleson KE, Billings DR, YC Chen J, Hancock PA, (2011) A model of human-robot trust theoretical model development. In: Proceedings of the human factors and Ergonomics society annual meeting, vol 55. Sage Publications, London, pp 1432–1436
97. Savla K, Frazzoli E (2012) A dynamical queue approach to intelligent task management for human operators. Proc IEEE 100(3):672–686
98. Sellner B, Heger FW, Hiatt LM, Simmons R, Singh S (2006) Coordinated multi-agent teams and sliding autonomy for large-scale assembly. Proc IEEE Spec Issue Multi-Robot Syst 94(7):1425–1444

99. Shah D, Schneider J, Campbell M (2012) A sketch interface for robust and natural robot control. Proc IEEE 100(3):604–622
100. Sheridan TB (1983) Supervisory control of remote manipulators, vehicles and dynamic processes: experiments in command and display aiding. Technical report, DTIC Document
101. Sheridan TB (1992) Telerobotics, automation, and human supervisory control. The MIT Press, Cambridge
102. Sheridan TB (ed) (2002) Humans and automation: system design and research issues. Wiley, New Jersey
103. Sheridan TB (2011) Adaptive automation, level of automation, allocation authority, supervisory control, and adaptive control: distinctions and modes of adaptation. IEEE Trans Syst Man Cybern Part A Syst Hum 41(4):662–667
104. Sheridan TB, Parasuraman R (2006) Human-automation interaction. In: Nickerson RS (ed) Reviews of human factors and ergonomics, vol 1. Human Factors and Ergonomics Society, Santa Monica, pp 89–129
105. Sheridan TB, Verplank WL (1978) Human and computer control of undersea teleoperators. Technical report, MIT Man-Machine Systems Laboratory, Cambridge, MA
106. Shi J, Jimmerson G, Pearson T, Menassa R (2012) Levels of human and robot collaboration for automotive manufacturing. In: Proceedings of the workshop on performance metrics for intelligent systems, PerMIS '12. ACM, New York, NY, USA, pp 95–100
107. Shneiderman B, Plaisant C, Cohen M, Jacobs S (2009) Designing the user interface: strategies for effective human-computer interaction, 5/e, New edn. Prentice Hall, New Jersey
108. Sholes E (2006) Evolution of a UAV autonomy classification taxonomy. IEEEAC paper, (1538)
109. Sloman A, Croucher M (1981) Why robots will have emotions. Proc IJCAI
110. Smith BW (2014) Automated vehicles are probably legal in the united states
111. Srivastava V (2012) Stochastic search and surveillance strategies for mixed human-robot teams. PhD dissertation, University of California at Santa Barbara
112. Trivedi MM, Hall B, Kogut G, Roche S (2000) Web-based teleautonomy and telepresence. In: International symposium on optical science and technology. International Society for Optics and Photonics, pp 81–85
113. Velásquez JD (1998) When robots weep: emotional memories and decision-making. In: AAAI proceedings
114. Wagner AR (2009) The role of trust and relationships in human-robot social interaction. PhD thesis, Georgia Institute of Technology
115. Wang X, Shi Z, Zhang F, Wang Y (2015) Dynamic real-time scheduling for human-agent collaboration systems based on mutual trust. In: Cyber-Physical Systems (CPS), Taylor and Francis, Milton Park, 2015
116. Wang Y (2016) Regret-based automated decision-making aids for domain search tasks using human-agent collaborative teams. IEEE Trans Control Syst Technol 99:1–16
117. Wegner R, Anderson J (2006) Agent-based support for balancing teleoperation and autonomy in urban search and rescue. Int J Robot Autom 21(2):120–128
118. Whetten J, Goodrich M, Guo Y (2010) Beyond robot fan-out: towards multi-operator supervisory control. In: IEEE international conference on systems man and cybernetics (SMC), pp 2008–2015
119. Wickens CD, Gordon SE, Liu Y (2004) An introduction to human factors engineering. Pearson Prentice Hall, Upper Saddle River, NJ
120. Wilcox R, Nikolaidis S, Shah J (2012) Optimization of temporal dynamics for adaptive human-robot interaction in assembly manufacturing. In: Proceedings of robotics science and systems, Sydney, Australia
121. Yagoda RE (2010) Development of the human robot interaction workload measurement tool (HRI-WM). In: Proceedings of the human factors and Ergonomics society annual meeting, vol 54. Sage Publications, London, pp 304–308

Chapter 2
Robust Shared-Control for Rear-Wheel Drive Cars

Jingjing Jiang and Alessandro Astolfi

2.1 Introduction

Nowadays vehicles are one of the major means of transportation. The driving safety depends highly on the driver's attention, experience and skills. According to the report given by the World Health Organization, more than one million fatalities are caused by traffic-related accidents per year all over the world. To keep the human beings out of the vehicle control loop and to relieve them from the task of driving, researchers started to study self-driving cars.

A local path planning algorithm has been introduced in [12], in which the path modification is performed using a Voronoi cell when a collision is detected, while two path planning methods and multiple PID controllers have been used in [19] to achieve auto-driving. The paper [14] has presented a robust controller composed of a longitudinal controller and a lateral controller for a self-driving car to track a given trajectory. In [2] an electric race car is driven by a high-level navigation system via a multi-threaded program and an embedded low-level control to achieve accurate tracking of a straight line and good turning speed comparable to the human driver. Model Predictive Control (MPC) is also widely used in auto-driving cars. The paper [10] has used simulation and experimental results to demonstrate that smooth trajectory tracking with small position errors is achievable with the application of MPC. Other control methods, such as fuzzy control laws [13, 22], adaptive controls [15, 17], and sliding mode controls [4], have also been utilized in vehicle auto-navigation.

J. Jiang (✉) · A. Astolfi
Dept. of Electrical and Electronic Engineering, Imperial College London, London SW7 2AZ, UK
e-mail: jingjing.jiang10@imperial.ac.uk

A. Astolfi
The DICII, University of Roma "Tor Vergata", Via Del Politecnico 1, 00133 Rome, Italy
e-mail: a.astolfi@imperial.ac.uk

Y. Wang and F. Zhang (eds.), *Trends in Control and Decision-Making for Human–Robot Collaboration Systems*, DOI 10.1007/978-3-319-40533-9_2

Even though the driverless cars, such as Google Cars, have already been successfully tested in some areas, they still have several limitations. For instance, much intricate preparations have to be made before Google Cars can be driven on a certain road. Data have to be studied carefully after the testing drive, meter by meter, by both computers and humans. Even though the sensors of the car can detect obstacles on the road, they may be unable to distinguish between a rock and a crumpled piece of paper. Therefore the car will drive around both of them. Compared with auto driving, the cooperative driving which does not abandon the human driver has several advantages: it provides additional information (beyond those sensed directly by vehicle-based sensors) and better quality of information (lower noise and smaller delays), and is able to enable direct negotiations about maneuvers (such as merging and creating gaps) [20]. A shared-control algorithm for a car has been proposed in [11] and [5]: the car is able to drive autonomously by following the path created by waypoints generated by the human operator, while a passive human–machine interface has been utilized in [16] to suggest control actions to the driver of a vehicle with n-trailers. A longitudinal driving-assistance system based on a self-learning algorithm for the driver modeling and an automatic braking activation (or collision warning) system has been studied in [21]. A human-in-the-loop design approach to the control of a smart car based on a supervisory control that determines the desired velocity for the vehicle has been presented in [1]. Differently from the traditional braking model, the paper [3] has proposed a new safety indicator, named as "time gap interval", for the safe distance to the car in front in car-following situations. The paper [8] has presented a shared-control algorithm for the kinematic model of a rear-wheel drive car to "correct" the driver's behaviors in "dangerous" situations. The design is based on the measurement of absolute positions. This chapter studies the human–robot cooperation, extends the results given in [8] to the cases in which no absolute positions are available and to the cases in which disturbances are taken into consideration, thus also extending [9]. The shared controller established in the chapter is designed to assist the human driver drive safely.

The rest of the chapter is organized as follows. In Sect. 2.2 the shared-control problem for the kinematic model of a rear-wheel drive car is formulated. The solutions to the problem with and without measurements of absolute positions are given in Sects. 2.3 and 2.5, respectively. Formal properties of the closed-loop system with shared control are presented in each section. To deal with model uncertainties, we consider the system with the presence of bounded disturbances as explained in Sect. 2.4. Three case studies showing the effectiveness of the shared-control algorithm are provided in Sect. 2.6. Finally, some conclusions and suggestions for future work are given in Sect. 2.7.

2.2 Problem Formulation, Definitions, and Assumptions

The chapter studies the kinematic model of a rear-wheel drive car, the dynamics of which, as explained in [8], are described by the equations

$$\begin{aligned}
\dot{x} &= v_s \cos\theta, \\
\dot{y} &= v_s \sin\theta, \\
\dot{\theta} &= \frac{v_s \tan\phi}{l}, \\
\dot{\phi} &= \omega_s,
\end{aligned} \tag{2.1}$$

where (x, y) are the coordinates of the middle point of the rear-axle, θ and ϕ are the inertial heading and the steering angle, respectively. l denotes the distance between the front-axle and the rear-axle. Furthermore, the shared-control inputs are the forward velocity v_s and the angular velocity of front wheels ω_s.

As explained in Sect. 2.1 the shared-control input $u_s = [v_s, \omega_s]^T$ discussed in this chapter is a combination of the human input $u_h = [v_h, \omega_h]^T$, named as *h-control*, and the feedback control input $u_f = [v_f, \omega_f]^T$, named as *f-control*. Figure 2.1 illustrates the architecture of the system with the shared controller, where "H" describes the human action. Define u_s as

$$u_s(t) = (1 - k(t))u_f(t) + k(t)u_h(t), \tag{2.2}$$

where $k(t) \in [0, 1]$ is named as *sharing function*, describing how the control authority is shared between the driver and the feedback controller. The value of k can provide feedback to the driver indicating how safe his/her behavior is. This is helpful to improve the driver's performance, especially for new drivers. Figure 2.1 also shows that the human driver receives more information than the feedback controller since he/she is able to directly perceive the environment besides reading data from the sensors.

In this chapter, we use the subscripts h, f, and s to denote the control inputs generated by the human operator, the feedback controller, and the shared controller, respectively.

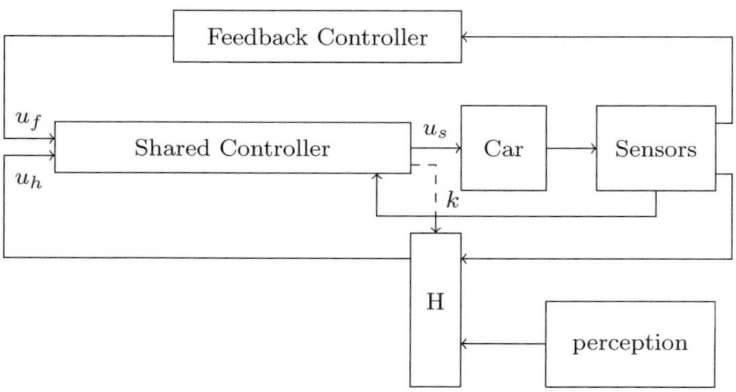

Fig. 2.1 The proposed architecture of the shared-control

Definition 2.1 Similar to [8], the names *s-closed-loop* and *h-closed-loop* are used to represent the closed-loop system with shared control u_s (described by (2.1)) and human control u_h, respectively.

Let $\mathcal{P}_a(t) \in \mathbb{R}^2$ be a compact set describing the admissible Cartesian configuration set[1] of the system (2.1) and u_h be a given h-control. Then the shared-control problem for the kinematic model of the car can be formulated as follows.

Given the system (2.1), a compact nonempty set $\mathcal{P}_a(t)$ and an h-control $u_h(t)$, find (if possible)

- an f-control u_f;
- a sharing function k;
- a safe set $\mathcal{R}_s(v_h, t) \triangleq \mathcal{P}_a(t) \times \mathcal{H}_s \times \mathcal{A}_s \subset \mathcal{P}_a(t) \times \mathcal{H} \times \mathcal{A} \triangleq \mathcal{R}(v_h, t)$;

where $\mathcal{H}$ and $\mathcal{A}$ are the sets of heading angles and steering angles, respectively, such that the s-closed-loop system (2.1) and (2.2) has the following properties.

(1) The (x, y)-position of the car remains in its feasible Cartesian configuration set, *i.e.,* $(x(t), y(t)) \in \mathcal{P}_a(t)$ for all $t \geq 0$.
(2) The shared controller does not change the aims of the human driver if he/she drivers safely, *i.e.,*

$$\Omega_s = \begin{cases} \Omega_h & \text{if } \Omega_h \subset \mathcal{R}_s(v_h, t), \\ \Pi_{\mathcal{R}_s}(\Omega_h) & \text{if } \Omega_h \not\subset \mathcal{R}_s(v_h, t), \end{cases}$$

where Ω_s and Ω_h are the Ω-limit sets of the s-closed-loop system and h-closed-loop system, respectively. $\Pi_{\mathcal{R}_s}(\Omega_h)$ is the projection of Ω_h into the safe subset $\mathcal{R}_s(v_h, t)$, which will be defined in Sects. 2.3.1 and 2.5.1, with and without measurements of absolute positions, respectively.
(3) $u_s(t) = u_h(t)$ for all $t \geq 0$ such that the system state belongs to $\mathcal{R}_s(v_h, t)$.

Note that for any fixed v_h, $\mathcal{H}_s$ and $\mathcal{A}_s$ denote the sets of all heading angles and steering angles such that the car is unable to hit the boundary of $\mathcal{P}_a$ within an arbitrary short period of time, respectively.

Assumption 2.1 We assume that the projection of the car in the (x, y)-plane is a rectangle with length l and width w, where l denotes the distance between the middle of the front-axle and the middle of the rear-axle and w denotes the distances between the centers of the two rear wheels (or the centers of the two front wheels).

[1]We use $(x(t), y(t))$ to denote the Cartesian coordinates of the center of the rear-axle at the time instant t. The admissible Cartesian configuration set $\mathcal{P}_a(t)$ is such that

$$(x(t), y(t)) \in \mathcal{P}_a(t), \text{ for all } t \geq 0.$$

2.3 Design of the Shared-Control Law with Measurements of Absolute Positions

In this section, we provide a shared-control algorithm for the kinematic model of a rear-wheel drive car with measurements of absolute positions.

Assumption 2.2 We assume that the admissible Cartesian configuration set $\mathcal{P}_a$ is defined by a group of linear inequalities given by

$$\mathcal{P}_a = \{p \in \mathbb{R}^2 \,|Sp + T \leq 0\}, \tag{2.3}$$

where $p = [x, y]^T$, $S = [s_1^T, s_2^T, \ldots, s_m^T]^T \in \mathbb{R}^{m\times 2}$ and $T = [t_1, t_2, \ldots, t_m]^T \in \mathbb{R}^m$. In addition, if $m > 2$, the matrices S and T are such that

$$\text{rank}(\begin{bmatrix} s_{r_1} \\ \vdots \\ s_{r_g} \end{bmatrix}) < \text{rank}(\begin{bmatrix} s_{r_1} & t_{r_1} \\ \vdots & \vdots \\ s_{r_g} & t_{r_g} \end{bmatrix}),$$

for all $g \in [3, m]$ and $r_1, r_2, \ldots, r_g \in \{1, 2, \ldots, m\}$.

2.3.1 *Design of the Feedback Controller*

Lemma 2.1 *Consider the set $\mathcal{P}_a$ in (2.3) and assume Assumption 2.2 holds. Then for any fixed v_h and θ no more than two constraints are active*[2] *at any given time instant.*

Proof If $m \leq 2$ then the claim trivially holds. Consider now the case $m \geq 3$. We prove the claim by contradiction. Suppose that at a given time instant, for a fixed v_h and θ, three constraints are active. Without loss of generality assume that these are the first 3 constraints. Then there exists a positive constant η such that

$$\begin{bmatrix} s_1 \\ s_2 \\ s_3 \end{bmatrix} \left(p + \eta \begin{bmatrix} v_h \cos\theta \\ v_h \sin\theta \end{bmatrix} \right) + \begin{bmatrix} t_1 \\ t_2 \\ t_3 \end{bmatrix} = 0.$$

By Assumption 2.2 the above equation does not have any solutions, hence the claim.

As detailed in [6] the m inequality constraints can be partitioned into N_c groups, where $N_c \leq \binom{m}{2}$. Each group contains two constraints. Relative to the ith group of

---

[2]For each human input v_h and heading angle θ, the ith constraint is active at position p if there exists a positive η such that $s_i \left(p + \eta \begin{bmatrix} v_h \cos\theta \\ v_h \sin\theta \end{bmatrix} \right) + t_i = 0$. As a results, the constraint is said to be "active" if the car is moving towards the constraint.

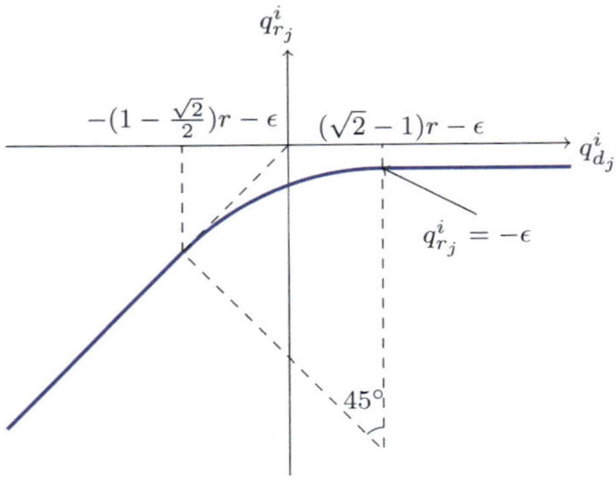

Fig. 2.2 Graph of the function $q^i_{r_j}$ given by (2.6) for $j \in \{1, 2\}$

constraints, *i.e.*,

$$q^i = S^i p + T^i \leq 0, \tag{2.4}$$

where S^i is invertible, we define a new variable $z^i = [z^i_1, z^i_2]^T$, with

$$z^i_j = \log \frac{q^i_j}{q^i_{r_j}}, \tag{2.5}$$

for all $j \in \{1, 2\}$, where $q^i_{r_j}$ is the reference signal relative to q^i_j and is defined by

$$q^i_{r_j} = \begin{cases} q^i_{d_j}, & \text{if } q^i_{d_j} \leq (1 - \frac{\sqrt{2}}{2})r - \epsilon, \\ -\epsilon, & \text{if } q^i_{d_j} \geq (\sqrt{2}-1)r - \epsilon, \\ -(r+\epsilon) + \sqrt{r^2 - [(\sqrt{2}-1)r - \epsilon - q^i_{d_j}]^2}, & \text{otherwise }, \end{cases} \tag{2.6}$$

where r is a user-selected positive constant, ϵ is a sufficiently small positive number, $q^i_d = [q^i_{d_1}, q^i_{d_2}]^T = S^i p_d + T^i$ and p_d describes the desired trajectory in the space $\mathcal{P}$. Note that the definition of $q^i_{r_j}$ is illustrated in Fig. 2.2 and $[{S^i}^{-1}(q^i_r - T^i)]$ is the projection of p_d into the admissible Cartesian configuration set $\mathcal{P}_a$ relative to the ith group of constraints.

It is clear that $q^i_{r_j}$ is a smooth function taking negative values. Therefore, $\dot{q}^i_{r_j}$ exists. Let $\alpha^i_r = [\alpha^i_{r_1}, \alpha^i_{r_2}]^T = {S^i}^{-1} \dot{q}^i_r$. Then[3]

[3]The definition of the function atg(·), given in [8], is close to that of the standard four quadrant arctan function except that its range equals to $(-\infty, +\infty)$ rather than $[-\pi, \pi)$. In addition, it is a smooth function with values in the range $(-\infty, +\infty)$. Therefore $\dot{\theta}^i_r$ always exist, which is a necessary condition for calculating $\dot{L}^i$, where L^i is given by (2.9).

$$
\begin{aligned}
p_r^i &= S^{i^{-1}}(q_r^i - T^i), \ \theta_r^i = \operatorname{atg}(\alpha_{r_2}^i, \alpha_{r_1}^i),\\
v_r^i &= \sqrt{{\alpha_{r_1}^i}^2 + {\alpha_{r_2}^i}^2}, \ \ \phi_r^i = \operatorname{atan}\left(\frac{\dot{\theta}_r^i l}{v_r^i}\right),\\
\omega_r^i &= \dot{\phi}_r^i.
\end{aligned}
\tag{2.7}
$$

Let $(p_d, \theta_d, \phi_d) \in \Omega_h$, then the projection of (p_d, θ_d, ϕ_d) into the safe subset $\mathcal{R}_s^i$ is defined as

$$
\Pi_{\mathcal{R}_s}^i(p_d, \theta_d, \phi_d) = (p_r^i, \theta_r^i, \phi_r^i),
$$

where $(p_r^i, \theta_r^i, \phi_r^i)$ is given by (2.7). Furthermore, the projection of Ω_h into the set $\mathcal{R}_s^i$ is defined as

$$
\Pi_{\mathcal{R}_s}^i(\Omega_h) = \{s \in \mathcal{R}_s | s = \Pi_{\mathcal{R}_s}^i(p_d, \theta_d, \phi_d)\}, \quad \forall (p_d, \theta_d, \phi_d) \in \Omega_h.
$$

System (2.1) controlled by the feedback controller ${u'}_f^i$ can be rewritten, using the variable z^i, as

$$
\begin{aligned}
\dot{z}_1^i &= \frac{{v'}_f^i \cos\theta^i}{e^{z_1^i} q_{r_1}^i} - \frac{v_r^i \cos\theta_r^i}{q_{r_1}^i},\\
\dot{z}_2^i &= \frac{{v'}_f^i \cos\theta^i}{e^{z_2^i} q_{r_2}^i} - \frac{v_r^i \cos\theta_r^i}{q_{r_2}^i},\\
\dot{\theta}^i &= \frac{{v'}_f^i \tan\phi^i}{l},\\
\dot{\phi}^i &= {\omega'}_f^i.
\end{aligned}
\tag{2.8}
$$

Let

$$
\begin{aligned}
\theta^{i*} &= \operatorname{atg}(e^{z_2^i}(v_r^i \sin\theta_r^i - \gamma_2 z_2^i), e^{z_1^i}(v_r^i \cos\theta_r^i - \gamma_1 z_1^i)),\\
\phi^{i*} &= \operatorname{atg}\left(\begin{matrix} l\dfrac{z_1^i}{e^{z_1^i} q_{r_1}^i} \sin\dfrac{\theta + \theta^{i*}}{2} \operatorname{sinc}\dfrac{\theta - \theta^{i*}}{2} + \dfrac{\theta^{i*} l}{{v'}_f^i} \\ -l\dfrac{z_2^i}{e^{z_2^i} q_{r_2}^i} \cos\dfrac{\theta + \theta^{i*}}{2} \operatorname{sinc}\dfrac{\theta - \theta^{i*}}{2} \end{matrix}, 1\right),
\end{aligned}
$$

where $\gamma_1 > 0$, $\gamma_2 > 0$.

Consider the Lyapunov function candidate relative to the ith group of constraints (2.4) given by

$$
L^i(z_1^i, z_2^i, \theta^i, \phi^i) = \frac{1}{2}[{z_1^i}^2 + {z_2^i}^2 + (\theta^i - \theta^{i*})^2 + (\tan\phi^i - \tan\phi^{i*})^2], \tag{2.9}
$$

and choose ${u'}_f^i = [{v'}_f^i, {\omega'}_f^i]^T$ such that $L^i < 0$ for all $(z_1^i, z_2^i, \theta^i, \phi^i) \neq (0, 0, \theta^{i*}, \phi^{i*})$. One such choice is given by

$$v_f^i = \sqrt{e^{2z_1^i}(v_r^i \cos\theta_r^i - \gamma_1 z_1^i)^2 + e^{2z_2^i}(v_r^i \sin\theta_r^i - \gamma_2 z_2^i)^2},$$
$$\omega_f^i = \cos^2\phi^i\left[-\frac{v(\theta^i - \theta^{i*})}{l} + \frac{\dot{\phi}^{i*}}{\cos^2\phi^{i*}} - \gamma_3(\tan\phi - \tan\phi^{i*})\right],$$

which yields

$$\dot{L}^i = \gamma_1 \frac{{z_1^i}^2}{q_{r_1}^i} + \gamma_2 \frac{{z_2^i}^2}{q_{r_2}^i} - \gamma_3(\tan\phi - \tan\phi^*)^2 \leq 0.$$

This can be "pulled back" into the (p, θ, ϕ) coordinates by

$$v_f^i = \sqrt{\left(\frac{q_1^i}{q_{r_1}^i}\right)^2\left(v_r^i\cos\theta_r^i - \gamma_1 \log\frac{q_1^i}{q_{r_1}^i}\right)^2 + \left(\frac{q_2^i}{q_{r_2}^i}\right)^2\left(v_r^i\sin\theta_r^i - \gamma_2 \log\frac{q_2^i}{q_{r_2}^i}\right)^2},$$
$$\omega_f^i = \cos^2\phi^i\left[-\frac{v(\theta^i - \theta^{i*})}{l} + \frac{\dot{\phi}^{i*}}{\cos^2\phi^{i*}} - \gamma_3(\tan\phi - \tan\phi^{i*})\right], \tag{2.10}$$

where

$$\phi^{i*} = \operatorname{atg}\left(\frac{l\dfrac{\log\frac{q_1^i}{q_{r_1}^i}}{q_1^i}\sin\dfrac{\theta+\theta^{i*}}{2}\operatorname{sinc}\dfrac{\theta-\theta^{i*}}{2} + \dfrac{\theta^{i*} l}{v_f^i}}{-l\dfrac{\log\frac{q_2^i}{q_{r_2}^i}}{q_2^i}\cos\dfrac{\theta+\theta^{i*}}{2}\operatorname{sinc}\dfrac{\theta-\theta^{i*}}{2}}, 1\right), \tag{2.11}$$
$$\theta^{i*} = \operatorname{atg}\left(\frac{q_2^i}{q_{r_2}^i}\left(v_r^i\cos\theta_r^i - \gamma_2\log\frac{q_2^i}{q_{r_2}^i}\right), \frac{q_1^i}{q_{r_1}^i}\left(v_r^i\cos\theta_r^i - \gamma_1\log\frac{q_1^i}{q_{r_1}^i}\right)\right),$$

and $q_j^i = s_j^i p + t_j^i$, $q_{r_j}^i = s_j^i p_r + t_j^i$ for all $j \in \{1, 2\}$.

2.3.2 Shared-Control Algorithm

The overall set $\mathcal{R}$, denoting the set of all feasible values of the system states, can be partitioned into three subsets: the safe subset $\mathcal{R}_s$, the hysteresis subset $\mathcal{R}_h$ and the dangerous subset $\mathcal{R}_d$. The set division is based on the distance between the current (x, y)-position and the boundary of the set $\mathcal{P}_a$ and the velocity towards the boundary. Relative to the ith group of constraints, the three subsets are defined by (12) given on the top of next page,

$$\tilde{\mathcal{R}}_s^i(v_h) = \left\{ \begin{array}{r} (q^i, \theta^i, \phi^i) \in \mathcal{Q}_a^i \times \mathcal{H} \times \mathcal{A} : m_j^i < \dfrac{1}{q_j^i + b_2} - \dfrac{1}{b_2} \text{ if } q_j^i \geq -b_2 \\ \text{for all } j \in \{1, 2\} \end{array} \right\}$$

$$\tilde{\mathcal{R}}_h^i(v_h) = \left\{ \begin{array}{r} (q^i, \theta^i, \phi^i) \in \mathcal{Q}_a^i \times \mathcal{H} \times \mathcal{A} : \quad m_j^i \leq \dfrac{1}{q_j^i + b_1} - \dfrac{1}{b_1} \text{ if } q_j^i \geq -b_1 \\ \text{for all } j \in \{1, 2\} \\ \text{and } \exists k \in \{1, 2\} \text{ such that} \\ m_k^i \geq \dfrac{1}{q_k^i + b_2} - \dfrac{1}{b_2} \text{ and } q_k^i \geq -b_2 \end{array} \right\} \tag{12}$$

$$\tilde{\mathcal{R}}_d^i(v_h) = \left\{ \begin{array}{r} (q^i, \theta^i, \phi^i) \in \mathcal{Q}_a^i \times \mathcal{H} \times \mathcal{A} : \qquad \exists j \in \{1, 2\} \text{ such that} \\ m_j^i > \dfrac{1}{q_j^i + b_1} - \dfrac{1}{b_1}, -b_1 \leq q_j^i < 0 \\ \text{or } \exists j \in \{1, 2\} \text{ such that} \\ m_j^i > \dfrac{1}{q_j^i + b_1} - \dfrac{1}{b_1}, q_j^i = 0 \end{array} \right\}$$

where[4] $\mathcal{Q}_a^i = S^i \mathcal{P}_a + T^i$, $b_2 > b_1 > 0$, $m_j^i = s_j^i[\cos\theta^i, \sin\theta^i]^T v_h$ and $m_k^i = s_k^i$ $[\cos\theta^i, \sin\theta^i]^T v_h$ with $k, j \in \{1, 2\}$.

Example 2.1 Assume that the admissible Cartesian configuration set $\mathcal{P}_a$ is a square described by (2.3), where $S = [a_1^T, a_2^T, a_3^T, a_4^T]^T$ and $T = [-3, -3, -3, -3]^T$ with $a_1 = [-1, 0]$, $a_2 = [1, 0]$, $a_3 = [0, -1]$ and $a_4 = [0, 1]$. Figure 2.3 on the top of next page provides graphical illustrations of the three subsets for different values of v_h and θ.

Note that this definition is given in the (q^i, θ^i, ϕ^i) coordinates and can be transferred to the (x, y, θ^i, ϕ^i) coordinates by the equations

$$\begin{aligned} \mathcal{R}_s^i(v_h) &= \mathrm{diag}(\mathrm{S^i}^{-1}, \mathrm{I})(\tilde{\mathcal{R}}_\mathrm{s}^\mathrm{i} - \mathrm{col}(\mathrm{T^i}, 0)), \\ \mathcal{R}_h^i(v_h) &= \mathrm{diag}(\mathrm{S^i}^{-1}, \mathrm{I})(\tilde{\mathcal{R}}_\mathrm{h}^\mathrm{i} - \mathrm{col}(\mathrm{T^i}, 0)), \\ \mathcal{R}_d^i(v_h) &= \mathrm{diag}(\mathrm{S^i}^{-1}, \mathrm{I})(\tilde{\mathcal{R}}_\mathrm{d}^\mathrm{i} - \mathrm{col}(\mathrm{T^i}, 0)), \end{aligned}$$

where $\mathrm{col}(\mathrm{T^i}, 0)$ is a column vector built by stacking the zero vector under the vector T^i.

[4]The notation $S^i \mathcal{P}_a + T^i$ with $S^i \in \mathbb{R}^{2\times 2}$, $\mathcal{P}_a \in \mathbb{R}^2$ and $T^i \in \mathbb{R}^2$, denotes the set defined by

$$\{x \in \mathbb{R}^2 | x = S^i y + T^i, y \in \mathcal{P}_a\}.$$

.

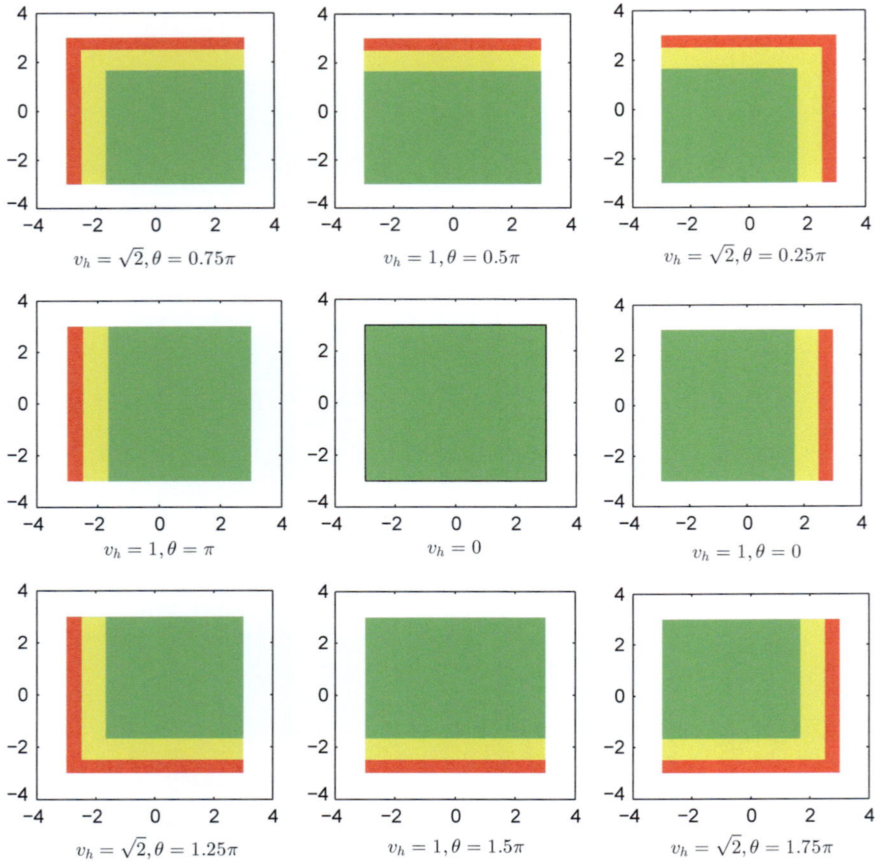

Fig. 2.3 Illustration of the safe (*green*), the hysteresis (*yellow*) and the dangerous subsets (*red*) for the case $n = 2$ ($b_1 = 1, b_2 = 2$)

The sharing function relative to the ith group of constraints can then be defined as

$$k^i(p, \theta, v_h) = \begin{cases} 0, & (p, \theta, \phi) \in \mathcal{R}_d^i(v_h) \text{ and } L^i = \min_{j=I_1}^{I_{n_c}} L^j, \\ l^i, & (p, \theta, \phi) \in \mathcal{R}_h^i(v_h), \\ 1, & \text{otherwise}, \end{cases} \tag{2.13}$$

where

$$l^i = \begin{cases} 1, & \text{if } k^i(t^-) = 0, \\ 0, & \text{if } k^i(t^-) = 1. \end{cases}$$

Note that in the cases in which two constraints are active, there is only one group of active constraints. However, if only one constraint is active, there could be n_c groups of active constraints, where $n_c \leq m - 1$. Therefore, the activation of the ith group of constraints does not imply the activation of the ith feedback controller u_f^i.

Suppose $I_1, I_2, \ldots, I_{n_c}$ groups of constraints are active at the time instant t, where $I_i \in \{1, 2, \ldots, N_c\}$ for all $i \in \{1, 2, \ldots, n_c\}$ and L^j is defined in Eq. (2.9).

Finally, the shared-control input u_s is given as

$$\begin{aligned} u_s(p, \theta, \phi, v_h) = &\sum_{i=1}^{N_c}[(1 - k^i(p, \theta, v_h))u^i_f(p, \theta, \phi, p_r, \theta_r, \alpha_r)] \\ &+ \min_{i=1}^{N_c} k^i(p, \theta, v_h)\, u_h(t). \end{aligned} \tag{2.14}$$

Proposition 2.1 *Consider the kinematic model of a rear-wheel car (2.1) with the shared-control algorithm (2.10)–(2.13) and (2.14). Let $\mathcal{P}_a$ be a given compact and nonempty admissible Cartesian configuration set described by (2.3) and u_h be a given h-control. Assume $(x(0), y(0)) \in \mathcal{P}_a$. Then there exists positive γ_1, γ_2, γ_3 and $b_2 > b_1 > 0$ such that the following properties hold for the s-closed-loop system.*

i) $(x(t), y(t)) \in \mathcal{P}_a$ *for all* $t \geq 0$.
ii) $\Omega_s = \Pi_{\mathcal{R}_s}(\Omega_h)$.
iii) $u_s(t) = u_h(t)$ *for all* $t \geq 0$ *such that* $(p(t), \theta(t), \phi(t)) \in \mathcal{R}_s(v_h(t))$.

Proof To begin with, as detailed in Sect. 2.3.1, the f-control is such that the Cartesian configuration of the system stays in the admissible set $\mathcal{P}_a$. In addition, as detailed in [7] any trajectory should enter the dangerous subset $\mathcal{R}_d$ before leaving $\mathcal{R}$ where u_f is active. Therefore, the set $\mathcal{R}$ is forward invariant and claim i) holds.

Claim ii) is a consequence of the general results in [18], and of the fact that $\Pi_{\mathcal{R}_s}(\Omega_h)$ is the Ω-limit set of both the h-closed-loop and the f-closed-loop systems (by definition, the former, and by Eqs. (2.5)–(2.9), the latter).

Finally, claim iii) is a direct consequence of the definition of the shared-control input u_s.

2.4 Disturbance Rejections

The model studied in Sect. 2.3 does not include uncertainties. In this section, we include additive disturbances to the system dynamics to make the model more flexible and then provide a solution to the underlying robust shared-control problem.

The dynamics of the kinematic model of a rear-wheel drive car with disturbances can be written as

$$\begin{aligned} \dot{x} &= v_s \cos\theta + d_x, \\ \dot{y} &= v_s \sin\theta + d_y, \\ \dot{\theta} &= \frac{v_s \tan\phi}{l} + d_\theta, \\ \dot{\phi} &= \omega_s + d_\phi, \end{aligned} \tag{2.15}$$

where d_x, d_y, d_θ and d_ϕ are disturbances on the dynamics of x, y, θ and ϕ, respectively.

Assumption 2.3 The disturbances are bounded, *i.e.*, $|d_x| \le d_1$, $|d_y| \le d_2$, $|d_\theta| \le d_3$ and $|d_\phi| \le d_4$, where d_1, d_2, d_3 and d_4 are positive constants known to the users.

We assume that Assumptions 2.2 and 2.3 hold for the rest of this section.

With the use of z^i defined in (2.5), system (2.15) with the feedbacks controller u^i_f can be rewritten as

$$\begin{aligned}
\dot{z}^i_1 &= \frac{s^i_{11}(v^i_f \cos\theta + d_x) + s^i_{12}(v^i_f \sin\theta + d_y)}{e^{z^i_1} q^i_{r_1}} - \frac{\dot{q}^i_{r_1}}{q^i_{r_1}},\\
\dot{z}^i_2 &= \frac{s^i_{21}(v^i_f \cos\theta + d_x) + s^i_{22}(v^i_f \sin\theta + d_y)}{e^{z^i_2} q^i_{r_2}} - \frac{\dot{q}^i_{r_2}}{q^i_{r_2}},\\
\dot{\theta} &= \frac{v^i_f \tan\phi}{l} + d_\theta,\\
\dot{\phi} &= \omega^i_f + d_\phi,
\end{aligned} \tag{2.16}$$

where q^i_r is defined by (2.6) and $S^i = [{s^i_1}^T, {s^i_2}^T]^T$ with $s^i_1 = [s^i_{11}, s^i_{12}]$ and $s^i_2 = [s^i_{21}, s^i_{22}]$.

Let

$$\tan\phi^{i*} = \frac{l}{v^i_f}\begin{bmatrix}
\dot{\theta}^{i*} - f(\theta-\theta^{i*})d_3 + \dfrac{z^i_1 v^i_f s^i_{11}}{m^i_{r_1} e^{z^i_1}} \sin\dfrac{\theta+\theta^{i*}}{2} \operatorname{sinc}\dfrac{\theta+\theta^{i*}}{2} \\
- \dfrac{z^i_1 v^i_f s^i_{12}}{m^i_{r_1} e^{z^i_1}} \cos\dfrac{\theta+\theta^{i*}}{2} \operatorname{sinc}\dfrac{\theta+\theta^{i*}}{2} \\
+ \dfrac{z^i_2 v^i_f s^i_{21}}{m^i_{r_2} e^{z^i_2}} \sin\dfrac{\theta+\theta^{i*}}{2} \operatorname{sinc}\dfrac{\theta+\theta^{i*}}{2} \\
- \dfrac{z^i_2 v^i_f s^i_{22}}{m^i_{r_2} e^{z^i_2}} \cos\dfrac{\theta+\theta^{i*}}{2} \operatorname{sinc}\dfrac{\theta+\theta^{i*}}{2}
\end{bmatrix},$$

$$S^i \begin{bmatrix} v^i_f \cos\theta^{i*} \\ v^i_f \sin\theta^{i*} \end{bmatrix} = \begin{bmatrix} s^i_{11} f(s^i_{11} z^i_1) d_1 + s^i_{12} f(s^i_{12} z^i_1) d_2 + e^{z^i_1}(\dot{m}^i_{r_1} + \gamma_1 z^i_1) \\ s^i_{21} f(s^i_{21} z^i_2) d_1 + s^i_{22} f(s^i_{12} z^i_2) d_2 + e^{z^i_2}(\dot{m}^i_{r_2} + \gamma_2 z^i_2) \end{bmatrix} = \begin{bmatrix} \beta \\ \zeta \end{bmatrix}, \tag{2.17}$$

where γ_1 and γ_2 are two user-selected positive constants. As stated in Sect. 2.3.1, S^i is invertible. Define $m^i = [m^i_1, m^i_2]^T$ as

$$\begin{bmatrix} m^i_1 \\ m^i_2 \end{bmatrix} = \begin{bmatrix} v^i_f \cos\theta^{i*} \\ v^i_f \sin\theta^{i*} \end{bmatrix} = {S^i}^{-1} \begin{bmatrix} \beta \\ \zeta \end{bmatrix}.$$

Then v^i_f and θ^{i*} can be calculated as[5]

[5]Recall that the definition of the function atg(·) is given in [8].

$$v_f^i = \sqrt{(m_1^i)^2 + (m_2^i)^2}, \quad \theta^{i*} = \operatorname{atg}(m_2^i, m_1^i). \tag{2.18}$$

Consider the Lyapunov function (2.9) with θ^{i*} and $\tan\phi^{i*}$ defined by (2.18) and (2.17), respectively. Select ω_f^i, together with v_f^i given by (2.18), such that the derivative of L^i satisfies

$$\dot{L}^i \leq -k_1(z_1^i)^2 - k_2(z_2^i)^2 - k_3(\tan\phi - \tan\phi^{i*})^2 + K,$$

where $k_i > 0$ for all $i \in \{1, 2, 3\}$, and K is a positive constant. One such selection is given by

$$\begin{aligned} \omega_f^i = & \cos^2\phi \left[\frac{\dot{\phi}^{i*}}{\cos^2\phi^{i*}} - \gamma_3(\tan\phi - \tan\phi^{i*}) - \frac{v_f^i}{l}(\theta - \theta^{i*}) \right] \\ & - d_4 \epsilon \tanh \frac{c(\tan\phi - \tan\phi^{i*})}{\epsilon}, \end{aligned} \tag{2.19}$$

yielding

$$\dot{L}^i \leq \frac{\gamma_1(z_1^i)^2}{q_{r_1}^i} + \frac{\gamma_2(z_2^i)^2}{q_{r_2}^i} - \gamma_3(\tan\phi - \tan\phi^{i*})^2 + K, \tag{2.20}$$

where

$$\begin{aligned} K = & \frac{-(\epsilon+1)}{2c(\epsilon-1)} \ln\frac{\epsilon+1}{\epsilon-1} \left(d_1 e^{\frac{\epsilon}{2c|s_{11}^i|}} + d_2 e^{\frac{\epsilon}{2c|s_{12}^i|}} \right) + \frac{\epsilon}{2c} d_3 \ln\frac{\epsilon+1}{\epsilon-1} \\ & - \frac{(\epsilon+1)}{2c(\epsilon-1)} \ln\frac{\epsilon+1}{\epsilon-1} \left(d_1 e^{\frac{\epsilon}{2c|s_{21}^i|}} + d_2 e^{\frac{\epsilon}{2c|s_{22}^i|}} \right). \end{aligned}$$

with $c > 0$ and $\epsilon > 1$ user-selected constants.

Note that by choosing proper values of ϵ and c, we can render K arbitrarily small. In addition, (2.20) indicates that

$$\lim_{t\to\infty} |z_1^i| \leq \sqrt{-K q_{r_1}^i / \gamma_1}, \quad \lim_{t\to\infty} |z_2^i| \leq \sqrt{-K q_{r_2}^i / \gamma_2}.$$

This discussion is summarized in the following statement.

Proposition 2.2 *Consider the system (2.15) with the feedback controller* $u_f^i = [v_f^i, \omega_f^i]^T$ *given by (2.18) and (2.19). In addition,* q_r^i, θ^{i*} *and* $\tan\phi^{i*}$ *are given by (2.6), (2.18) and (2.17), respectively. Assume that the admissible configuration set of the car* $\mathcal{P}_a$ *is defined by (2.3) and* $(x(0), y(0)) \in \mathcal{P}_a$. *Then the closed-loop system has the following properties.*

i) $(x(t), y(t)) \in \mathcal{P}_a$ *for all* $t \geq 0$.
ii) *Define the tracking errors as* $e^i(t) = [e^i_x(t), e^i_y(t)]^T$ *with*

$$e^i_x(t) = \left| |\frac{x(t)}{x^i_r(t)}| - 1 \right|, \quad e^i_y(t) = \left| |\frac{y(t)}{y^i_r(t)}| - 1 \right|.$$

Then $\lim_{t\to\infty} e^i(t)$ *is bounded and the bound can be controlled by tuning the parameters,* $\gamma_1, \gamma_2, \epsilon$ *and* c*, of the feedback controller.*

Remark 2.1 According to (2.17) $\tan\phi^{i*}$ is not defined when $v^i_f = 0$. In this case, we redesign the feedback controller as

$$\omega^i_f = \omega^i_r - \text{sign}(\phi - \phi^i_r)d_4 - \gamma_4(\phi - \phi^i_r), \tag{2.21}$$

with $\gamma_4 > 0$.

Consider the Lyapunov function[6] $L^i = \frac{1}{2}(\phi - \phi^i_r)^2$ with ω^i_f given by (2.21), then the derivative of L^i is calculated as

$$\begin{aligned}\dot{L}^i &= (\phi - \phi^i_r)(\omega^i_f + d_\phi - \omega^i_r) = (\phi - \phi^i_r)[d_\phi - \text{sign}(\phi - \phi^i_r)d_4 - \gamma_4(\phi - \phi^i_r)] \\ &\leq -\gamma_4(\phi - \phi^i_r)^2 \leq 0.\end{aligned}$$

2.5 Design of the Shared Control Without Measurements of Absolute Positions

In this section, we discuss how to design the shared controller for the kinematic model of a rear-wheel drive car without absolute positioning while satisfying all the properties given in Sect. 2.2 for any nonempty admissible configuration set. In the rest of this section, we assume that the measurements are the distance to the obstacle along the current forward direction D, the angular difference between the real and the reference heading angle θ_e, the wheel angle ϕ and the distances to the obstacles along and orthogonal to the reference forward direction, d_1 and

$$d_2 = \begin{cases} d_{2l}, & \text{if } |\log \dfrac{d_{2l} - \tilde{d}_2}{d_{r_{2l}} - \tilde{d}_2}| \leq |\log \dfrac{d_{2r} - \tilde{d}_2}{d_{r_{2r}} - \tilde{d}_2}|, \\ d_{2r}, & \text{if } |\log \dfrac{d_{2l} - \tilde{d}_2}{d_{r_{2l}} - \tilde{d}_2}| > |\log \dfrac{d_{2r} - \tilde{d}_2}{d_{r_{2r}} - \tilde{d}_2}|, \end{cases}$$

where d_{2l} and d_{2r} denote the distances to the obstacles on the left-hand side and right-hand side orthogonal to the reference forward direction, respectively. In addition, $\tilde{d}_2$ is given in Sect. 2.5.1.

[6]Note that this is a Lyapunov function only for the ϕ system.

2.5.1 Design of the Feedback Controller

According to the definition of $d_2, d_2 = d_{2l}$ or $d_2 = d_{2r}$. Without loss of generality, we provide the design of the shared controller only for the case in which $d_2 = d_{2l}$. The dynamics of the closed-loop system with the feedback controller $u_f = [v_f, \omega_f]^T$ can be written, using the variables d_1, d_2, θ_e and ϕ, as

$$\begin{aligned} \dot{d}_1 &= -\, v_f \cos\theta_e, \\ \dot{d}_2 &= -\, v_f \sin\theta_e, \\ \dot{\theta}_e &= \frac{v_f \tan\phi - v_r \tan\phi_r}{l}, \\ \dot{\phi} &= \omega_f. \end{aligned} \tag{2.22}$$

In addition, for any nonempty admissible configuration set $\mathcal{P}_a$, the constraint $(x(t), y(t)) \in \mathcal{P}_a$ can be rewritten as

$$d_i(t) \geq \tilde{d}_i, \quad \forall i \in \{1, 2\}, \tag{2.23}$$

where $\tilde{d}_i$ is a positive constant. One choice for $\tilde{d}_1$ and $\tilde{d}_2$ is $\tilde{d}_1 = \tilde{d}_2 = \sqrt{l^2 + (\frac{w}{2})^2}$.

Assumption 2.4 The signals d_{d_1} and d_{d_2}, describing the desired distances to the relative obstacles (the obstacles along and orthogonal to the reference forward direction), are continuous.

The previous assumption holds for the rest of this section.

Define the variable $z = [z_1, z_2]^T$, based on the measurements of distances, as

$$z_i = \log\frac{d_i - \tilde{d}_i}{d_{r_i} - \tilde{d}_i}, \forall i \in \{1, 2\}, \tag{2.24}$$

where d_{r_1} and d_{r_2} are the feasible reference signal relative to d_1 and d_2, respectively and are defined as

$$d_{r_i} = \begin{cases} d_{d_i}, & \text{if } d_{d_i} \geq (1 - \frac{\sqrt{2}}{2})r + \epsilon + \tilde{d}_i, \\ \epsilon + \tilde{d}_i, & \text{if } d_{d_i} \leq (1 - \sqrt{2})r + \epsilon + \tilde{d}_i, \\ m_i, & \text{otherwise}\ , \end{cases} \tag{2.25}$$

where $m_i = r + \epsilon + \tilde{d}_i - \sqrt{r^2 - [(\sqrt{2} - 1)r - \epsilon + d_{d_i} - \tilde{d}_i]^2}$ and ϵ is a sufficiently small positive constant. Note that d_{r_i} is a smooth continuous function, indicating that $\dot{d}_{r_i}$ exists.

Suppose $(d_{d_1}, d_{d_2}, \theta_d, \phi_d)$ belongs to the Ω-limit set for the h-closed-loop system, *i.e.*, $(d_{d_1}, d_{d_2}, \theta_d, \phi_d) \in \Omega_h$, then the projection of it into the safe subset $\mathcal{R}_s$ is given by

$$\Pi_{\mathcal{R}_s}(d_{d_1}, d_{d_2}, \theta_d, \phi_d) = (d_{r_1}, d_{r_2}, \theta_d, \phi_d),$$

where d_{r_1} and d_{r_2} are defined by (2.25). Hence, the projection of Ω_h into the safe subset $\mathcal{R}_s$ is defined as

$$\Pi_{\mathcal{R}_s}(\Omega_h) = \left\{ s \in \mathcal{R}_s \,\middle|\, s = \Pi_{\mathcal{R}_s}(d_{d_1}, d_{d_2}, \theta_d, \phi_d), \quad \forall (d_{d_1}, d_{d_2}, \theta_d, \phi_d) \in \Omega_h \right\}.$$

With the variable z given by (2.24), system (2.22) can be written as

$$\begin{aligned}
\dot{z}_1 &= \frac{v_r}{d_{r_1} - \tilde{d}_1} - \frac{v_f \cos\theta_e}{d_1 - \tilde{d}_1}, \\
\dot{z}_2 &= -\frac{v_f \sin\theta_e}{d_2 - \tilde{d}_2}, \\
\dot{\theta}_e &= \frac{v_f \tan\phi - v_r \tan\phi_r}{l}, \\
\dot{\phi} &= \omega_f.
\end{aligned}$$

Let[7]

$$\theta_e^* = \operatorname{atg}\left(\gamma_2 (d_2 - \tilde{d}_2) z_2,\, (d_1 - \tilde{d}_1)\left(\frac{v_r}{d_{r_1} - \tilde{d}_1} + \gamma_1 z_1 \right) \right), \tag{2.26a}$$

$$\xi = \sqrt{(\gamma_2 (d_2 - \tilde{d}_2) z_2)^2 + \left[(d_1 - \tilde{d}_1)\left(\frac{v_r}{d_{r_1} - \tilde{d}_1} + \gamma_1 z_1 \right) \right]^2}, \tag{2.26b}$$

$$\phi^* = \operatorname{atg}\left(\begin{array}{c} \dfrac{l z_2}{d_2 - \tilde{d}_2} \cos\dfrac{\theta_e + \theta_e^*}{2} \operatorname{sinc}\dfrac{\theta_e - \theta_e^*}{2} + \dfrac{v_r \tan\phi_r}{\xi} \\ -\dfrac{l z_1}{d_1 - \tilde{d}_1} \sin\dfrac{\theta_e + \theta_e^*}{2} \operatorname{sinc}\dfrac{\theta_e - \theta_e^*}{2} + \dfrac{l \dot{\theta}_e^*}{\xi} \end{array}, 1 \right), \tag{2.26c}$$

where $\gamma_1 > 0$, $\gamma_2 > 0$.

Consider the Lyapunov function

$$L(z_1, z_2, \theta_e, \phi, t) = \frac{1}{2}[z_1^2 + z_2^2 + (\theta_e - \theta_e^*)^2 + (\tan\phi - \tan\phi^*)^2], \tag{2.27}$$

and choose v_f and ω_f such that $\dot{L}(t) < 0$ for all $t \geq 0$ and $(z_1, z_2, \theta_e, \tan\phi) \neq (0, 0, 0, \tan\phi_r)$. One such a choice is given by

$$\begin{aligned}
v_f &= \xi, \\
\omega_f &= \cos^2\phi \left[\frac{\dot{\phi}^*}{\cos^2\phi^*} - \frac{v(\theta_e - \theta_e^*)}{l} - \gamma_3 (\tan\phi - \tan\phi^*) \right],
\end{aligned}$$

[7] Recall that the definition of the function atg(·) is given in [8].

where γ_3 is a positive constant. This can be transfered into the $(d_1, d_2, \theta_e, \phi)$ coordinates by

$$\upsilon_f = \sqrt{(d_1 - \tilde{d}_1)^2\left(\frac{\upsilon_r}{d_{r_1} - \tilde{d}_1} + \gamma_1 \log\frac{d_1 - \tilde{d}_1}{d_{r_1} - \tilde{d}_1}\right)^2 + \gamma_2^2 (d_2 - \tilde{d}_2)^2 \log^2\frac{d_2 - \tilde{d}_2}{d_{r_2} - \tilde{d}_2}},$$
$$\omega_f = \cos^2\phi\left[\frac{\dot{\phi}^*}{\cos^2\phi^*} - \frac{\upsilon(\theta_e - \theta_e^*)}{l} - \gamma_3(\tan\phi - \tan\phi^*)\right], \tag{2.28}$$

where

$$\theta_e^* = \text{atg}\left(\gamma_2(d_2 - \tilde{d}_2)\log\frac{d_2 - \tilde{d}_2}{d_{r_2} - \tilde{d}_2}, (d_1 - \tilde{d}_1)\left(\frac{\upsilon_r}{d_{r_1} - \tilde{d}_1} + \gamma_1\log\frac{d_1 - \tilde{d}_1}{d_{r_1} - \tilde{d}_1}\right)\right),$$
$$\phi^* = \text{atg}\left(\begin{array}{c} \frac{l}{d_2 - \tilde{d}_2}\log\frac{d_2 - \tilde{d}_2}{d_{r_2} - \tilde{d}_2}\cos\frac{\theta_e + \theta_e^*}{2}\text{sinc}\frac{\theta_e - \theta_e^*}{2} + \frac{\upsilon_r \tan\phi_r}{\upsilon_f} \\ -\frac{l}{d_1 - \tilde{d}_1}\log\frac{d_1 - \tilde{d}_1}{d_{r_1} - \tilde{d}_1}\sin\frac{\theta_e + \theta_e^*}{2}\text{sinc}\frac{\theta_e - \theta_e^*}{2} + \frac{l\dot{\theta}_e^*}{\upsilon_f} \end{array}, 1\right), \tag{2.29}$$

and $d_r = [d_{r_1}, d_{r_2}]^T$ is given by (2.25).

2.5.2 Shared-Control Algorithm

For any given υ_h, the safe, hysteresis, and dangerous subsets can be defined by (30) at the bottom of this page, where D is the measured distance to the obstacles along the current forward direction, b_1 and b_2 are user-selected parameters and $b_2 > b_1 > 0$.

$$\mathcal{R}_s(\upsilon_h) = \left\{(d_1, d_2, \theta_e, \phi) \in \mathbb{R}^+ \times \mathbb{R}^+ \times \mathbb{S} \times \mathbb{S} : \upsilon_h < \frac{1}{b_2 - D} - \frac{1}{b_2} \text{ if } D \le b_2\right\}$$

$$\mathcal{R}_h(\upsilon_h) = \left\{\begin{array}{r} (d_1, d_2, \theta_e, \phi) \in \mathbb{R}^+ \times \mathbb{R}^+ \times \mathbb{S} \times \mathbb{S} : \upsilon_h \le \frac{1}{b_1 - D} - \frac{1}{b_1} \\ \text{and } \upsilon_h \ge \frac{1}{b_2 - D} - \frac{1}{b_2} \\ \text{and } D \le b_1 \\ \text{or } \upsilon_h \ge \frac{1}{b_2 - D} - \frac{1}{b_2} \\ \text{and } b_1 \le D \le b_2 \end{array}\right\}$$

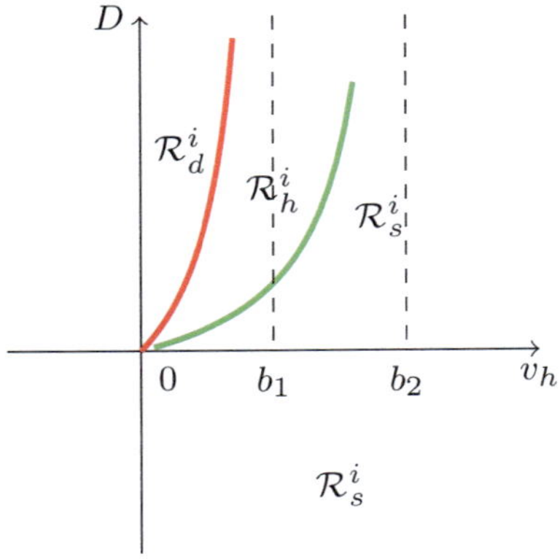

Fig. 2.4 Illustration of the safe, the hysteresis and the dangerous sets with measurements of D and v_h

$$\mathcal{R}_d(v_h) = \left\{ \begin{array}{r} (d_1, d_2, \theta_e, \phi) \in \mathbb{R}^+ \times \mathbb{R}^+ \times \mathbb{S} \times \mathbb{S} : v_h > \dfrac{1}{b_1 - D} - \dfrac{1}{b_1} \\ \text{and } 0 \leq D \leq b_1 \end{array} \right\} \tag{2.30}$$

Figure 2.4 illustrates how the overall set is divided into three subsets based on v_h and D. Basically, the state belongs to the dangerous set if D is close to zero (*i.e.*, $0 \leq D \leq b_1$) and v_h is large (*i.e.*, (D, v_h) is located above the red curve); the state belongs to the safe subset if D is large (*i.e.*, $D > b_2$) or v_h is small (*i.e.*, (D, v_h) is located below the green curve).

The sharing function k can then be defined as

$$k(D, v_h) = \begin{cases} 1, & (d_1, d_2, \theta_e, \phi) \in \mathcal{R}_s(v_h), \\ l, & (d_1, d_2, \theta_e, \phi) \in \mathcal{R}_h(v_h), \\ 0, & (d_1, d_2, \theta_e, \phi) \in \mathcal{R}_d(v_h), \end{cases} \tag{2.31}$$

where

$$l = \begin{cases} 1, \text{ if } (d_1, d_2, \theta_e, \phi) \text{ enters } \mathcal{R}_h(v_h) \text{ from } \mathcal{R}_s(v_h), \\ 0, \text{ if } (d_1, d_2, \theta_e, \phi) \text{ enters } \mathcal{R}_h(v_h) \text{ from } \mathcal{R}_d(v_h). \end{cases}$$

Finally, the shared-control input of the system (2.22) is given as

$$u_s = (1 - k(D, v_h))u_f(d, d_r, \theta_e, \phi, v_r) + k(D, v_h)u_h. \tag{2.32}$$

The following statement summarizes the discussions in this section without proof.

Proposition 2.3 *Consider the kinematic model of a rear-wheel drive car controlled by the shared-control law (2.28)–(2.31) and (2.32) without absolute positioning, the dynamics of which can be described by (2.22). Assume that the admissible Cartesian configuration set of the car is nonempty and the initial position of the car is feasible,* i.e., $(x(0), y(0)) \in \mathcal{P}_a, d_1(0) > \tilde{d}_1, d_2(0) > \tilde{d}_2$. *Let u_h be a given h-control. Then there exists positive $\gamma_1, \gamma_2, \gamma_3$, and $b_2 > b_1 > 0$ such that the closed-loop system has the following properties.*

i) $d_1(t) > \tilde{d}_1$ *and* $d_2(t) > \tilde{d}_2$ *for all* $t \geq 0$;
ii) $\Omega_s = \Pi_{\mathcal{R}_s}(\Omega_h)$;
iii) $u_s(t) = u_h(t)$ *for all t such that* $(d_1(t), d_2(t), \theta_e(t), \phi(t)) \in \mathcal{R}_s$.

Remark 2.2 In the cases in which $\tilde{d}_1 = \tilde{d}_2 = \sqrt{l^2 + (\frac{w}{2})^2}$, the car is unable to reach the boundary of its admissible configuration set (such as parking next to the curb). However, this can be easily solved by modifying the values of $\tilde{d}_1$ and $s\tilde{d}_2$ to l and $\frac{w}{2}$, respectively when θ_e is close to zero.

Remark 2.3 According to the shared-control input (2.28)–(2.31) and (2.32), the feedback controller is active only if the position of the car is "close" to the boundary of $\mathcal{P}_a$. Therefore, d_i can be changed to $\hat{d}_i = \min(d_i, \mathcal{B})$ for both $i \in \{1, 2\}$, where $\mathcal{B}$ is a positive user-selected constant.

As a result one can design the feedback controller only for the cases in which $d_i \leq \mathcal{B}$. This is helpful in real applications, especially when the obstacle is exactly in parallel or orthogonal to the required direction, *i.e.,* $d_1 = \infty$ or $d_2 = \infty$.

2.6 Case Studies

This section provides three case studies to illustrate how the shared-control algorithm works and demonstrates that it is effective in guaranteeing the safety of the car in various situations. The first case study deals with the situation in which absolute positions are not available, while the other two show how the vehicle controlled by the shared controller behaves in the presence of disturbances.

2.6.1 Case I: Turning Without Absolute Positioning

Consider the kinematic model of a rear-wheel drive car described by the Eq. (2.22) and controlled by the shared-control law (2.28)–(2.31) and (2.32). Assume that the absolute positions are not measurable and that the controlled car is approaching a junction with busy traffic on the main road. The driver wants to turn left and continue his/her journey on the main road but does not check the traffic carefully. Therefore, he/she makes the wrong decision at the junction and puts himself/herself in a dangerous situation.

Simulation results are given in Fig. 2.5, where the controlled car changes its color from green to red indicating that the feedback controller is active and gains the control authority from the driver to keep him/her safe. With the help of the shared controller,

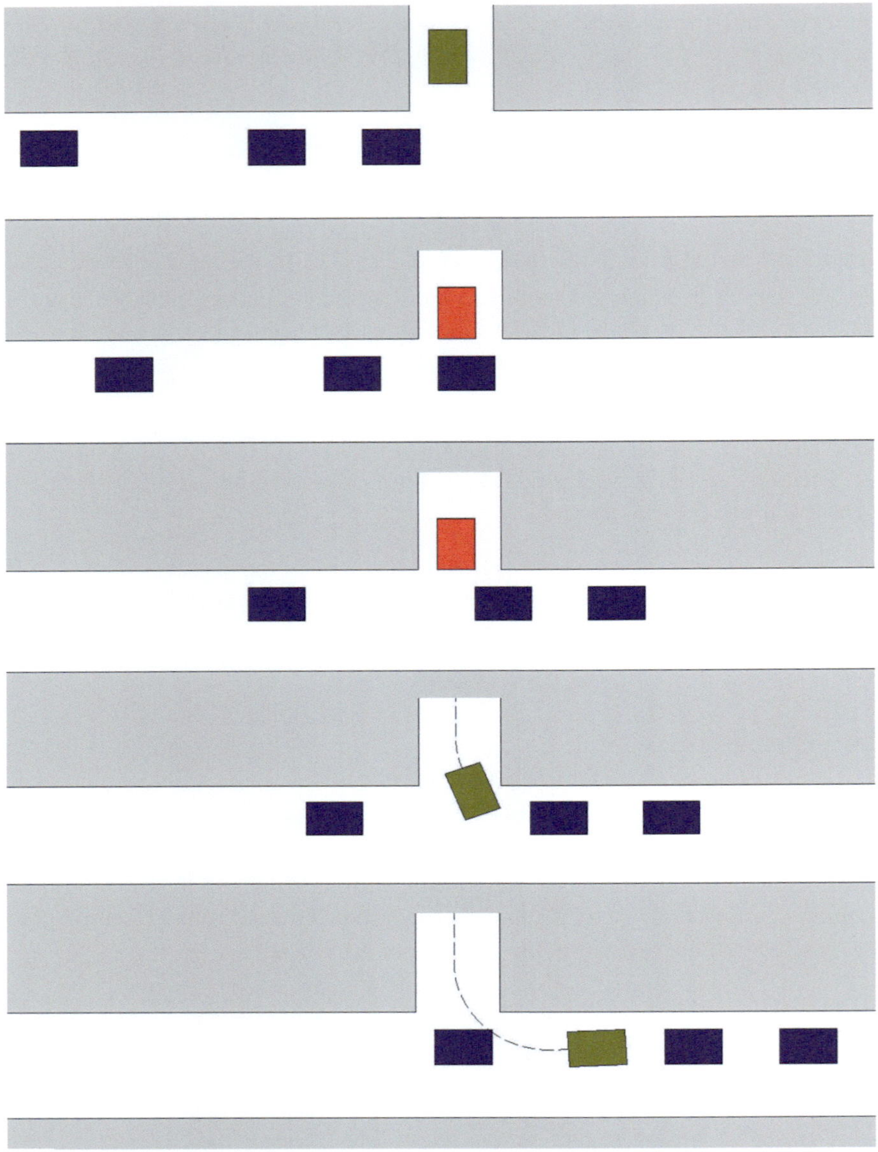

Fig. 2.5 Snapshots of the path of the car with the shared control in the (x, y)-plane for the set $\mathcal{P}_a$ represented by the *white* area. *Green* car: the controlled car (the feedback controller is not active). *Red* car: the controlled car (the feedback controller is active). *Blue* cars: the other vehicles on the road. Dashed line: (x, y)-trajectory of the controlled car

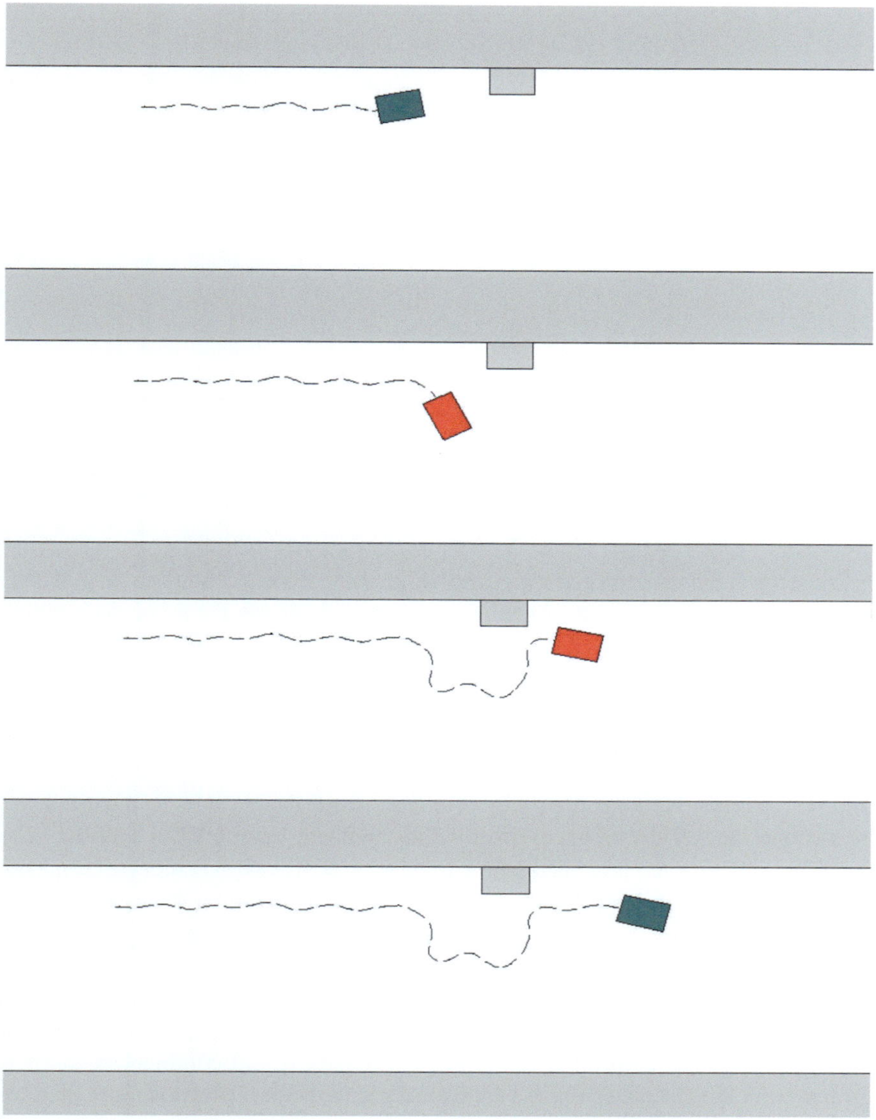

Fig. 2.6 Snapshots of the path of the car with the shared control in the (x, y)-plane for the set $\mathcal{P}_a$ represented by the *white* area. *Green* car: the controlled car (the feedback controller is not active). *Red* car: the controlled car (the feedback controller is active). *Grey* car: the parked car. Dashed line: (x, y)-trajectory of the controlled car

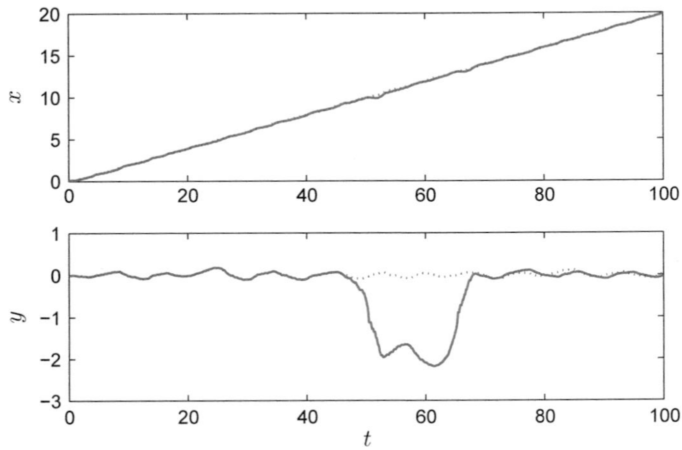

Fig. 2.7 Time histories of the variables x and y for the h-closed-loop system (*red*, dotted) and for the s-closed-loop system (*green*, solid)

the car stops at the junction and waits until there is a sufficiently long gap to make the turn. Finally, the car leaves the junction and turns left joining the main road safely.

2.6.2 Case II: Driving on a Road with Parked Cars

This case studies the shared control of a rear-wheel drive car with disturbances. We assume that all disturbances are random numbers between -0.1 and 0.1. The controlled car is driven on a road with parked cars. Simulation results are displayed in Figs. 2.6 and 2.7. Figure 2.6 shows that the car stays in the feasible configuration set $\mathcal{P}_a$ represented by the white areas, demonstrating the effectiveness of the shared controller. Figure 2.7 shows that the (x, y)-trajectory of the car is not smooth and is full of oscillations. This is because of the unknown disturbances. However, even though the model is not perfect and the driver does not recognize the parked car on his/her lane, the controlled car is able to bypass it and continue its journey. Furthermore, the distance from the controlled car to the parked car is more than we expect because we add additional space to compensate the bounded tracking error when designing the feedback controller.

2.6.3 Case III: Emergency Breaking

Similar to Case II, we assume that the model is imperfect and $d_1 = d_2 = d_3 = d_4 = 0.1$. We also assume that the controlled car is driven on the road, while the car in front of it (represented by the blue rectangle in Fig. 2.8) brakes suddenly. However, the

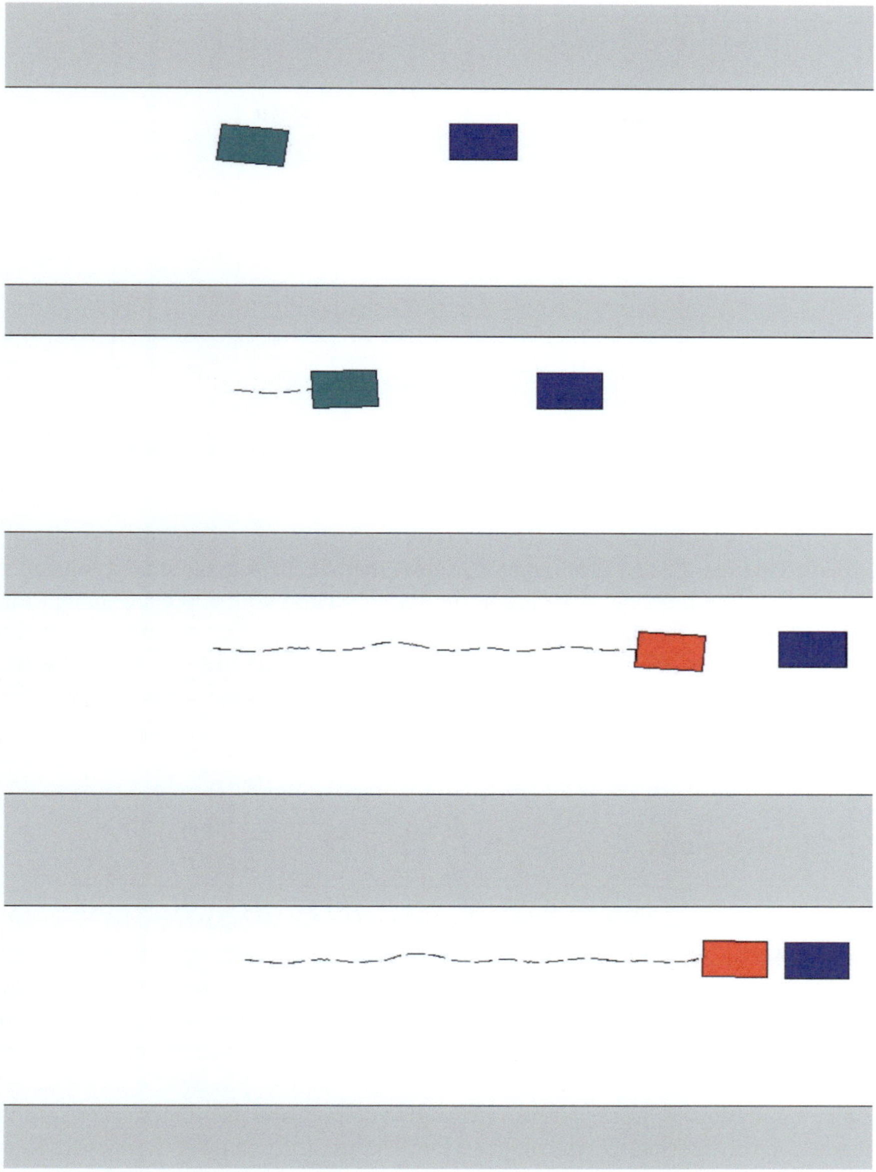

Fig. 2.8 Snapshots of the path of the car with the shared control in the (x, y)-plane for the set $\mathcal{P}_a$ represented by the *white* area. *Green* car: the controlled car (the feedback controller is not active). *Red* car: the controlled car (the feedback controller is active). *Blue* cars: other vehicles on the road. Dashed line: (x, y)-trajectory of the controlled car

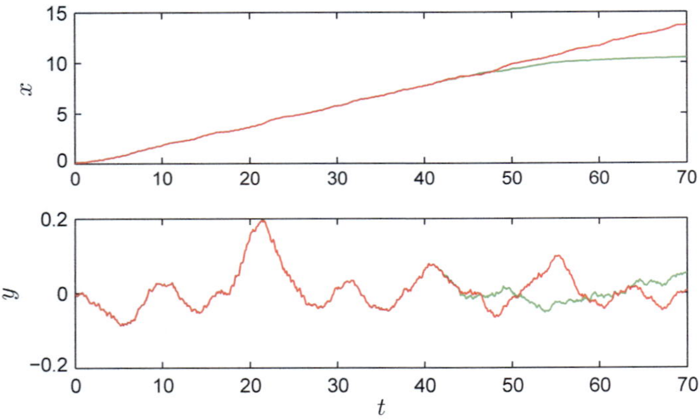

Fig. 2.9 Time histories of the variables x and y for the h-closed-loop system (*red*, dotted) and for the s-closed-loop system (*green*, solid)

driver sitting in the controlled car does not respond to it in time. Simulation results are displayed in Figs. 2.8 and 2.9. Figure 2.8 shows that the car with the shared controller stops behind the blue car at a certain distance to avoid collision even though the driver keeps pressing the gas pedal, hence, demonstrating the effectiveness of the shared-control law. Figure 2.9 indicates that the (x, y)-trajectories of the car for the h-closed-loop system and the s-closed-loop system are the same when $t < 44$ as the controlled car is far away from the car in front.

2.7 Conclusions

This chapter studies the shared-control problem for the kinematic model of a rear-wheel drive car with and without absolute positioning. We present solutions to the problem with static and dynamic admissible configuration sets. In addition, we study the system in the presence of bounded disturbances and prove that the tracking error is bounded. The sharing function, describing how the control authority is shared between the driver and the feedback controller, is defined as a hysteresis switch to reduce oscillations. Three examples are used to demonstrate the effectiveness of the shared-control algorithm. In the future, we will focus on the study of the cooperations between cars and of systems with measurement errors.

References

1. Chiang HH, Wu SJ, Perng JW, Wu BF, Lee TT (2010) The human-in-the-loop design approach to the longitudinal automation system for an intelligent vehicle. IEEE Trans Syst Man Cybern 40(4):708–720

2. Drage T, Kalinowski J, Braunl T (2014) Integration of drive-by-wire with navigation control for a driverless electric race car. IEEE Intell Transp Syst Mag 6(4):23–33
3. Fadilah SI, Shariff ARM (2014) A time gap interval for safe following distance (TGFD) in avoiding car collision in wireless vehicular networks (VANET) environment. In: Proceedings of international conference on intelligent systems, modelling and simulation, pp 683–689
4. Ferrara A, Librino R, Massola A, Miglietta M, Vecchio C (2008) Sliding mode control for urban vehicles platooning. In: Proceedings of IEEE intelligent vehicles symposium, pp 877–882
5. Gnatzig S, Schuller F, Lienkamp M (2012) Human-machine interaction as key technology for driverless driving - a trajectory-based shared autonomy control approach. In: Proceedings of IEEE conference on RO-MAN, Paris, pp 913–918
6. Jiang J, Astolfi A (2013) Shared-control for fully actuated linear mechanical systems. In: Proceedings of IEEE conference on decision and control, Italy
7. Jiang J, Astolfi A (2014) Shared-control for the kinematic model of a mobile robot. In: Proceedings of IEEE conference on decision and control, USA
8. Jiang J, Astolfi A (2015) Shared-control for the kinematic model of a rear-wheel drive car. In: Proceedings of american control conference, USA
9. Jiang J, Astolfi A (2016) Shared-control for typical driving scenarios. In: submitted to european control conference, Denmark
10. Kianfar R, Augusto B, Ebadighajari A, Hakeem U, Nilsson J, Raza A, Tabar RS, Irukulapati NV, Englund C, Falcone P, Papanastasiou S, Svensson L, Wymeersch H (2012) Design and experimental validation of a cooperative driving system in the grand cooperative driving challenge. IEEE Trans Intell Transp Syst 13(3):994–1007
11. Kim JS, Ryu JH (2013) Shared teleoperation of a vehicle with a virtual driving interface. In: Proceedings of IEEE international conference on control, automation and systems, Gwangju, pp 51–857
12. Lee U, Yoon S, Shim HC, Vasseur P (2014) Local path planning in a complex environment for self-driving car. In: Proceedings of IEEE annual international conference on cyber technology in automation, control, and intelligent systems, Hong Kong, pp 445–450
13. Lilly JH (2007) Evolution of a negative-rule fuzzy obstacle avoidance controller for an autonomous vehicle. IEEE Trans Fuzzy Syst 15(4):718–728
14. Massera Filho C, Wolf DF, Grassi V, Osorio FS (2014) Longitudinal and lateral control for autonomous ground vehicles. In: Proceedings of IEEE symposium on intelligent vehicles, pp 588–593
15. McGann C, Py F, Rajan K, Ryan J, Henthorn R (2008) Adaptive control for autonomous underwater vehicles. In: Proceedings of AAAI conference on artificial intelligence, pp 1319–1324
16. Michalek M, Kielczewski M (2013) Helping a driver in backward docking with n-trailer vehicles by the passive control-assistance system. In: Proceedings of IEEE international conference on intelligent transportation systems, pp 1993–1999
17. Mohajerpoor R, Dezfuli SS, Bahadori B (2013) Teleoperation of an unmanned car via robust adaptive backstepping control approach. In: Proceedings of IEEE/ASME international conference on advanced intelligent mechatronics, pp 1540–1545
18. Prieur C (2001) Uniting local and global controllers with robustness to vanishing noise. Math Control Signals Syst 14:143–172
19. Thrun S (2010) Toward robotic cars. Commun ACM 53(4):99–106

20. Wang J, Zhang L, Zhang D, Li K (2009) Cooperative (rather than autonomous) vehicle-highway automation systems. IEEE Intell Transp Syst Mag 1(1):10–19
21. Wang J, Zhang L, Zhang D, Li K (2012) An adaptive longitudinal driving assistance system based on driver characteristics. IEEE Trans Intell Transp Syst 14(1):1–12
22. Zhou SH, Yasunobu S (2006) A cooperative auto-driving system based on fuzzy instruction. In: Proceedings of 7th international symposium on advanced intelligent systems, pp 300–304

Chapter 3
Baxter-On-Wheels (BOW): An Assistive Mobile Manipulator for Mobility Impaired Individuals

Lu Lu and John T. Wen

3.1 Introduction

According to the National Spinal Cord Injury Statistical Center, the population with spinal cord injury (SCI) in the US is estimated to be around 270,000 with 12,500 new injuries every year [3]. Over 40% of SCI patients are quadriplegic, i.e., with paralysis in all four limbs. SCI primarily affects young adults (age 16–30) due to vehicular accidents, falls, violence, sports, etc. This issue is not confined to the U.S. – there are estimated 2.5 million SCI patients [4] around the world. There are also other causes of paraplegia (lower limb paralysis) and quadriplegia, including stroke, cerebral palsy, multiple sclerosis, muscular dystrophy, amyotrophic lateral sclerosis (Lou Gehrig's Disease), etc. The cost of their lifetime treatment and care can be quite significant [3, 5, 6]. Beyond the initial hospital and continuing rehabilitation, much of the cost is for in-home assistants to help with daily tasks. Many homebound individuals, including SCI patients or other people with disability, are cognitively sound and capable of speech and hearing, but mobility and dexterity challenged. Even routine daily tasks, such as grasping items from shelves, picking up objects from floor, opening cabinet doors, etc., require the help of a human assistant. Such dedicated human help is not only expensive but often not needed for long stretches of time. A mobile robotic manipulator that could work collaboratively with individuals with disability to perform some of these tasks under the human user command would increase user independence, improve the quality of life, and potentially lower the cost of care by reducing the need for a dedicated human assistant.

This chapter was developed and enhanced from earlier papers published as [1, 2] © 2015 IEEE.

L. Lu (✉)
New Jersey Institute of Technology, Newark, NJ, USA
e-mail: lulu@njit.edu

J. T. Wen
Rensselaer Polytechnic Institute, Troy, NY, USA
e-mail: wenj@rpi.edu

Y. Wang and F. Zhang (eds.), *Trends in Control and Decision-Making for Human–Robot Collaboration Systems*, DOI 10.1007/978-3-319-40533-9_3

Various types of robotic manipulation aids have been proposed to work with persons with disability [7–9]. Configurations range from a robot arm mounted on the user wheelchair [10–12], a robot arm mounted in a fixed location in the workspace [13], and a robot arm mounted on a mobile base [9, 14–17]. Among commercial systems, Manus (by Exact Dynamics) and Rapter (by Applied Resources Corporation) are the most successful [10, 12]. Both of these systems offer users the ability to directly control the movement of the robot arm to interact with the physical world around them. Mounting a robotic arm onto an existing wheelchair designed for human transport poses significant limitations. The modified wheelchair may be too wide to navigate doorways, impede the user's ability to see objects or people around them, obstruct the user's entry or exit of the wheelchair, and render the wheelchair less stable [11]. Willow Garaged developed a mobile dual-arm robot, PR2, and has shown limited demonstration of its use in the assistive context [16]. The main focus of robot is on autonomous operation under high-level user direction, rather than the shared autonomy in our approach. The system is also costly and is not currently available. Another dual-arm mobile robot HERB [18, 19] has been proposed for home and assistive use. The focus has been on autonomous operations, but its application to unstructured home environments remains unfulfilled. Several mobile manipulators have been developed for persons with mobility challenges, including a wheelchair-based robot arm KARES-II [14, 15] and a portable mobile assistive manipulator ASIBOT that could climb stairs [17, 20]. They are technologically sophisticated but the redundancy resolution of such mobile manipulators has not been extensively studied.

There is a large body of literature on teleoperation, where the human user completely determines the robot motion, typically through the use of a master robot (to control the remote "slave" robot) [21–23]. Such scheme is impractical for disabled users, as the user could typically command only a limited number of degrees of freedom (dof). In between the fully autonomous control, as in industrial robots, and completely human-based control, as in teleoperation, is *shared autonomy* [24]. In this common human–robot collaboration approach, the human user commands certain aspects of the robot motion, and the robot controller incorporates the human input and other considerations such as obstacle avoidance, accurate positioning, or disturbance rejection, for the actual execution of the robot motion. Shared autonomy has been used in numerous applications involving humans and robots working together. Examples include cooperative object handling [25, 26], assistance to disabled or elderly people [27–29], mobile robot and vehicle control [30, 31], multi-agent control [32] and tele-surgery [33]. Recent studies have focused on human intention modeling and prediction [34–36], robot/human task partition [37], and trust building [38, 39].

This chapter describes our implementation of shared autonomy on a mobile manipulator for assisting mobility impaired users with activities of daily living. We have developed a prototype system as a testbed for this research. This system involves the integration of a human-friendly dual-arm robot (the Baxter by Rethink Robotics), a commercial motorized wheelchair (3-dof base with independently driven wheels), a multitude of sensors for collision avoidance and contact sensing (Kinect sensors and robot wrist force sensors), and various input devices, including a breath-operated sip-puff device, various types of joysticks, and voice command [40]. The focus of this chapter is on the integration of human command, sensor inputs, and redundancy resolution to achieve effective dexterous manipulation under the human user guidance.

A user with disability is only able to provide a restricted set of input commands using a part of the body or voice. A key issue in assistive robotics is to translate this restricted input to the complete robot motion in a natural and intuitive way. We adopt the resolved velocity redundancy resolution approach for its simplicity and ease to include various control objectives and constraints. The general frame involves the human user commanding a part of the desired motion or force of the system through some input device. The baseline optimization problem resolves the instantaneous robot joint velocities such that the human input command is maximally realized, while certain equality and inequality constraints are satisfied. These constraints may involve physical requirements of the system, e.g., collision avoidance and joint limits, and human user consideration, such as intuitiveness of the combined arm and base motion. It may also involve sensor-based motion, e.g., vision guidance or force control. The constraint optimization problem is strictly convex and may be efficiently solved as a quadratic program. We tailor the baseline optimization for two operating scenarios: free-space motion and constrained motion (with the arm in contact with the environment). We demonstrate the algorithm through two tasks: (1) Pick up a bottle from the shelf and place it on a table while avoiding obstacles; (2) Erase a whiteboard while maintaining certain contact force level (to ensure proper cleaning of the board).

Though tele-supervisory control has long been studied for remote operations, application to assistive robotics presents new challenges. The novelty and uniqueness of our approach involve the following:

1. *Efficient Collaborative Redundancy Resolution.* The optimization-based resolved velocity approach allows a user with disability to use a low-dof input device to effectively command a high-dof system to perform complex tasks. The user can add or remove constraints to affect the behavior of the robot, e.g., keeping the approach direction, tugging in elbows, etc., and incorporate sensors, e.g., keep the gripper pointing to the target.
2. *Seamless Incorporation of Force Control.* Force control has rarely been addressed for users with disability. We apply impedance control to combine motion and force control, which also fits in the same optimization-based resolved velocity framework.
3. *Intuitive Base-Arm Coordination.* A constraint is added to the optimization algorithm to ensure that the base and arm motion does not result in nonintuitive opposite directions.
4. *Integration of Off-the-Shelf Components.* The demonstration platform, BOW, is based on an industrial robot Baxter integrated with a retrofitted power wheelchair. This allows a low-cost, flexible, and reconfigurable testbed.

In summary, the proposed algorithm and mobile manipulator system, BOW, provides an effective tool for a user with disability to manipulate the motion and force of the robot end-effector while achieving an intuitive and safe internal motion. With some training, the user would be able to perform certain activities of daily living independently.

3.2 System Description

3.2.1 Experimental Platform: BOW

Our mobile assistive robot experimental testbed is the Baxter-on-Wheel (BOW) shown in Fig. 3.1. It consists of a dual-arm robot, Baxter by Rethink Robotics [41], mounted on a converted power wheelchair. The arms and wheelchair may be controlled at the same time for coordinated operations. The initial design of the system has been described in [40] with improved sensing and control algorithms incorporated since then. Such combination of industrial robot with wheelchair has also been proposed in [42] for industrial applications. Our focus here is to empower mobility impaired individuals to gain more independence in daily living. Though we have chosen to use the Baxter, it may be substituted by other "human-friendly" robots such as Universal UR5, Kinova Jaco, or ABB Yumi. The dual 7-dof arms of Baxter are attractive for certain tasks, e.g., reaching the top shelf, picking up a large tray, opening and closing drawers, etc.

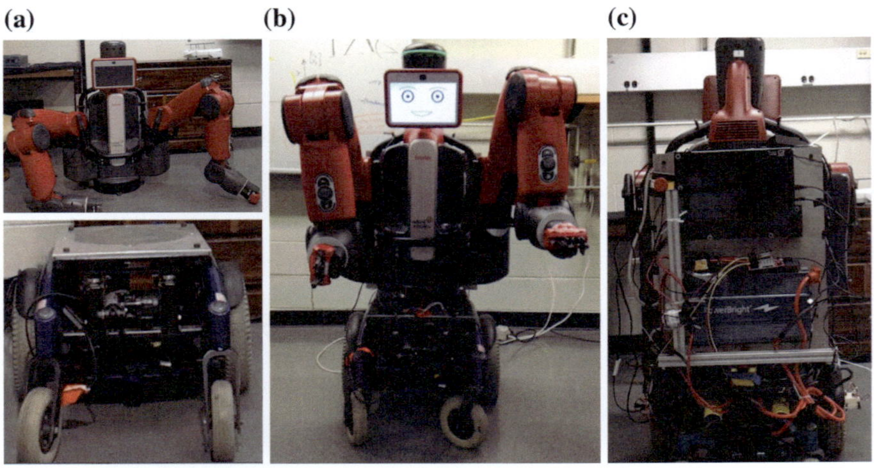

Fig. 3.1 **a** The Baxter-on-Wheels integrates the Baxter without the stand and a stripped-down wheelchair with a custom adapter plate. **b** *Front view*: Note that the BOW is sufficiently low to reach the ground to pick up objects (© 2016 IEEE. Reprinted, with permission, from [1]). **c** *Back view*: A custom backpack houses a laptop computer and inverter to allow the wheelchair battery to power both the wheelchair and the Baxter

3.2.1.1 Mobile Base

The mobile base is a commercially available wheelchair with two independently driven wheels, and two small caster wheels for balance. Smart wheelchair projects such as [43, 44] use the joystick port for the computer control of the wheelchair. This was also our initial approach, but we found that the analog signals in joystick control is susceptible to noise, especially from the inverter powering the Baxter. We have therefore replaced the joystick-based control with a Roboclaw motor controller to allow direct digital computer control. The wheel rotational velocities are measured by an inertial measurement units (IMU) consisting of a gyroscopic sensor and accelerometer mounted on each wheel. The signals are sampled at 100 Hz by an Arduino microcontorller and communicated to the computer via a Bluetooth shield. At the present, only the gravitational acceleration is used for wheel velocity sensing. More sophisticated sensor fusion scheme may be used to improve wheelchair localization, but has not been implemented yet. Four IR sensors and a Kinect sensor are mounted at the front and sides of the wheelchair to detect obstacles.

3.2.1.2 On-Board Manipulator

The electronics of the dual-arm Baxter robot are entirely contained in the torso above the waist. It could therefore be mounted on the seat of the stripped-down wheelchair through a custom adapter plate. The robot has two articulated arms, each with 7 rotational joints. The arms are designed with plastic casing and low inertia. The joints are elastic and the moving speeds are limited. Thus, human can stay within the workspace without excessive safety concerns. Baxter is equipped with three cameras, one at each wrist and one on the head which may be used to detect the targets and obstacles. Several parallel jaw grippers with different opening widths come with the Baxter. The gripper fingers may be servoed and there are infrared sensors to detect the distance between an object and the gripper fingers. The Baxter operates under Robot Operating System (ROS) and allows joint- and task-level position and velocity control. Torque control is also possible, but we use primarily joint-level velocity control. The joint positions and velocities are published by ROS at 100 Hz. There are additional services in the low-level Baxter control system to prevent self-collision, compensate for gravity, and maintain safety by halting the servos during an unexpected impact. The joint torques can also be accessed at 100 Hz. We use these values to estimate the contact force and torque for force control.

The wheelchair with batteries weighs 118.8 kg while the Baxter robot weighs an additional 74.8 kg, similar to a normal adult and well within the payload range of the wheelchair. Mounted on the wheelchair base, the Baxter's working envelop ranges from the ground to a maximum working height of 1.63 m off the ground. Each arm can extend 1.04 m out to either side of the robot and 0.68 m in front of the wheelchair. The combined footprint of the system is 0.914 m in length and 0.673 m in width, which is only 2 cm wider than the wheelchair itself and equally as long. This is important as the BOW can operate through doors and hallway in any wheelchair-friendly facilities. A rear mounting surface is designed to accommodate the inverter, control laptop, and any future hardware.

3.2.1.3 Power

The wheelchair comes with two 12 V/73 Ah gel-cell batteries connected in series. We use the same batteries to power the Baxter, laptop, and other sensors on the BOW as well. The Baxter and the laptop require 120 V AC while the wheelchair needs a 24 V DC line. A Power Bright 24 V 1000 W Inverter is used to generate the AC supply. The Baxter and other AC loads draw about 720 W maximum. The wheelchair draws 250 W in continuous usage. The batteries supply $12\,\mathrm{V} \times 73\,\mathrm{Ah} \times 2 = 1752\,\mathrm{W\text{-}h}$. This means that the BOW can run continuously for around 2 h. Addition of fuel cell and supercapacitors can increase the run time and performance.

3.2.1.4 Human Input Device

Many input devices are available to disabled users, depending on the level of disability. Some of these choices include electronic pointing, eye tracking, facial feature tracking, head movement, arm gesture tracking, sip-and-puff, wands and sticks, joysticks, trackballs, touch screen, electromyoraphy (EMG), electroencephalography (EEG), and voice recognition. We will consider users with severe mobility impairment, so will only assume limited degrees of freedom available as the input command. As an example, we focus on 3-dof input devices such as joysticks, e.g., the Microsoft Xbox controller, and sip-puff type of devices (commonly used by individuals with quadriplegia as a computer interface). The operation of the sip-puff device only requires the movement of the head and mouth. An example of a breath-operated sip-puff device is Jamboxx. Similar to input devices with joysticks and buttons, Jamboxx senses 3-dof input from user: left-right sliding of the nozzle, up-down motion of the nozzle, and sip-puff action to the nozzle detected with a pressure sensor. We have also implemented simple verbal commands for user speech inputs using the CMU Sphinx library [45], an open source speech recognition toolkit. In particular, we use PocketSphinx variant for speed and portability.

3.2.1.5 System Architecture

The overall system architecture is shown in Fig. 3.2. The BOW control is performed by a Linux laptop computer mounted on the back of the Baxter. This on-board computer relays commands (e.g., desired velocity, position, or torque) and reads sensor input (joint angles and torques) to and from the Baxter through ROS. It also interfaces with the Arduino microcontroller to send motor current command to

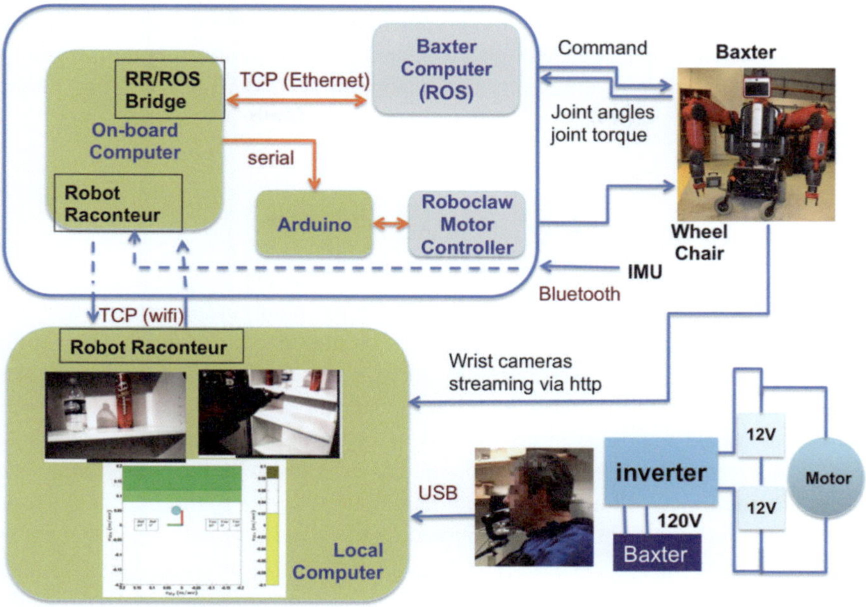

Fig. 3.2 The overall BOW architecture includes the integration of Baxter ROS-based control system, wheelchair control, sensors, user interface, display, and power plant. We use the middleware Robot Raconteur developed at RPI for the overall coordination. The control algorithm is implemented in MATLAB on the local computer

the wheelchair through the Roboclaw motor controller and reads the IMU modules mounted on each wheel axle via Bluetooth.

A separate local computer provides the interface to the user including receiving commands from the user input and shows user and robot operations through a graphical display and robot wrist cameras streaming feed. The robot control based on the user input and robot feedback is performed in the local computer as well which communicates with the on-board computer through a wireless TCP connection.

The orchestration of the various heterogeneous components in the system is conducted by Robot Raconteur (RR), which is an object-oriented client-service- based distributed control middleware originally developed in our lab by John Wason [46]. RR communicates with ROS on the Baxter robot using our RR/ROS Bridge [47]. There are several RR services on the on-board computer. They provide ROS interface, interface to the wheelchair, and the interface to the on-board sensor data. On the local computer, an RR service provides interface to the user input device. An RR client on the local computer accesses this multitude of services for the Baxter configuration, wheelchair measurements, sensor data, and user input. The information is processed based on the algorithm that will be described in the rest of this chapter, to provide driving commands to the Baxter and the wheelchair motors. The client may be implemented in various languages, e.g., MATLAB, Python, C++, etc. For the ease of prototyping, we choose to implement the control algorithm in MATLAB, with some sacrifice of speed. Though the system is not real time, communication in the local area network is sufficiently fast to support around 50 ms sampling period on average.

3.2.2 System Kinematics

Figure 3.3 shows a schematics of the system at the configuration where all joint angles are defined as zero, and the wheelchair is centered at the origin and facing the positive x-direction of the inertia frame. Only the left arm is labeled; the right arm is identical to the left, symmetrical with respect to the x-z plane of the base frame. The link parameters of the arms are given by a, b, c, d. The displacement of the base of the arms from the wheelchair coordinate center is given by the vector p_{BO_L} and p_{BO_R}. The numerical values of the robot parameters may be found in [2]. The joint angles of the left and right arms are denoted by the vectors q_L and q_R, respectively, where q_L consists of $\{q_{L_1}, \cdots, q_{L_7}\}$ and q_R consists of $\{q_{R_1}, \cdots, q_{R_7}\}$.

Assuming no wheel slippage, the kinematics of the wheelchair is given by the standard unicycle model:

$$\underbrace{\begin{bmatrix} \dot{\theta}_B \\ \dot{x}_B \\ \dot{y}_B \end{bmatrix}}_{\dot{q}_B} = \underbrace{\begin{bmatrix} 1 & 0 \\ 0 & \cos\theta_B \\ 0 & \sin\theta_B \end{bmatrix}}_{J_B} \begin{bmatrix} \omega_B \\ v_B \end{bmatrix} \tag{3.1}$$

where (x_B, y_B) is the wheelchair center with respect to inertia frame, θ_B is the orientation of the wheelchair, and (ω_B, v_B) are the turning and linear velocities of the wheelchair, respectively.

The differential kinematics of the left arm/base system (similarly for the right arm) in the mobile base frame is given by

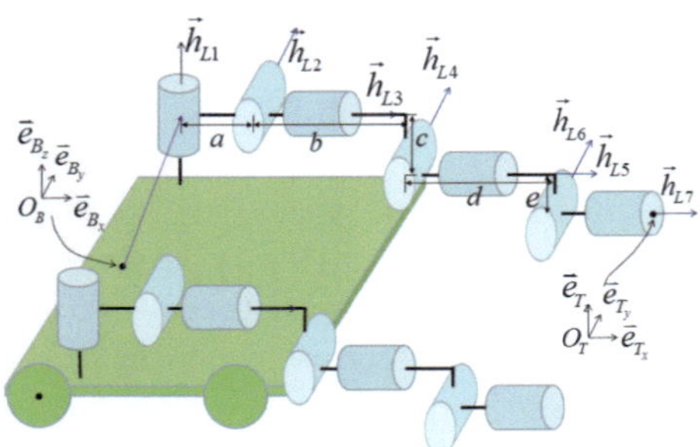

Fig. 3.3 Kinematics of the mobile manipulator system

$$V_T = J_{T_L}\dot{q}_L + \begin{bmatrix} e_z\omega_B \\ -(p_{BO_L} + p_{O_LT_L}) \times e_z\omega_B + e_x v_B \end{bmatrix} \tag{3.2}$$

where $V_T = [\omega_T, v_T]^T$ is the spatial velocity (angular and linear) of the task frame at the end-effector, J_{T_L} is the corresponding arm Jacobian, $p_{O_LT_L}$ is the arm base to end-effector vector, $e_x = [1, 0, 0]^T$ and $e_z = [0, 0, 1]^T$ are the unit vectors in the base frame.

Combining (3.1)–(3.2), we can write the overall kinematics in a compact form:

$$V_T = J_T(q)u = \begin{bmatrix} J_{T_L} & \begin{matrix} e_z \\ -(p_{BO_L} + p_{O_LT_L}) \times e_z \end{matrix} & \begin{matrix} 0 \\ e_x \end{matrix} \end{bmatrix} u \tag{3.3a}$$

$$\dot{q} = J_q(q)u = \begin{bmatrix} I & 0 \\ 0 & J_B \end{bmatrix} u \tag{3.3b}$$

where $q = \begin{bmatrix} q_L \\ q_B \end{bmatrix}$, $u = \begin{bmatrix} \dot{q}_L \\ \omega_B \\ v_B \end{bmatrix}$.

3.3 Control Algorithm

3.3.1 Baseline Shared-Control Algorithm

The basic algorithm is based on resolved velocity: given the user commanded task velocity, v_H, find the joint velocity u to best match the command while satisfying specified constraints. The constraints include:
(a) Joint or task position equality constraint, $h_{E1}(q) = 0$, to hold certain joints at specified position (e.g., fixed elbow angle, maintaining wrist heading).
(b) Joint or task velocity equality constraint, $h_{E2}(u) = 0$, to prevent unwanted motion (e.g., stationary base, stationary elbow).
(c) Joint or task position inequality constraint, $h_{I1}(q) \geq \eta_1 > 0$, to avoid collision and joint stops.
(d) Joint or task velocity inequality constraint, $h_{I2}(u) \geq \eta_2 > 0$, to avoid excessive velocity in certain directions.

We convert the position constraints into velocity constraints using Jacobian-based velocity control to steer the robot toward the constraint surface:

$$\frac{dh_{E1}}{dt} = \underbrace{\frac{\partial h_{E1}(q)}{\partial q} J_q(q)}_{J_{E1}(q)} u = -k_{E1} h_{E1}(q). \tag{3.4}$$

For the inequality position constraint, we use the external penalty function approach of moving the robot away from the constraint:

$$\frac{dh_{I1i}}{dt} = \underbrace{\frac{\partial h_{I1i}}{\partial q} J_q(q)}_{J_{I1}(q)} u \geq \sigma_i, \; i = 1, \ldots, m_{I1}. \tag{3.5}$$

The function σ_i is positive if the inequality constraint is nearly violated, $h_{I1i} > \eta_{I1i}$, where η_{I1i} is a small negative number; it is negative if the inequality constraint is satisfied, to allow the h_{I1i} to be smaller if necessary. The velocity-based constraints, (b) and (d), are assumed to be linear in u with lower bound for the inequality constraint given by constants c_I. The overall optimization problem at each time then becomes:

$$\min \|J_T u - v_H\|^2 + \lambda \|S_R u\|^2 \tag{3.6}$$

subject to $\dot{q} = J_q(q)u$ and

(a) $J_{E1}(q)u = -k_{E_1} h_{E1}(q)$, **(b)** $J_{E2}(q)u = 0$, **(c)** $J_{I1}(q)u \geq \sigma_i$, **(d)** $J_{I2}(q)u \geq c_I$.

Note that a penalty term on u (in the direction of S_R) is added to ensure strict convexity of the optimization problem and avoid unnecessary base motion. Sensor-based feedback may be incorporated into constraints (a) and (c). For example, if the robot end-effector is required to satisfy the constraint, $h_E(p_{OT}) = 0$ (such as keeping the heading angle or keep a target in the center of the wrist camera image), the corresponding equality constraint is

$$\left(\frac{\partial h_E}{\partial p_{OT}} J_T\right) u = -k_E h_E(p_{OT}) \tag{3.7}$$

where $h_E(p_{OT})$ may be obtained based on the camera measurement. For force control along the direction given by S_F, a generalized damper foce control law may be implemented as an equality constraint as

$$S_F^T J_T u = -k_F(S_F^T F_m - F_H) \tag{3.8}$$

where F_m is the measured spatial force and F_H is the desired force set by the user (see the force control section below) or specified by the task.

For dual-arm operation, v_H may specify the desired velocity of a frame that is fixed with respect to the left and right arms. In this case, the optimization problem will find the left and arm joint velocities and the base velocity to best match with v_H:

$$\min_u \left\|J_{T_L} u - v_H\right\|^2 + \left\|J_{T_R} u - v_H\right\|^2 + \lambda \|S_R u\|^2 \tag{3.9}$$

subject to the constraint force requirement (3.8) (with F_H as the specified squeeze force).

3.3.2 *Free-Space Mode and Contact Mode*

The baseline shared-control algorithm applies to both free-space mode and contact mode (by incorporating force control as an equality constraint). We will separately consider the two modes and design different constraints and objective functions to achieve specific requirements. The measured contact force will be used to switch between these two modes.

3.3.2.1 Free-Space Mode

In the free-space mode, the user commands the end-effector motion. The internal degrees of freedom of the arm and base need to be resolved to be intuitive to the human user while satisfying all the physical constraints. The "intuitiveness" of the motion is subjective. Based on our experience working with disabled uses, we impose the following requirements in an attempt to quantify this notion.

1. When the user does not issue a command (v_H is zero) and all equality and inequality constraints are satisfied ($h_{E1}(q) = 0$ and $h_{I1}(q) \geq \eta_1$), the arm and the wheelchair do not move. This requirement is obviously necessary to avoid any unintended surprise motion.
2. When the arm is inside the interior of the workspace and the desired user command can be fully realized through arm motion, only the arm moves but not the wheelchair. This requirement is to avoid excessive base motion when arm alone could follow the user command.
3. When the arm is at the workspace boundary, and the user commanded velocity is the direction that would move beyond the arm workspace, then the wheelchair should move to achieve the desired user motion commands. This requirement allows base motion only when it is necessary to follow the user command.
4. When the arm is at the workspace boundary, the base motion should result in the end-effector motion in the same direction as the user commanded motion. This requirement will avoid the unnatural motion where the base seems to go against the user command (with the arm also moving to compensate for the undesired base motion) in the transient phase.
5. The un-commanded end-effector directions will be autonomously controlled via the equality constraint. For example, if the user commands the translational motion only, then the end-effector orientation is fixed. If the user wants to approach along a fixed axis, then the translation orthogonal to the approach axis is also fixed. In visual servoing, the object position measured the image plane could also be imposed as an equality constraint.
6. To avoid colliding with the environment, the wheelchair speed is reduced when it is near an obstacle (as sensed by the on-board proximity sensors), and eventually comes to a stop as a specified minimum distance to the obstacle is reached. This requirement overrides the user command if the specified velocity is into the obstacle.

7. The joint angles are within their limits and self-collision avoidance must be guaranteed. The joint range may need to be artificially restricted for more intuitive motion. For example, since the arms will mostly operate in front of the robot instead of on the side, the shoulder joint angles may need to be constrained to a narrower range. This would also allow the user to more effectively move the base to pushing the arm against the shoulder joint limits.
8. The magnitudes of the velocity inputs are bounded to ensure the validity of the kinematics-based control.

There may be other considerations such as minimizing the number of joints that move. This may entail ℓ_0 minimization which we do not address in this paper.

We now describe our approach to design the constraints and objective function to satisfy the above requirements. Requirement (1) follows from the fact that when $v_H = 0$ and all constraints are satisfied, $u = 0$ is the solution of the optimization problem. The additional penalty term on u in (3.6) will avoid unnecessary base motion, to satisfy requirement (2) and (3). When the arm is within the workspace boundary and its motion satisfies the desired end-effector motion v_H, the optimal solution will have zero base velocity in order to minimize the objective function. When the arm reaches the workspace boundary and user continues to command the end-effector beyond the arm workspace, the arm motion alone cannot fully realize v_H due to fact that the inequality constraint. In this case the base will move to realize the desired end-effector velocity command, while minimizing the objective function. To satisfy requirement (4), we notice that when the user commands the end-effector to move in certain direction and the arm reaches its boundary, it is more natural to have the base follow the desired direction of motion (projected onto the x-y plane), rather than having other combined base/arm motion. In this case, we include an additional equality constraint as

$$J_{E2}u = e_z^T(v_H \times (\omega_B \times p_{OT} + v_B)) = 0. \tag{3.10}$$

The equality constraint is only enforced in the e_z direction since we are only interested in the x-y components of v_H. To satisfy (5), we let h_{E1} be the difference between the desired and actual configuration of the end-effector in the direction not directly controlled by the user. For example, if the human user directly controls the translational velocity of the end-effector, $h_{E1}(q)$ can be selected as any three-parameter representation of the orientation error (e.g., deviation from the initial orientation). In visual servoing case, $h_{E1}(q)$ may be based on the image-plane error of a target from its desired configuration. To satisfy (6) and (7), we divide the inequality constraints $h_{I1}(q) \geq \eta_1$ into three parts: $h_{I1i}(q) \geq \eta_{1i}, i = 1, 2, 3$. For the first part, $h_{I11}(q)$ denotes the minimum distance between the robot to the obstacle detected through onboard sensing devices like IR or Kinect sensors, with η_{11} as the lower bound on the minimum distance below which the wheelchair speed must be reduced for collision avoidance. The second part, $h_{I12}(q)$, specifies the minimum distance between the arm and body of the robot with η_{12} as the lower bound. The third part, $h_{I13}(q) - \eta_{13} \geq 0$, represents all the joint angles limit constraints. For example, if $q_{Li} \in [q_{Limin}, q_{Limax}]$, then the corresponding constraint inequalities are $q_{Li} - q_{Limin} \geq 0$ and $-q_{Li} + q_{Limax} \geq 0$. To satisfy (8), we design the inequality constraints $h_{I2}(u) > \eta_2$ to be consisting of all the constraints on the input magnitudes.

3.3.2.2 Contact Mode

In contact mode, it is necessary to first identify the constrained direction S_F appropriate for the task. The user should be able to either adjust the contact force setpoint or control the movement in unconstrained directions while maintaining contact. These additional requirements are summarized as follows:

1. The constrained direction S_F is known or identified using onboard sensors.
2. The user can choose either to adjust the contact force setpoint F_H or command the motion in complementary unconstrained directions v_H, but not simultaneously.

An intuitive method to identify the constraint direction is to use the collection of the measured contact force, i.e., $F_m/||F_m||$, as the constrained direction. To remove the effect of frictional and inertial forces, we use the actual velocity measurement of the end-effector to determine the constraint force direction (orthogonal to motion). The force component in the motion direction is then removed.

To satisfy requirement (2), we define two cases of user intention based on the input v_H. Consider the two regions Ω_1 and Ω_2 as shown in Fig. 3.4. If v_H is in Ω_2, v_H is interpreted as the command to increase or decrease the desired contact force. If v_H is into the constraint, then the force set point increases. If v_H is backing away from the constraint, then the desired contact force will decrease, eventually breaking off the contact. If v_H is in Ω_1, then v_H is interpreted as the motion command of the end-effector along unconstrained direction while keeping the contact force around its desired value (set in Ω_2).

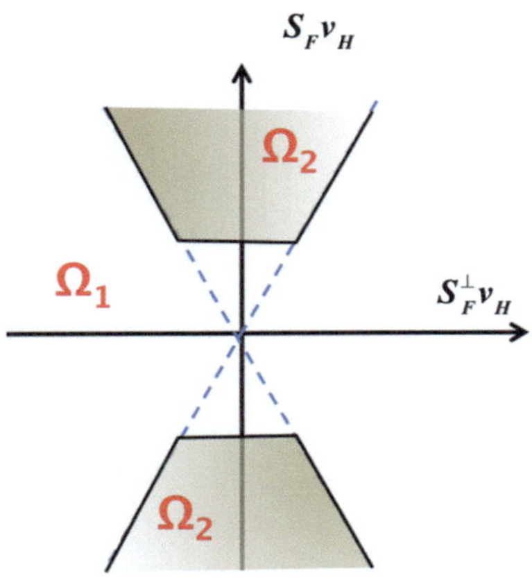

Fig. 3.4 The regions Ω_1 and Ω_2

3.3.2.3 Mode Switching

To perform tasks like cleaning and door opening, the end-effector needs to be operated at free-space mode first, approaching the constrained object, and then switched to contact mode once contact has been detected. The detection of contact can be achieved by measuring the end-effector spatial force F_m, and setting a threshold on its magnitude for transition between free-space and contact mode. Since the measurement is noisy, we use a hysteresis function for mode transition to prevent chattering between the two modes: When the system is in free-space mode, then $||F_m|| > F_{Th}$ triggers the transition to contact mode, when the system is in contact mode, then $||F_m|| < F_{Tl} < F_{Th}$ triggers the transition back to free-space mode.

3.4 Application to the BOW

In this section, we describe the application of the optimization-based kinematic control algorithm (3.6) to the BOW testbed. To illustrate our approach, we consider two special cases: In the free-space mode, the user controls the translational motion of the end-effector, with orientation stabilized around a desired configuration. The task involves the human user commanding the BOW to pick up a bottle from the shelf and place it on the table. In the contact mode, the user controls the force into a flat surface and then wiping motion along the surface while maintaining the desired contact force. The task involves the human user commanding the BOW approach the whiteboard, press down with a specified contact force, and clean the letters on the board.

3.4.1 User Interface

User interface plays a critical role in terms of the usability of the system. It allows the user to command the robot to interact with the environment and visualize sensor information to determine the appropriate response. Figure 3.5a shows a prototype interface constructed in MATLAB to provide the following functionalities:

1. Displays the magnitude of the user input command.
2. Allows user to switch between operating modes.
3. Allows user to change system configuration.

The sip-puff of the Jamboxx nozzle controls the robot gripper in/out motion and the left-right sliding of the nozzle controls the side motion of the gripper. The blue dot in the left figure indicates the magnitude of the horizontal commanded velocity of the gripper. When the user gives a large v_H command, the assumed intent is to primarily move the base for large-range motion. When the user gives a small v_H command, the assumption is that the intent is for fine arm movement. In the first case, the arm should be folded with $q_{L4_{\min}} = \pi/2$. In the second case, the arm can stretch out with $q_{L4_{\min}} = 0$. The dark-green region corresponds to the first case, the white region for the second case, and the light-green region the transition region that switches between the two regions using a hysteresis function to avoid chattering. The bar on the right-hand shows the magnitude of v_H in the z direction, the up/down velocity of the end-effector commanded by the Jamboxx pitch motion. When the user moves in and out of the brown region, the system toggles into and out of the orientation mode. In the orientation mode, the user can select a fixed amount of rotation about the wrist. In the prototype interface, we include a set of fixed yaw and roll angles to facilitate grasping in tight spaces. To return to the linear velocity mode, the user simply moves in and out of the brown region again. Note that the user interface is completely customizable to tailor to individual preference and capabilities. In terms of information display, the Baxter has three cameras, one at the wrist of each arm and one on the torso. Only two of the three could be streamed at a time. Figure 3.5b shows the left and right arm camera views. When one arm is commanded to perform a task, such as grasping an object from the shelf, the second arm may serve as a roving eye, especially when the first arm camera is occluded. The contact force information is directly plotting on the user screen.

3.4.2 *Object Pick-Up and Placement Task*

In the object pick-and-place task, the user needs to first direct the end-effector of the robot toward the shelf, command the robot to pick up the bottle from shelf, back up the robot and turn, and move to the table and drop the bottle. The results shown here is based on the operation of the Jamboxx by the first author. The overall task execution is completely in 80sec. Figure 3.6 shows the key stages during the task execution: (1) initial configuration (2) arm reaching boundary and wheelchair about to move (3) reaching and grasping the bottle on the shelf (4) arm holding the bottle and wheelchair ready for backing up and turning (5) completion of wheelchair turning around (6) placement of bottle on the table.

The shoulder and elbow joint angles q_{L_1} to q_{L_4} are shown in Fig. 3.7a. The minimum distance to obstacles and self-collision distance $p_{OT}^{(x)}$ are shown in Fig. 3.7b. The commanded velocity v_H and the actual robot velocity $J_T u$ are compared in Fig. 3.7d. The base motion ω_B and v_B are shown in Fig. 3.7c. As can be seen from the plots, all joint angles are within their upper and lower bounds throughout the motion. The human input command is realized most of the time, except for the time period when the wheelchair is too close to obstacle and the human still drives it forward. From Fig. 3.7b, it can be seen that the distance to obstacle is kept outside of the bounding box of wheelchair to ensure collision avoidance. At about $t = 50$ s, the human command still tries to push the wheelchair forward but the wheelchair does not move due the inequality constraint. In this case, human user backs up the wheel-

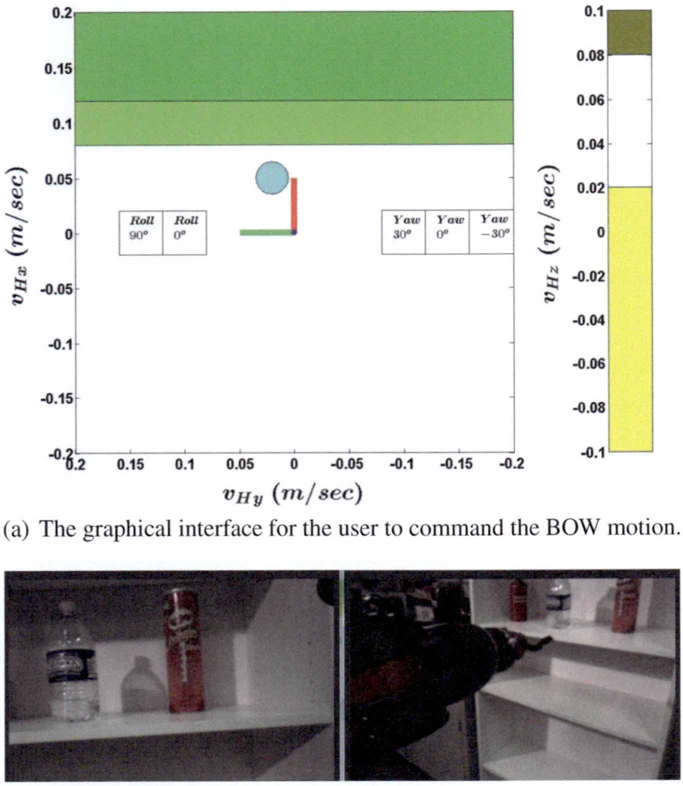

(a) The graphical interface for the user to command the BOW motion.

(b) Left-hand and right-hand camera views.

Fig. 3.5 Graphical user interface for the operation of the BOW. **a** shows the user command for the BOW motion **b** shows the views from BOW hand cameras

chair to continue with the operation. The plot also show the self-collision distance $p_{OT}^{(x)}$ larger than its minimum allowable value for all time. From Fig. 3.7c, before $t = 6$ s, the human command can be fully realized by arm movement only, so that the wheelchair speed is zero. After 6sec, the joint angle q_{L_4} reaches its lower bound and the arm cannot stretch further. The wheelchair then starts to move to compensate for the desired motion. This demonstrates the intended combined arm-base motion proposed in theory. Since the wheelchair subsystem has time delay and nonlinearities (only open loop command is implemented at this time), the velocity commands sent to the wheelchair do not exactly equal the actual velocities, causing some discrepancies between v_H and $J_T u$ even if they should be equal ideally. However, it can be seen that v_H and $J_T u$ largely match each other, showing a good realization of human commands.

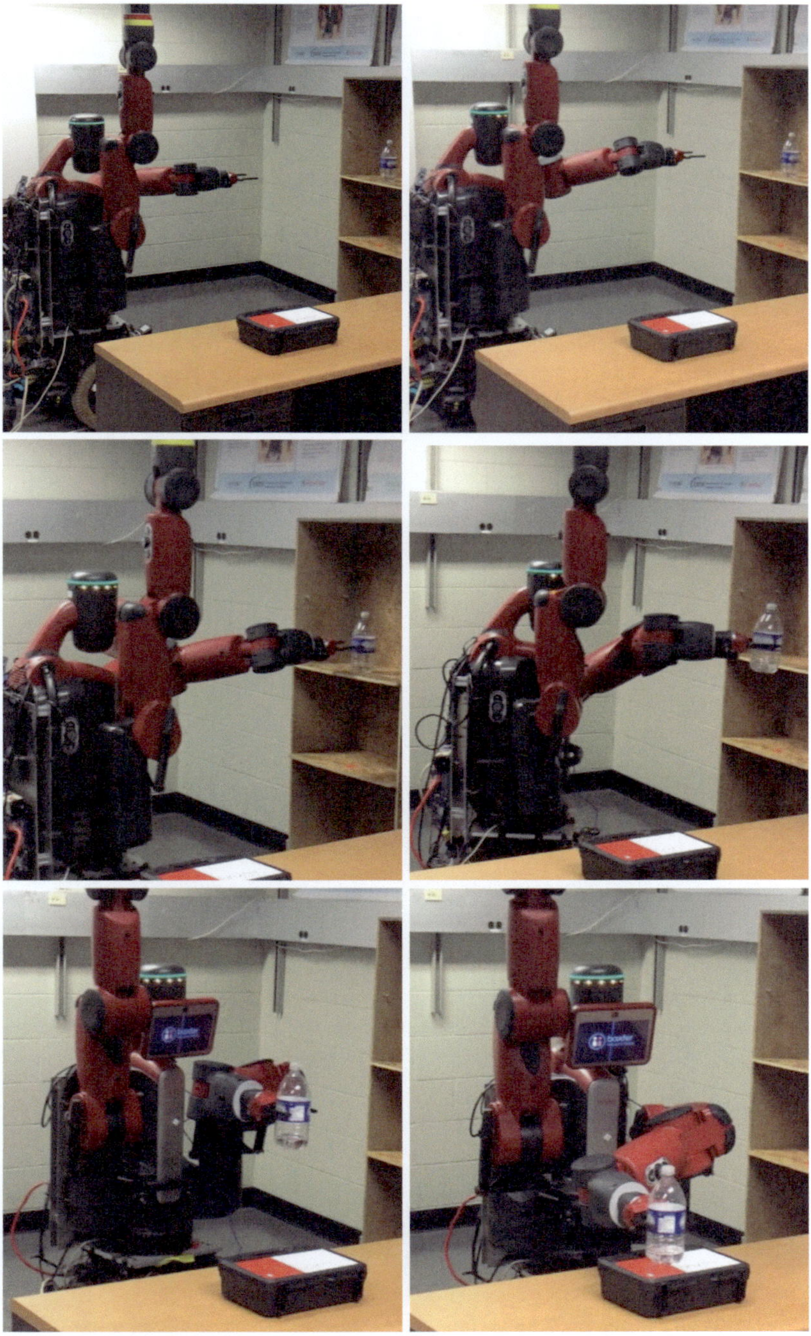

Fig. 3.6 Pictures for different stages of the task execution: (1) initial configuration (2) arm at boundary of workspace and wheelchair about to move (3) approach bottle for grasp (4) arm holding the bottle and wheelchair about to back up and turn (5) completion of wheelchair turn toward table (6) bottle placement

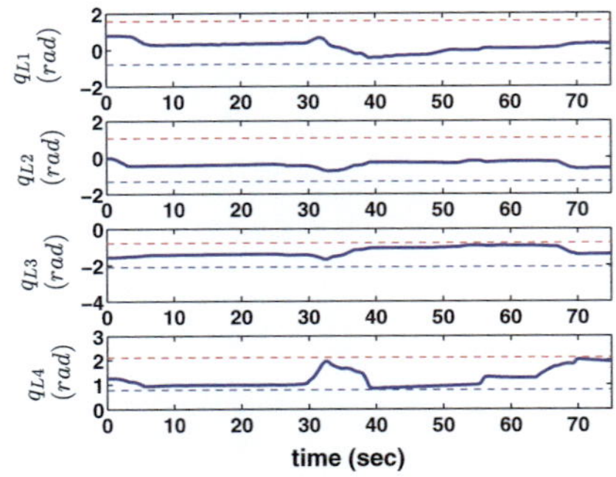

(a) Joint angles q_{L_1} to q_{L_4} in experiment are constrained to be within the joint limits indicated by the dashed lines.

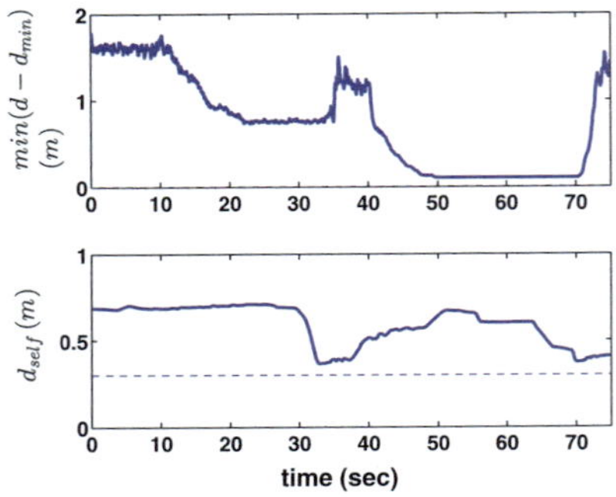

(b) The minimum distance from the bounding box of wheelchair to the obstacles (shelf and table), and the self-collision distance $p_{OT_L}^{(x)}$ in the experiment both satisfy the constraint throughout the experiment.

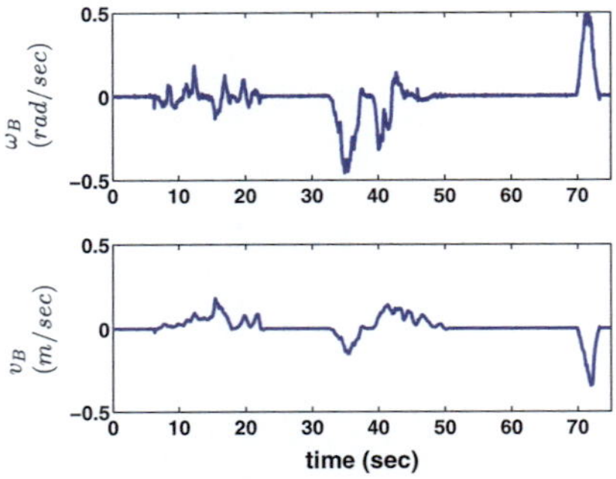

(c) The wheelchair velocities ω_B and v_B in the free space experiment.

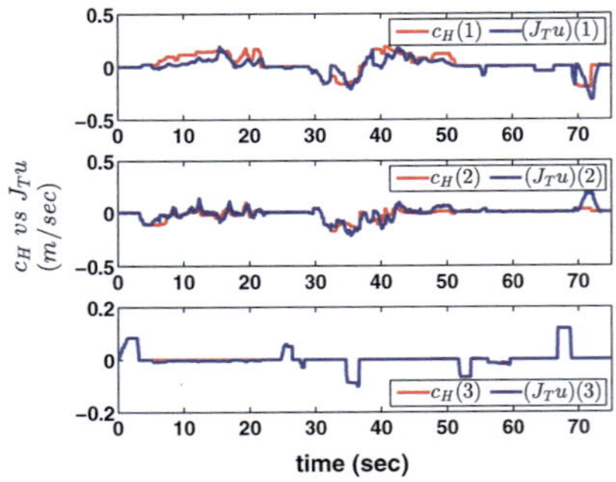

(d) The human commanded end-effector velocity, v_H, and the actual end-effector velocity, $J_T u$. Although the wheelchair control faces time delay and nonlinearities, the human commands are largely satisfied.

Fig. 3.7 Performance of the free- space pick-and-place experiment with the human user commanding the translational velocity of the BOW *left* arm end-effector motion

3.4.3 Board Cleaning Task

The board cleaning task involves both free-space and contact modes and the transition between them. In this case, the human user employs a joystick to provide 3-dof command signals to drive the arm toward the board, adjust the desired contact force, erase the board and then departs from the board. This work has been presented in part in [1]. The human input v_H versus the actual end-effector velocity $J_T u$ are plotted in Fig. 3.8a, and the measured end-effector force magnitude versus the desired contact force are plotted in Fig. 3.8b.

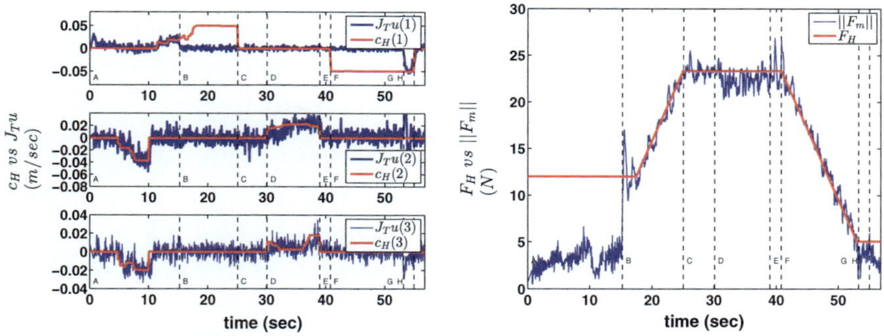

(a) The human input v_H versus the actual end-effector velocity $J_T u$.

(b) The measured end-effector force magnitude versus the desired contact force.

Fig. 3.8 Motion and force responses for the board cleaning experiment (© 2016 IEEE. Reprinted, with permission, from [2])

In the plots, we highlight the following key phases of the task execution:

- Phase A: The system is in the free- space mode, and the end-effector is commanded to move towards the board.
- Phase B: The eraser makes contact with the whiteboard, and the measured contact force jumps above F_{Th}. The system immediately switches to the contact mode, after which the user pushes v_H into the board to increase the desired contact force.
- Phase C: The user finishes adjusting the desired contact force, which is maintained at $24.2N$.
- Phase D: The user starts wiping up letters on the board.
- Phase E: The user finishes wiping up the board.
- Phase F: The user starts to command the arm away from the board (with $S_F^T v_H < 0$) to reduce the desired contact force.
- Phase G: The contact force F_m falls below F_{Tl}, and the system switches to the free-space mode.
- Phase H: The final phase of the system when the user finishes backing up the arm and the arm is several centimeters away from the board.

The system configurations at phase A, phase B-D, phase E-G, and phase H are shown in Fig. 3.9.

From the end-effector velocity plot Fig. 3.8a, it is seen that in the free-space mode, $J_T u$ largely tracks v_H. In the contact mode, $J_T u$ tracks v_H closely in the unconstrained directions (y and z). The difference in the x-direction results in contact force control. From the end-effector force plot Fig. 3.8b, the measured contact force shows reasonably good tracking of the desired force. The force tracking error could be reduced by incorporating real-time velocity information into the normal direction estimation, canceling the effect of friction when the eraser is in motion. For human-directed contact control tasks in everyday environment, the force tracking accuracy is not a primary consideration, as long as the contact is maintained and the force

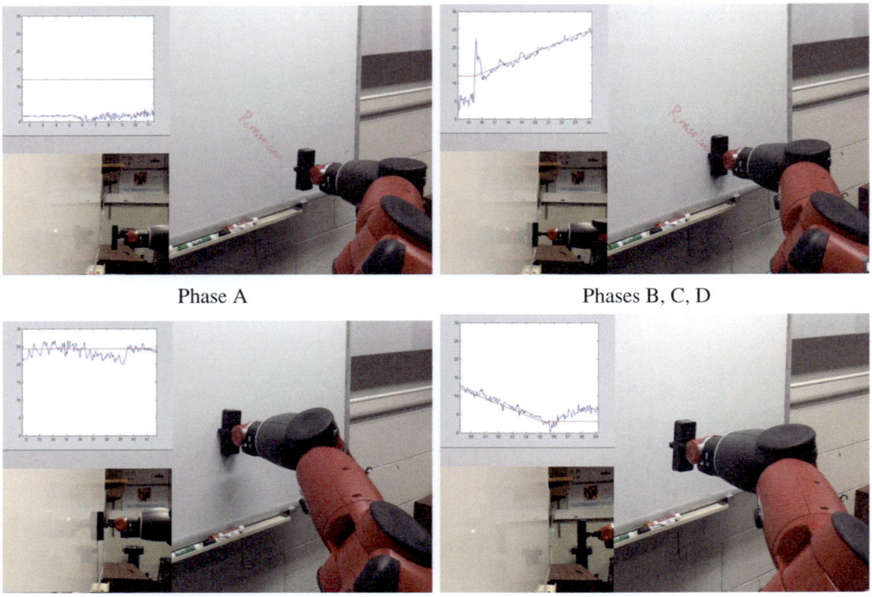

Fig. 3.9 The system configurations at phase A, phase B-D, phase E-G, and phase H

magnitude is within a reasonably large neighborhood of the desired force. In our example, if we choose F_H to be sufficiently large, the eraser is able to maintain contact throughout the erasing motion.

3.5 Conclusion

Assistive robots hold promise of improving the quality of life of severely mobility-impaired individuals. We have presented our shared-control implementation of an mobile assistive robotic manipulator, the BOW, combining a human-friendly industrial robot, and a standard power wheelchair. By posing the robot motion control problem as a kinematics-based optimization, we are able to combine the limited user input of the end-effector motion with sensor and environment information to produce intuitive and effective motion. This approach has been demonstrated in simulation and experimentation for navigation, grasping, and force control. Our current plan is to work with disabled population to improve the efficacy of the user interface, and allow the robot controller to learn from the human guidance through experience.

Acknowledgements We would like to thank Andrew Cunningham and Jingyu Su for the wheelchair control system implementation, Dan Kruse for the Robot Raconteur Baxter Bridge, John Wason for the development and sharing of the Robot Raconteur software, Will Keddy-Hector and

Utkarsh Sinha for the initial development of the BOW, and Jonas Braasch and the Center for Cognition, Communication, and Culture (CCC) for the support of this endeavor. The enthusiasm, encouragement and feedback from David Whalen is a continuing source of inspiration. This work is supported in part by the National Science Foundation Smart Lighting Engineering Research Center under NSF Cooperative Agreement No. EEC-0812056 and in part by the Center for Automation Technologies and Systems (CATS) under a block grant from the New York State Empire State Development Division of Science, Technology and Innovation (NYSTAR).

References

1. Lu L, Wen JT (2015) Human-directed robot motion / force control for contact tasks in unstructured environments. In: IEEE Conference on automation science and engineering, Sweden, pp 1165–1170. doi:10.1109/CoASE.2015.7294255
2. Lu L, Wen JT (2015) Human-robot cooperative control for mobility impaired individuals. In: American control conference, Chicago, July-2015, pp 447–452. doi:10.1109/ACC.2015.7170776
3. Spinal Cord Injury (SCI) facts and figures at a glance. National Spinal Cord Injury Statistical Center (NSCISC). https://www.nscisc.uab.edu/Public/Facts2015.pdf (2015). Last Accessed 10 Mar 2016
4. General Information: Internation Campaign for Cures of Spinal Cord Injury Paralysis (ICCP). http://www.campaignforcure.org. Last Accessed 10 Mar 2016
5. Cao Y, Chen Y, DeVivo M (2011) Lifetime direct costs after spinal cord injury. Topics Spinal Cord Inj Rehabil 16(4):10–16. doi:10.1310/sci1604-10
6. DeVivo M, Chen Y, Mennemeyer S, Deutsch A (2011) Costs of care following spinal cord injury. Topics Spinal Cord Inj Rehabil 16(4):1–9. doi:10.1310/sci1604-1
7. Argall BD (2015) Turning assistive machines into assistive robots. In: Proceedings of SPIE - The International Society for Optical Engineering, vol 9370, pp 1–12. doi:10.1117/12.2085352
8. Goodrich MA, Schultz AC (2007) A survey. Found. Trends Hum-Comput Interact 1(3):203–275. doi:10.1561/1100000005
9. Haigh KZ, Yanco H (2002) Automation as caregiver: a survey of issues and technologies. In: AAAI-02 Workshop on automation as caregiver: the role of intelligent technology in elder care, pp 39–53
10. Capille JW Jr (2010) Kinematic and experimental evaluation of commercial wheelchair-mounted robotic arms. Ph.D. Thesis, University of South Florida
11. Hillman M, Hagan K, Hagan S, Jepson J, Orpwood R (2002) The Weston wheelchair mounted assistive robot-the design story. Robotica 20(2):125–132
12. Mahoney RM (2001) The raptor wheelchair robot system. Integration of assistive technology in the information age. IOS Press, Netherlands, pp 135–141
13. Meal Body Assistive Feeder. http://www.richardsonproducts.com. Last Accessed 19 Mar 2016
14. Bien Z, Chung MJ, Chang PH, Kwon DS, Kim DJ, Han JS, Kim JH, Kim DH, Park HS, Kang SH, Lee K, Lim SC (2004) Integration of a rehabilitation robotic system (KARES II) with human-friendly man-machine interaction units. Auton Robots 16(2):165–191. doi:10.1023/B:AURO.0000016864.12513.77
15. Bien Z, Kim DJ, Chung MJ, Kwon DS, Chang PH (2003) Development of a wheelchair-based rehabilitation robotic system (KARES II) with various human-robot interaction interfaces for the disabled. In: IEEE/ASME international conference on advanced intelligent mechatronics, AIM, vol 2(Aim), pp 902–907. doi:10.1109/AIM.2003.1225462
16. Chen T, Ciocarlie M, Cousins S, Grice P, Hawkins K, Hsiao K, Kemp C, King CH, Lazewatsky D, Leeper A, Nguyen H, Paepcke A, Pantofaru C, Smart W, Takayama L (2013) Robots for humanity: using assistive robotics to empower people with disabilities. IEEE Robot Autom Mag 20(1):30–39

17. Huete AJ, Victores JG, Martinez S, Giminez A, Balaguer C (2012) Personal autonomy rehabilitation in home environments by a portable assistive robot. IEEE Trans Syst Man Cybern Part C: Appl Rev 42(4):561–570. doi:10.1109/TSMCC.2011.2159201
18. Srinivasa S, Berenson D, Cakmak M, Romea AC, Dogar MR (2012) HERB 2.0: lessons learned from developing a mobile manipulator for the home. In: Proceedings of the IEEE, pp 1–19
19. Srinivasa SS, Ferguson D, Helfrich CJ, Berenson D, Collet A, Diankov R, Gallagher G, Hollinger G, Kuffner J, Weghe MV (2010) HERB: a home exploring robotic butler. Auton Robot 28(1):5–20. doi:10.1007/s10514-009-9160-9
20. Jardón A, Gil ÁM, de la Peña AI, Monje C, Balaguer C (2011) Usability assessment of ASIBOT: a portable robot to aid patients with spinal cord injury. Disabil Rehabil Assist Technol 6(4):320–330. doi:10.3109/17483107.2010.528144
21. Anderson RJ, Spong MW (1989) Bilateral control of teleoperators with time delay. IEEE Trans Autom Control 34(5):494–501
22. Lawrence DA (1993) Stability and transparency in bilateral teleoperation. IEEE Trans Robot Autom 9(5)
23. Vertut J (2013) Teleoperation and robotics: application and technology. Springer, Netherlands
24. Sheridan TB (1992) Telerobotics, automation, and human supervisory control. MIT Press, Cambridge
25. Kruse D, Wen JT, Radke RJ (2015) A sensor-based dual-arm tele-robotic system. IEEE Trans Autom Sci Eng 12(1):4–18
26. Stückler J, Behnke S (2011) Following human guidance to cooperatively carry a large object. In: 11th IEEE-RAS international conference on humanoid robots, Slovenia
27. Kim H, Biggs SJ, Schloerb DW, Carmena JM, Lebedev MA, Nicolelis MAL, Srinivasan MA (2006) Continuous shared control for stabilizing reaching and grasping with brain-machine interfaces. IEEE Trans Biomed Eng 53(6):1164–1173
28. Lacey G, MacNamara S (2000) Context-aware shared control of a robot mobility aid for the elderly blind. Int J Robot Res 19(11):1054–1065
29. Wang H, Liu XP (2014) Adaptive shared control for a novel mobile assistive robot. IEEE/ASME Trans Mechatron 19(6):1725–1736
30. Jiang J, Astolfi A (2015) Shared-control for the kinematic model of a rear-wheel drive car. In: Proceedings of 2015 American control conference, Chicago
31. Storms JG, Tilbury DM (2014) Blending of human and obstacle avoidance control for a high speed mobile robot. In: American control conference (ACC), USA
32. Franchi A, Secchi C, Ryll M, Bulthoff HH, Giordano PR (2012) Shared control: balancing autonomy and human assistance with a group of quadrotor UAVs. IEEE Robot Autom Mag 19(3):57–68
33. Nudehi SS, Mukherjee R, Ghodoussi M (2005) A shared-control approach to haptic interface design for minimally invasive telesurgical training. IEEE Trans Control Syst Technol 13(4):588–592
34. Dragan AD, Srinivasa SS (2013) A policy-blending formalism for shared control. Int J Robot Res 32(7):790–805
35. Storms JG, Vozar S, Tilbury DM (2014) Predicting human performance during teleoperation. In: ACM/IEEE international conference on human-robot interaction, Bielefeld University, Germany
36. Wang Z, Mülling K, Deisenroth MP, Amor HB, Vogt D, Schölkopf B, Peters J (2013) Probabilistic movement modeling for intention inference in human-robot interaction. Int J Robot Res 32(7):841–858
37. Cai H, Mostofi Y (2015) To ask or not to ask: a foundation for the optimization of human-robot collaborations. In: Proceedings of 2015 American control conference, Chicago
38. Argall BD, Murphey TD (2014) Computable trust in human instruction. In: AAAI fall symposium series, pp 31–32
39. Wang X, Shi Z, Zhang F, Wang Y (2015) Mutual trust based scheduling for (semi) autonomous multi-agent systems. In: American control conference (ACC), pp 459–464

40. Cunningham A, Keddy-Hector W, Sinha U, Whalen D, Kruse D, Braasch J, Wen J (2014) Jamster: a mobile dual-arm assistive robot with Jamboxx control. In: Conference on automation science and engineering (CASE), pp 509–514
41. Baxter: Redefining robotics and manufacturing (2014). http://www.rethinkrobotics.com/products/baxter. Last Accessed 14 Mar 2014
42. Venator E, Lee GS, Newman W (2013) Hardware and software architecture of ABBY: an industrial mobile manipulator. In: 2013 IEEE international conference on automation science and engineering (CASE), IEEE, pp 324–329
43. Madigan EA, Newman WS (2012) What do users want from "smart wheelchairs". In: NI 2012: Proceedings of the 11th international congress on nursing informatics, 2012. American Medical Informatics Association
44. Pineau J, Atrash A (2007) SmartWheeler: a robotic wheelchair test-bed for investigating new models of human-robot interaction. In: AAAI spring symposium: multidisciplinary collaboration for socially assistive robotics, pp 59–64
45. CMU Sphinx. http://cmusphinx.sourceforge.net. Last Accessed 25 Jan 2016
46. Wason Technology: Robot Raconteur (2014). https://robotraconteur.com Accessed 28 July 2015
47. Kruse D (2014) Baxter_RR_Bridge. https://github.com/ChopChopperDan/baxter_RR_bridge Accessed 28 July 2015

Chapter 4
Switchings Between Trajectory Tracking and Force Minimization in Human–Robot Collaboration

Yanan Li, Keng Peng Tee and Shuzhi Sam Ge

4.1 Introduction

Human–robot collaboration is becoming a popular research topic due to two major reasons: first, robots are anticipated in many applications where human–robot interactions are inevitable; second, there still exists a big gap between state-of-the-art technologies of autonomous robotics and real-world requirements. A lot of effort has been made by scientists and engineers in designing safe, reliable, and natural human–robot interface to integrate complementary capabilities of humans and robots [17, 18]. One of the most straightforward ways to realize human–robot collaboration is through physical interactions, which allows humans and robots to exchange force and position information of each other efficiently [1].

A typical scenario of this type of human–robot collaboration is sketched in Fig. 4.1, where a human physically interacts with a robot. There is a force/torque sensor at the interaction point, by which the robot is able to measure the interaction force (including force and torque) exerted by the human and moves accordingly. Figure 4.1 also shows an application example to illustrate the necessity of human–robot collaboration: in a normal situation, the robot is able to follow a predefined trajectory and finishes a task (e.g., pick and place) alone; in another possible situation, an obstacle appears

---

This chapter was developed and enhanced from an earlier paper published as [20] ©2015 IEEE.

Y. Li (✉)
Imperial College London, SW7 2AZ London, London, UK
e-mail: hit.li.yn@gmail.com

K.P. Tee
Institute for Infocomm Research, 1 Fusionopolis Way, Singapore 138632, Singapore
e-mail: kptee@i2r.a-star.edu.sg

S.S. Ge
National University of Singapore, 21 Lower Kent Ridge Road,
Singapore 119077, Singapore
e-mail: samge@nus.edu.sg

Y. Wang and F. Zhang (eds.), *Trends in Control and Decision-Making for Human–Robot Collaboration Systems*, DOI 10.1007/978-3-319-40533-9_4

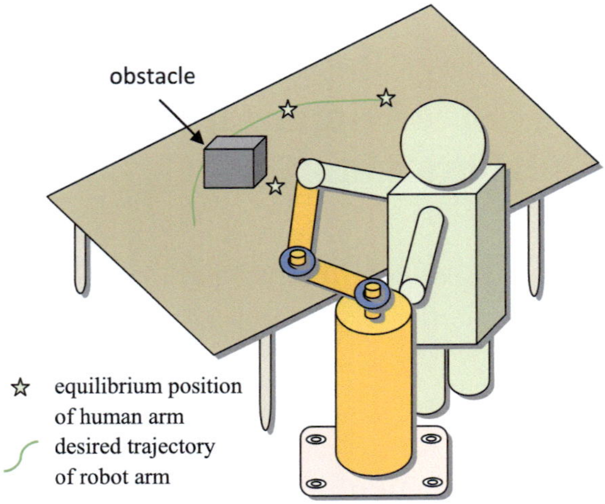

Fig. 4.1 A typical scenario of human–robot collaboration. The equilibrium position of the human arm and the desired trajectory of the robot arm will be explained later in system modeling

in halfway of the robot's desired trajectory or the human has several position points or areas of interest due to a certain other reason; in this situation, the human expects to fine-tune the robot's motion by jumping into the robot's control loop on the fly; when the human intervention disappears, the robot is able to resume its unfinished task and continue to follow its desired trajectory.

One natural idea to achieve the above objectives is to integrate trajectory tracking and force minimization (such that the human can easily move the robot without much effort), and switch between them when necessary. In particular, tracking control can be adopted when there is no human intervention, and force or impedance control is adopted when there is human intervention. In the literature, tracking control has been studied in the case of non-interaction to achieve trajectory tracking, including proportional-derivative control, adaptive control [26], learning control [2], etc. In the case of physical interaction, force control or impedance control is employed to realize force regulation or a trade-off between trajectory tracking and force regulation. Impedance control is originally derived from optimal control by considering such a trade-off [13]. In the field of human–robot collaboration through physical interactions, it is natural to take force regulation as a control objective because human's feeling (interaction force between the robot and human, in the case of physical interaction) is one of the most important indexes. To regulate the interaction force, modeling of the human intention has been studied in several works: motion characteristics of the human arm is investigated to generate a point-to-point movement for the robot to collaborate with the human [5]; the human intention is observed by the change of the interaction force [7]; haptic data (position and force) are used to recognize the human intention so that the robot is able to provide assistance to the human [25]; a model is obtained using learning by demonstration to predict the human intention [22]; the walking direction of the users of a crane robot is estimated using a Kalman filter [29]; the human intention is estimated by employing a human arm model [17] and minimizing the interaction force [18], respectively.

Tracking control takes the external force as a disturbance while force or impedance control regulates the robot's motion according to the external force. Therefore, a hard switching between them will lead to a sudden change of the control signal and thus oscillation and even instability. Another option to achieve the above objectives is to employ impedance control alone, but it has been shown in many studies that a fixed impedance model can only achieve either trajectory tracking, minimization of human's force, or a trade-off between these two.

The scenario described in Fig. 4.1 represents a general case where the robot needs to change its control objectives according to its circumstances. This research topic has attracted a lot of attention recently. In [14], it is pointed out that the robot should have an adjustable role when it interacts with a human, such that different desired performances can be achieved in corresponding situations. In [8], an abstract homotopy model is proposed to describe the interactive behavior of a dyad, which sheds lights on the control design for the robot in interaction with a human. In [23], several methods are developed to change the robot's role on the fly in a cotransporting task, and they are compared with both objective and subjective measures. In [4], shared control of human and robot is investigated for a tracking problem with a terminal constraint, and the robot updates its control strategy by taking human's control input into account. The task considered in [4, 23] includes a common goal which is known to both human and robot. Instead, in the scenario described in Fig. 4.1, human and robot may have their own goals which are unknown to each other.

In this work, we employ impedance control and propose a general framework which is able to change the robot's impedance and reference trajectory intuitively, and eventually realize switchings between trajectory tracking and force minimization. To achieve trajectory tracking and force minimization in two separate situations, variable impedance control with different impedance parameters can be developed: high impedance when trajectory tracking is expected while low impedance when force minimization is expected [19]. Besides this, adaptation of the robot's reference trajectory also has to be taken into account because typically the human's desired trajectory is different from the predefined reference trajectory of the robot. Reference adaptation will help in improving the transient performance during switchings between two cases. Part of this work has been presented in [20], and more details about selection of cost functions are given with new simulation results in this work.

The rest of this work is organized as follows. In Sect. 4.2, modeling of the system under study is discussed. In Sect. 4.3, the proposed framework is introduced, selection of cost functions elaborated with examples and the employed dynamic programming explained. Simulation results under the proposed method are presented in Sect. 4.4, and concluding remarks are given in Sect. 4.5.

4.2 Dynamic Models

The system under study includes a robot arm and a human arm, of which the dynamic models and their properties are introduced in this section.

4.2.1 Robot Model

The kinematics of the robot arm are

$$x(t) = \phi(q) \tag{4.1}$$

where $x(t) \in \mathbb{R}^n$ is position in the Cartesian space and $q \in \mathbb{R}^n$ is coordinate in the joint space. Differentiating Eq. (4.1) with respect to time leads to

$$\dot{x}(t) = J(q)\dot{q} \tag{4.2}$$

where $J(q) \in \mathbb{R}^{n\times n}$ is the Jacobian matrix. Further differentiating Eq. (4.2) with respect to time leads to

$$\ddot{x}(t) = \dot{J}(q)\dot{q} + J(q)\ddot{q} \tag{4.3}$$

The dynamic model of the robot arm in the joint space is

$$M(q)\ddot{q} + C(q,\dot{q})\dot{q} + G(q) = \tau + J^T(q)f \tag{4.4}$$

where $M(q) \in \mathbb{R}^{n\times n}$ is the inertia matrix, $C(q,\dot{q})\dot{q} \in \mathbb{R}^n$ the Coriolis and Centrifugal force, $G(q) \in \mathbb{R}^n$ the gravitational force, $\tau \in \mathbb{R}^n$ the control input and $f \in \mathbb{R}^n$ the interaction force applied by the human arm. Substituting (4.1)–(4.3) into the dynamic model (4.4), we have the dynamic model of the robot arm in the Cartesian space

$$M_R(q)\ddot{x} + C_R(q,\dot{q})\dot{x} + G_R(q) = u + f \tag{4.5}$$

where

$$\begin{aligned} M_R(q) &= J^{-T}(q)M(q)J^{-1}(q) \\ C_R(q,\dot{q}) &= J^{-T}(q)(C(q,\dot{q}) - M(q)J^{-1}(q)\dot{J}(q))J^{-1}(q) \\ G_R(q) &= J^{-T}(q)G(q),\ u = J^{-T}(q)\tau \end{aligned} \tag{4.6}$$

For feasibility of the control design, we rewrite Eq. (4.5) as

$$\ddot{x} = -M_R^{-1}(q)C_R(q,\dot{q})\dot{x} - M_R^{-1}(q)G_R(q) + M_R^{-1}(q)f + M_R^{-1}(q)u \tag{4.7}$$

4.2.2 Human Arm Model

In a physical interaction system, a human arm can be modeled by a system that

- applies a force to control the state of the robot arm, and simultaneously

- changes its own state according to the state of the robot arm and the force applied by the robot arm.

In the literature, equilibrium point control hypothesis is proposed to model the human arm [9], as below

$$M_H\ddot{x} + C_H\dot{x} + K_H(x - x_H) = -f \tag{4.8}$$

where M_H, C_H and K_H are unknown inertia, damping and stiffness matrices, respectively, and x_H is the equilibrium position which indicates the rest position of the human arm when there is no interaction force. Note that M_H, C_H and K_H are not inherent physical parameters of the human arm. Instead, they are the results of the human arm dynamics under human control. In particular, humans may control their muscles to result in different M_H, C_H and K_H when necessary. For feasibility of the control design, we neglect the effect of the inertial component in the above model, i.e.,

$$C_H\dot{x} + K_H(x - x_H) = -f \tag{4.9}$$

This simplification is supported by the literature which shows that C_H and K_H dominate the dynamics of the human arm in physical human–robot interaction [6, 28]. Furthermore, C_H and K_H have the following property:

Property 4.1 *C_H and K_H are functions of x and $\dot{x}$.*

The above property is easy to understand: first, C_H describes damping and Coriolis effects which is thus dependent on the position or/and velocity of the human arm; second, K_H is the stiffness component which has been shown to be also dependent on the position or/and velocity of the human arm [24, 27].

Lemma 4.1 *Given a vector $\xi \in \mathbb{R}^n$, we have*

$$\begin{aligned} \dot{C}_H\xi &= C_{H1}(x, \xi)\dot{x} + C_{H2}(\dot{x}, \xi)\ddot{x} \\ \dot{K}_H\xi &= K_{H1}(x, \xi)\dot{x} + K_{H2}(\dot{x}, \xi)\ddot{x} \end{aligned} \tag{4.10}$$

where $C_{H1}(x, \xi) \in \mathbb{R}^{n\times n}$, $C_{H2}(\dot{x}, \xi) \in \mathbb{R}^{n\times n}$, $K_{H1}(x, \xi) \in \mathbb{R}^{n\times n}$ and $K_{H2}(\dot{x}, \xi) \in \mathbb{R}^{n\times n}$ are defined in the proof.

Proof See Appendix.

By taking the derivative of Eq. (4.9), we obtain

$$\dot{C}_H\dot{x} + C_H\ddot{x} + \dot{K}_H(x - x_H) + K_H\dot{x} = -\dot{f} \tag{4.11}$$

By considering Eqs. (4.9), (4.11) and Property 4.1, we obtain

$$\begin{aligned}\dot{f} &= -\dot{K}_H(x - x_H) - K_H\dot{x} - \dot{C}_H\dot{x} - C_H\ddot{x} \\ &= -K_{H1}(x, x - x_H)\dot{x} - K_{H2}(\dot{x}, x - x_H)\ddot{x} - K_H\dot{x} \\ &\quad -C_{H1}(x, \dot{x})\dot{x} - C_{H2}(\dot{x})\ddot{x} - C_H\ddot{x}\end{aligned} \tag{4.12}$$

By denoting $K_{H1}(x, x - x_H), K_{H2}(\dot{x}, x - x_H), C_{H1}(x, \dot{x})$ and $C_{H2}(\dot{x})$ as K_{H1}, K_{H2}, C_{H1} and C_{H2}, respectively, we have

$$\dot{f} = -K_{H1}\dot{x} - K_{H2}\ddot{x} - K_H\dot{x} - C_{H1}\dot{x} - C_{H2}\ddot{x} - C_H\ddot{x} \tag{4.13}$$

4.2.3 Unified Model

In this subsection, we combine the robot model and the human model in a unified one. In particular, we denote the state $z = [z_1, z_2, z_3]^T$ where $z_1 = x$, $z_2 = \dot{x}$ and $z_3 = f$. Then, the robot model (4.7) and the human model (4.13) can be combined into

$$\dot{z} = A(z)z + B(z)u + D(z) \tag{4.14}$$

where

$$\begin{aligned}A(z) &= L^{-1}\begin{bmatrix}\mathbf{0}_n & I_n & \mathbf{0}_n \\ \mathbf{0}_n & -M_R^{-1}(q)C_R(q, \dot{q}) & M_R^{-1}(q) \\ \mathbf{0}_n & -K_{H1} - K_H - C_{H1} & \mathbf{0}_n\end{bmatrix} \\ B(z) &= L^{-1}\begin{bmatrix}\mathbf{0}_n \\ M_R^{-1}(q) \\ \mathbf{0}_n\end{bmatrix}, \quad D(z) = L^{-1}\begin{bmatrix}\mathbf{0}_n \\ -M_R^{-1}(q)G_R(q) \\ \mathbf{0}_n\end{bmatrix} \\ L &= \begin{bmatrix}I_n & \mathbf{0}_n & \mathbf{0}_n \\ \mathbf{0}_n & I_n & \mathbf{0}_n \\ \mathbf{0}_n & C_{H2} + C_H + K_{H2} & I_n\end{bmatrix}\end{aligned} \tag{4.15}$$

with $\mathbf{0}_n \in \mathbb{R}^{n\times n}$ and $I_n \in \mathbb{R}^{n\times n}$ being zero and unit matrices, respectively.

Remark 4.1 Note that A, B and D are functions of z because $q = \phi^{-1}(x), \dot{q} = J^{-1}\dot{x}$, $x - x_H = K_H^{-1}(f + C_H\dot{x})$ and C_H and K_H are functions of x and $\dot{x}$ (Property 4.1). Also note that it is easy to show that L is non-singular.

Remark 4.2 Different from existing works about physical human–robot interaction, in this work, the human model is taken into consideration and the interaction force f is deemed as a state in the unified model. In this way, the effect of the states of human and robot to each other can be investigated. This is one of the main contributions of this work.

4.2.4 Trajectory Tracking

In the last step of modeling, we consider the transformation between regulation and trajectory tracking in optimal control [12].

Denote the desired trajectory of the robot arm as x_d. It can be obtained by developing the following system:

$$\dot{x}_d = F(x_d, z) \tag{4.16}$$

where F is a known vector function. Thus, the complete system under study becomes

$$\dot{\bar{z}} = \bar{A}(\bar{z}) + \bar{B}(\bar{z})u \tag{4.17}$$

where

$$\bar{z} = [z^T \ x_d^T]^T, \ \bar{A}(\bar{z}) = \begin{bmatrix} A(z)z + D(z) \\ F(x_d, z) \end{bmatrix}, \ \bar{B}(\bar{z}) = \begin{bmatrix} B(\bar{z}) \\ \mathbf{0}_l \end{bmatrix} \tag{4.18}$$

4.3 Control Design

4.3.1 Control Objective

Recalling the scenario described in Fig. 4.1, we expect two control objectives in two situations: trajectory tracking and force minimization. Therefore, it is natural to consider the following cost function:

$$\Gamma(t) = \int_0^\infty c(v)\mathrm{d}v \tag{4.19}$$

where $c(t)$ is the instant cost defined as

$$c(t) = (x - x_d)^T Q_1 (x - x_d) + \dot{x}^T Q_2 \dot{x} + f^T Q_3 f + u^T R u \tag{4.20}$$

$Q_1 \in \mathbb{R}^{n\times n} \geq 0$, $Q_2 \in \mathbb{R}^{n\times n} \geq 0$, $Q_3 \in \mathbb{R}^{n\times n} \geq 0$ and $R \in \mathbb{R}^{n\times n} > 0$. This cost function is similar to that in optimal robot control: trajectory tracking is achieved by including the position and velocity errors while the control input is also constrained. This is one objective in the scenario described in Fig. 4.1. The difference is that the interaction force f also appears in the cost function. This lies in an assumption that the interaction force f is generated due to the conflict between human's destination and the current position, which is in line with the human model (4.9). By penalizing it, the robot can be moved to the rest position of the human arm so the interaction force becomes 0, which is the other objective in the scenario described in Fig. 4.1.

Therefore, two control objectives in two situations can be achieved by adjusting the costs, e.g., $c(t) = (x - x_d)^T Q_1 (x - x_d)$ for trajectory tracking and $c(t) = f^T Q_3 f$ for force minimization. Moreover, the trade-off between the two objectives can be also realized by choosing a cost that includes both the tracking error and the interaction force.

4.3.2 Selection of Cost Functions

Following the previous subsection, we discuss how to choose the cost function according to different circumstances. This is task-dependent and we cannot provide a general framework in this work. Therefore, we still focus on the scenario in Fig. 4.1 and propose two update laws for illustration.

First, we may update the cost function by changing the weights Q_1, Q_2, Q_3, and R. For example, it is expected that trajectory tracking is more important when the actual position of the robot arm is near the prescribed desired trajectory. Therefore, we may divide the work space of the robot arm $\mathscr{WS}$ into two subspaces: $\mathscr{WS}_1 = \{x \mid \|x - x_d\| < d\}$ where d is a designed distance, and $\mathscr{WS}_2 = \mathscr{WS} - \mathscr{WS}_1$. In $\mathscr{WS}_1$, trajectory tracking is more important while in $\mathscr{WS}_2$, force minimization is more important. This is achieved by the following update law:

$$\begin{cases} Q_1^{i,i} \gg 1 \text{ and } Q_3^{i,i} \ll 1, \text{ if } x \in \mathscr{WS}_1; \\ Q_1^{i,i} \ll 1 \text{ and } Q_3^{i,i} \gg 1, \text{ if } x \in \mathscr{WS}_2. \end{cases} \tag{4.21}$$

where $Q_1^{i,i}$ and $Q_3^{i,i}$ stand for the i-th diagonal element of Q_1 and Q_3, respectively. It is assumed that Q_1, Q_2, Q_3 and R are diagonal, and $Q_2^{i,i} = 1$ and $R^{i,i} = 1$ with normalization.

Second, we can update the cost function by adapting the desired trajectory x_d. According to the human model (4.9), the interaction force f becomes 0 when the robot arm reaches the equilibrium position of the human arm x_H. Therefore, if $x_d \neq x_H$ (which is a typical case), f will be generated. Then, $Q_1 \neq 0$ and $Q_3 \neq 0$ indicate a trade-off between trajectory tracking and force minimization, even if they can be made very small. Moreover, the above update law of weights is a hard switching which is undesired as discussed in Introduction. Therefore, we propose to replace the instant cost by $c(t) = (x - x_r)^T Q_1 (x - x_r) + \dot{x}^T Q_2 \dot{x} + f^T Q_3 f + u^T R u$ and update x_r as follows:

$$\dot{x}_r = \begin{cases} \dot{x}_d + \alpha(x_d - x_r), & \text{if } f = 0; \\ \beta f, & \text{if } f \neq 0. \end{cases} \tag{4.22}$$

where x_r is the new reference trajectory of the robot arm, and $\alpha > 0$ and $\beta > 0$. Interpreting the above update law (4.22), if $f \neq 0$, x_r is adapted until f becomes 0 which indicates that $x \to x_H$; if $f = 0$, x_r is updated until $x_r = x_d$ and the control objective downgrades to trajectory tracking.

From the point of view of impedance control, the first update law (4.21) is to adapt the impedance parameters while the second one (4.22) is to adapt the reference trajectory [19]. By combing the above two update laws, it is possible to simultaneously adapt both impedance and reference trajectory, like what humans do [31].

4.3.3 Optimal Control

The last step of the control design is to find a control policy to minimize the cost function (4.19) subject to the system dynamics (4.17). Since the cost function changes on the fly, we need to find an online method which is able to provide an optimal control in real time. Dynamic programming is a natural option for this purpose, which mimics the way that biological systems interact with environments [15, 16, 30]. It usually includes an actor to generate actions according to the environment stimuli and a critic to evaluate the results of the actions.

In [21], a dynamic programming method is proposed for continuous systems with completely unknown dynamics. Besides its model-free feature, this method is employed in this work to solve the formulated optimal control problem due to its synchronized form, i.e., the updates of critic and actor networks take effect simultaneously. It will be briefly introduced in the following three steps.

First of all, the following system observer is developed to observe the unknown state $\dot{\bar{z}}$:

$$\dot{\hat{\bar{z}}} = \bar{A}_0\hat{\bar{z}} + \hat{G}(\hat{\bar{z}}, u) + L(\bar{z} - \hat{\bar{z}}) \tag{4.23}$$

where $\bar{A}_0$ is a designed Hurwitz matrix, $\hat{G}(\hat{\bar{z}}, u)$ the estimate of $G = \bar{A}(\bar{z}) + \bar{B}(\bar{z})u - \bar{A}_0\hat{\bar{z}}$ and $\hat{\bar{z}}$ the estimate of the system state $\bar{z}$. L is a positive definite matrix and is chosen to make $\bar{A}_0 - L$ a Hurwitz matrix. $\hat{G}(\hat{\bar{z}}, u)$ is approximated by using the following neural networks (NN):

$$\begin{aligned}
&\hat{G}(r) = \hat{W}S(\hat{V}r),\\
&S(\hat{V}r) = [s_1(\hat{V}r),\ s_2(\hat{V}r), \ldots, s_p(\hat{V}r)]^T,\\
&s_i(\hat{V}r) = \exp[\frac{-(\hat{V}r - \mu_i)^T(\hat{V}r - \mu_i)}{\eta_i^2}],\ i = 1, 2, \ldots, p
\end{aligned} \tag{4.24}$$

where $r = [\hat{\bar{z}}^T\ u^T]^T$ is the NN input, p the number of nodes in the hidden layer, $\mu_i = [\mu_{i,1}, \mu_{i,2}, \ldots, \mu_{i,k}]^T$ the center of the receptive field, η_i the width of the Gaussian function, and $\hat{W} \in \mathbb{R}^{(3n+l)\times p}$ and $\hat{V} \in \mathbb{R}^{k\times(4n+l)}$ estimates of optimal weights W and V, respectively. To obtain $\hat{W}$ and $\hat{V}$, the following update laws are developed:

$$\dot{\hat{W}} = -k_1(\bar{A}_0 - L)^{-T}(\bar{z} - \hat{\bar{z}})S^T(\hat{V}r) - k_2\|\bar{z} - \hat{\bar{z}}\|\hat{W}$$
$$\dot{\hat{V}} = -k_3(I_p - \text{diag}(s_i^2(\hat{V}r)))^T\hat{W}^T(\bar{A}_0 - L)^{-T}(\bar{z} - \hat{\bar{z}})\,\text{sgn}^T(r) - k_4\|\bar{z} - \hat{\bar{z}}\|\hat{V} \quad (4.25)$$

where k_i are positive scalars for $i = 1, 2, 3, 4$ and $\text{sgn}(r)$ is a vector in which the element is the sign of the corresponding element in r.

Second, define the value function corresponding to the cost function Γ as Υ, and it is approximated by a critic network

$$\hat{\Upsilon}(\hat{\bar{z}}) = \hat{W}_1^T S_1(V_1^T\hat{\bar{z}}) \quad (4.26)$$

where $\hat{W}_1$ is the estimate of the optimal weight W_1, V_1 a designed matrix and $S_1(V_1^T\hat{\bar{z}})$ the activation function that is defined similarly to $S(\hat{V}r)$ in Eq. (4.24). To obtain $\hat{W}_1$, the following update law is developed:

$$\dot{\hat{W}}_1 = -k_5\frac{\xi}{(\xi^T\xi + 1)^2}(\xi^T\hat{W}_1 + \hat{c}) \quad (4.27)$$

where k_5 is a positive scalar and the estimated instant cost is

$$\hat{c} = (\hat{x} - \hat{x}_d)^T Q_1(\hat{x} - \hat{x}_d) + \hat{\dot{x}}^T Q_2\hat{\dot{x}} + \hat{f}^T Q_3\hat{f} + u^T Ru \quad (4.28)$$

$\hat{x}$, $\hat{x}_d$, $\hat{\dot{x}}$ and $\hat{f}$ are estimates of x, x_d, $\dot{x}$ and f, respectively, which are obtained from the observer in the first step. ξ is defined as below

$$\xi = \left[\frac{\partial S_1^T(V_1^T\hat{\bar{z}})}{\partial\hat{\bar{z}}}\right]^T\dot{\hat{\bar{z}}}$$
$$= \left[-\frac{2}{\eta_{1,1}^2}(V_1^T\hat{\bar{z}} - \mu_{1,1})s_{1,1}\ \ldots\ -\frac{2}{\eta_{1,p_1}^2}(V_1^T\hat{\bar{z}} - \mu_{1,p_1})s_{p_1}\right]^T V_1^T\dot{\hat{\bar{z}}} \quad (4.29)$$

where $\eta_{1,j}$, $\mu_{1,j}$, $j = 1, 2, \ldots, p_1$, and p_1 are NN parameters that are similar as in Eq. (4.24).

The third step is to estimate the optimal control which minimizes the cost function (4.19) subject to the system dynamics (4.17). This estimated control is proposed as

$$u = -\frac{1}{2}R^{-1}\left(\frac{\partial\hat{G}(\hat{\bar{z}}, u)}{\partial u}\right)^T\hat{\Upsilon}_{\hat{\bar{z}}} \quad (4.30)$$

where $\hat{\Upsilon}_{\hat{\bar{z}}}$ is obtained as

$$\hat{\Upsilon}_{\hat{\bar{z}}} = \frac{\partial\hat{\Upsilon}(\hat{\bar{z}})}{\partial\hat{\bar{z}}} = \frac{\partial S_1^T(V_1^T\hat{\bar{z}})}{\partial\hat{\bar{z}}}\hat{W}_1 \quad (4.31)$$

and $\frac{\partial \hat{G}(\hat{\bar{z}},u)}{\partial u}$ is obtained as

$$\begin{aligned}
\frac{\partial \hat{G}(\hat{\bar{z}}, u)}{\partial u} &= \hat{W}\frac{\partial S(\hat{V}r)}{\partial u} \\
&= \hat{W}\frac{\partial S(\hat{V}r)}{\partial(\hat{V}r)}\frac{\partial(\hat{V}r)}{\partial u} \\
&= \hat{W}\frac{\partial S(\hat{V}r)}{\partial(\hat{V}r)}\hat{V}\frac{\partial r}{\partial u} \\
&= \hat{W}\left[-\frac{2}{\eta_1^2}(\hat{V}r - \mu_1)s_1 \ \ldots \ -\frac{2}{\eta_p^2}(\hat{V}r - \mu_p)s_p\right]\hat{V}\begin{bmatrix}\mathbf{0}_{(3n+l)\times n} \\ I_n\end{bmatrix}
\end{aligned} \tag{4.32}$$

According to [21], the control (4.30) with the observer (4.23) and update laws (4.25) and (4.27) minimizes the cost function (4.19) subject to the unknown system dynamics (4.17). Interested readers may refer to [21] for detailed derivations, stability analysis and proof. As discussed in Sect. 4.3.2, different control objectives can be defined by updating the weights Q_1, Q_2, Q_3 and R, and the reference x_d in the cost function. Subsequently, the control objective discussed in Introduction and Sect. 4.3.1 is achieved.

4.4 Simulations

4.4.1 Simulation Settings

In this section, we conduct the simulation study based on the scenario described in Fig. 4.1. The robot arm has two revolute joints and operates in a planar space. Its physical parameters are: $m_1 = m_2 = 20.00\,\text{kg}$, $l_1 = l_2 = 0.30\,\text{m}$ and $I_{z1} = I_{z2} = 0.45\,\text{kgm}^2$, where m_i, l_i and I_{zi}, $i = 1, 2$, represent the mass of the i-th link, length, and inertia about the z-axis that comes out of the page passing through the center of mass, respectively.

The impedance parameters of the human arm are assumed to be diagonal, and their values are determined from [3]: (21 ± 20) Ns/m and (201 ± 200) N/m, respectively. In particular, we set the following parameters according to Property 4.1:

$$\begin{aligned}
C_H &= \text{diag}\{[21 - 20\sin(\dot{x}_X), 21 - 20\sin(\dot{x}_Y)]\} \\
K_H &= \text{diag}\{[201 - 200\sin(\dot{x}_X), 201 - 200\sin(\dot{x}_Y)]\}
\end{aligned} \tag{4.33}$$

where $\dot{x}_X$ and $\dot{x}_Y$ stand for two elements of the velocity $\dot{x}$ in X direction and Y direction, respectively. We specify the impedance parameters of the human arm as above to simulate that the human arm becomes more compliant when the velocity

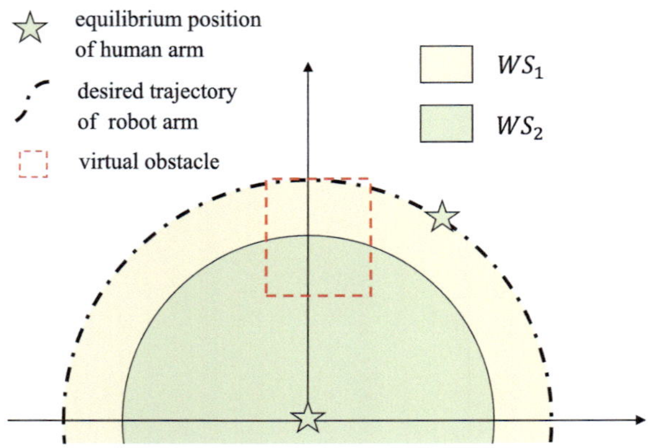

Fig. 4.2 Simulation settings. The workspaces $\mathscr{W}\mathscr{S}_1$ and $\mathscr{W}\mathscr{S}_2$ are introduced in Sect. 4.3.2

increases. The human intervention period is from $\frac{100}{3}$s to $\frac{200}{3}$s, i.e., $C_H = K_H = \mathbf{0}_2$ for $t < \frac{100}{3}$s and $t > \frac{200}{3}$s.

The desired trajectory of the robot arm is half a circle with a radius of 0.2 m, from $[-0.2\ 0]^T$ for $t = 0$ s to $[0.2\ 0]^T$ for $t = 100$ s. In particular, the desired trajectory $x_d = [-0.2\cos(\frac{\pi}{100}t)\ 0.2\sin(\frac{\pi}{100}t)]^T$ is generated by Eq. (4.16). The virtual obstacle is a square with the following four edges: $[-0.05\ 0.2]^T$, $[0.05\ 0.2]^T$, $[0.05\ 0.1]^T$, and $[-0.05\ 0.1]^T$. In order to avoid this obstacle, the human arm has the following two equilibrium positions: $[0\ 0]^T$ for $t = \frac{100}{3}$ s and $[0.1\ 0.1\sqrt{3}]^T$ for $t = \frac{200}{3}$ s. The initial coordinate of the robot arm in the joint space is $[-\frac{\pi}{2}\ \pi]^T$. The simulation settings are illustrated in Fig. 4.2.

The Hurwitz matrix $\bar{A}_0$ in Eq. (4.23) is given as $\bar{A}_0 = \text{diag}([-I_6, E])$, $L = 10I_8$, and the initial values in Eq. (4.23) are: $u(0) = [0\ 0]^T$, $\hat{\bar{z}}(0) = [-0.2\ 0\ 0\ 0\ 0\ 0\ -\frac{\pi}{100}\ 0]^T$, $\hat{W}(0) = \mathbf{1}_{8\times 10}$, $\hat{V}(0) = I_{10}$, $\hat{W}_1(0) = \mathbf{0}_{10\times 1}$ and $V_1 = I_8$, where $\mathbf{1}_{8\times 10}$ stands for a matrix with all elements as 1. The NN parameters are: $p = 10$, $p_1 = 10$, $\eta_i = 1$, $\mu_i = -1 + 0.2i$ for $i = 1, 2, \ldots, p$ and $\eta_{1,j} = 1$, $\mu_{1,j} = -1 + 0.2j$ for $j = 1, 2, \ldots, p_1$. The parameters in the update laws (4.25) and (4.27) are: $k_i = 1$ for $i = 1, \ldots, 5$. A sweeping frequency signal is added into u to satisfy the persistent excitation condition.

4.4.2 Change of Weights

To achieve switchings between trajectory tracking and force minimization, the method to update the weights in Sect. 4.3.2 is evaluated. In particular, two subspaces are defined: $\mathscr{W}\mathscr{S}_1 = \{x \mid \|x - x_d\| < 0.01\}$ in which the robot arm leads, and $\mathscr{W}\mathscr{S}_2 = \mathscr{W}\mathscr{S} - \mathscr{W}\mathscr{S}_1$ in which the human arm leads. The weights are updated as follows: $R = 0.01I_2$ when $x \in \mathscr{W}\mathscr{S}_1$ and $R = 10I_2$ when $x \in \mathscr{W}\mathscr{S}_2$, and $Q_1 = 500I_2$, $Q_2 = 0.1I_2$ and $Q_3 = 0.001I_2$ in both subspaces. Another two cases are considered for comparison: $R = 0.001 I_2$ fixed for both subspaces and $R = 10 I_2$ fixed for both cases.

The simulation results in three cases are shown in Figs. 4.3, 4.4 and 4.5, respectively. In Fig. 4.3, it is shown that trajectory tracking is always the control objective of the robot arm, so it does not comply to the human's control objective. This can be seen from two aspects: first, the actual trajectory of the robot arm tries to follow the desired trajectory of the robot arm, even when the human wants to move it to his/her equilibrium position; second, a large interaction force is resulted which indicates that more effort is required from the human arm to move the robot arm. Conversely, in Fig. 4.4, it can be found that force minimization is always the control objective of the robot arm, and the results include a better tracking of the human's equilibrium position, a small interaction force but a worse tracking of the robot's desired trajectory when the human intervention disappears. Therefore, it is essential to consider switchings between different cases, which is achieved by the proposed method with updated weight: Fig. 4.5 illustrates that a better tracking of the human's equilibrium position and a small interaction force are resulted during the human intervention, and tracking of the robot's desired trajectory is also guaranteed when the human intervention disappears. Note that the employed optimal control requires a period for convergence, which is roughly from the beginning of the task to $t = 20$ s. Besides, there exist obvious overshoots when the human intervention appears and disappears. In practice, human can be required to gradually change the force that is applied to the robot arm, and thus the issue of overshoot may be addressed.

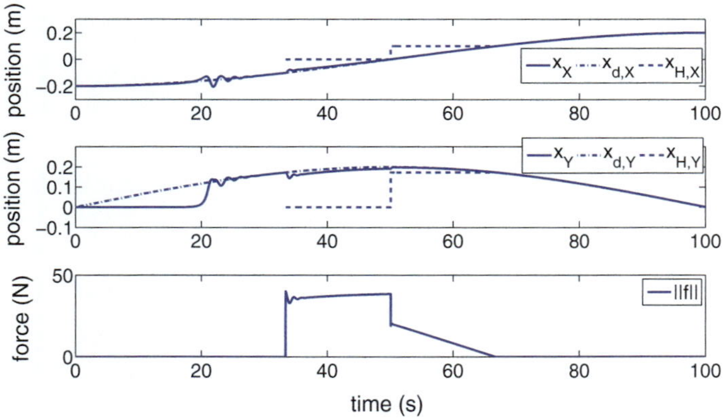

Fig. 4.3 Positions in X and Y directions and norm of interaction forces, under the condition of fixed weight $R = 0.001 I_2$

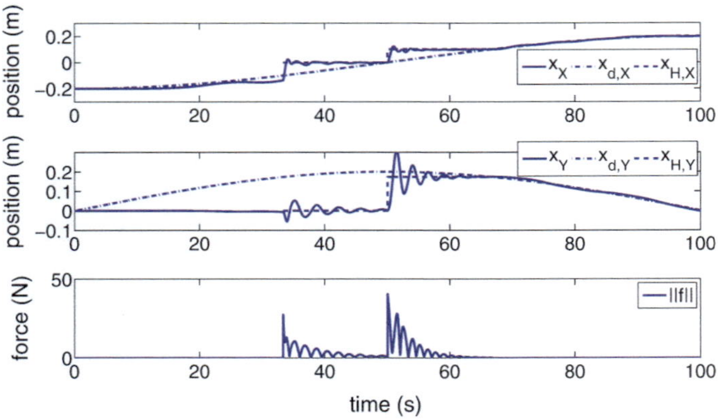

Fig. 4.4 Positions in X and Y directions and norm of interaction forces, under the condition of fixed weight $R = 10I_2$

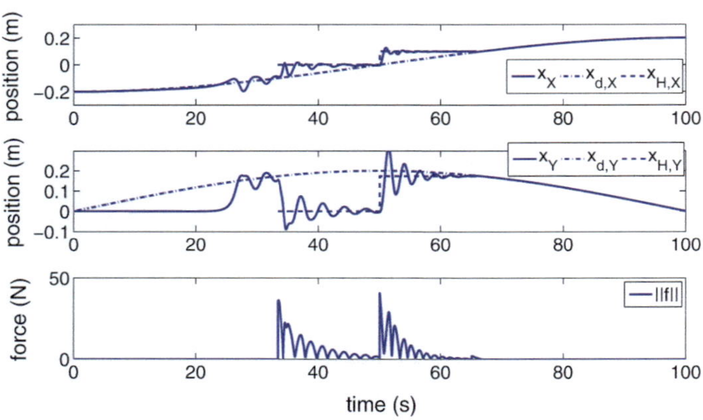

Fig. 4.5 Positions in X and Y directions and norm of interaction forces, under the condition of updated weight

4.4.3 Adaptation of Desired Trajectory

In the last part of the simulation, we verify the validity of reference adaptation in Sect. 4.3.2 for switchings of trajectory tracking and force minimization. In Eq. (4.22), $\alpha = \beta = 0.1$ while the values of the weights are the same as those under the condition of $R = 0.001$ in the previous subsection. Results are shown in Fig. 4.6. It is clearly seen that the expected performance for switchings is achieved: force minimization when the human arm intervenes and trajectory tracking when the intervention disappears. The adaptation process of the employed optimal control and the overshoots when the human intervention appears and disappears exist, similarly as in the previous subsection.

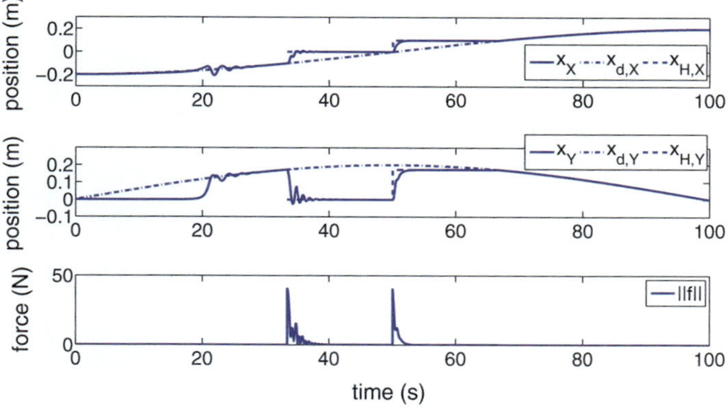

Fig. 4.6 Positions in X and Y directions and norm of interaction forces, under the condition of reference adaptation

4.5 Conclusions

A framework has been developed for switchings between trajectory tracking and force minimization in a typical human–robot collaboration scenario. Human dynamics have been considered in system modeling and a cost function has been proposed to include both control objectives of the robot and the human. Under the developed framework, two examples have been introduced to achieve the switchings. Dynamic programming has been employed to solve the formulated optimal control problem. At last, the validity of the proposed method has been verified by simulation studies, which have illustrated achievement of the expected switching performance.

In above simulation studies, weight update and reference adaptation have been implemented independently. It is interesting to investigate the control performance when they are applied simultaneously. This is relevant to simultaneous adaptation of impedance and reference trajectory in human motor control [10, 11], as briefly discussed in Sect. 4.3.2. Examples of weight update and reference adaptation are given but they are not applicable to a general case. Therefore, they will be further studied for other applications and tasks.

Appendix: Proof of Lemma 4.1

Proof Denote $\rho = \dot{C}_H \xi$, the vector composed by elements at the i-th row of C_H as C_i, and the element of C_H at the i-th row and j-th column as c_{ij}. According to Property 4.1, we have $\dot{c}_{ij} = \frac{\partial c_{ij}}{\partial x}\dot{x} + \frac{\partial c_{ij}}{\partial \dot{x}}\ddot{x}$.

Consider the i-th element of ρ as below

$$\begin{aligned}\rho_i &= \Sigma_{j=1}^{n}\dot{c}_{ij}\xi_j \\ &= \Sigma_{j=1}^{n}\left[\left(\frac{\partial c_{ij}}{\partial x}\dot{x}+\frac{\partial c_{ij}}{\partial \dot{x}}\ddot{x}\right)\xi_j\right] \\ &= \Sigma_{j=1}^{n}\Sigma_{k=1}^{n}\left[\left(\frac{\partial c_{ij}}{\partial x_k}\dot{x}_k+\frac{\partial c_{ij}}{\partial \dot{x}_k}\ddot{x}_k\right)\xi_j\right] \\ &= \Sigma_{k=1}^{n}\Sigma_{j=1}^{n}\left[\left(\frac{\partial c_{ij}}{\partial x_k}\dot{x}_k+\frac{\partial c_{ij}}{\partial \dot{x}_k}\ddot{x}_k\right)\xi_j\right] \\ &= \Sigma_{k=1}^{n}\left(\frac{\partial C_i}{\partial x_k}\xi\dot{x}_k+\frac{\partial C_i}{\partial \dot{x}_k}\xi\ddot{x}_k\right) = \frac{\partial C_i}{\partial x^T}\xi\dot{x}+\frac{\partial C_i}{\partial \dot{x}^T}\xi\ddot{x} \end{aligned} \tag{4.34}$$

From the above equation, we find that the i-th rows of $C_{H1}(x,\xi)$ and $C_{H2}(\dot{x},\xi)$ are $\frac{\partial C_i}{\partial x^T}\xi$ and $\frac{\partial C_i}{\partial \dot{x}^T}\xi$, respectively. Therefore, $\dot{C}_H\xi = C_{H1}(x,\xi)\dot{x} + C_{H2}(\dot{x},\xi)\ddot{x}$. Similarly, it can be shown that $\dot{K}_H\xi = K_{H1}(x,\xi)\dot{x} + K_{H2}(\dot{x},\xi)\ddot{x}$, of which the details are omitted.

References

1. Argall BD, Billard AG (2010) A survey of tactile human-robot interactions. Robot Auton Syst 58(10):1159–1176
2. Arimoto S (1990) Learning control theory for robotic motion. Int J Adapt Control Signal Process 4(6):543–564
3. Buerger SP, Hogan N (2007) Complementary stability and loop shaping for improved human-robot interaction. IEEE Trans Robot 23(2):232–244
4. Chipalkatty R, Droge G, Egerstedt MB (2013) Less is more: Mixed-initiative model-predictive control with human inputs. IEEE Trans Robot 29(3):695–703
5. Corteville B, Aertbelien E, Bruyninckx H, De Schutter J, Van Brussel H (2007) Human-inspired robot assistant for fast point-to-point movements. In: IEEE international conference on robotics automation. Roma, Italy, pp 3639–3644
6. Duchaine V, Gosselin C (2009) Safe, stable and intuitive control for physical human-robot interaction. In: IEEE International conference on robotics and automation, pp 3676–3681
7. Erden MS, Tomiyama T (2010) Human-intent detection and physically interactive control of a robot without force sensors. IEEE Trans Robot 26(2):370–382
8. Evrard P, Kheddar A (2009) Homotopy switching model for dyad haptic interaction in physical collaborative tasks. In: 3rd joint eurohaptics conference and symposium on haptic interfaces for virtual environment and teleoperator systems. Salt Lake City, Utah, USA, pp 45–50
9. Feldman AG (1966) Functional tuning of the nervous system with control of movement or maintenace of a steady posture, ii: Controllable parameters of the muscles. Biophysics 11:565–578
10. Ganesh G, Albu-Schaeffer A, Haruno M, Kawato M, Burdet E (2010) Biomimetic motor behavior for simultaneous adaptation of force, impedance and trajectory in interaction tasks. In: Proceedings of the IEEE international conference on robotics and automation (ICRA). Anchorage, Alaska, USA, pp 2705–2711
11. Ganesh G, Jarrasse N, Haddadin S, Albu-Schaeffer A, Burdet E (2012) A versatile biomimetic controller for contact tooling and haptic exploration. In: Proceedings of the IEEE international conference on robotics and automation (ICRA), pp 3329–3334

12. Ge SS, Li Y, Wang C (2014) Impedance adaptation for optimal robot-environment interaction. Int J Control 87(2):249–263
13. Hogan N (1985) Impedance control: an approach to manipulation-Part I: Theory; Part II: Implementation; Part III: Applications. Trans ASME J Dyn Syst Meas Control 107(1):1–24
14. Jarrasse N, Sanguineti V, Burdet E (2014) Slaves no longer: review on role assignment for human-robot joint motor action. Adapt Behav 22(1):70–82
15. Lewis FL, Vrabie D (2009) Reinforcement learning and adaptive dynamic programming for feedback control. IEEE Circuits Syst Mag 9(3):32–50
16. Lewis FL, Vrabie D, Vamvoudakis KG (2012) Reinforcement learning and feedback control: using natural decision methods to design optimal adaptive controllers. IEEE Circuits Syst Mag 32(6):76–105
17. Li Y, Ge SS (2014) Human-robot collaboration based on motion intention estimation. IEEE/ASME Trans Mechatron 19(3):1007–1014
18. Li Y, Ge SS (2016) Force tracking control for motion synchronization in human-robot collaboration. Robotica 34(06):1260–1281
19. Li Y, Tee KP, Chan WL, Yan R, Chua Y, Limbu DK (2015a) Continuous role adaptation for human-robot shared control. IEEE Trans Robot 31(3):672–681
20. Li Y, Tee KP, Yan R, Limbu DK, Ge SS (2015) Shared control of human and robot by approximate dynamic programming. In: Proceedings of the 2015 American control conference (ACC). IL, USA, Chicago, pp 1167–1172
21. Liu D, Huang Y, Wang D, Wei Q (2013) Neural-network-observer-based optimal control for unknown nonlinear systems using adaptive dynamic programming. Int J Control 86(9):1554–1566
22. Medina JR, Lawitzky M, Mortl A, Lee D, Hirche S (2011) An experience-driven robotic assistant acquiring human knowledge to improve haptic cooperation. In: IEEE/RSJ international conference on intelligent robots and systems, pp 2416–2422
23. Mortl A, Lawitzky M, Kucukyilmaz A, Sezgin M, Basdogan C, Hirche S (2012) The role of roles: physical cooperation between humans and robots. Int J Robot Res 31(13):1656–1674
24. Mussa-Ivaldi FA, Hogan N, Bizzi E (1985) Neural, mechanical, and geometric factors subserving arm posture in humans. J Neurosci 5(10):2732–2743
25. Passenberg C, Stefanov N, Peer A, Buss M (2011) Enhancing task classification in human-machine collaborative teleoperation systems by real-time evaluation of an agreement criterion. In: IEEE world haptics conference, pp 493–498
26. Slotine JJE, Li W (1987) On the adaptive control of robotic manipulators. Int J Robot Res 6(3):1
27. Tsuji T, Morasso PG, Goto K, Ito K (1995) Human hand impedance characteristics during maintained posture. Biol Cybern 72(6):475–485
28. Tsumugiwa T, Yokogawa R, Hara K (2002) Variable impedance control based on estimation of human arm stiffness for human-robot cooperative calligraphic task. In: IEEE international conference on robotics and automation. DC, USA, Washington, pp 644–650
29. Wakita K, Huang J, Di P, Sekiyama K, Fukuda T (2013) Human-walking-intention-based motion control of an omnidirectional-type cane robot. IEEE/ASME Trans Mechatron 18(1):285–296
30. Werbos PJ (2009) Intelligence in the brain: a theory of how it works and how to build it. Neural Netw 22(3):200–212
31. Yang C, Ganesh G, Haddadin S, Parusel S, Albu-Schaeffer A, Burdet E (2011) Human-like adaptation of force and impedance in stable and unstable interactions. IEEE Trans Robot 27(5):918–930

Chapter 5
Estimating Human Intention During a Human–Robot Cooperative Task Based on the Internal Force Model

Ehsan Noohi and Miloš Žefran

5.1 Introduction

Despite the significant challenge that physical interaction with humans introduces to robots, successful strategies have been proposed in the literature that demonstrate effective cooperation between humans and robots [3–6, 17, 18, 25]. The key factor in the success of these strategies is their ability to estimate human intent. While the haptic channel is believed to convey the information about the human intent, there has always been an ambiguity in interpretation of haptic clues [6, 29]. The internal force is one of the sources of this ambiguity.

Different approaches have been taken for both estimating the human intent and resolving the ambiguity. For example, Medina et al. [17] proposed a prediction framework, based on Hidden Markov Models (HMMs), to estimate human's intended trajectory. The model employs a position-based impedance controller to map the haptic observations to the estimated trajectory. They proposed a cooperation strategy for an assistive robot, engaged in a (heavy) object manipulation task. To deal with the ambiguity, they embed their HMM in a stochastic optimal control framework [18]. The goal of the optimal controller is to minimize the human's contribution to the task wrt. the intention-estimation error. In other words, the human is only required to steer the motion while the physical work is carried out by the robot. They extended their work to the case where the task model is unknown to the robot and proposed a model-free assistance scheme, relying only on the measured forces [19]. They combined this

This chapter was developed and enhanced from the following two earlier papers published as [24, 27] © 2016 IEEE.

E. Noohi · M. Žefran (✉)
Robotics Lab, Electrical and Computer Engineering Department,
University of Illinois at Chicago, Chicago, IL, USA
e-mail: mzefran@uic.edu

E. Noohi
e-mail: enoohi2@uic.edu

Y. Wang and F. Zhang (eds.), *Trends in Control and Decision-Making for Human–Robot Collaboration Systems*, DOI 10.1007/978-3-319-40533-9_5

model-free controller with their model-based controller, using a strategy-selection decision maker. The goal of the controller is again to reduce human efforts. This approach is not applicable to the tasks that require active contribution of both human and the robot, such as lifting a large object.

As another example, Dumora et al. [6] proposed a classifier to map the haptic measurements to the human intents. The classifier was embedded in a compliance control scheme; referred to as virtual mechanism by the authors. They proposed a cooperation strategy for an assistive robot, engaged in a (large) object manipulation task. To deal with the ambiguity, the task was decomposed into a sequence of rotation and translation (primitive) motions and a virtual mechanism was defined for each primitive motion. Sequencing to the primitive motions resulted in a nonsmooth and unnatural motion trajectory.

In addition to the position and force, researchers have also considered object's velocity and exchanged energy to design algorithms for human–robot cooperation. For example, Bussy et al. [3] proposed a velocity- based controller to estimate the human intent during a dyadic object transportation task. By studying the cooperation behavior of human dyads, they observed a velocity-based control law that can be embedded in an impedance controller. The controller has been tested on a HRP-2 humanoid robot and demonstrated the desired performance in performing elementary primitive motions. Unfortunately, the velocity-based methods are unable to deal with the ambiguity of internal forces (force cancellation).

As another example, Donner et al. [5] proposed an energy monitoring scheme to estimate the human intent. The variation of energy was calculated based on the haptic information and then was associated with the human intent. That is, the robot injected (released) energy into (from) the system whenever the human did so. The proposed controller was tested in a collaborative goal-directed object swinging task and was shown that when the robot is leading (following) the task, its performance is comparable with a human leader (follower). While the flow of energy reveals important information about the interaction, the energy-based approaches cannot detect human's force (torque) if it does not change the energy. As a result, the method is unable of resolving the ambiguity when a big portion of human efforts are canceled out as internal force.

In a different approach, researchers have studied the process of role-exchange during a cooperative collaboration, in order to estimate the human intent. For instance, Mörtl et al. [20] compared a dynamic role allocation strategy with constant role allocation strategies (such as fixed leader or follower roles). In the dynamic strategy, the robot estimates the human applied force and torques and an impedance controller responds to the estimated human intent, accordingly. They observed that while the human subjects preferred strategies with constant roles for the robot, the objective performance measures were higher in dynamic role allocation schemes. In other words, the robot sacrifices the human comfort (natural responses) to the higher performance (synthetic optimization goals). In a similar work, Groten et al. [13] introduced a dominance measure for haptic interaction tasks. Unlike many other researchers, they suggest that a natural human–robot interaction should not fall into the leader/follower modes. They observed that the average individual dominance values for humans are between 0.4 and 0.6 (in 95% of the trials). They also reported the observation that humans prefer a difference in dominance values (instead of both values being equal). In other words, robot and the human should always balance their efforts (dominance values around 0.5), instead of taking the leader or the follower roles (values of zero and one) or performing exactly half of the work (value of 0.5). As another example, Evrard and Kheddar [7] proposed a homotopy (interpolation) switching controller and showed that it describes a large number of physical interaction behaviors. The homotopy function assigns a weighted role to the robot (between two extreme behaviors: the leader behavior and the follower behavior). They tested their controller on an avatar, performing a cooperative obstacle avoidance task with a human partner.

While the role of the internal force in the ambiguity of haptic information is neglected in many works, it has been recognized in several others [3, 7, 13, 20, 23, 35]. For example, while studying the dominance distribution in dyadic haptic collaboration, Groten et al. [13] took the ad-hoc assumption that the internal force satisfies the minimum-energy constraint. Or, while proposing a homotopy model for the role taking in a dyadic physical interaction, Evrard and Kheddar [7] speculate that the homotopy variable may be correlated with the internal force. Noohi and Žefran [23] suggested that human perception of the cooperativeness of a dyadic collaboration is linked to the internal force. Mörtl et al. [20] related the haptic measurements to the exchange of the leader/follower roles between the human and the robot. They observed that dynamic role assignments resulted in a considerably larger internal force.

None of these works have investigated the features of the internal force and its relation to the human intent. In this work, we study a human–human physical interaction and explore the characteristics of the desired internal force. Taking advantage of the observed features in the internal force, we propose an internal-force-aware cooperation policy that can elevate the performance of the existing human–robot cooperation strategies.

The rest of the paper is organized as follows. By briefly reviewing some related theories in the field of human movement science, we propose a computational model for the internal force during a dyadic object manipulation task in Sect. 5.2. To validate the model and investigate its properties, a human study with 22 healthy adult subjects was conducted. The study is described in Sect. 5.3. The results of the human study are presented in Sect. 5.4. To validate our model, we compare the proposed model with several alternative models in Sect. 5.5 and show that the proposed model better explains human motions. Next, we further investigate the properties of the internal force signals in Sect. 5.6 and explore statistical characteristics of these forces. We show that a strong correlation exists between the object's velocity and a component in the internal force. As a result, we propose a cooperation policy for the robot in Sect. 5.7. We show that this internal-force-aware policy can be embedded in existing robot controllers and enhance their performance. We conclude our work in Sect. 5.8 and discuss extensions of and future directions for our work.

5.2 Internal Force Model

In this section we first introduce the standard formulation of the internal force for a dyadic collaborative task. Then, some of the well-recognized internal force models in the literature are introduced. Next, we introduce our proposed internal force model and discuss its variations and its relation to the existing models.

5.2.1 Problem Formulation

Let us consider a dyadic object manipulation task. Let f_1 and f_2 be the two forces applied to the manipulated object. It is common to decompose the applied forces into the effective forces (f_1^e and f_2^e) and the internal force (F^i):

$$\begin{aligned} f_1 &= f_1^e + F^i \\ f_2 &= f_2^e - F^i \end{aligned} \tag{5.1}$$

Note that any component of the applied forces that cancels out is part of the internal force and the remaining components are effective forces. Also note that the manipulation task is performed by the total force $F_{sum} = f_1 + f_2 = f_1^e + f_2^e$. As a result,

$$\begin{aligned} f_1^e &= \left(\tfrac{1}{2} + \delta\right) F_{sum} \\ f_2^e &= \left(\tfrac{1}{2} - \delta\right) F_{sum} \end{aligned} \tag{5.2}$$

where δ is called the *cooperation index*, see [23]. The effective forces are hypothesized to be linked with different features of the collaboration. As mentioned earlier, taking the leader or the follower role during a dyadic collaboration is related to a homotopy function between these forces [7]. Also, a human study shows that the human perception of the performance of a collaboration is linked to the effective forces [23]. In another study [13], effective forces are explored to characterize the dominance in a haptic collaborative task.

Note that (5.1) is an under determined system of equations. Assuming the applied forces are available (through measurement), to obtain the effective forces one should calculate the internal force (or equivalently the cooperation index). Since the internal force cancels out in the equations of motion, it does not influence the motion trajectory. However, it introduces potentially undesirable push–pull forces that affect the cooperativeness of the collaboration. To further study the internal force model, let us introduce another decomposition for the applied forces, this time into two components parallel with and perpendicular to the F_{sum}:

$$\begin{aligned} f_1 &= f_1^{\parallel} + f_1^{\perp} = \gamma_1 F_{sum} + \eta_1 f^{\perp} \\ f_2 &= f_2^{\parallel} + f_2^{\perp} = \gamma_2 F_{sum} + \eta_2 f^{\perp} \end{aligned} \tag{5.3}$$

where $f^{\perp} \cdot F_{sum} = 0$. Recalling that $f_1 + f_2 = F_{sum}$, it is easy to show that $\gamma_1 + \gamma_2 = 1$ and $\eta_1 + \eta_2 = 0$. As a result, the applied forces can simply be decomposed as:

$$\begin{aligned} f_1 &= \alpha F_{sum} + F^{\perp} \\ f_2 &= (1-\alpha) F_{sum} - F^{\perp} \end{aligned} \tag{5.4}$$

where $F^{\perp} = \eta_1 f^{\perp} = -\eta_2 f^{\perp}$ and $\alpha = \gamma_1$. Comparing (5.4) with (5.1) and (5.2) results in:

$$F^i = F^{\perp} + \beta F_{sum} \tag{5.5}$$

where $\beta = \alpha - \left(\frac{1}{2} + \delta\right)$. Equation (5.4) reveals an important property of the internal force: it not only contains all force components that are perpendicular to F_{sum}, but may also contain components parallel to it. While the perpendicular component ($F^{\perp}$) is easy to calculate, the components of the internal force that are parallel with F_{sum} are not easy to obtain. In fact, these components are the source of the aforementioned ambiguity in the haptic clues. Indeed, the main challenge in modeling the internal force lies in the calculation of β.

We claim that a natural (human-like) human–robot cooperation requires the internal force to be similar to that of a human–human cooperation, what we call the *desired* internal force. We will discuss different models for this internal force in the following sections. We first introduce some existing models in the literature and then present our proposed model. Later, we will discuss how these models are different compared to each other and how they are related to the above formulation.

5.2.2 Existing Models

To calculate the *desired* internal force, researchers have taken different approaches. In this section we briefly discuss two of those models that have received the most attention in the literature. The first model is inspired by the physical characteristics of objects when several forces are exerted on them. The second model is inspired by the energy efficiency of the human when performing tasks.

5.2.2.1 Physics Inspired

In the first model, the assumption is that the effective forces would not contribute to the mechanical strain of the object (also known as engineering strain). In case of a dyadic object manipulation, this assumption requires the effective forces to be equal, i.e., $f_1^e = f_2^e$. Applying this equation in (5.1) results in:

$$\begin{aligned} F^i &= \tfrac{1}{2}(f_1 - f_2) \\ f_1^e &= f_2^e = \tfrac{1}{2}(f_1 + f_2) \end{aligned} \tag{5.6}$$

Since this model is related to the model of the mechanical internal force [12], it has received attention from many researchers, see [28] as an example.

5.2.2.2 Human Inspired

In the second model the assumption is that humans do not apply unnecessary internal forces to the object. The assumption is based on the hypothesis that humans are energy efficient agents and thus, they apply forces only when the task demands it (such as securing grasp stability). Considering (5.4) and (5.5), this assumption results in $\beta = 0$ for $0 < \alpha < 1$, which means that the internal force consists only of the perpendicular force components, $F^{\perp}$. In other words, since effective forces do not cancel each other when $0 < \alpha < 1$, no component of the internal force appears in the direction of F_{sum} and thus, $\beta = 0$. On the other hand, when $\alpha \leq 0$ ($\alpha \geq 1$) the model suggests that f_1 ($-f_2$) is only contributing to the internal force, not the effective force. In such cases, the model takes $\beta = \alpha$ ($\beta = \alpha - 1$). In other words:

$$\begin{cases} F^i = f_1 & \alpha \leq 0 \\ F^i = -f_2 & \alpha \geq 1 \\ F^i = F^{\perp} & \text{otherwise} \end{cases} \tag{5.7}$$

An example of employing this model is [13], where the dominance distribution in a dyadic haptic collaboration is investigated.

5.2.3 *Proposed Model*

We claim that for the internal force to be the *desired* internal force, it should contain some information about the object's motion trajectory. In this section, first we review some important properties of human movements and then, we will describe how our model relates the object's motion profile to the *desired* internal force.

5.2.3.1 Human Movement Properties

Human movements have been studied extensively in the literature and many interesting theories have been established [10, 11, 15, 21, 34]. Here, we briefly review some of the related theories to set the stage for the discussions in the following sections.

Minimum-Jerk Movement

It has been shown that many human movements, in particular single arm reaching movement, follow the minimum-jerk trajectory [10]. The minimum-jerk trajectory can be obtained by optimizing the jerk function along the movement path. That is:

$$x^*(t) = \underset{x(t)}{\operatorname{argmin}} \left(\frac{1}{2} \int_{t_s}^{t_e} \left\| \frac{\mathrm{d}^3 x(t)}{\mathrm{d}t^3} \right\|^2 \mathrm{d}t \right) \tag{5.8}$$

where $x(t)$ is the position trajectory, t_s is the movement start time and t_e is the end time. Using the calculus of variations, it has been shown [33] that the solution to (5.8) is in the form of a 5th order polynomial. That is:

$$x^*(t) = \sum_{k=0}^{5} a_k \, t^k \tag{5.9}$$

To be able to determine the a_k coefficients, the boundary conditions should be taken into consideration. If we assume that the motion starts and ends at rest (zero boundary conditions), the minimum-jerk trajectory would be:

$$x^*(\tau) = x_s + L(6\tau^5 - 15\tau^4 + 10\tau^3) \tag{5.10}$$

where $\tau = (t - t_s)/(t_e - t_s)$ is the normalized time and $L = \|x_e - x_s\|$ is the traveled distance. Here, $x_s = x(t_s)$ and $x_e = x(t_e)$ are the start and the end positions of the movement, respectively.

Constrained Movement

Many studies show that humans can adapt to the external force field and his/her motion trajectory will return to the minimum-jerk trajectory after enough learning trials [9, 16, 32]. That is, when human movements are constrained, the motion trajectory still follows the bell-shape velocity profile and the motion trajectory is robust to the external disturbances. For instance in a dyadic object manipulation, each person introduces a disturbance force to the other person's hand. However, based on the above theory and after the familiarization phase, both persons learn to cooperatively move the object on a minimum-jerk trajectory. We have recently shown that the motion trajectory of the object during a cooperative object manipulation task is strongly correlated with the minimum-jerk trajectory [26].

It should be noted that the learning performance of a constraint movement is related to the ability of the person to build an internal model for the constraint. If the constraint is predictable, the learning phase would be fast and effective. That is why people are very good in cooperative manipulation with another person. However, if the constraint model changes or it is not easily predictable, then the cooperation would fail (or would be ineffective). That is one reason why cooperation between a human and a robot is challenging. In this work we show that the internal force is a key to understanding the human intent. In turn, a reliable model of the internal force is crucial to achieving successful cooperation.

5.2.3.2 Description of the Model

Let us note the fact that the internal force can only deform (compress or stretch) the object, but not altering the object's motion trajectory. As a result, to model the internal force, the motion-descriptor variables (position and velocity) are not suitable. Instead, a model for the deformation of the object would lead us to a model for the internal force. Since elastic deformation appears in many different forms (such as compression, stretch, sheer, and twist), we take the most general form of linear relation between the internal force and the deformation strain ε, expressed as:

$$F^i = \sum_{k=0}^{N} b_k \frac{\mathrm{d}^k \varepsilon}{\mathrm{d}t^k} \tag{5.11}$$

The deformation strain is defined as $\varepsilon(t) = (l(t) - L)/L$, where L is the original length of the object and $l(t)$ is length of the deformed object at time t. The order of the model is N (where $0 \leq N \leq \infty$). For instance, if the object is a linear spring the order is $N = 0$ and (5.11) would be simplified as $F^i = b_0\ \varepsilon$.

Now let us consider a dyadic object manipulation task. The length of the deformed object can be obtained as:

$$l(t) = x_2^*(t) - x_1^*(t) \tag{5.12}$$

where $x_1^*(t)$ and $x_2^*(t)$ are the movement trajectories of the subjects' hands. As we discussed earlier in Sect. 5.2.3.1, both humans follow a minimum-jerk motion profile (as a result of a learning process). That is, both $x_1^*(t)$ and $x_2^*(t)$ are minimum-jerk trajectories and can be represented by 5th order polynomials; see (5.9). Plugging (5.9) into (5.12) and the result into (5.11) would reveal that:

$$F^i(t) = \sum_{k=0}^{5} c_k t^k \tag{5.13}$$

In order to determine the coefficients in (5.13), the boundary conditions should be enforced. The boundary conditions correspond to the time instances $t^\dagger$ when $F_{sum}(t^\dagger) = 0$. At such time instances, $f_1^e(t^\dagger) = f_2^e(t^\dagger) = 0$ and $f_1(t^\dagger) = -f_2(t^\dagger)$. Consequently, $F^i(t^\dagger) = f_1(t^\dagger)$, see (5.1). As discussed in Sect. 5.2.3.1, we have recently shown [26] that F_{sum} follows a minimum-jerk profile and thus, it has exactly one zero-crossing point in the middle, namely at time t_m. Taking t_s and t_e as the start time and the end time of the motion, $F_{sum}(t^\dagger) = 0$ for $t^\dagger \leq t_s$, $t^\dagger = t_m$ and $t^\dagger \geq t_e$. This results in the following set of boundary conditions:

$$F^i(t_s) = f_1(t_s) \qquad F^i(t_m) = f_1(t_m) \qquad F^i(t_e) = f_1(t_e) \tag{5.14}$$

The smoothness of the applied forces imposes additional constraints on the internal force: the derivatives of the applied forces should be continuous. Note that if we ignore the smoothness constraint in our model, we obtain two variations of the proposed model: the minimum-variation model (MV) and the quadratic model (QM). The minimum-variation model (MV) hypothesizes that the internal force is either constant or changes linearly. Therefore, the internal force in MV is a piece-wise linear interpolation between the boundary conditions in (5.14). On the other hand, the quadratic model (QM) suggests that the internal force is a quadratic function that passes through the boundary conditions in (5.14). Both models satisfy (5.13) and (5.14), but assume different smoothness constraints.

5.2.4 *Discussion*

To better understand the differences between the proposed model and the existing models, we need to compare them with the original definition of the internal force, provided by (5.1) and (5.2). Let us first consider the physics inspired model. Comparing (5.6) with (5.1) and (5.2) reveals that this model basically suggests that $\delta = 0$, in all situations. Having a fixed cooperation index regardless of the manipulation type (bimanual versus dyadic) is problematic and will be discussed in more detail in Sect. 5.5. Now let us consider the human inspired model. Comparing (5.7) with (5.1) and (5.2) shows that this model can be simplified to $\delta = \alpha - 0.5$, when $0 < \alpha < 1$, and to $\delta = \pm 0.5$, otherwise. In contrast, our proposed model does not constrain δ. That is, regardless of the value of α, the value of δ is obtained indirectly by employing the model described in (5.13) and (5.14).

While our proposed model is based on the theories and well-recognized models for human movements, the alternative models take ad-hoc assumptions that are not supported by the results of human studies. We will compare the performance of our proposed model with the existing models in the following sections. To be able to compare these models with each other and evaluate our proposed model, we designed a human study. In the following section we describe the details of the method we used for this study and discuss the results in the later sections.

5.3 Method

To be able to study the role of the internal force in the cooperation policies, we consider both bimanual and dyadic manipulation. In bimanual manipulation, we ask the subjects to grasp an object with both hands and move it horizontally. For dyadic manipulation, we grouped the subjects into pairs and ask them to perform dyadic manipulation, cooperatively. In both cases, the grasps are power-grasp and the subjects can apply independent forces and torques to the object.

5.3.1 Apparatus

We choose a pot as the object to be manipulated. The subjects' applied forces are measured using two SI-65-5 ATI Gamma force sensors [2]. The force sensors are placed in between each pot handle and the pot container. Two PCI-6034E NI data acquisition boards [22] sample the force signals at 1 KHz frequency. To transform all signals in the same frame (the earth reference frame), we use a 9DOF-Sensor-Stick SparkFun IMU and measure the pot's orientation [31]. The IMU works at the sampling frequency of 100 Hz. An Arduino Mega microcontroller board [1] is used to interface the IMU with the computer. All devices are managed through a Matlab GUI that we develop. Figure 5.1 details the system components.

5.3.2 Procedure

To study bimanual and dyadic manipulation, we consider three different cooperation scenarios. In the first scenario (bimanual mode), each subject performs the task alone. This scenario is referred to as single-person bimanual (**SPB**) scenario. In the next scenario (dyadic mode), a leader role (and a follower role) is assigned to each subject (and his/her partner). The leader is the person who initiates all the movements, and

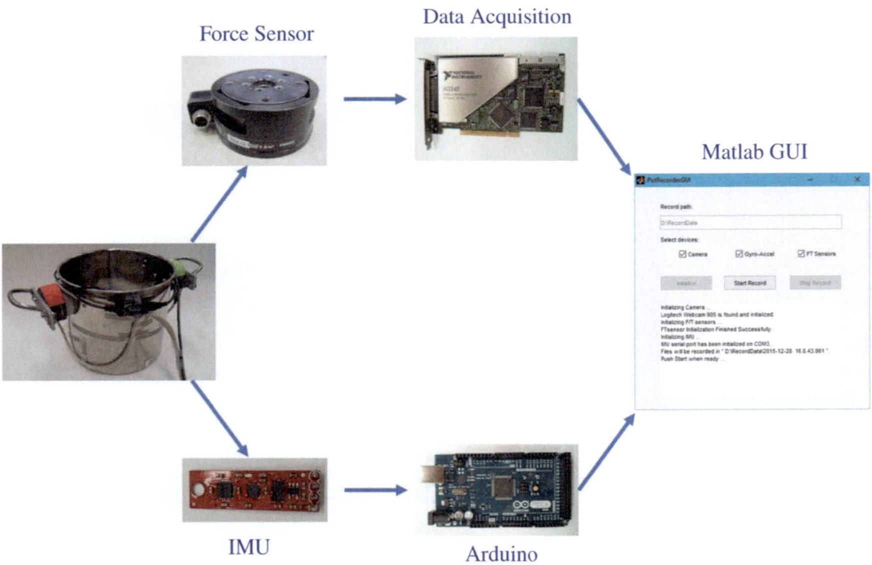

Fig. 5.1 Experimental setup components. The force sensors are installed between the pot handles and the pot container. The IMU is interfaced with an Arduino board and placed in a box that is glued to the inside bottom of the pot. The sampled data is collected by a GUI in Matlab

the follower is the one who follows his/her lead. We will refer to this scenario as the leader-follower (**L/F**) scenario. In the last scenario (dyadic mode), a beep is played by the software at certain points in time and the subjects are asked to perform their movements immediately after hearing the beep. In this scenario the computer synchronizes the subjects and thus we will refer to this scenario as the synchronized (**Sync**) scenario. Since the start time of the movement is known for both subjects by the beep, no time lag exists between the subjects at the beginning of the movement (unlike the L/F scenario). Thus, comparing Sync scenario with L/F scenario is helpful in understanding the role of the time lag between the leader and the follower in L/F scenario.

Each trial of the experiment consists of lifting the pot from the table at the start point (point 1), moving the pot horizontally toward the destination point (point 2), and putting the pot down on the table at the end point. In this work we only study the horizontal movements and discard the first and last movement in each trial. The configurations of the start points and the end points are designed such that two distinct types of horizontal motions exist. In the first type, the direction of the motion is perpendicular to the direction of the grasp force (the line connecting the handles). As a result, the grasp force has small components in the direction of motion. We will refer to these horizontal motions as type 1 motions. In the other type, the direction of the motion is parallel with the grasp force (the line connecting the handles). As a result, the grasp force has dominant components in the direction of motion. We will refer to these horizontal motions as type 2 motions. Figure 5.2 depicts these motion types. We also consider both short-range and long-range motions. In the short-range motions, the distance between the start point and the end point is selected to be 28 cm, while in the long-range motions it is 83 cm.

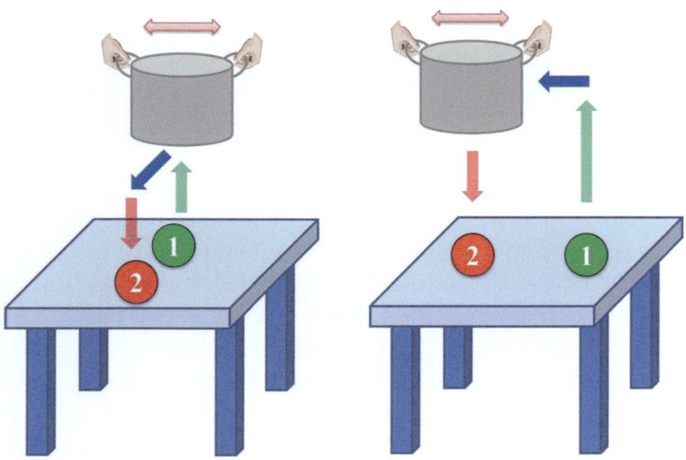

(a) Type 1 motion (b) Type 2 motion

Fig. 5.2 Motion types: **a** Type 1 motion. **b** Type 2 motion

The subjects are given as many familiarization trials as they need. Then, three trials of SPB scenario are executed by each subject, including a short-range type 1 motion, a short-range type 2 motion, and a long-range type 2 motion. The long-range type 1 motion cannot be executed, because it is out of the range of normal human-arm reachable space. Next, the subjects are paired and each pair performs three trials in Sync scenario, excluding the short-range type 2 motion. Finally, each pair performs two trials in L/F scenario. That is, we assign the leader role to a subject and ask the pair to perform a long-range type 1 motion. Then the subjects switch the leading role and repeat the task. In all of the trials, no repeated measurements are collected. To eliminate the high frequency noise, a low pass FIR smoothing filter is applied to the signals. It is a 7th-degree polynomial Savitzky–Golay filter [30] with the frame size of 201 and the cutoff frequency of 12.5 Hz.

5.4 Results

A human study was conducted with 22 adult subjects, consisting of 12 men and 10 women between 19 and 35 years of age. Discarding the corrupted measurements, we collected 58 trials in SPB scenario, 31 trials in Sync scenario, and 21 trials in L/F scenario (a total of 52 trials in dyadic mode). Using the models explained in Sect. 5.2.3, the internal forces, $F^i(t)$, were calculated for all trials. In all of the graphs in this section, red dashed line represents f_1 (or f_1^e), green solid line represents f_2 (or f_2^e), and black dotted line represents F_{sum}. Also, the start time ($t = 0$) in all graphs is associated with the time that the horizontal motion begins (excluding the vertical motion).

To validate our model, we first consider the SPB manipulation. In type 1 motions, where the internal force is very small, we expect that both hands of each subject apply similar forces and the hand's collaboration is close to ideal cooperation (i.e., $f_1^e = f_2^e = \frac{1}{2}F_{sum}$ or $\delta = 0$). The results showed that all subject exhibit high degree of cooperation in this type of motion. Figure 5.3 depicts the applied forces and the effective forces for subject-13 as an example.

In type 2 motions, the internal force is larger and our model plays a more influential role. Again we observed good cooperation between each subject's hands, as expected (meaning $\delta \approx 0$). Panels (a) and (b) of Fig. 5.4 show forces of subject-1 as an example. More than 90% of the subjects applied similar forces as depicted in this figure. However, 2 subjects performed the task differently. Panels (c) and (d) of Fig. 5.4 show subject-18's forces. The graph reveals that the subject's hands operate differently: one hand generates a coarse trajectory for the whole task while the other hand only introduces fine tuning to the path. The specialization of human hands and the asymmetric division of labor in human bimanual actions has been reported previously [14].

Now, let us consider dyadic manipulation. Here, the synchronization between subjects is not as good as during SPB manipulation. As a result, only 55% of the groups demonstrated good cooperation. Panels (a) and (b) of Fig. 5.5 depict forces of pair-6

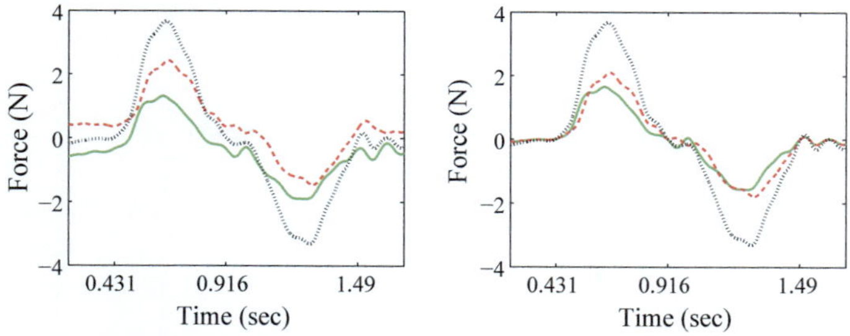

(a) Applied forces (S13) (b) Effective forces (S13)

Fig. 5.3 Force interaction in SPB manipulation, type 1 motion

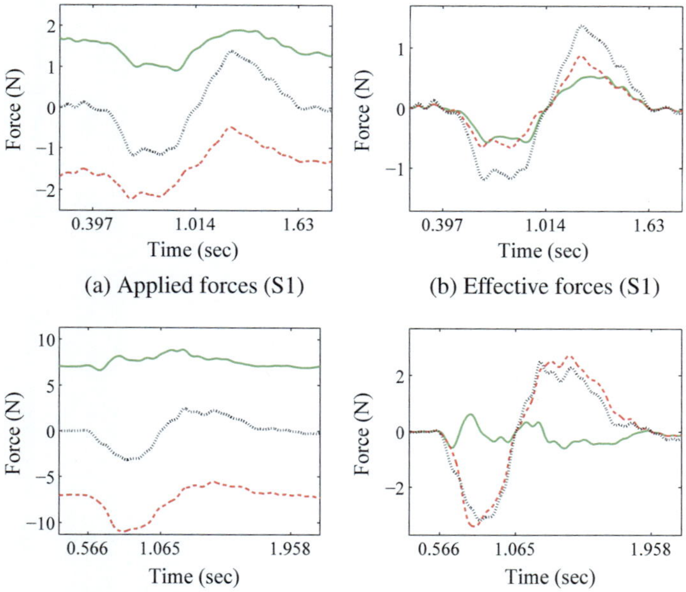

(a) Applied forces (S1) (b) Effective forces (S1)

(c) Applied forces (S18) (d) Effective forces (S18)

Fig. 5.4 Force interaction in SPB manipulation, type 2 motion. Forces of subject-1 (**a**, **b**) and subject-18 (**c**, **d**) are shown

as an example of a good cooperation. The other 45% of the groups developed some kind of specialization. Panels (c) and (d) of Fig. 5.5 show an example of this case, in which pair-7 started the manipulation cooperatively and in the middle, switched to the leader/follower mode. Interestingly, pair-3 extended their specialization and developed roles. As illustrated in panels (e) and (f) of Fig. 5.5, one subject only accelerated the object and the other subject decelerated it. The acceleration/deceleration role in collaborative tasks has also been reported in the literature [29].

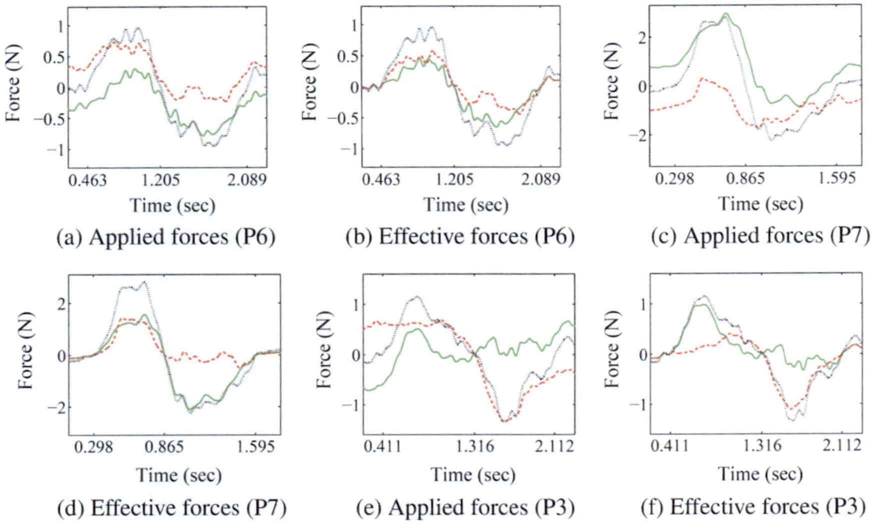

Fig. 5.5 Force interaction in dyadic manipulation, type 1 motion. Pairs 3, 6, and 7 are shown

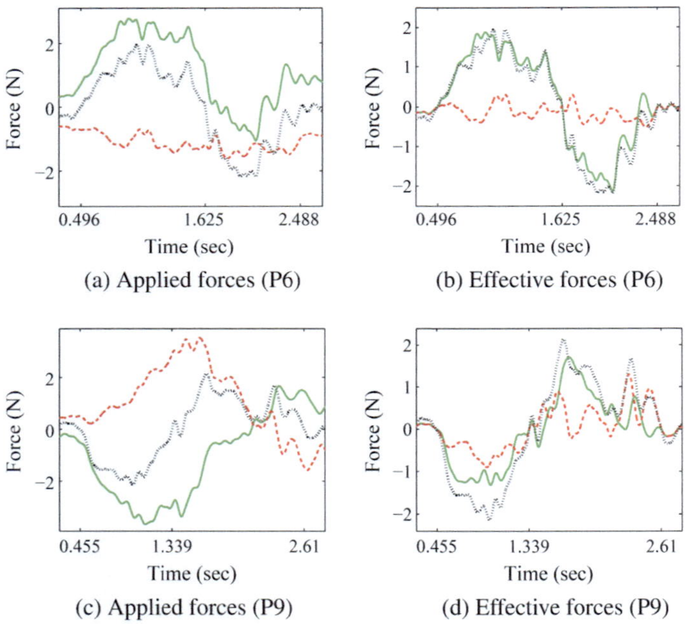

Fig. 5.6 Force interaction in dyadic manipulation, type 2 motion. Pairs 6 and 9 are shown

Finally, in type 2 motion of dyadic manipulation, groups demonstrate a wide range of collaboration types. Figure 5.6 depicts two examples of these collaborations. While pair-6 showed a very successful specialization (panels (a) and (b) of Fig. 5.6), pair-9 performed more cooperatively (panels (c) and (d) of Fig. 5.6).

5.5 Validation of the Model

In the previous section, we showed that employing our proposed model generates signals that are in agreement with the existing theories on human movements. In this section we compare the proposed model with a few alternative models and show that they contradict with established theories in human movement science. By studying the behavioral characteristics of the internal force and the effective forces for different models, we provide a qualitative assessment of the validity of the models.

As observed in the previous section, almost all bimanual manipulation trials were highly cooperative, the effective forces for each subject were very similar and $\delta \approx 0$ in each trial. These results strongly support the physics inspired model that suggests identical effective forces and $\delta = 0$. In other words, in case of bimanual manipulation, our proposed model and the physics inspired model agree with one another. However, in case of the dyadic manipulation, while our proposed model reveals a variety of collaboration types, the physics inspired model still suggests identical effective forces (perfect cooperation). Based on the human assessments, it has been shown [23] that subjects' perceptions of effective forces are not similar in dyadic tasks. Since the physics inspired model is unable to detect any differences in the effective forces, we exclude this model from our comparison.

The first alternative model that we discuss here is the human inspired model. Figure 5.7 presents a comparison between the human inspired model (Hu) and our proposed model (PM). The signals belong to an SPB manipulation in type 2 motion and therefore, we expect to observe smooth internal and effective forces (recall that in SPB only one brain controls both hands). As illustrated in this figure, the variations of the internal force in Hu are too large and too fast for a natural hand movement. That is, when a single person manipulates an object, he/she would not experience such push/pulls during the task. Furthermore, the effective forces in Hu suggest several role changing (leader to follower and vice versa), which is not expected in a single-person bimanual manipulation task. In contrast, PM introduces a smooth variation in the internal force and also a cooperative behavior in effective forces.

The second alternative model is the minimum-variation for the internal force. Recall that the minimum-variation model (MV) hypothesizes that the internal force is either constant or changes linearly. We chose the quadratic model (QM) as the third alternative model (see Sect. 5.2.3.2). In type 1 motion, the differences between the internal forces generated by PM, MV, and QM are negligible. For instance in SPB manipulation, the maximum difference between all internal forces over the time, over all subjects, and over all models is 0.057 N. However, in type 2 motion, the effect of different models is more evident. Figure 5.8 shows the results of these models

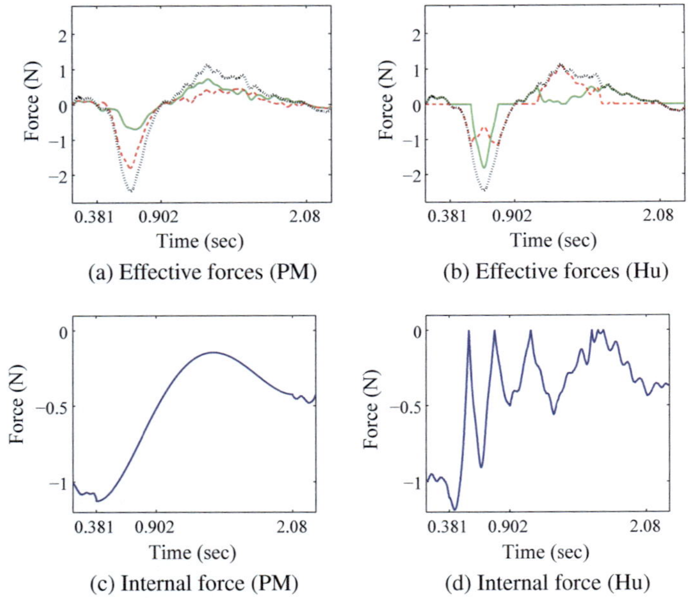

Fig. 5.7 Comparison between our proposed model (PM) and human inspired model (Hu). Forces of subject-9 are shown (type 2 motion in SPB scenario)

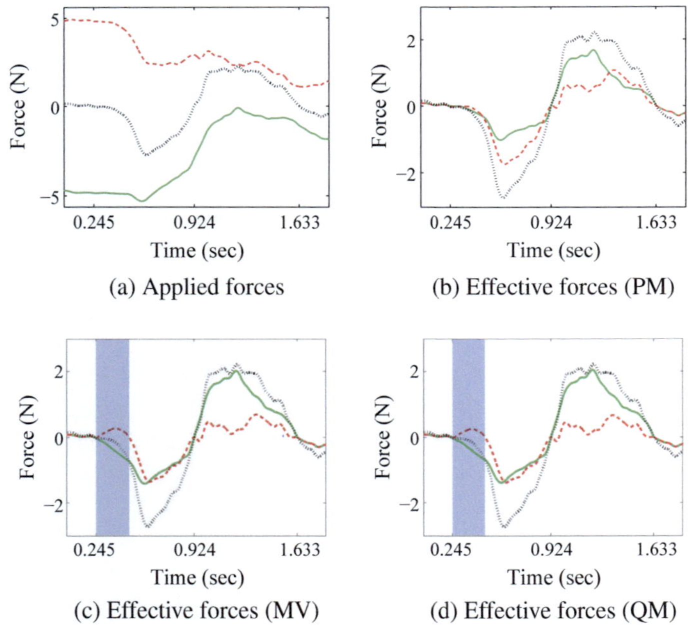

Fig. 5.8 Comparison between our model and alternative models. Forces of subject-2 are shown (SPB manipulation, type 2 motion)

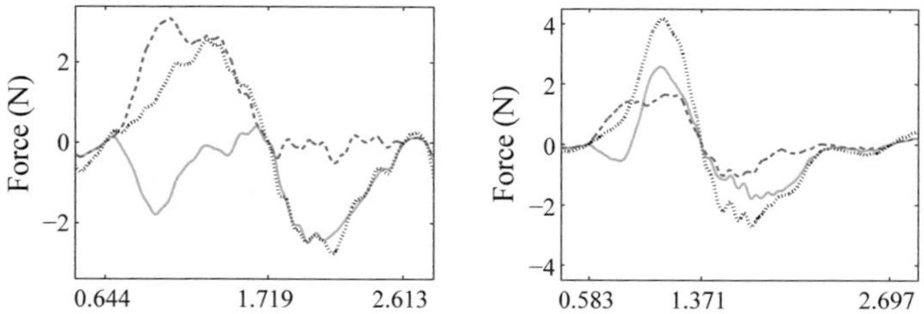

(a) Effective forces (PM, P8) (b) Effective forces (PM, P9)

Fig. 5.9 Force interaction in a leader/follower, type 1 motion

on an SPB manipulation. While the signals are very similar, a subtle difference exists between PM and the alternative models. Both MV and QM suggest that at the beginning of the motion, first hand opposes the second hand's motion for around 250 ms (see the shaded areas in panels (c) and (d)). This behavior is well-known in the leader/follower scheme (no synchronization). Figure 5.9 shows two examples of leader/follower scheme. We turned the beep off and asked one subject to follow the motions of the other subject. Due to the delay between the leader's motion and the follower's perception of the motion, follower resists the motion for a short period of time. This behavior is not expected in an SPB manipulation, when both hands are controlled by the same brain. It shows that while the proposed model provides the expected signals, other alternative models do not.

According to the above qualitative assessment, while the proposed model is in full agreement with the established theories of human movements, the alternative models contradict them. For instance, alternative models suggest jerky push–pulls that humans would not usually experience in a natural bimanual task. For a brief review on related theories on human movements, see Sect. 5.2.3.1. More specifically, refer to [10, 14, 23, 29, 32]. In summary:

Corollary 5.1:
Unlike the alternative models, the proposed model respects the well-established theories of human movement and can better explain the nature of a dyadic collaboration.

5.6 Statistical Analysis of the Internal Force Features

In this section we study the statistical properties of the internal force during a dyadic manipulation task. We show that while many properties vary between different scenarios, a certain component is strongly correlated with the object velocity. We will

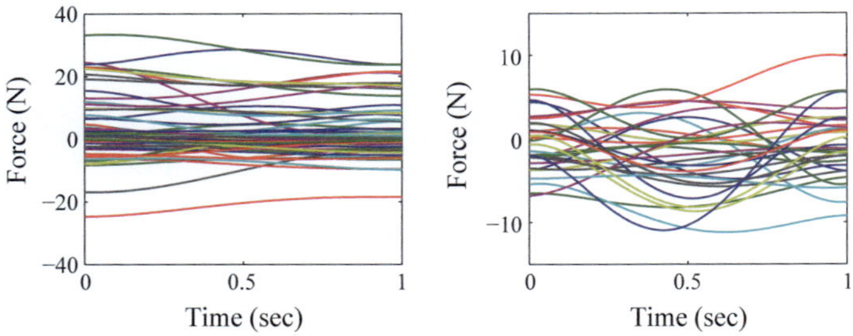

(a) SPB scenario (b) Sync scenario

Fig. 5.10 Internal forces exemplified with two scenarios

use this component as an abstract model for the human intent and accordingly will propose a cooperation strategy in the following section.

For the signals to be comparable in different motion types and in different scenarios, we normalized them in both time and magnitude. That is, motion duration is scaled to be exactly 1 s and the traveled distance is also scaled to be exactly 1 m. The normalization scaling factors are calculated for F_{sum} and then applied to f_1, f_2 and F^i. We used G*Power software [8] to calculate the statistical power. The number of trials were high enough to provide a power of $1 - \beta > 0.95$ for all scenarios. All other statistical figures were obtained using Matlab. Following the model explained in Sect. 5.2.3, we first calculated the internal forces, $F^i(t)$, for all trials. Then we studied the variations and magnitude of the internal forces among different scenarios. Figure 5.10 shows the internal forces in SPB and Sync scenarios as two examples. As exemplified in this figure, the internal forces in bimanual mode are larger than in dyadic mode. More specifically, the magnitude of the forces, both at the beginning and at the end, in bimanual mode is larger than in Sync mode. The results of the statistical analysis of these signals appear in the following sections.

5.6.1 Initial Grasp Force Magnitude

Before the movement is initiated, the internal force is equivalent to the grasp force. Our statistical analysis shows that the magnitude of the initial grasp force ($\| F^i(0) \|$) is significantly larger in bimanual mode compared with the dyadic mode. For a significance level of $\alpha = 0.05$, we observe that $\mu_{bimanual} > \mu_{dyadic}$ with a p-value of $p = 0.0001$. The 95% confidence interval for $\mu_{bimanual}$ is $CI = [5.42,\ 9.71]$ and for μ_{dyadic} is $CI = [2.34,\ 3.89]$. The same trend exists in all scenarios, $\mu_{SPB} > \mu_{Sync}$ ($p = 0.0008$) and $\mu_{SPB} > \mu_{L/F}$ ($p = 0.019$). The variations of the initial grasp force are also significantly larger in bimanual mode compared with the dyadic mode. That is, $\sigma_{bimanual} > \sigma_{dyadic}$ with a p-value of $p < 0.0001$ for $\alpha = 0.05$.

5.6.2 Final Grasp Force Magnitude

Immediately after the movement is finished, the applied forces sum up to zero and therefore, the internal force would be the grasp force again. Similar results are observed for the magnitude of the final grasp force ($\| F^i(1) \|$). That is, the final grasp force (and its variations) is significantly larger in bimanual mode compared with the dyadic mode. For a significance level of $\alpha = 0.05$, we observe that $\mu_{bimanual} > \mu_{dyadic}$ with a p-value of $p = 0.004$. Also $\sigma_{bimanual} > \sigma_{dyadic}$ with a p-value of $p < 0.0001$. The same trend exists in all scenarios, $\mu_{SPB} > \mu_{Sync}$ ($p = 0.01$) and $\mu_{SPB} > \mu_{L/F}$ ($p = 0.05$). The 95% confidence interval for $\mu_{bimanual}$ is $CI = [4.25,\ 7.77]$ and for μ_{dyadic} is $CI = [2.4,\ 4.12]$.

5.6.3 Internal Force Energy

Similarly, the average energy of the internal force signals ($\int_0^1 \| F^i(t) \|^2 dt$) are significantly higher in bimanual mode compared with the dyadic mode. For $\alpha = 0.05$, we observe that $\mu_{bimanual} > \mu_{dyadic}$ with a p-value of $p = 0.001$. Also $\sigma_{bimanual} > \sigma_{dyadic}$ with a p-value of $p < 0.0001$ and the power of $1 - \beta = 0.999$. The same trend exists in all scenarios, $\mu_{SPB} > \mu_{Sync}$ ($p = 0.008$) and $\mu_{SPB} > \mu_{L/F}$ ($p = 0.04$). The 95% confidence interval for $\mu_{bimanual}$ is $CI = [47.4,\ 137.2]$ and for μ_{dyadic} is $CI = [10.5,\ 30.2]$.

A Discussion on the Grasp Force

Above observations can be combined and expressed as following:

Grasp Force Observation 1:
Humans apply larger grasp forces in bimanual mode compared to the dyadic mode. The variations of the grasp forces are also larger in bimanual mode.

While we showed statistical evidence in favor of the above observation, it seems completely counter intuitive. In bimanual mode, both hands belong to the same person and both are controlled with the same brain. So, the movement is expected to be smoother compared to the dyadic mode and the internal force is expected to be negligible. However, the results show that, in fact, the average energy of the internal force signals are significantly higher in bimanual mode than in dyadic mode. We speculate that the higher energy is an artifact of the higher initial (final) grasp force. In other words, if these grasp forces are compensated, the remaining signals have less energy in bimanual mode than in dyadic mode.

To test this hypothesis, we need to study the variations of the internal force in more detail. We first consider the difference between the initial grasp force and final grasp force, i.e., $\Delta F = F^i(1) - F^i(0)$. Then the initial grasp force is canceled, i.e., $F^i_c = F^i - F^i(0)$. Finally, both initial and final grasp forces will be canceled.

5.6.4 Difference Between Initial and Final Grasp Forces

In this test, no statistically significant difference is observed for $\Delta F = F^i(1) - F^i(0)$. That is, no strong evidence exists to reject the null hypothesis that $\mu_{bimanual} = \mu_{dyadic}$ and the p-value of the test is $p = 0.19$. The same trend exists for all scenarios, $\mu_{SPB} = \mu_{Sync}$ ($p = 0.39$) and $\mu_{SPB} = \mu_{L/F}$ ($p = 0.23$). In addition, the data fails to reject the null hypothesis that the $\Delta F = 0$. Particularly, the 95% confidence interval for $\mu_{bimanual}$ is $CI = [-2.26,\ 0.48]$ and for μ_{dyadic} is $CI = [-0.8,\ 1.28]$. These observations can be summarized as following:

Grasp Force Observation 2:
Regardless of the cooperation scenario (in both dyadic and bimanual modes), the initial and final grasp forces are not significantly different.

5.6.5 Internal Force Variation

In order to study the internal force variations, first the initial grasp force is canceled, $F^i_c = F^i - F^i(0)$. Figure 5.11 shows F^i_c in SPB and Sync scenarios as two examples. The statistical analysis shows that the average energy of F^i_c is not significantly different in bimanual mode compared with the dyadic mode. That is, no strong evidence exists to reject the null hypothesis that $\mu_{bimanual} = \mu_{dyadic}$ and the p-value of the test is $p = 0.53$. The 95% confidence interval for $\mu_{bimanual}$ is $CI = [5.27,\ 25.80]$ and for μ_{dyadic} is $CI = [7.98,\ 15.99]$. The same trend exists in all scenarios, $\mu_{SPB} = \mu_{Sync}$ ($p = 0.79$) and $\mu_{SPB} = \mu_{L/F}$ ($p = 0.49$).

On the contrary, the variation of average energy is significantly larger in bimanual mode compared with the dyadic mode. For a significance level of $\alpha = 0.05$, we observe that $\sigma_{bimanual} > \sigma_{dyadic}$ with a p-value of $p < 0.0001$. The same trend exists in all scenarios, $\sigma_{SPB} > \sigma_{Sync}$ ($p < 0.0001$) and $\sigma_{SPB} > \sigma_{L/F}$ ($p < 0.0001$).

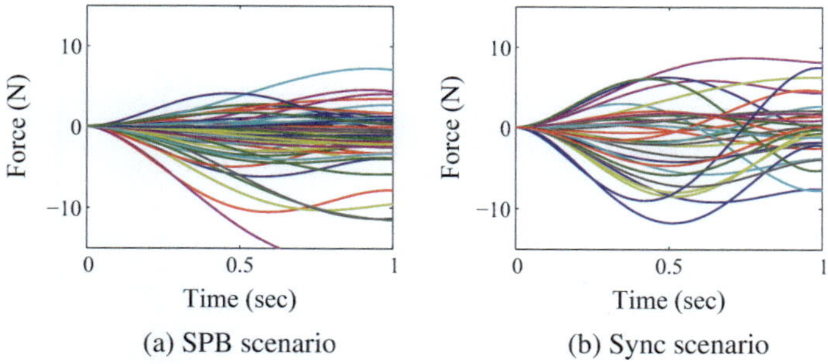

Fig. 5.11 Internal forces variations exemplified with two scenarios

5.6.6 Negotiation Force

In order to cancel both the initial grasp force and the final grasp force, we propose to decompose the internal force into two components. According to the model discussed in Sect. 5.2.3, the internal force is obtained by satisfying the boundary conditions. These boundary conditions are the results of the negotiation of two hands on the timing of the reaching movement, see (5.14). The initial and final grasp forces are two of these boundary conditions. If there had been no other boundary conditions, the internal force would have followed the minimum-jerk trajectory. In other words, if $\dot{F}^i(0) = \dot{F}^i(1) = 0$ and $\ddot{F}^i(0) = \ddot{F}^i(1) = 0$, then the internal force in (5.13) can be expressed [33] as a smooth minimum-jerk function of time (similar to (5.10)):

$$F_s^i = F^i(0) + (F^i(1) - F^i(0)) \times (6t^5 - 15t^4 + 10t^3) \tag{5.15}$$

If we take this smooth force as the first component of the internal force, the remaining force would be the negotiation component, that is:

$$F_n^i = F^i - F_s^i \tag{5.16}$$

Figure 5.12a exemplifies the decomposition of two signals and Fig. 5.12b, c show the negotiation components for the SPB and Sync scenarios as two examples. As it is illustrated, both initial and final grasp forces have been canceled in F_n^i.

As expected, the average energy of the negotiation component is significantly lower in bimanual mode compared with the dyadic mode. For $\alpha = 0.05$, we observe that $\mu_{bimanual} < \mu_{dyadic}$ with a p-value of $p = 0.0015$. Also $\sigma_{bimanual} < \sigma_{dyadic}$ with a p-value of $p < 0.0001$. The same trend exists in all scenarios, $\mu_{SPB} < \mu_{Sync}$ ($p = 0.0004$) and $\mu_{SPB} < \mu_{L/F}$ ($p = 0.03$). The 95% confidence interval for $\mu_{bimanual}$ is $CI = [1.23,\ 3.22]$ and for μ_{dyadic} is $CI = [4.39,\ 11.51]$. These observations can be summarized as following:

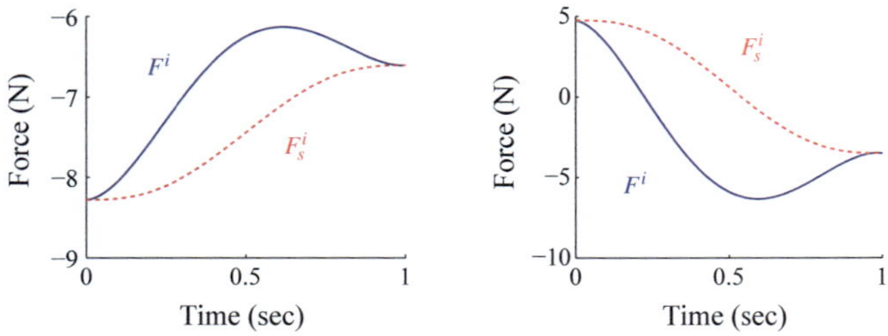

(a) Internal force decomposition examples

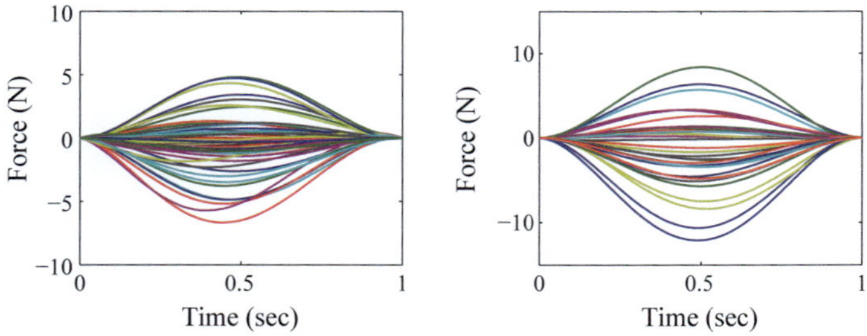

(b) SPB scenario (c) Sync scenario

Fig. 5.12 Negotiation forces exemplified with two scenarios

Negotiation Force Observation 1:
The average energy of the negotiation force is significantly smaller in bimanual mode compared with the dyadic mode.

5.6.7 Negotiation Force Versus Object Velocity

A closer look at Fig. 5.12 reveals that the negotiation forces are very similar to the velocity profile in the reaching movement. As an accepted practice [16, 32], we use Pearson correlation coefficient, r, to measure this similarity. The correlation between the normalized object velocity and the negotiation force is calculated for each trial. The values of r show a strong correlation between the two signals. For instance, the average value of r for the negotiation forces in Fig. 5.12c is $\bar{r} = 0.91$ and its standard deviation is $s = 0.1$. More specifically, the 95% confidence interval for $\mu_{bimanual}$ is $CI = [0.75,\ 0.87]$ and for μ_{dyadic} is $CI = [0.84,\ 0.93]$. The results strongly support the finding that:

Negotiation Force Observation 2:
The negotiation force is strongly correlated with the velocity of the object.

5.7 Proposed Cooperation Policy

In this section we propose a cooperation policy for a robot, collaborating with a human in a dyadic reaching movement. The robot needs to apply forces to the object in such a way that the human perceives a natural (human-like) physical interaction. An important component of this physical interaction is the internal force. While the internal forces do not contribute to the motion trajectory of the object, they play an important role in how natural the interaction is.

The internal force model discussed in Sect. 5.2.3 cannot be implemented in a real-time robot controller. For instance, the model needs the value of the grasp force in between the movement to construct the internal force. This kind of information is available only after the movement is over. Here, we propose a cooperation policy that eliminates the need for this information and generates a natural internal force. The proposed policy is based on the two-component decomposition of the internal force, discussed in Sect. 5.6.6. Accordingly to (5.16), $F^i = F^i_s + F^i_n$.

To fully determine F^i_s, the values of initial and final grasp forces ($F^i(0)$ and $F^i(1)$) are required, see (5.15). As discussed in Sect. 5.6.4, strong evidence exists suggesting that $\Delta F = 0$, or equivalently $F^i(0) = F^i(1)$. Therefore, we propose to select the same value for both the initial grasp force and the final grasp force. Let F_G refer to this value. By plugging $F^i(0) = F^i(1) = F_G$ in (5.15), the first component of the internal force is obtained as, $F^i_s = F_G$.

For the second component, F^i_n, we exploit the observation in Sect. 5.6.7. As discussed there, F^i_n is strongly correlated with the normalized object velocity. That is, we can approximate the negotiation force as $F^i_n = \kappa \cdot v$, in which κ is a scaling factor. As a result,

$$F^i = F_G + \kappa \cdot v \tag{5.17}$$

As discussed in Sects. 5.6.1 and 5.6.2, the 95% confident interval of the initial and final grasp forces in dyadic mode are $CI = [2.34,\ 3.89]$ and $CI = [2.4,\ 4.12]$, respectively. Thus, F_G belongs to the intersection of these ranges, i.e., $F_G \in [2.4,\ 3.89]$. The exact value of F_G would be obtained during the initial grasp. On the other hand, the value of κ is obtained by calculating the energy of the signal. According to the observation discussed in Sect. 5.6.6, the 95% confident interval of the average energy of F^i_n in dyadic mode is $CI = [4.39,\ 11.51]$. The energy of

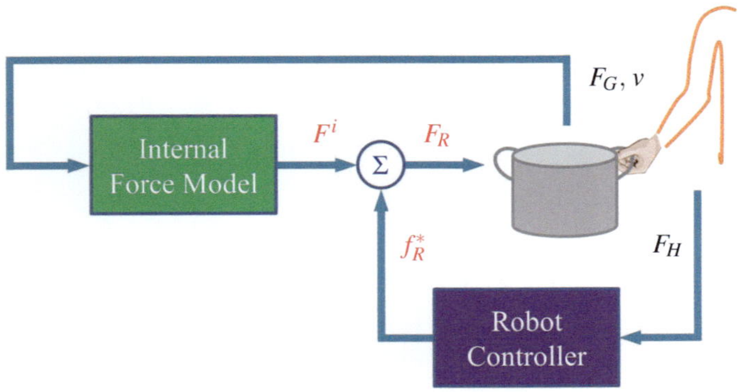

Fig. 5.13 Sketch of the control diagram for the proposed policy. The internal force block receives the velocity of the pot and the grasp force signals. The robot controller receives the human applied force

the normalized velocity profile for a minimum-jerk trajectory[1] is 1.414. As a result $\kappa \in [1.76, 2.85]$. The value of $\kappa = 2.3$ would be considered as a safe choice.

Considering (5.17) and (5.1), the robot's contribution to the dyadic task would be:

$$F_R = f_R^* + F_G + \kappa \cdot v \tag{5.18}$$

where f_R^* is the effective force provided by the robot. As discussed in the Introduction section, interesting methods have been proposed in the literature for planning f_R^*. Our proposed policy takes advantage of such strategies and augments them with a model for internal force (the last two terms in (5.18)).

Figure 5.13 shows the block diagram of the proposed policy. The "Robot Controller" block in this figure represents one of the existing strategies in the literature. The "Internal Force Model" block implements (5.17). By appending the internal force model into the controller, our cooperation policy can resolve the ambiguity of the haptic channel and introduce a more natural (human-like) internal force. Merging this policy with the existing cooperation strategies will increase the performance of those strategies, as well. The integrated policy reduces the undesirable push–pull forces and introduces a smoother human–robot interaction.

Discussion

In many applications, the cooperative manipulation task consists of a sequence of point-to-point ballistic motions. Reaching movement is an effective model for such movements. In particular, the concept of reaching movement with via-points considers complex movements (such as drawing) that comprise a sequence of point-to-point movements. As a result, the proposed cooperation policy can effectively be utilized for this type of applications. On the other hand, in some specific applications the goal of cooperation is not just to move an object, but also to reorient it. For instance, let

[1]A dyadic/bimanual reaching movement has been shown to follow minimum-jerk trajectory [26].

us assume that a human and a robot cooperatively assemble a piece of furniture, say a table. They need to put the tabletop upside down, attach the legs, lift the table, and then turn it over its legs. The last sub-task does not demand a large displacement, but requires substantial reorientation. Since our proposed policy is constructed based on the motion characteristics of the object, it cannot be utilized in this sub-task.

To be able to extend the proposed policy, the characteristics of the internal torque during human–human cooperative object reorientation need to be studied. We speculate that object's angular velocity may play a role in coordination of the cooperation between humans and therefore, it may appear in the generalized cooperation policy. We believe that further investigation of this conjecture is an interesting extension to this work.

5.8 Conclusion

In this paper we studied the role of the internal force in the estimation of the human intent. We first proposed a new model for the internal force that can better explain the interaction between dyads. To validate our model, we set up a human study and collected the applied forces during human–human cooperative object manipulation. Using these signals, we compared the proposed model with several existing models for the internal force. We showed that, unlike the alternative models, the proposed model respects the well-established theories on human movements. After justifying the validity of the model, we explored different features and regularities in the internal force signals.

We observed that, while the initial and final grasp forces were both significantly larger in bimanual mode, the difference between these forces (in each trial) were negligible. We also observed that no significant difference existed between bimanual and dyadic mode regarding the average energy of the variations of the internal force. However, when the internal force is decomposed into the negotiation force and a smooth force, the average energy of the negotiation component was significantly higher in dyadic mode, as expected. Furthermore, we observed that the negotiation force is strongly correlated with the object's velocity.

The last observation was particularly important. Based on this observation, we proposed a cooperation policy for a human–robot physical collaboration. It was discussed that by merging this policy with existing cooperation strategies, a smoother human–robot interaction would be achieved. Quantifying the performance gain of such integration is the next step for this work. Also, in the proposed cooperation policy, we assume that the robot's effective force is available (using existing cooperation strategies). An extension to this work is to study the interplay between the robot's effective force, the internal force, and the human intent.

Acknowledgements This work has been supported by the National Science Foundation under Grants IIS-0905593, CNS-0910988 and CNS-1035914.

References

1. Arduino Mega (2014) http://arduino.cc/en/Main/arduinoBoardMega
2. ATI (2014) ATI industrial automation, F/T sensor: gamma, force/torque sensor. http://www.ati-ia.com/products/ft/ft_models.aspx?id=Gamma
3. Bussy A, Kheddar A, Crosnier A, Keith F (2012) Human-humanoid haptic joint object transportation case study. In: IEEE/RSJ international conference on intelligent robots and systems (IROS), 2012, pp 3633–3638
4. Chen L, Javaid M, Di Eugenio B, Žefran M (2015) The roles and recognition of haptic-ostensive actions in collaborative multimodal human-human dialogues. Comput Speech Lang 34:201–231
5. Donner P, Mortl A, Hirche S, Buss M (2013) Human-robot cooperative object swinging. In: 2013 IEEE international conference on, robotics and automation (ICRA), IEEE, pp 4343–4349
6. Dumora J, Geffard F, Bidard C, Fraisse P (2013) Towards a robotic partner for collaborative manipulation. In: HRI workshop on collaborative manipulation, pp 1–6
7. Evrard P, Kheddar A (2009) Homotopy switching model for dyad haptic interaction in physical collaborative tasks. In: World haptics, pp 45–50
8. Faul F, Erdfelder E, Buchner A, Lang AG (2009) Statistical power analyses using g*power 3.1: tests for correlation and regression analyses. Behav Res Methods 41:1149–1160
9. Flash T, Gurevich I (1997) Models of motor adaptation and impedance control in human arm movements. Adv Psychol 119:423–481
10. Flash T, Hogan N (1985) The coordination of arm movements: an experimentally confirmed mathematical model. J Neurosci 5(7):1688–1703
11. Garvin G, Žefran M, Henis E, Kumar V (1997) Two-arm trajectory planning in a manipulation task. Biol Cybern 76(1):53–62
12. Gere J, Goodno B (2012) Mechanics of materials. Nelson Education
13. Groten R, Feth D, Goshy H, Peer A, Kenny D, Buss M (2009) Experimental analysis of dominance in haptic collaboration. In: The 18th ieee international symposium on robot and human interactive communication, 2009. RO-MAN 2009, pp 723–729
14. Guiard Y (1987) Asymmetric division of labor in human skilled bimanual action: the kinematic chain as a model. J Mot Behav 19(4):486–517
15. Harris CM, Wolpert DM (1998) Signal-dependent noise determines motor planning. Nature 394:780–784
16. Hwang EJ, Donchin O, Smith MA, Shadmehr R (2003) A gain-field encoding of limb position and velocity in the internal model of arm dynamics. PLoS Biol 1(2):209–220. doi:10.1371/journal.pbio.0000025
17. Medina JR, Lawitzky M, Mortl A, Lee D, Hirche S (2011) An experience-driven robotic assistant acquiring human knowledge to improve haptic cooperation. In: 2011 IEEE/RSJ international conference on intelligent robots and systems (IROS), IEEE, pp 2416–2422
18. Medina JR, Lee D, Hirche S (2012) Risk-sensitive optimal feedback control for haptic assistance. In: 2012 IEEE international conference on robotics and automation (ICRA), IEEE, pp 1025–1031
19. Medina JR, Lawitzky M, Molin A, Hirche S (2013) Dynamic strategy selection for physical robotic assistance in partially known tasks. In: 2013 IEEE international conference on robotics and automation (ICRA), IEEE, pp 1180–1186
20. Mörtl A, Lawitzky M, Kucukyilmaz A, Sezgin M, Basdogan C, Hirche S (2012) The role of roles: physical cooperation between humans and robots. Int J Robot Res 31(13):1656–1674
21. Nakano E, Imamizu H, Osu R, Uno Y, Gomi H, Yoshioka T, Kawato M (1999) Quantitative examinations of internal representations for arm trajectory planning: minimum commanded torque change model. J Neurophysiol 81(5):2140–2155
22. NI (2014) National instruments, NI PCI-6034E (Legacy) data acquisition board. http://sine.ni.com/nips/cds/view/p/lang/en/nid/11916

23. Noohi E, Žefran M (2014) Quantitative measures of cooperation for a dyadic physical interaction task. In: 2014 14th IEEE-RAS international conference on humanoid robots (Humanoids), pp 469–474
24. Noohi E, Žefran M (2016) Modeling the interaction force during a haptically-coupled cooperative manipulation. In: 2016 25th IEEE international symposium on robot and human interactive communication (RO-MAN), pp 119–124
25. Noohi E, Parastegari S, Žefran M (2014) Using monocular images to estimate interaction forces during minimally invasive surgery. In: 2014 IEEE/RSJ international conference on intelligent robots and systems, pp 4297–4302
26. Noohi E, Parastegari S, Žefran M (2015) Computational model for dyadic and bimanual reaching movements. In: IEEE world haptics conference (WHC), pp 260–265
27. Noohi E, Žefran M, Patton JL (2016) A model for human-human collaborative object manipulation and its application to human-robot interaction. In: IEEE transactions on robotics, 32(4):880–896
28. Reed KB, Peshkin M, Hartmann MJ, Colgate JE, Patton J (2005) Kinesthetic interaction. In: 9th International conference on rehabilitation robotics, 2005. ICORR 2005. IEEE, pp 569–574
29. Reed K, Peshkin M, Hartmann MJ, Grabowecky M, Patton J, Vishton PM (2006) Haptically linked dyads: are two motor-control systems better than one? Psychol Sci 17(5):365–366
30. Schafer R (2011) What is a Savitzky-Golay filter? [lecture notes]. IEEE Signal Process Mag 28(4):111–117. doi:10.1109/MSP.2011.941097
31. Sparkfun (2014) 9 Degrees of freedom - sensor stick, (accelerometer, gyroscope and magnetometer). https://www.sparkfun.com/products/10724
32. Shadmehr R, Mussa-Ivaldi FA (1994) Adaptive representation of dynamics during learning of a motor task. J Neurosci 14(5):3208–3224
33. Shadmehr R, Wise SP (2005) The computational neurobiology of reaching and pointing: a foundation for motor learning. MIT Press, Cambridge
34. Uno Y, Kawato M, Suzuki R (1989) Formation and control of optimal trajectory in human multijoint arm movement. Minimum torque-change model. Biol Cybern 61:89–101
35. Žefran M, Kumar V, Yun X (1994) Optimal trajectories and force distribution for cooperating arms. In: Proceedings of the 1994 IEEE international conference on robotics and automation, 1994, vol 1, pp 874–879

Chapter 6
A Learning Algorithm to Select Consistent Reactions to Human Movements

Carol Young and Fumin Zhang

6.1 Introduction

As robots move away from the factory floor they encounter environments that are no longer well structured in time or space. This provides motivation for them to be equipped with real-time algorithms that tolerate noise from their environment and adapt to it. These challenges are compounded when humans come in close contact with a robot [1, 2], where robots should behave in predictable ways [3, 4] in response to human activities. Since human safety is an important concern, any choice of reaction that a robot makes should occur in real time. A learning algorithm is usually employed by the robot to make predictions of the human action, and then select safe reactions. Since humans have individual differences and preferences, the learning algorithms need to maintain a balance between adaptiveness and consistency, which is the challenge that will be addressed in this chapter.

Figure 6.1 illustrates a common structure of online learning algorithms. Typically, a learning algorithm interacts with a hidden mapping of parameters to output, commonly referred to as a concept. An algorithm uses observed parameters to make a selection of a control law. Then when the expected output is known it will be compared to the actual output from the system. If these outputs match then we say that the selection is correct, otherwise an error occurs, and the estimation of concept is updated. The online learning method handles each parameter-output pair and update

---

The research work is supported by ONR grants N00014-10-10712 (YIP) and N00014-14-1-0635; and NSF grants OCE-1032285, IIS-1319874, and CMMI-1436284, and an ARCS Scholar Award.

C. Young · F. Zhang (✉)
School of Electrical and Computer Engineering, Georgia Institute of Technology, Atlanta, GA 30332, USA
e-mail: fumin@gatech.edu

Y. Wang and F. Zhang (eds.), *Trends in Control and Decision-Making for Human–Robot Collaboration Systems*, DOI 10.1007/978-3-319-40533-9_6

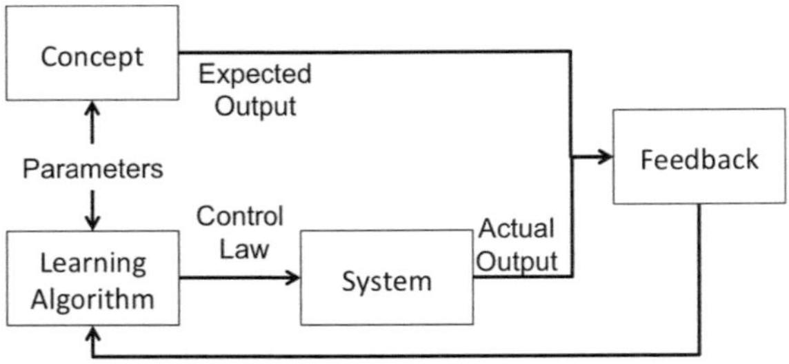

Fig. 6.1 How learning algorithms work

sequentially and only uses each pair once, hence only requiring limited memory and processing time [5]. This makes online learning ideal for implementation on robotic systems. This chapter deals with learning algorithms that choose between two control laws. Each of these control laws generate a unique output, only one of which is the output expected by the concept. At any point in time we define the output value with the highest probability of being expected by the concept for given parameters as a *preference*, while the other output value is defined as a *deviation* from the preference.

The particular online learning paradigm we are examining is ensemble learning [5, 6]. Ensemble leaning uses multiple experts, each of which represent a different estimation of unknown quality of the concept. These experts are then combined into the algorithm's estimation of the concept. A notable example of online ensemble learning is the weighted majority algorithm [7]. This algorithm can be used to take the Winnow algorithm [8] online [9]. It also has inspired popular boosting algorithms such as AdaBoost [10] and its derivative Brown Boost [11].

These online ensemble methods can follow drift, which is a change in the concept, e.g., mapping from parameters to outputs, over time [12]. When binary choices are considered, drift is a change in preference over time. Multiple online methods have been implemented to adapt to drift, including the dynamic weighted majority where new experts are added and deleted over time [13]. Other methods for tracking drift include using regret minimization [14], and tracking the average of online parameter-output pairs [15].

These methods can also tolerate uncertainty in the concept. Robustness measures how well an algorithm can tolerate this uncertainty. Robustness has been considered for online learning methods [7, 9]. The novel approach that this chapter makes is to analyze a specialized form of robustness, the consistency. Consistency is a measure of how easily the selection made by an algorithm changes when deviations are observed. If an algorithm never changes its selection, then the algorithm will be the most consistent, but it cannot follow drift; hence it is not adaptive. On the other hand, if an algorithm changes its selection every time a deviation occurs, then the algorithm is adaptive, but it is not consistent because it cannot tolerate uncertainty. The notion of consistency is important to analyze whether a learning algorithm meets predictability requirements for easy human–robot interaction [3], and prevents chatter in motion.

The goal of this chapter is not to create an algorithm that is the most adaptive or the most consistent, but rather an algorithm that can manage the tradeoff between the two. A learning algorithm called the dual expert algorithm (DEA) is developed in this paper to select between two possible control laws, and each generates a reaction to one of two choices a human will make. We show that the DEA has bounded errors in selection between two reactions, improving consistency as the number of data points increases, while retaining its ability to adapt to concept drift. We compare the DEA with two similar algorithms (weighted majority algorithm (WMA) [7] and the online Winnow algorithm [9]).

We justify the conclusion and demonstrate real-life applications in human–robot interaction (HRI) by implementing the algorithm onto a moving robot avoiding a known user approaching it in the hallway. Because pedestrians tend to pass on a predetermined side [16, 17] with parameters that are similar to pedestrians passing other human [18], the existing methods to solve this problem have the robot stop [19] or treat the humans as obstacles and plan a path around them [20–22]. Our method allows the robot to move around a human in a socially aware manor without necessitating large path planing costs.

In the experiment the concept is the internal beliefs of a human that determines which wall they turn towards when encountering a robot moving towards them in the hallway. The two control laws that the robot has to select from are for passing the human by moving toward either the right or the left side of the hallway. The expected output from the robot is that the human and the robot move toward opposite walls to avoid each other. The expected output is then compared to the actual output that is which side of the hallway the human moves toward. If the two outputs match each other, then the selection made by the learning algorithm is a success, otherwise the selection is an error. The learning algorithm then uses this success or error to improve its next prediction.

The layout of this chapter is as follows. Section 6.2 presents the problem formulation. Section 6.3 explains the dual expert algorithm and Sect. 6.4 introduces the expanded dual expert algorithm. Section 6.5 shows the simulation results of both algorithms. Section 6.6 presents the results of the experiment, and Sect. 6.7 presents the conclusion.

6.2 Background

6.2.1 Expert-Based Learning

An expert in an algorithm represents mapping from a set of input to a specific output. In the case of learning binary choices, two experts can be employed: one expert always make the first choice, and the other expert always make the second choice. We can assume that at least one expert will preform as good or better than random guessing. An expert-based learning algorithm in this case will try to select the best expert to believe in. This idea can be generalized to multiple experts. Of all the experts considered there will be one that makes the minimum amount of errors, which we will call the best expert. This best expert is unknown before hand. The learning algorithm will need to discover the best expert by making a number of trials. Errors will be made during these trials, and the algorithm learns from these errors. Thus a feasible learning algorithm must make a bounded number of errors before finding the best expert.

We define *consistency* as how well the algorithm resists different outputs from the same parameters. In the binary case, suppose the first choice is predicted by the best expert, but the concept exhibits the second choice, then a consistent algorithm should not switch its best expert immediately. The switching should only happen when the second choice appears repeatedly. Each time an algorithm changes its prediction, we say a switch occurs. The number of switches is an important parameter for an algorithm used for human–robot interaction because it affects the human's perception on the consistency of the robot. A learning algorithm that learns quickly but without consistency will be regarded as unpredictable by human even if the error bound for the algorithm is low.

We define *adaptiveness* as how quickly the algorithm can adapt when the concept that it is attempting to learn drifts. In the binary case, suppose the concept changes its preference from the first choice to the second choice, then the algorithm should switch its prediction from the first choice to the second choice after small number errors. In general, for an algorithm to be considered adaptive, it needs to learn the new preference quickly within a minimal number of errors.

6.2.2 Binary Learning Algorithms

The weighted majority algorithm (WMA) [7] and the Winnow algorithm [8, 9] are the two basic expert-based learning algorithms that can be used to make selections between two control laws. The WMA using two experts is shown in Algorithm 1. Each expert takes in the parameters of the system and uses them to make their selection. Each expert is assigned a weight that is adjusted by the algorithm. The weights of the experts are then compared, and the expert with the highest weight is chosen. More specifically, the algorithm computes two positive weights $\tilde{W}_1$ and $\tilde{W}_2$ that gauge the likelihood of which choice will be correct in each iteration. A selection is made based on comparing the two weights. If $\tilde{W}_1 > \tilde{W}_2$, then the algorithm chooses expert 1, and if $\tilde{W}_1 < \tilde{W}_2$, then it chooses expert 2. If $\tilde{W}_1 = \tilde{W}_2$, a selection is made randomly with equal probability. If there is an error in the prediction, then the weight corresponded to the expert is reduced by half. The Winnow algorithm is very similar to the WMA but has a deviation in line 9 where a correct choice also results in the weight of the corresponding expert being increased by a factor of two.

Both WMA and Winnow have bounded error [7, 8]. And WMA is able to adapt easily to drift; however, it is not consistent, since one deviation is enough to cause it to switch its selection. The additional doubling of weights allows Winnow to be consistent but at the cost of its adaptiveness, since now a switch in selection can only be caused when there are an equal number of deviations as there are of the preferred output.

By extending these algorithms, we would like to design an algorithm that is more consistent than the WMA, and more adaptive than the Winnow. It has bounded error

Algorithm 6.1 Weighted Majority Algorithm (using two "experts" with opposite selections)

Set $\tilde{W}_1 = \tilde{W}_2 = 0.5$
Choose selection a from argmax $\left(\tilde{W}_1, \tilde{W}_2\right)$
if $\tilde{W}_1 = \tilde{W}_2$ **then**
  Choose a randomly from $\{1, 2\}$ with equal probability
end if
if Error **then**
  $\tilde{W}_a = \frac{W_a}{2}$
else if (Winnow) **then**
  $\tilde{W}_a = 2\tilde{W}_a$
end if

like both the WMA and Winnow while maintaining a balance between consistency and adaptiveness. The dual expert algorithm (DEA) is shown in Algorithm 2. The setup is similar to WMA with the increasing weights of Winnow; however, there is one key difference in line 12. When a correct selection is made and the weight is over 0.25, line 12 increases the weight by taking its square root, which keeps the value $W_a \leq 1$. This bound is the key difference that allows for the DEA to maintain a balance between consistency and adaptiveness.

Algorithm 6.2 Dual Expert Algorithm

1: Set $W_1 = W_2 = 0.5$
2: Choose selection a from argmax (W_1, W_2)
3: **if** $W_1 = W_2$ **then**
4:   Choose a randomly from $\{1, 2\}$ with equal probability
5: **end if**
6: **if** Error **then**
7:   $W_a = \frac{W_a}{2}$
8: **else if** Correct **then**
9:   **if** $W_a \leq 0.25$ **then**
10:     $W_a = 2W_a$
11:   **else**
12:     $W_a = \sqrt{W_a}$
13:   **end if**
14: **end if**

6.3 Analysis

Suppose the DEA algorithm has learned the preference of C. Let δ be the number of iterations when C disobeys its preferred option and makes a "deviation." Let the total number of iterations be N, and we require that $\delta < \frac{N}{2} - 1$.

6.3.1 Performance

If the concept makes a deviation from its preference, then it will trigger the learning algorithm to make a "wrong" selection based on its past knowledge. The following claim can be made about the error in selection.

Proposition 6.1 *The maximum number of selection errors that Algorithm 6.2 generates is* $1 + 2\delta$.

Proof Let W_p be the weight of the preferred option of the concept C, and let W_d be the weight of the deviation option. When $\frac{W_p}{W_d} > 1$ the preferred selection is chosen. When $\frac{W_p}{W_d} < 1$ the deviation is chosen. And when $\frac{W_p}{W_d} = 1$, which is the starting condition, a random choice is made.

When $\frac{W_p}{W_d} \geq 1$ a deviation can cause an error, leading to a total of δ possible errors. Each deviation can decrease the ratio $\frac{W_p}{W_d}$ by at most a factor of 2. Since the ratio starts at 1 the minimum that the ratio can ever be is $2^{-\delta}$. Thus to return this ratio back to 1 it must be increased by a factor of 2^{δ}. This requires there are δ times that the preferred selection is chosen by C while $\frac{W_p}{W_d} < 1$. Hence, the algorithm can make at most δ additional selection errors when $\frac{W_p}{W_d} < 1$.

Therefore after considering all deviations, $\frac{W_p}{W_d} \geq 1$. And since the deviation can be chosen when $\frac{W_p}{W_d} = 1$, but this error will then raise $\frac{W_p}{W_d} > 1$, there can be at most one additional error. Thus the total number of errors is less than $\delta + \delta + 1$, making the error bound of Algorithm 6.2 as $E = 1 + 2\delta$.

The error bound $1 + 2\delta$ is on the same order of magnitude as the weighted majority algorithm which has an error bound of $E = 2.4 + 2.4\delta$ [7].

6.3.2 Consistency

We show that the selection of the DEA is consistent to deviations made by C. This means that the number of switches between selections of the algorithm is kept small, by ignoring deviations. We say the DEA becomes more consistent as the number of iterations increases the ability of the algorithm to ignore deviations increases.

First we show that for the weighted majority algorithm, the two weights will always be within a factor of two from each other.

Proposition 6.2 *Let* $\tilde{W}_1$ *and* $\tilde{W}_2$ *be the two weights in the weighted majority algorithm, then* $\max\{\frac{\tilde{W}_1}{\tilde{W}_2}, \frac{\tilde{W}_2}{\tilde{W}_1}\} = 1$ *or* 2.

Proof The weighted majority algorithm starts with $\tilde{W}_1 = \tilde{W}_2$. So that initially $\max\{\frac{\tilde{W}_1}{\tilde{W}_2}, \frac{\tilde{W}_2}{\tilde{W}_1}\} = 1$. At any time, when $\frac{\tilde{W}_1}{\tilde{W}_2} = \frac{\tilde{W}_2}{\tilde{W}_1} = 1$ a random choice between the selections is made. And if $\frac{\tilde{W}_1}{\tilde{W}_2} > 1$ then selection 1 is chosen, likewise if $\frac{\tilde{W}_2}{\tilde{W}_1} > 1$ then selection 2 is chosen. If an error happens in the selection when $\frac{\tilde{W}_1}{\tilde{W}_2} = \frac{\tilde{W}_2}{\tilde{W}_1} = 1$, $\max\{\frac{\tilde{W}_1}{\tilde{W}_2}, \frac{\tilde{W}_2}{\tilde{W}_1}\}$ will be increased to 2. When $\max\{\frac{\tilde{W}_1}{\tilde{W}_2}, \frac{\tilde{W}_1}{\tilde{W}_2}\} = 2$, then if an error is made then the largest weight is reduced by half. The ratio $\frac{\tilde{W}_1}{\tilde{W}_2}$ will be reduced to 1.

Now due to this fact, if C makes a deviation, then the weighted majority algorithm will change the weights to be equal because the larger weight will be reduced by half. The maximum number of switchings an algorithm can make is 2δ where δ is the number of deviations C makes. No deviation is ignored.

Next we will argue that using the dual expert algorithm, the algorithm will make less switchings in its selection because it can ignore deviations made by C. Let us define $R = \max\{\frac{W_1}{W_2}, \frac{W_2}{W_1}\}$. First, we argue that in a number of situations, R will be increased to a value that is greater than 2.

Claim Starting from $W_1 = W_2 = 0.5$, if the DEA makes one error followed by one correct selection, then $R > 2$.

Proof The weight that is updated is always the larger weight, since a weight must be chosen to be modified, and a weight must be larger to be chosen. This means that when there is an error the maximum weight will decrease by a factor of two, in this case making $\min(W_1, W_2) = 0.25$. Then the $\max(W_1, W_2) = 0.5$. If the next selection is correct then according to Algorithm 6.2 $\max(W_1, W_2) = \sqrt{(0.5)} > 0.5$, which leads to $R = \sqrt{(0.5)}/0.25 > 2$.

Claim Starting from $\max\{W_1, W_2\} < 0.25$ and $R \le 2$, if DEA makes two consecutive correct selections, then $R > 2$.

Proof R is the maximum of two positive reciprocal values, thus $R \ge 1$. A single correct selection when $\max\{W_1, W_2\} \le 0.25$ increases the correct weight by a factor of 2, meaning that R is increased by a factor of 2. Thus $R \ge 2$, and $\max\{W_1, W_2\} < 0.5$. An additional correct selection will then increase the $\max\{W_1, W_2\}$, thus increasing R so that $R > 2$.

Claim Starting from $\max\{W_1, W_2\} < 0.5$ and $R \le 2$, if DEA makes one error followed by one correct selection, then $R > 2$.

Proof A single error would decrease the max weight by a factor of 2 so that $\min\{W_1, W_2\} \le 0.25$ and $1 < R \le 2$. If $\max\{W_1, W_2\} < 0.25$ then a single correct selection would increase the max weight by a factor of 2 so that $2 < R \le 4$. And if $\max\{W_1, W_2\} \ge 0.25$ then a single correct selection would increase the max weight such that $\max\{W_1, W_2\} \ge 0.5$. Then $R = \max\{W_1, W_2\}/\min\{W_1, W_2\} > 0.5/0.25 = 2$.

Claim By making consecutive correct selections, the value of R can be increased until $R = \frac{1}{\min\{W_1, W_2\}}$.

Proof Each correct selection only affects the maximum weight. And while $\max\{W_1, W_2\} < 0.25$, $\max\{W_1, W_2\}$ increases by a factor of 2. Once $\max\{W_1, W_2\} \geq 0.25$, $\max\{W_1, W_2\}$ is square rooted to form the new maximum. Thus $\max\{W_1, W_2\}$ converges to 1 as the number of consecutive correct selections increase. Since $R = \max\{W_1, W_2\}/\min\{W_1, W_2\}$, and the minimum weight is unchanged, R converges to $\frac{1}{\min\{W_1,W_2\}}$.

The number of deviations the dual expert algorithm can ignore depends on the actual ratio between the weights.

Claim If $R > 2^K$ where $K > 1$, then $(K - 1)$ consecutive deviations can be ignored.

Proof Each deviation reduces the ratio R by a factor of 2, because the larger weight is decreased by a factor of 2. Since $R > 2^K$, the larger weight can be decreased by 2 for a total of $K - 1$ times, and still be the larger weight, thus allowing the algorithm to ignore $K - 1$ deviations.

Therefore, if C makes deviations, the DEA will be more robust than the WMA. Since the weighted majority algorithm does not ignore deviations and the DEA does.

6.3.3 Adaptiveness

If we consider drift to be the change in preferred output of the concept C from p to $\tilde{p}$, then the adaptiveness of the DEA can be shown to be directly linked to the amount of deviations, which have been encountered before drift occurred.

Claim R is bounded above by a finite number.

Proof We have previously shown that the maximum of R is $R = \frac{1}{\min\{W_1,W_2\}}$.

Only an error can decrease weights in DEA and Proposition 6.1 showed that the number of error is bounded by the finite number of deviations. Thus $\min\{W_1, W_2\}$ is bounded below. Therefore, $R = \frac{1}{\min\{W_1,W_2\}}$ is bounded above by a finite number.

Claim For any given $2^{K-1} \leq R < 2^K$ after output $\tilde{p}$ is observed for K consecutive iterations, DEA will change from its selection p to the alternate selection $\tilde{p}$.

Proof Each time the output is $\tilde{p}$ and DEA selects p there is an error. This error decreases the maximum weight, and thus R is reduced by a factor of 2. Then after $K - 1$ consecutive outputs of $\tilde{p}$, $1 \leq R < 2$.

If $1 \leq R < 2$ then an addition output of $\tilde{p}$ will decrease the weight corresponding to p to be less than the weight corresponding to $\tilde{p}$. This will change the selection from p to $\tilde{p}$. Thus a maximum of K consecutive C outputs of $\tilde{p}$ are needed to change the selection of the DEA.

These two claims imply that the number of errors needed to change selections is limited by the number of deviations leading up to the change. This can be compared to Winnow algorithm where the 1 in $R = \frac{1}{\min\{W_1,W_2\}}$ is replaced by $\max\{W_1, W_2\}$, which can be very large. This allows Winnow to be consistent but not adaptive.

6.3.4 Tie Breaking

In DEA as well as WMA and Winnow, it can be easily seen that the algorithm cannot issue a prediction when the weights of the experts are tied. A tie will force the algorithm to randomly choosing an expert, which is not preferred. In the following, we show that the DEA has reduced number of ties compared to WMA or Winnow.

Proposition 6.3 *If a tie is created by DEA, then* $\delta = \frac{N}{2}$ *must be satisfied.*

Proof Define η as the number of predictions in the preferred choice such that $\eta + \delta = N$. Define r_η as the ratio between the weight of the expert making the preferred choice and the weight of the expert making the other choice. And define $r_\delta = 1/r_\eta$, so that $R = \max\{r_\eta, r_\delta\}$.

When $\max\{W_1, W_2\} < 0.25$ each deviation decreases r_η by a factor of two because each deviation will either decrease the weight of the expert making the preferred choice or increase the weight of the expert making the other choice. Likewise each correct prediction increases r_η by a factor of two. Therefore when $\eta = \delta$ the ratio between the maximum and minimum weights is identical.

To raise $\max\{W_1, W_2\} \geq 0.25$ a correct prediction is needed. When $\max\{W_1, W_2\} \geq 0.25$, each correct prediction square roots $\max\{W_1, W_2\}$, so that $0.5 \leq \max\{W_1, W_2\} < 1$ for any finite number of correct predictions. And because the weights are always positive only the positive square root result is considered, this means that for each number between 0.25 and 1 that is square rooted there will be a unique solution.

Because the maximum weight is the only weight that is ever changed, the minimum weight can only decrease. This means that if one weight goes below 0.25 then all potential ties can only take place when the other weight is below 0.25. Thus for a tie to occur each weight would have to be lowered below 0.25, each time the weight increases above 0.25 due to a correct prediction, an incorrect prediction would be needed to lower it again below 0.25. Thus $\eta = \delta$.

This means that for tied weights, $\eta = \delta$ while $\max\{W_1, W_2\} \geq 0.25$ and the ratio must remain the same when below 0.25 which implies $\eta = \delta$ while $\max\{W_1, W_2\} < 0.25$. Therefore $\eta = \delta$, which is equivalent to $\delta = \frac{N}{2}$, must be true for there to be a tie.

Note that even if $\delta = \frac{N}{2}$, there may not be a tie. For example starting at $W_1 = W_2 = 0.5$ a correct prediction on the preferred direction will raise $\max\{W_1, W_2\} = \sqrt{0.5}$, and then an incorrect prediction, caused by a single δ, would lower that weight to $\sqrt{0.5}/2 \approx 0.3536$, while leaving the other weight unchanged at 0.5. Since this example uses $\delta = 1$ and $N = 2$, and $0.5 \neq 0.3536$, this serves as an example to show that $\delta = \frac{N}{2}$ may not lead to a tie.

We note that the weighted majority algorithm will create a tie whenever one correct prediction is followed by an incorrect prediction, regardless of the relationship between δ and N. In addition the Winnow algorithm will create a tie whenever $\delta = \frac{N}{2}$. Therefore, we can say that DEA has a lower chance of generating a tie compared to the weighted majority algorithm and the Winnow algorithm.

6.4 Expanded Dual Expert Algorithm

The DEA uses two weights to select potential outcomes. However, an algorithm can encounter situations where the preference depends on a parameter. If the parameter is smaller than a threshold then output 1 is preferred, and if the parameter is larger than a threshold then output 2 is preferred. An example of this type of parameter would be the relative position that a human starts in a hallway. Starting along the left wall would imply that the human would tend to stay along the left wall to pass, and starting along the right wall would imply that the human would tend to stay along the right wall to pass. However somewhere between the walls the preference of the wall to pass along changes. We propose the extended dual expert algorithm (EDEA) to learn these preferences using weights that are function of the parameter.

The parameter is discretized into M values. We call each value of the parameters a section and each section is indexed by n. Let us assume that there are two edges $n = 1$ and $n = M$, and that output 1 is the preference of edge $n = 1$, while output 2 is the preference at edge $n = M$. And the two weights are now represented as two M-dimensional vectors W_1^i and W_2^i indexed by $i = 1, 2, \ldots, M$.

The EDEA algorithm in Algorithm 6.3 updates its weights based on whether a correct or an incorrect selection is made. Line 8 shows that when an incorrect selection is made the weights between n and the edge that prefers the observed output are decreased. And when a correct selection is made the weights between n and the edge that prefers the observed output are increased, as shown in line 10.

Algorithm 6.3 Expanded Dual Expert Algorithm

1: Set $W_1^i = W_2^i = 0.5\ \forall i = 1 : M$
2: The parameter is in section n
3: Choose selection a from $\text{argmax}(W_1^n, W_2^n)$
4: **if** $W_1^n = W_2^n$ **then**
5: Choose selection randomly
6: **end if**
7: **if** Error **then**
8: $W_a^i = {}^{W_a^i}/_2,\ \forall i = \begin{cases} n : M & \text{if } a = 1 \\ 1 : n & \text{if } a = 2 \end{cases}$
9: **else if** Correct **then**
10: $W_a^i = \begin{cases} 2W_a^i & \text{if } W_a^i < 0.25 \\ \sqrt{W_a^i} & \text{if } W_a^i \geq 0.25 \end{cases}\ \forall i = \begin{cases} 1 : n & \text{if } a = 1 \\ n : M & \text{if } a = 2 \end{cases}$
11: **end if**

6.4.1 Performance Analysis

In order to discuss the performance of the algorithm we will make the following additional assumptions. We assume that there is a section index m so if $n \leq m$, then the preferred output is 1, and if $n > m$, then the preferred output is 2. Thus consider that the preference is dependent on the section. Since "deviations" are defined as when the output varies from the concept preference, we must allow deviations to be dependent on sections.

Define δ_n as the number of deviations in section n. Thus the total number of deviations denoted as Δ is $\Delta = \sum_{n=1}^{M} \delta_n$. Under this assumption, the error bound of the EDEA is given below:

Proposition 6.4 *The upper bound on the total error for the EDEA algorithm is* $E = (M+1)\Delta + M$.

Proof There are two potential sources of errors in the EDEA algorithm. Consider section n. An error that occurs when a deviation occurs will decrease the ratio of $\frac{W_{\mathrm{p}}^n}{W_{\mathrm{d}}^n}$ by a factor of 2 in the section where the error took place. Each error that occurs without a deviation will increase the ratio of $\frac{W_{\mathrm{p}}^n}{W_{\mathrm{d}}^n}$ by a factor of 2 in the section where the error took place. Each error that occurs, regardless of source, can affect the ratio in a maximum of M sections and a minimum of one section.

The maximum number of errors that can be caused by deviations is the total number of deviations Δ. Since each error due to a deviation can effect a maximum of M sections in the worst case, the ratio of $\frac{W_{\mathrm{p}}^i}{W_{\mathrm{d}}^i}$ for $i = 1 \ldots M$ would decrease by a factor of 2^{Δ}. In order to increase this ratio to be greater than 1, the ratio in each section must have an increase $\Delta + 1$ times.

In the worst case, each of these increases would be caused by an error made without a deviation. And because the minimum number of sections that an error can affect is 1, the worst case is that the ratio in each section is brought to be greater than 1 individually. Since there are M sections this means that there are at most $M\Delta + M$ errors that occur without a deviation. Therefore the bound on the error is $E = (M+1)\Delta + M$.

This error bound can be tightened if the section m where the preference changes is known.

Proposition 6.5 *With a known m the error bound for the EDEA algorithm is* $E = (\max\{m, M-m\} + 1)\Delta + M$.

Proof The error bound given in the previous proof, $(M+1)\Delta + M$, assumes that one deviation can decrease the ratio of the preferred weight over the deviation weight in M sections. However the deviation is dependent on the section. So a deviation can only decrease the weight of the preferred output in $\max\{m, M-m\}$ sections. This means that the error bound of Algorithm 6.3 can be tightened to be $(\max\{m, M-m\} + 1)\Delta + M$.

6.4.2 Consistency and Adaptiveness

In EDEA, because weights within sections are updated in the same way as weights are updated in DEA, the ratio of weights in a section determines how many deviations in that section can be ignored.

The consistency also has a dependence related to parameter. Suppose that the same selection is preferred in sections w and s.

Proposition 6.6 *If* $|m - w| > |m - s|$ *then* $\frac{W_p^w}{W_d^w} \geq \frac{W_p^s}{W_d^s}$.

Proof From EDEA, if W_p^s is increased, then W_p^w is also increased. And if W_d^s is decreased then W_d^w is also decreased. But an increase of W_p^w does not guarantee an increase of W_p^s and a decrease of W_d^w does not guarantee a decrease of W_d^s. Thus W_p^w/W_d^w increases if W_p^s/W_d^s increases, but is not guaranteed to decrease if W_p^s/W_d^s decreases. Since $W_p/W_d = 1$ in all sections before the first iteration. It is true that $\frac{W_p^w}{W_d^w} \geq \frac{W_p^s}{W_d^s}$.

Because the increase of weights within each section is bounded similar to the bound in DEA, there will be the same bound on the number of errors needed to change selections. Proposition 6.6 showed that if $|m - w| > |m - s|$ then $\frac{W_p^w}{W_d^w} \geq \frac{W_p^s}{W_d^s}$. This also implies that adaptiveness has a dependence on the parameter, and sections near m will be more adaptive than sections near $n = 1$ or $n = M$.

6.5 Simulation

6.5.1 Dual Expert Algorithm

6.5.1.1 Setup

The learning of a concept C by the DEA, WMA, and Winnow algorithms was simulated. C was created to choose one of two outputs in each iteration of a trial. This output was randomly generated using a constant probability of choosing the deviation. This percentage deviation represents the difference between the concept and the best expert, and was used to simulate the effect of noise in the system. After 50% of the iterations in a trial, drift was added by changing which output was the deviation while keeping the probability of making a deviation constant.

6.5.1.2 Results

Figure 6.2 shows the average total rate of errors by DEA, WMA, and Winnow. DEA always kept the rate of errors smaller than the WMA. Before drift occurred Winnow

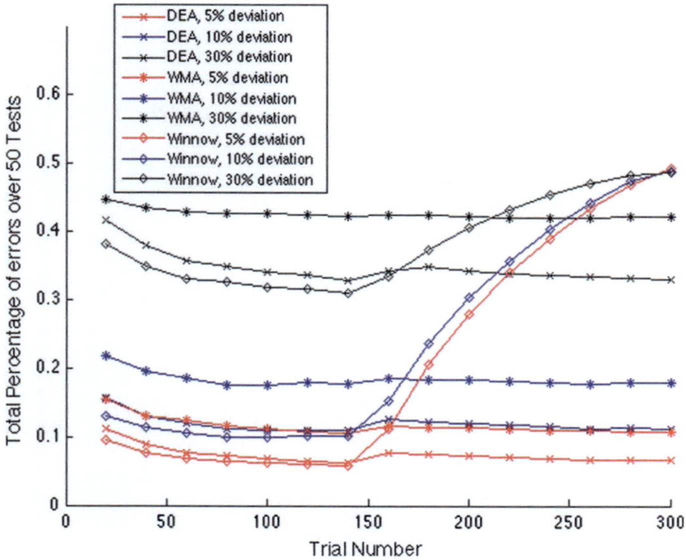

Fig. 6.2 Total percentage of errors, averaged over 50 Tests for dual expert algorithm, weighted majority algorithm, and the Winnow algorithm when the percentages of deviations are 5, 10, and 30%, and drift takes place at 150 trials (© 2016 IEEE. Reprinted, with permission, from [23])

had a slightly lower error rate than DEA; however, after drift occurred, Winnow's error rate increased dramatically while WMA and DEA's remained relatively consistent. This shows that DEA offers error rate comparable to Winnow, while maintaining a drift resistance comparable to WMA.

Figure 6.3 shows that the rate of switches for DEA is lower than WMA and close to Winnow, except for a spike that happens when drift occurs. In addition the number of switches in all cases for DEA and Winnow decreases towards 0, both before and after drift occurs, while the number of switches in the WMA remains relatively constant, as the number of trials increase.

This shows that DEA manages the balance between consistency and adaptiveness better than the two expert WMA and Winnow.

6.5.2 *Expanded Dual Expert Algorithm*

6.5.2.1 Setup

The learning of a concept C by five different algorithms was simulated. At the start of the trial the section m, where the preference changes, was selected. Each iteration could take place in one of 10 sections, randomly selected with equal probability. And the output of C was randomly chosen using the preference of the selected section and

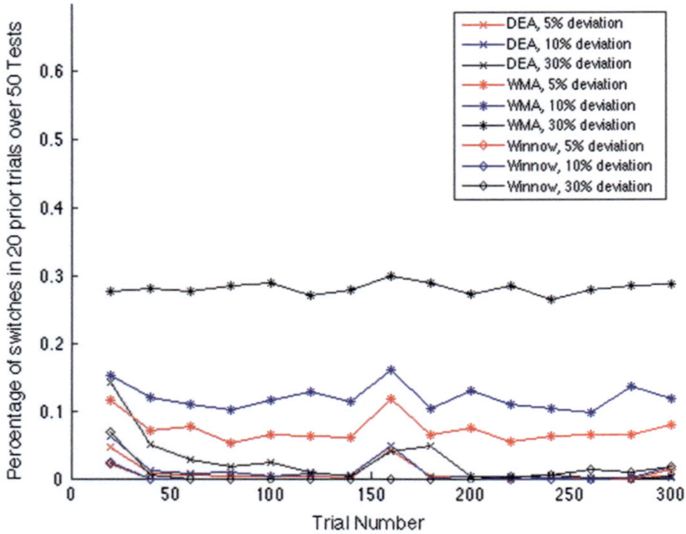

Fig. 6.3 Percentage of switches for each group of 20 trials, averaged over 50 tests for duel expert algorithm, weighted majority algorithm, and the Winnow algorithm when the percentage of deviations is 5, 10, and 30%, and drift takes place at 150 trials (© 2016 IEEE. Reprinted, with permission, from [23])

a constant probability of deviation. After half of the iterations drift was simulated by changing m.

Here is the list of algorithms used to learn concept C.

1. Expanded dual expert algorithm
2. The weighted majority algorithm
3. The Winnow algorithm
4. DEA applied to each section individually
5. DEA applied with no regards to sections
6. Multi-expert algorithm, which is, DEA modified to use more than two experts.

Both the Winnow and WMA are created according to their traditional multi-expert use [8, 9]. And the multi-expert algorithm uses experts in the same way as the WMA and Winnow; however, the update of weights is identical to the update used in the dual expert algorithm, in which it is limited by the weight of the selection that is made.

6.5.2.2 Results

Figure 6.4 shows the average total rate of errors. The inclusion of sections noticeably reduces the error rate for DEA. In addition before drift occurs EDEA can be seen to be decreasing quickly, approaching the lowest error rate algorithm, Winnow,

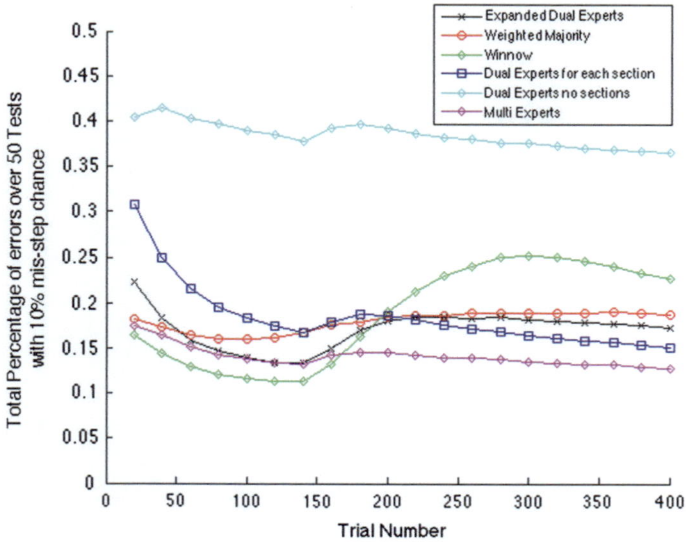

Fig. 6.4 Total percentage of incorrect selections, averaged over 50 tests where the percentage of deviations is 10% and drift occurs at 150 iterations, changing m from 7 to 3 (©2016 IEEE. Reprinted, with permission, from [23])

along with the multi-expert algorithm. While after drift occurs Winnow's error rate dramatically increases while WMA, multi-expert, and dual expert in each section stay relatively constant. EDEA's error rate also increases, but it still remains less than WMA. This shows that EDEA offers error rate approaching Winnow and the multi-expert algorithms, and drift resistance that keeps it preforming better than WMA.

Figure 6.5 shows the number of switches. EDEA and Winnow both have a small rate of switches that decrease towards 0 as time increases before drift. After the drift, the rate of EDEA switches spikes but then resumes descending towards 0. Winnow's switch rate also spikes to about the same height but this is after the same number of trials that were done predrift, the multi-expert case spikes at the same time as the EDEA, but to a switching rate approximately four times as large.

The continued good performance both before and after drift in the EDEA, and the limited spike in switching at the time of drift supports the fact that it manages the balance between consistency and adaptiveness well.

6.6 Experiment

To test the applicability of the EDEA in an embodied robot we implemented it onto a TurtleBot passing a human pedestrian in the hallway. The parameter was the distance the human was from the right wall. This would affect the concept, here representing

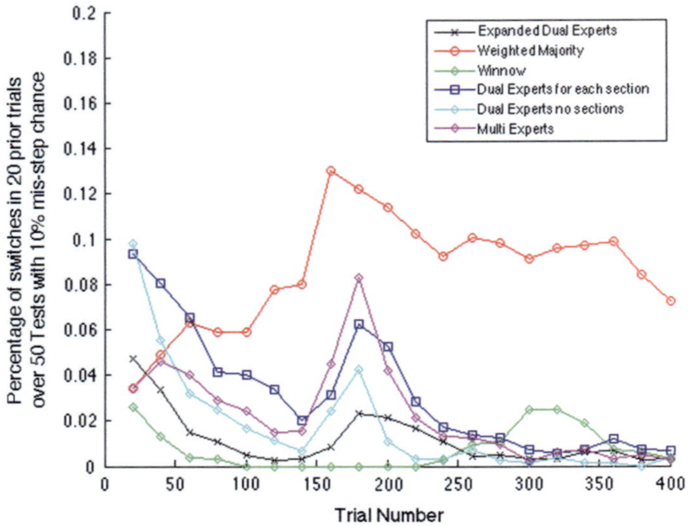

Fig. 6.5 Percentage of switches for each group of 20 trials, averaged over 50 tests where the percentage of deviations is 10% and drift occurs at 150 iterations, changing m from 7 to 3 (© 2016 IEEE. Reprinted, with permission, from [23])

the human's internal preference to deflect to either the right or left wall. The experts also used this parameter to predict which wall the human would deflect towards. It would then implement a control law that would allow the TurtleBot to deflect towards the wall opposite the one predicted.

The TurtleBot was chosen for this experiment since it satisfied the following conditions.

1. The physical dimension of the mobile base occupies a footprint that is similar to the footprint of an averaged human. Hence, the robot is able to pass by the bystander in a typical hallway setting.
2. The robot can move at a speed that is comparable to the averaged casual walking speed of a human.
3. The robot is equipped with a Microsoft Kinect sensor and necessary software that can identify an approaching object and its movements to the left or to the right.
4. The robot can be equipped with obstacle avoidance behaviors that can be modified to adjust the avoidance direction.

6.6.1 Setup

The experiments were performed using a single TurtleBot that started centered in the hallway. The TurtleBot continued down the center of the hallway until it detected a human approaching. It detected the human using a depth point cloud, looking for a

Table 6.1 Averaged error and switch rate of experiment and simulation

Test	Error rate (%)	Switch rate (%)
Experiment average	35.0	12.5
Simulation 5% error rate	21.8	9.8
Simulation 10% error rate	27.2	10.7
Simulation 30% error rate	42.2	12.1

roughly human-shaped blob approaching, and noted the relative position within the hallway of the human. Then the TurtleBot implemented an obstacle avoidance control law to avoid the human. The direction that the TurtleBot turned when implementing this avoidance was determined by the learning algorithm.

While the robot was moving towards the wall, it continued to track the human to see if its selection was correct. It determined this by tracking what side of the TurtleBot the human was on as he/she left the field of view of the sensor. If the human was between the TurtleBot and the wall it was deflecting towards, an error will be recorded, otherwise the trial was a success. The TurtleBot then used this information to update the Algorithm 6.3.

Some limitations were found during the implementation of the program. Most notably the fact that the noise from the depth-based human detection made it difficult to identify human error quickly if there is a simple jerky motion. It was however usually able to identify errors when the human continued moving towards their intended wall.

Each test was set up with 10 iterations. The human initial location was arbitrarily chosen to get a large coverage of initial positions within tests and different orders of iterations between tests. We placed a marker on the floor so that the human could determine if they started in one of the 6 lanes with a preference towards the left or one of the 4 lanes with a preference towards the right. The tests were repeated for 4 times (Fig. 6.6).

6.6.2 *Results*

The first row of Table 6.1 compiles the averaged result over all four tests. It includes the rate of errors and switches. The next three rows of the table presents simulation results for 5, 10, and 30% error rates, each were 10 trials long and averaged over 50 tests.

The experimental data in the first row was best matched by the simulation results with the 30% error rate in the fourth row. Note the average number of switching was only slightly higher than the 30% error bound, this shows even with the difficulties inherent in human perception capability of the robot, the EDEA could still have a bounded error and switch rate.

Fig. 6.6 TurtleBot approaching human starting on the left side of the preference dividing line

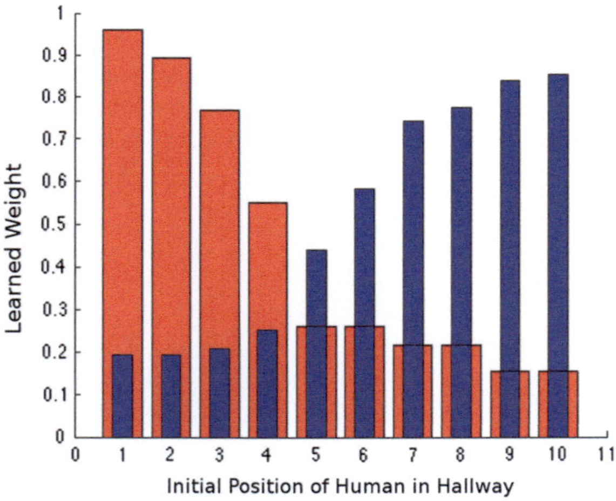

Fig. 6.7 Weights for selection of human turn direction. Hallway position is noted along the x-axis. The *red* bars represent the weight for the human passing along the *left wall* while the *blue* bars represent the weight for the human passing along the *right wall* (© 2016 IEEE. Reprinted, with permission, from [23])

Figure 6.7 shows the averaged weights computed by the robot. For each initial human position, the TurtleBot will turn in the direction of the larger weight. The increased difference in weights near the walls can be seen in Fig. 6.7. This increased difference requires more errors for the TurtleBot to change its output and is thus more resistant to noise. This increased noise resistance is the key to EDEA's consistency. In addition weights near the switch point were closer and thus less errors were required to change the TurtleBot's output there, which is key for the EDEA's adaptiveness.

Overall the robot did learn the human's preference which was consistent throughout the tests. The number of errors and switches is consistent with simulation, and the weight ratio displays the same spatial dependence as selected. This indicates that EDEA transfers well to physical implementation.

6.7 Conclusions

This paper has shown that a balance between adaptiveness and consistency can and should exist in robotic systems. And we show that the weight adjustment of the DEA and EDEA allow those algorithms to have a better balance between the two than the WMA and online Winnow. Increasing the applicability of these algorithms could be accomplished by expanding them to consider additional parameters. And as seen with the EDEA expansion, increasing the number of variables could also help with decreasing the noise of the system. These algorithms can also be more tailored to robotic systems by increasing the number of selections that the robot can chose from, more closely resembling a behavior-based robotic system.

Acknowledgements The research work is supported by ONR grants N00014-10-10712 (YIP) and N00014-14-1-0635; and NSF grants OCE-1032285, IIS-1319874, and CMMI-1436284 and an ARCS Scholar Award.

And the authors would like to thank Aida Khosroshahi of San Jose State University and Chris Paucar of Georgia Institute of Technology for their assistance in the implementation of these algorithms onto the TurtleBot test bed.

References

1. Huttenrauch H, Green A, Norman M, Oestreicher L, Eklundh KS (2004) Involving users in the design of a mobile office robot. IEEE Trans Syst Man Cybern Part C: Appl Rev 34(2):113–124
2. Mutlu B, Forlizzi J (2008) Robots in organizations: the role of workflow, social, and environmental factors in human-robot interaction. In: 3rd ACM/IEEE international conference on human-robot interaction (HRI). IEEE, pp 287–294
3. Breazeal C (2004) Social interactions in hri: the robot view. IEEE Trans Syst Man Cybern Part C: Appl Rev 34(2):181–186
4. Dautenhahn K, Woods S, Kaouri C, Walters ML, Koay KL, Werry I (2005) What is a robot companion-friend, assistant or butler? In: IEEE/RSJ international conference on intelligent robots and systems. IEEE, pp 1192–1197
5. Kuncheva L (2004) Classifier ensembles for changing environments. In: Roli F, Kittler J, Windeatt T (eds) Multiple classifier systems, ser. Lecture notes in computer science, vol 3077. Springer Berlin, Heidelberg, pp 1–15

6. Polikar R (2012) Ensemble learning. In: Zhang C, Ma Y (eds) Ensemble machine learning. Springer, US, pp 1–34
7. Littlestone N, Warmuth MK (1989) The weighted majority algorithm. In: (1989) 30th annual symposium on foundations of computer science, vol 1989. IEEE, pp 256–261
8. Littlestone N (1988) Learning quickly when irrelevant attributes abound: a new linear-threshold algorithm. Mach Learn 2(4):285–318
9. Blum A (1997) Empirical support for winnow and weighted-majority algorithms: results on a calendar scheduling domain. Mach Learn 26(1):5–23
10. Freund Y, Schapire RE (1997) A decision-theoretic generalization of on-line learning and an application to boosting. J Comput Syst Sci 55(1):119–139
11. Freund Y (2001) An adaptive version of the boost by majority algorithm. Mach Learn 43(3):293–318
12. Tsymbal A (2004) The problem of concept drift: definitions and related work. Computer Science Department, Trinity College Dublin 106:
13. Kolter JZ, Maloof MA (2007) Dynamic weighted majority: an ensemble method for drifting concepts. J Mach Learn Res 8:2755–2790
14. Crammer K, Mansour Y, Even-Dar E, Vaughan JW (2010) Regret minimization with concept drift. In: COLT, pp 168–180
15. Ross GJ, Adams NM, Tasoulis DK, Hand DJ (2012) Exponentially weighted moving average charts for detecting concept drift. Pattern Recognit Lett 33(2):191–198
16. Pacchierotti E, Christensen HI, Jensfelt P (2006) Design of an office-guide robot for social interaction studies. In: IEEE/RSJ international conference on intelligent robots and systems. IEEE, pp 4965–4970
17. Pacchierotti E, Christensen HI, Jensfelt P (2006) Evaluation of passing distance for social robots. In: The 15th IEEE international symposium on robot and human interactive communication. IEEE, pp 315–320
18. Lauckner M, Kobiela F, Manzey D (2014) Hey robot, please step back!-exploration of a spatial threshold of comfort for human-mechanoid spatial interaction in a hallway scenario. In: The 23rd IEEE international symposium on robot and human interactive communication. IEEE, pp 780–787
19. Severinson-Eklundh K, Green A, Hüttenrauch H (2003) Social and collaborative aspects of interaction with a service robot. Robot Auton Syst 42(3):223–234
20. Chung S-Y, Huang H-P (2010) A mobile robot that understands pedestrian spatial behaviors. In: IEEE/RSJ international conference on intelligent robots and systems (IROS), pp 5861–5866
21. Chung S-Y, Huang H-P (2010) Predictive navigation by understanding human motion patterns. Int J Adv Robot Syst 8(1), 52–64
22. Sisbot EA, Marin-Urias LF, Alami R, Simeon T (2007) A human aware mobile robot motion planner. IEEE Trans Robot 23(5):874–883
23. Young C, Zhang F (2016) A learning algorithm to select consistent reactions to human movements. In: Proceedings of 2016 American control conferences, pp 6152–6157

Chapter 7
Assistive Optimal Control-on-Request with Application in Standing Balance Therapy and Reinforcement

Anastasia Mavrommati, Alex Ansari and Todd D. Murphey

7.1 Introduction

This chapter develops and explores the benefits of a specific form of control—referred to as *burst control*—to human–robot collaboration settings. Burst control laws consist of a continuous nominal control mode (NC) that switches to alternate, assistive control modes (AC) lasting for short but finite durations (see Fig. 7.1). One of the simplest way robots can collaborate with humans is to aid in the completion of physical tasks. More often than not, this assistive setting requires that control authority is shared between the collaborators. Where traditional shared control strategies work in conjunction with or filter user input commands (see [8, 10, 20, 37, 54]), burst control strategies can provide for a different form of shared control following a *control-on-request* (COR) paradigm. Potentially working along with existing shared controllers, burst control signals can be applied on request to implement quick changes, assisting with control goals as human tasks become too challenging to maintain. Due to their short duration, the key advantage of these signals is that they allow computers to take control authority and implement changes before users notice the loss in control. Through interactive simulation studies, we demonstrate how this COR paradigm can be implemented using simple interface components (e.g. buttons, toggles, or

A. Mavrommati · A. Ansari · T.D. Murphey (✉)
Department of Mechanical Engineering, Northwestern University,
2145 Sheridan Road, Evanston, IL 60208, USA
e-mail: t-murphey@northwestern.edu

A. Mavrommati
e-mail: stacymav@u.northwestern.edu

A. Ansari
e-mail: alexanderansari2011@u.northwestern.edu

Y. Wang and F. Zhang (eds.), *Trends in Control and Decision-Making for Human–Robot Collaboration Systems*, DOI 10.1007/978-3-319-40533-9_7

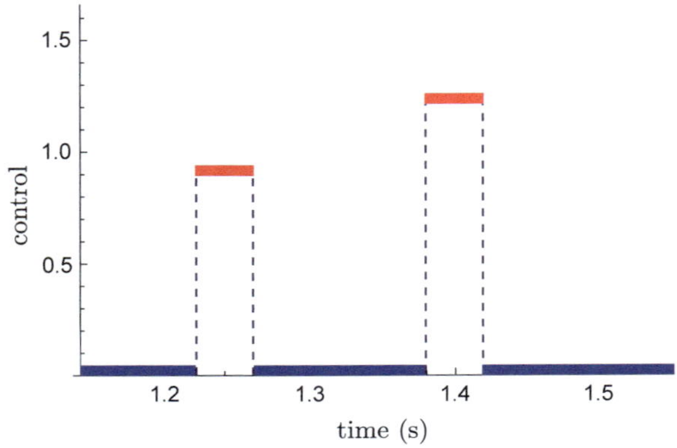

Fig. 7.1 An example burst control law. Continuous NC mode $u(t) = 0$ (*blue curve*) switches at $t = 1.22$ s and $t = 1.38$ s to alternate control modes (*red*) that are applied for short time horizons (0.04 s) (© 2015 IEEE. Reprinted, with permission, from [3])

switches), providing users the ability to collaborate with robotic systems by quickly requesting bursts of assistance.

The challenge in implementing these on-demand, quick-assist controllers is that optimal control responses for nonlinear systems require constrained iterative optimization and so are infeasible to compute on-the-fly. To address this concern, the first part of this chapter provides the theoretical foundation that allows for rapid generation of optimal burst control laws. First, the process computes the control action[1] to apply at any time to which a system's tracking objective is optimally sensitive. As Sect. 7.2.1 proves, this action is guaranteed to exist under standard assumptions. Planning for infinitesimal activation durations, we show optimal actions can be computed extremely rapidly from a simple algebraic expression of state and co-state (no iterative optimization). Next, Sect. 7.2.2 provides a simple line search that is guaranteed to find a finite application duration for any selected activation time, τ_m, that will provide a desired improvement in trajectory. By quickly computing optimal actions and resolving their finite application durations, the approach calculates assistive burst controls on-demand.

The remainder of this chapter explores the potential of the proposed COR interface in enhancing human–robot collaboration in the field of rehabilitation. We present two studies, each of which aims to address different aspects of collaboration between humans and robotic platforms. Our first study focuses on the element of *human–robot interaction* and regards rehabilitative devices as assistant robots. In particular, we demonstrate that COR interfaces embedded in robotic lower-limb exoskeletons can benefit standing balance therapy by promoting real-time collaboration between the therapist and the robotic device. To illustrate our approach, consider the case where a standing patient attempts to maintain an upright posture while wearing a robotic lower-limb exoskeleton. In this scenario, a monitoring physical therapist equipped with a COR interface has the option to either assist the patient manually or activate a corrective exoskeleton response on-demand—hence actively collaborating

[1]In this chapter a control action is a control vector with constant values applied for a (possibly infinitesimal) duration of time.

with the robot to improve and facilitate therapy. In the event of COR activation, a burst of optimal assistance is calculated in real time and applied for a very short duration, so that immediately after, full control authority is ceded back to the patient/therapist. Using the Robot Operating System (ROS), Sect. 7.3 shows that this therapist-robot collaboration paradigm can *ensure patient safety* while promoting *active patient participation*.

However, there is one central aspect of collaboration that is not addressed in this first study, and that is *human–robot communication*. Past results suggest that a robot needs to be able to produce nonverbal communication cues to be an effective collaborative partner [19]. Diverging from commonly employed communication channels, our second—short—study proposes that the robot uses vibrotactile feedback to communicate COR-generated intents. As opposed to the first study where the therapist is the epicenter, Sect. 7.4 examines how the patient can actively collaborate with the robot in achieving a desired upright posture, by interpreting and following tactile cues.

Following this Introduction, Sect. 7.2 derives a technique that takes advantage of the structure of burst control signals to compute an algebraic optimal control solution for systems with nontrivial dynamics. Section 7.3 discusses the application of COR interfaces to robot-assisted balance therapy and provides promising simulation results. Finally, Sect. 7.4 explores the potential of employing COR-generated vibrotactile signals to achieve posture reinforcement.

7.2 Assistive Control Synthesis

This section presents a method to compute burst AC laws that attain specified levels of improvement in trajectory tracking cost functionals. Section 7.2.1 leverages the short intended switching duration of burst AC signals to avoid dynamic constraints and iterative optimization in solving for a schedule of infinitesimal optimal actions to apply at any time. These infinitesimal actions specify the magnitude and direction of the burst control. The following Sect. 7.2.2 provides a line search process that returns a finite duration to apply each action to produce a burst control law that achieves a desired reduction in tracking cost over a specified horizon.

7.2.1 Calculating a Schedule of Optimal Infinitesimal Actions

The type of systems addressed in this chapter are assumed to follow continuous trajectories, $(x(t), u(t))$, such that

$$\dot{x}(t) = f(x(t), u(t))\,. \tag{7.1}$$

The dynamics vector, $f(x(t), u(t))$, can be nonlinear with respect to the state vector, $x(t) \in \mathbb{R}^n$, but is assumed to be linear (or has been linearized) with respect to control vector, $u(t) \in \mathbb{R}^m$. With these assumptions, (7.1) can be written in control-affine form

$$f(x(t), u(t)) = g(x(t)) + h(x(t))\, u(t)\,. \tag{7.2}$$

The goal of the proposed COR controllers is to predict the trajectory of the system and compute an optimal AC that produces a specified level of improvement in this trajectory when activated. To accomplish this, a cost functional is used to compare the performance of different trajectories. The state tracking cost functional, [2]

$$\begin{aligned} J_1 &= \int_{t_0}^{t_f} l_1(x(t))\, dt + m(x(t_f)) \\ &= \frac{1}{2}\int_{t_0}^{t_f} \|x(t) - x_d(t)\|_Q^2\, dt + \frac{1}{2}\,\|x(t_f) - x_d(t_f)\|_{P_1}^2\,, \end{aligned} \tag{7.3}$$

serves this purpose. For control derivations, J_1 need only obey general form (7.3). However, the quadratic form is applied in the implementation described in this chapter, i.e., $l_1(x(t)) := \frac{1}{2}\|x(t) - x_d(t)\|_Q^2$ and $m(x(t_f)) := \frac{1}{2}\,\|x(t_f) - x_d(t_f)\|_{P_1}^2$. For this case, $Q = Q^T \geq 0$ defines the metric on incremental state tracking error and $P_1 = P_1^T \geq 0$ defines the metric on state error at terminal time t_f. Initial time, t_0, is assumed to be the activation time of the COR interface to select controls that improve the future system trajectory.

Assume the system can be in one of two dynamic modes at a given time that differ only in control. Under nominal conditions, the system applies NC law, $u(t)$, and dynamics, $f_1 = f(x(t), u(t))$, result. Upon activating the COR interface, the dynamics switch from NC mode f_1 to an alternate AC dynamic mode $f_2 = f(x(t), u_2(t))$ for a short duration, λ^+, before switching back. Consider the extreme case where the system evolves according to f_1, and AC $u_2(t)$ is applied for an infinitesimal duration before switching back to the NC law, $u(t)$. In this case the mode insertion gradient,

$$\frac{dJ_1}{d\lambda^+} = \rho(t)^T[f_2 - f_1]\,, \tag{7.4}$$

measures the resulting change in cost functional (7.3) (see [13, 14, 48]). The term, $\rho(t) \in \mathbb{R}^n$, is the adjoint (co-state) variable calculated from the current system trajectory based on the differential equation,[3]

$$\dot{\rho}(t) = -D_x l_1(x(t))^T - D_x f_1^T \rho(t)\,, \tag{7.5}$$

---

[2]The notation, $\|\cdot\|_M^2$, indicates a norm on the argument where matrix, M, provides the metric (i.e., $\|x(t)\|_Q^2 = x(t)^T\, Q\, x(t)$).

[3]$D_x f(\cdot)$ denotes the partial derivative $\frac{\partial f(\cdot)}{\partial x}$.

such that at the final time $\rho(t_f) = D_x m(x(t_f))^T$. The control duration, λ^+, is evaluated infinitesimally, as $\lambda^+ \to 0$.

Through the course of the trajectory, any time, τ_m, the mode insertion gradient is negative, the state tracking cost functional can be reduced if $u_2(\tau_m)$ is activated for some duration at that time. The magnitude of the mode insertion gradient provides a first-order model of the change in trajectory cost that would result relative to the duration of application of this control.

To produce a desired degree of improvement in a system trajectory, each applied burst of AC needs to improve cost (7.3) by a specified amount. In other words, control actions, $u_2^*(\tau_m)$, need to be computed that drive the mode insertion gradient to a desired negative value, $\alpha_d \in \mathbb{R}^-$. However, there is generally a cost associated with the application of control authority. As such, a trade-off must be made in tracking the desired value of the mode insertion gradient, α_d, relative to control effort. Following a trajectory optimization approach, these competing goals can be encoded into the cost functional,

$$J_2 = \int_{t_0}^{t_f} l_2(x(t), u(t), u_2(t), \rho(t))\, dt \qquad \text{where}$$

$$l_2(\cdot) := \frac{1}{2}\left[\frac{dJ_1}{d\lambda^+} - \alpha_d\right]^2 + \|u_2(t)\|_R^2 := \frac{1}{2}[\rho(t)^T(f_2 - f_1) - \alpha_d]^2 + \|u_2(t)\|_R^2 . \tag{7.6}$$

Matrix R allows the designer to encode the cost of control relative to tracking α_d. The continuous schedule[4] of control actions, $u_2^*(t)$, that minimizes (7.6), optimizes this trade-off.

The remainder of the chapter assumes quadratic norms in (7.6) with $R = R^T > 0$. Proved in [5], quadratic functionals and the space of positive semi-definite / definite cones is convex. Because convexity is preserved for nonnegative weighted sums and integration, this choice of $R > 0$ provides convexity of (7.6). Additionally, NC is often (indicated by case) assumed to be null, $u(t) = 0$, over the control planning horizon. This choice allows easier interpretation of $u_2^*(\tau_m)$ as the optimal action at τ_m relative to doing nothing (allowing the system to drift for a horizon into the future).

Theorem 7.1 *Define* $\Lambda \triangleq h(x(t))^T \rho(t) \rho(t)^T h(x(t))$. *The schedule of controls,* $u_2(t)$, *that optimizes* (7.6) *based on dynamics* (7.2) *and state tracking cost functional* (7.3) *is given by the algebraic expression,*

$$u_2^*(t) = (\Lambda + R^T)^{-1} [\Lambda\, u(t) + h(x(t))^T \rho(t)\, \alpha_d] . \tag{7.7}$$

Proof See [3] for a detailed proof derivation.

---

[4] At any specified application time, τ_m, of the COR interface, $u_2^*(\tau_m)$ represents the optimal action that balances control authority and drives the mode insertion gradient to α_d if activated around that time. Thus, $u_2^*(t)$ is a schedule of optimal actions that can be switched to from NC, $u(t)$, to produce a desired change in mode insertion gradient at τ_m.

Though the principles applied to derive the schedule of infinitesimal optimal control actions (7.7) are reasonable, they are also nontraditional. To provide intuition regarding the properties of these solutions, the following proposition proves that the optimization posed to derive these controls is equivalent to a well-studied class of Tikhonov regularization problems (see [7, 9, 18, 21]).

Proposition 7.1 *Assume u, u_2, ρ, $h \in \mathcal{H}$ where $\mathcal{H}$ is an infinite dimensional reproducing kernel Hilbert function space (RKHS).*[5] *With appropriate change of variables, minimization of* (7.6) *obeys the structure of generalized continuous-time linear Tikhonov regularization problem*[6]

$$\min_{z} \|\Gamma z - b\|^2 + \kappa^2 \|L(z - z_0)\|^2 , \tag{7.8}$$

and (7.7) *obeys the structure of associated solution*

$$z^* = (\Gamma^T \Gamma + \kappa^2 L^T L)^{-1}(\Gamma^T b + \kappa^2 L^T L z_0) . \tag{7.9}$$

Above, Γ and L are bounded linear operators on $\mathcal{H}$, vectors z and $z_0 \in \mathcal{H}$, and b and $\kappa \in \mathbb{R}$. See [7, 9, 18, 21] for more detail on (7.8) *and* (7.9).

Proof See [3] for a detailed proof derivation.

As the following corollary indicates, because minimization of (7.6) can be posed as a Tikhonov regularization problem, solutions (7.7) inherit useful properties that regularization solutions obey.

Corollary 7.1 *With the assumptions stated in Proposition 7.1, solutions* (7.7) *for minimization of* (7.6) *exist and are unique.*

Proof The proof follows from Proposition 7.1, which shows the minimization can be formulated as a Tikhonov regularization problem with convex $\mathcal{L}_2$ error norm, $\|\Gamma z - b\|$. These problems are guaranteed to have solutions that exist and are unique. A proof is provided in [9].

Globally, optimal control actions (7.7) inherit properties of Tikhonov regularization solutions. However, the following corollary shows that near equilibrium points, solutions (7.7) simplify to linear state feedback laws.

Corollary 7.2 *Assume system* (7.2) *contains equilibrium point $x = 0$, the state and co-state are continuous, and tracking cost* (7.3) *is quadratic.*[7] *There exists a neighborhood around the equilibrium and nonzero time horizon for which optimal actions* (7.7) *are equivalent to linear feedback regulators.*

---

[5]Practically, this is not a very stringent requirement. Most spaces of interest are RKHS. Refer to [7] for more detail.

[6]For equivalence, $\|\cdot\|$ refers to the $\mathcal{L}_2$ norm and $\mathcal{H}$ is additionally assumed to be an $\mathcal{L}_2$ space.

[7]A quadratic cost is assumed so that resulting equations emphasize the local similarity between burst control and LQR [2].

Proof At final time, $\rho(t_f) = P_1 x(t_f)$. Due to continuity, this linear relationship between the state and co-state must exist for a nonzero neighborhood around t_f such that

$$\rho(t) = P(t)\, x(t)\,. \tag{7.10}$$

Applying this relationship, (7.7) can formulated as

$$\begin{aligned} u_2^*(t) = &(h(x(t))^T P(t)\, x(t)\, x(t)^T P(t)^T h(x(t)) + R^T)^{-1} \\ &[h(x(t))^T P(t)\, x(t)\, x(t)^T P(t)^T h(x(t))\, u(t) \\ &+ h(x(t))^T P(t)\, x(t)\, \alpha_d]\,. \end{aligned}$$

This expression contains terms quadratic in $x(t)$. In the neighborhood of the equilibrium these quadratic terms go to zero faster than the linear terms, and controls converge to

$$u_2^*(t) = R^{-T} h(x(t))^T\; P(t)\, x(t)\, \alpha_d\,. \tag{7.11}$$

By continuity, in a sufficiently small neighborhood of the equilibrium the system dynamics can be approximated as LTV system, $\dot{x}(t) = A(t)\, x(t) + B(t)\, u(t)$, (where $A(t)$ and $B(t)$ are linearizations about the equilibrium). Applying this assumption and differentiating (7.10) produces

$$\begin{aligned} \dot{\rho}(t) &= \dot{P}(t)\, x(t) + P(t)\, \dot{x}(t) \\ &= \dot{P}(t)\, x(t) + P(t)\, (A(t)\, x(t) + B(t)\, u(t))\,. \end{aligned}$$

Inserting relation (7.5) yields

$$\begin{aligned} -D_x l_1(\cdot)^T - A(t)^T P(t)\, x(t) = \dot{P}(t)\, x(t) \;+\; P(t) \\ (A(t)\, x(t) + B(t)\, u(t))\,, \end{aligned}$$

which can be rearranged such that

$$0 = (Q + \dot{P}(t) + A(t)^T P(t) + \; P(t)A(t))\, x(t) + P(t)\, B(t)\, u(t).$$

When nominal control $u(t) = 0$, this reduces to

$$0 = Q + A(t)^T P(t) + P(t)\, A(t) + \dot{P}(t). \tag{7.12}$$

Note the similarity to a Lyapunov equation. As mentioned, this relationship must exist for some nonzero neighborhood of t_f. Therefore, by continuity of $\rho(t)$, there must exist a finite time horizon and neighborhood of the equilibrium where (7.7) simplifies to linear feedback regulator (7.11) and $P(t)$ can be computed from (7.12) subject to $P(t_f) = P_1$ (Fig. 7.2).

Algorithm 7.1 Burst Control Synthesis

Initialize α_d, minimum change in cost ΔJ_{min} from (1.13), current time t_{curr}, default control duration Δt_{init}, scale factor $\beta \in (0,1)$, time horizon T, and the max line search iterations i_{max}.
if COR interface activated **then**
  $(t_0, t_f) = (t_{curr}, t_{curr} + T)$
  Simulate $(x(t), \rho(t))$ for $t \in [t_0, t_f]$ from f_1
  Compute initial cost $J_{1,init}$
  Specify α_d
  Compute $u_2^*(t)$ from $(x(t), \rho(t))$ using Theorem 1
  $\tau_m = \frac{\Delta t_{init}}{2} + t_0$
  Initialize $i = 0$, $J_{1,new} = \infty$
  while $J_{1,new} - J_{1,init} > \Delta J_{min}$ **and** $i \leq i_{max}$ **do**
    $\lambda^+ = \beta^i \times \Delta t_{init}$
    $(\tau_0, \tau_f) = (\tau_m - \frac{\lambda^+}{2}, \tau_m + \frac{\lambda^+}{2})$
    Re-simulate $x(t)$ applying f_2 for $t \in [\tau_0, \tau_f]$
    Compute new cost $J_{1,new}$
    $i = i + 1$
  end while
end if
return $(u_2^*(\tau_m), \tau_0, \tau_f)$

Fig. 7.2 The algorithm above includes an optional line search phase which re-simulates the state $x(t)$ and trajectory cost until an appropriate duration for AC application of $u_2^*(\tau_m)$ can be found. This is only required if the approximation provided by the mode insertion gradient is not accurate over the default control interval chosen

7.2.2 Computing the Control Duration

Theorem 7.1 provides a means to compute a schedule of open-loop optimal AC actions, $u_2^*(t)$. When implemented infinitesimally around any time, τ_m, $u_2^*(\tau_m)$ is the needle variation (see [55]) in $u(\tau_m)$ that optimizes control authority in driving the mode insertion gradient, $\frac{dJ_1}{d\lambda^+}$, to α_d at that time. This value of the mode insertion gradient reflects the achievable sensitivity of cost (7.3) to application of $u_2^*(\tau_m)$ for infinitesimal duration. However, by continuity of the adjoint and mode insertion gradient as $\lambda^+ \to 0$, there exists an open, nonzero neighborhood, V, around $\lambda^+ \to 0$ for which the mode insertion gradient models this sensitivity to first order (see [6, 14]). Hence the mode insertion gradient can be used to model the change in cost achievable by application of $u_2^*(\tau_m)$ over a finite duration $\lambda^+ \in V$ as

$$\Delta J_1 \approx \left.\frac{dJ_1}{d\lambda^+}\right|_{\lambda^+ \to 0} \lambda^+. \tag{7.13}$$

As $u_2^*(\tau_m)$ regulates $\frac{dJ_1}{d\lambda^+} \approx \alpha_d$, (7.13) becomes $\Delta J_1 \approx \alpha_d \lambda^+$. Thus the choice of λ^+ and α_d allow the control designer to specify the desired degree of change in (7.3) provided by each $u_2^*(\tau_m)$. Also note that for a given $\frac{dJ_1}{d\lambda^+}$ and any choice of λ^+, (7.13) can be applied to test if $\lambda^+ \in V$ or that it at least provides a ΔJ_1 that is known to be achievable for $\lambda^+ \in V$.

Though the neighborhood where (7.13) provides a reasonable approximation varies depending on system, in practice it is fairly straightforward to select a $\lambda^+ \in V$. The easiest approach is to select a single conservative estimate for λ^+. This is analogous to choosing a small, fixed time step in finite differencing or Euler integration. However, to avoid a-priori selection of a $\lambda^+ \in V$ and unnecessarily conservative step sizes, we use a line search with a descent condition to select a $\lambda^+ \in V$ or one that provides a minimum change in cost (7.3). Described in [53], the line search iteratively reduces the duration from a default value. By continuity, the process is guaranteed to find a duration that produces a change in cost within tolerance of that predicted from (7.13). In implementation, we use an iteration limit to bind the algorithm in time. Note that failing to find an acceptable λ^+ within the iteration limit is not usually a concern because (7.7) provides a schedule of control values so a duration can be sought for any nearby time if the system is overly sensitive at the current time.

For a more thorough discussion on the results, see [3]. Next, we perform real-time simulation studies using ROS to show how Algorithm 7.1 can benefit human–robot collaboration in a rehabilitation setting.

7.3 Human–Robot Interaction in Assisted Balance Therapy

In the previous section, we provided the theoretical foundation for the calculation of optimal burst control signals (i.e., their magnitude and direction u_2^*, and their duration λ^+) when assistance is requested. Here, we consider the application of COR interfaces that use these signals in the field of rehabilitation. Stroke and spinal cord injuries (SCI) are the main causes of motor disability and severely affect the ability to control everyday tasks, such as standing, walking, reaching, etc. [17, 36, 40]. During the past decades, there has been a substantial increase in the range, type, and number of robotic trainers for rehabilitation of motor impairments that follow stroke and SCI. Most of the robot-driven trainers currently in use, serve to augment the therapeutic tools available to the physical therapists and physicians, by guiding the patient's limbs through movement [4, 24, 47, 49]. However, the amount of assistance provided by the robot can have a large effect on the patient's recovery. To facilitate motor-neuron recovery, it is essential that patients actively participate in the training task [23]. It is for this reason that assist-as-needed algorithmic paradigms have shown promise for robotic therapy, inherently promoting patient participation by limiting assistance to the minimum the patient requires.

In the particular case of stroke rehabilitation, one of the main objectives is to improve posture stability [50]. In this section, we use the COR interface as an assist-as-requested controller that allows for an alternative robot-assisted training scheme where collaboration between the therapist and the robot is encouraged. In particular, therapist's guidance is integrated in real time with the assistive control, hence ensuring that therapy objectives are attained all while the therapist's work load and interference remains low. See Sect. 7.3.1 for an analysis of how our approach differs from existing assist-as-needed techniques.

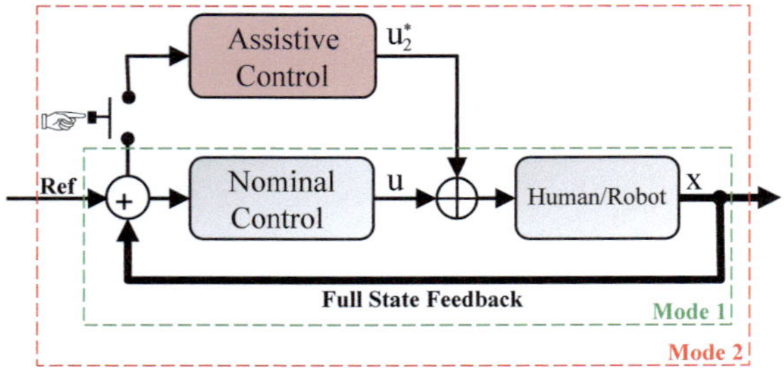

Fig. 7.3 Schematic representation of the COR interface used in balance therapy, consisting of the AC and NC (© 2015 IEEE. Reprinted, with permission, from [33])

Section 7.3.2 tests the real-time performance of a COR interface acting on a simulated standing figure, using the Robot Operating System (ROS), a standard platform for real robotic applications [39]. COR is shown to be suited for implementation on an embedded platform. In particular, our example verifies that real-time COR activation is feasible and requires only modest computational resources. In addition, it demonstrates how COR interfaces are effective even in the absence of force/pressure sensors which are usually required for assist-as-needed techniques [4, 25]. Note that a modified version of the optimal burst control in (7.7) is employed here, where AC is set to act in conjunction with (and not instead of) NC when activated (as shown in Fig. 7.3). For a detailed description of this derivation, see [33].

7.3.1 *Related Work: Assist-as-Needed Techniques*

Most existing assist-as-needed techniques in rehabilitation therapy provide performance based assistance, i.e., use task performance metrics as thresholds to initiate robot assistance. Krebs et al. [26, 27] propose an algorithm that uses performance-triggered impedance control to guide the patient through task completion. Speed, time, or EMG thresholds have been tested as performance metrics. Impedance control, systematized by Hogan [22] in 1985, is particularly effective when applied to contact tasks. However, this approach usually requires force feedback and also constitutes a sustained control strategy where the robot is continuously applying support (after the specified threshold has been exceeded). Banala et al. [4] follow a similar approach where the tangential forces of the controller are reduced as the patient's performance improves.

Wolbrecht et al. in [52] establish an assist-as-needed paradigm using a standard model-based adaptive controller and adding a novel force reducing term to the adaptive control law, which decays the force output from the robot when errors in task execution are small. The performance metric used in this technique is the kinematic error between the hand and goal location during reaching tasks.

Emken et al. in [16] formulate the "assist-as-needed" principle as an optimization problem. In particular, the robotic movement trainer is assumed to minimize a cost function that is the weighted sum of robot force and patient movement error. The authors find the controller that minimizes this cost function for the case in which motor recovery is modeled as a process of learning a novel sensory motor transformation. Here, step height achieved during walking is used as performance criterion. Although the approach has the advantage of being optimal, the resulting error-base controller is precalculated based on a fixed training model which does not allow for real-time model updates. For that, and because the controller is constantly active, possible inaccuracies of the selected model can have a negative impact on the long-term controller's performance as far as active patient participation is concerned.

Compared to the aforementioned control approaches that need formal performance metrics to be activated and remain active for a long time period, we follow an alternative approach that promotes real-time collaboration between the therapist and the robotic trainer. We can allow a therapist in the loop because the triggered optimal actions are only applied for a short duration of time and are calculated to assist the system as best as possible in the limited time frame available. This concept of shared control through intermittent bursts of optimal support comprises a unique assist-as-needed approach in that it completely avoids the issues of sustained control strategies where the interaction between two or more controllers acting on the same system simultaneously proves problematic.[8] An interactive simulation study provides more information about COR interfaces in the context of balance training in the following section.

7.3.2 Interactive Simulation Study

This section assesses the performance of the COR interface (shown in Fig. 7.3) in shared balance control of a simulated standing, humanoid figure allowed to move in the sagittal plane. Here, we use a double inverted pendulum to model the humanoid. This is a common abstraction used extensively to examine the dynamics of humanoid or human balance [11, 28, 30, 32, 34, 42]. Alternative methods commonly control this system using simplified center of mass (COM) and center of pressure (COP) models [1, 35]. In this work we demonstrate results directly controlling the nonlinear double pendulum model for two main reasons. First, it demonstrates the scalability of the proposed COR approach. In particular, we show that the controller is fast and applies to nonlinear dynamics in a way that directly generalizes to models with different number of links and actuator placement. Secondly, the choice to control to the double pendulum dynamics demonstrates flexibility. While the COR approach can be applied equally well to COP models, these models require specific force/pressure sensing hardware for feedback. Controlling to the double pendulum demonstrates how balance control can be achieved using alternative feedback based on joint angles and velocities.

[8] Interestingly, this scheme of intermittent instead of sustained control has been suggested to be natively used by the central nervous system (CNS) in human posture control in [31, 32].

The system model is illustrated in Fig. 7.4. Specifically, with both feet firmly planted to the ground and the knees locked in place, we allow 2 degrees of freedom, with the two pendulum pivot points corresponding to the hips and ankles respectively. The pendulum may be controlled through direct application of torque to the hip and ankle joints. Thus, the system has $n = 4$ states, i.e., the two joint angles $\boldsymbol{\theta}^{2\times 1}$ and their velocities $\dot{\boldsymbol{\theta}}^{2\times 1}$, and $m = 2$ inputs, i.e., the torques at the joints. The mass, inertia and length of the model segments were determined from anthropometric data provided by [51]. The pendulum equations of motion take the form in (7.2) and can be derived using Lagrangian Mechanics and the Euler–Lagrange equation. The derivation is straightforward and will not be shown here.

Our performance objective in this application is to drive the system to the upright position, i.e., to keep the center of gravity (COG) at the center of the base of support. As there are no particular constraints on our choice for the tracking cost function, we employ a quadratic performance metric of the form in (7.3), using the vertical body position as an angle reference so that $x_d = \mathbf{0}$.

For performance verification, we consider the example case where a human with impaired balance ability is in a standing position wearing a lower-limb robotic exoskeleton. In this scenario, the main system corresponds to the biomechanical model of a human augmented by a transparent overlying suit, while the NC plays the role of the central nervous system (CNS) attempting to maintain human balance. In literature, the CNS has been implemented both as a PID [34] and an optimal LQR controller [28]. Here, we chose to implement it as a PD controller providing feedback control on each of the joints. The AC is assumed to be embedded in the robotic suit. We emulate a scenario where the PD gains of the NC are insufficiently tuned and as a result, the provided torques fail to hold the figure upright.

The Robot Operating System (ROS), available online [39], lies in the center of the simulation setup, coordinating the real-time communication between the running processes. The humanoid balance model provided by the double inverted pendulum is simulated using the open-source software package called `trep` [44]. The system simulation initiates at time $t = 0$ given an initial state $(\boldsymbol{\theta}_0, \dot{\boldsymbol{\theta}}_0)$. At each time step, the program checks whether assistive control has been requested and enters one of the modes described next and shown in Fig. 7.3.

Mode 1: With no assistance requested, the process simulates the system (7.1) forward according to the NC input $u(t)$, and outputs the next system configuration.

Mode 2: If assistance is requested, the module solves for the state and co-state trajectories and evaluates the control formula (7.7) to compute the assistive controller (AC) output $u_2^*(t)$ (see Algorithm 7.1). It applies the control for duration λ^+, before switching back to the nominal mode.

The simulation runs indefinitely, allowing the user to activate the AC in real time as needed by pressing a button on the computer screen. For more information on the experimental setup, see [33]. Figure 7.4 demonstrates a sequence of snapshots where

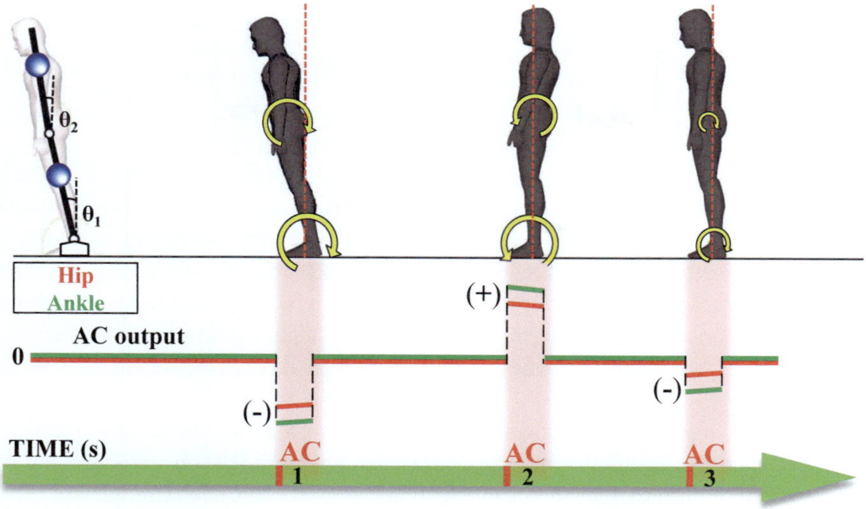

Fig. 7.4 Example of real-time standing balance control using the simulation setup. Assistive control (AC) is applied by pressing the button three times sequentially, eventually driving the system to the upright position (© 2015 IEEE. Reprinted, with permission, from [33])

the button is pushed three times sequentially and the figure is driven to the upright position.[9]

7.3.2.1 Results Based on Therapy Objectives

We now present three simulation examples that illuminate the benefits of using a COR interface in robot-assisted balance training therapy. As an informal measure of performance, we specify approximate angle safety constraints at $\theta = \pm 0.2$. It is assumed that if the angles advance beyond these values, the Center of Gravity leaves the base of support and a fall occurs. Furthermore, we quantify the magnitude of nominal control effort as the energy E_{NC} of the signals $u^{(1)}(t)$ and $u^{(2)}(t)$, one for each joint. Specifically,

$$E_{NC} = \int_0^{T_f} \|u(t)\|^2 dt \tag{7.14}$$

where T_f is either the final time of simulation T in Algorithm 7.1 or the time when the safety constraints are violated and the figure falls.

The COR parameters in Algorithm 7.1 were tuned as follows: the time horizon $T = 3.0$ s, the default control duration $\Delta\tau_{init} = 0.03$ s and the desired tracking cost reduction $\alpha_d = -200$. The weight matrices in (7.6) and (7.3) were set as

[9]As shown in Sect. 7.2, it is possible to provide parameters that guarantee local stability and convergence in the case when the Assistive Controller is continuously activated.

$$Q = \begin{pmatrix} 0.8 & 0 & 0 & 0 \\ 0 & 0.8 & 0 & 0 \\ 0 & 0 & 1.0 & 0 \\ 0 & 0 & 0 & 1.0 \end{pmatrix} \quad \text{and} \quad R = \begin{pmatrix} 0.1 & 0 \\ 0 & 0.1 \end{pmatrix}. \tag{7.15}$$

The higher weights on the velocities ensure a more realizable AC performance where fast motions are avoided. The parameter α_d defines how aggressive the assistance should be, i.e., either a "gentle" or a "strong push" toward the balanced position. Therefore, by setting the absolute value of α_d relatively low, we allow for generation of more realizable and gentle controls for application to humans. Finally, by tuning the time horizon value T, we define how long the assistive control has to affect the change in cost. A value of $3.0s$ was selected, as it resulted in smooth angle trajectories (see Fig. 7.5).

In all cases described below, once the button was pressed, it took approximately 0.1 s for the computation of the optimal assistive control in Python, on a laptop with an Intel Core i7 chipset. This confirms the feasibility of COR activation in real time.

Case 1 - No assistance: Consider the case where a standing patient is allowed to balance without assistance, i.e., only the NC is active (Fig. 7.5a). With no interference from the robot ($u_2^*(t) = 0$), the joint angles advance beyond the safety constraints indicating a fall. Therefore, although the training goal of active patient's involvement

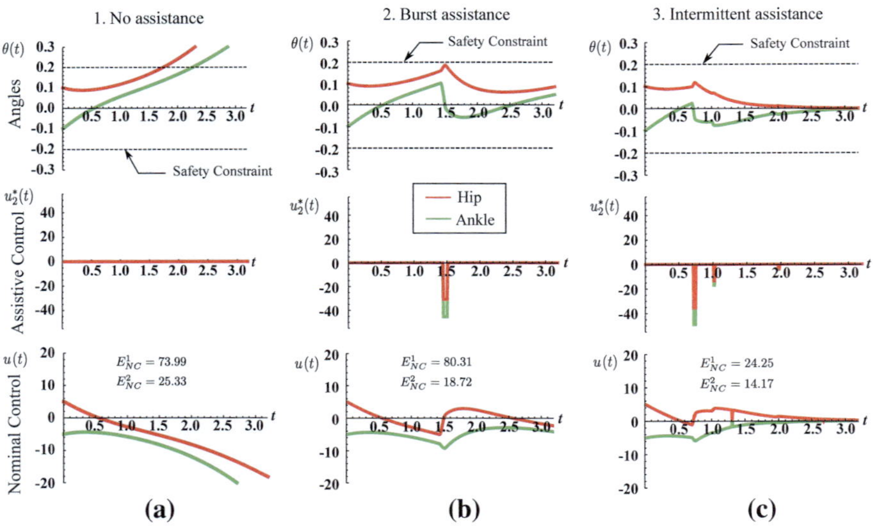

Fig. 7.5 Three examples demonstrate the optimal COR benefits in robot-assisted balance therapy. u_2^* indicates robot-applied torques and u corresponds to human–applied torques. E_{NC} indicates total nominal control effort. **a** With no assistance, safety constraints are violated. **b** With burst assistance, balance is maintained with high energy contribution from the NC (preferred implementation). **c** With intermittent assistance, balance is ensured with low energy contribution from the NC (© 2015 IEEE. Reprinted, with permission, from [33])

is achieved ($E^1_{NC} = 73.99$, $E^2_{NC} = 25.33$), there is *insufficient automated support* to prevent falls.

Case 2 - Burst assistance: In the second example, a monitoring physical therapist/observer is allowed to apply optimal control-on-request, to provide online optimal assistance when deemed necessary. The effect of the assistive controller activation is illustrated in Fig. 7.5b. In the previous case the unassisted model violated the safety constraints, now the patient is safely "pushed" back toward the upright position and a potential fall is prevented. What's interesting is that the patient is still actively applying torques ($u(t) > 0$) and thus applies almost the same amount of control effort as in the first case ($E^1_{NC} = 80.31$, $E^2_{NC} = 18.72$). The robot interferes with the human effort only for a very short duration, before ceding full control authority back to the NC. Thus, in this case, both training goals are achieved: *the patient is actively involved and balance is maintained.* Notice that this concept is similar to the action taken by a conventional therapist while assisting a person to maintain balance by gently "pushing" them to the right direction as needed. However, here, almost no effort is required from the side of the therapist (automated assistance).

Case 3 - Intermittent assistance: Our third example considers the case where the COR interface is activated multiple times intermittently (Fig. 7.5c). It can be observed that not only does the state remain inside the safety constraints but also it successfully reaches the origin. In practice, this means that a fall has been prevented and also that the upright posture has been achieved. However, in comparison to the second case, this approach is not ideal because *the patient is essentially held in the upright position by the robot* without using their own power ($E^1_{NC} = 24.25$, $E^2_{NC} = 14.17$).

Therefore, the second case is the most effective application of the optimal COR interface and indicates how an experienced therapist should take advantage of this control scheme to get the best therapy outcomes out of automated balance training.

7.4 Human–Robot Communication in Posture Reinforcement: A Short Study

In addition to employing COR interfaces to enhance therapist-robot collaboration, in this section we also explore its efficacy as a means for patient–robot communication in posture reinforcement. For this purpose, COR feedback is translated into vibrotactile cues to communicate intents through sensory augmentation. Previous studies by T. Murphey's group have provided evidence that task-based sensory augmentation yields nearly a factor of four improvement in time-to-failure for virtual balance tasks [45]. The idea of combining robotics and sensory substitution [15, 38] may be a viable alternative to conventional rehabilitation approaches. Several comparison studies have shown that neurologically intact people provided with tactile feedback can perform better than those acting on similar forms of auditory feedback; there are also cases in which tactile feedback is at least equally effective as vision, if not

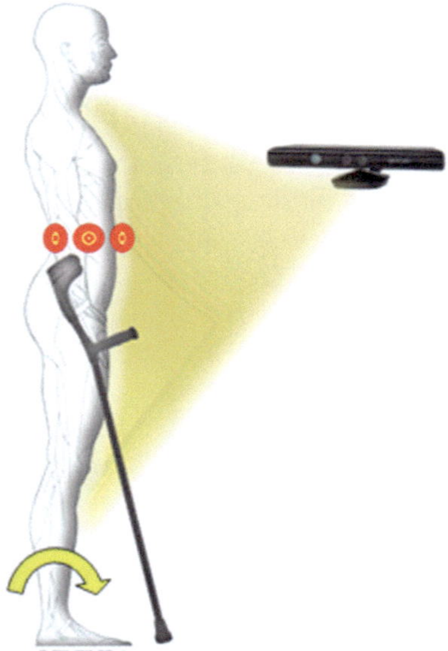

Fig. 7.6 An illustration of the subject's position during trials. The subject wears a vest with eight symmetrically positioned vibration motors around their torso (*red circles*) and uses crutches to help maintain balance while achieving different target postures

more so [41, 43]. A number of studies [12, 29, 46] conclude that tactile feedback can indeed be effective in promoting desired behaviors.

In this short study, it is hypothesized that optimal COR-generated tactile feedback on the torso can facilitate standing balance and prevent falls by providing cues for the correct placement of torso in real time. A preliminary experiment was designed to assess how a person perceives the optimal vibrotactile feedback generated by the COR interface and whether they are able to interpret it correctly to achieve a desired posture. The study was conducted with four healthy subjects. In the trials, a standing subject is wearing a vest with eight symmetrically positioned vibration motors on their chest and uses crutches to adjust their posture, moving back and forth (along the sagittal plane) and left and right (along the coronal plane). A Microsoft Kinect is used to capture sensory information about the subject's posture in real time. ROS collects and processes data from the Kinect at approximately 30 Hz. For an illustration of the subject's position during trials, see Fig. 7.6. The subject is instructed to hold their torso in a constant angle with respect to their hips so as to closely mimic the movement of a person wearing a hip-actuated exoskeleton.

During the main session, the subject is instructed to track eight subsequent random targets as fast as possible. Each target refers to a target ankle angle on the sagittal plane and a target ankle angle on the coronal plane, considering the upright position as the reference (zero) angle for both planes. The target may also be perceived as a target body Center of Gravity (CoG) with these two definitions used interchangeably. A new target is assigned as soon as the previous target has been reached. There is no other indication that the target has been reached, other than the vibration stopping.

A sample response from a single trial is illustrated in Fig. 7.7. It was observed that the subjects were adequately familiarized to the vibrational signals after a short

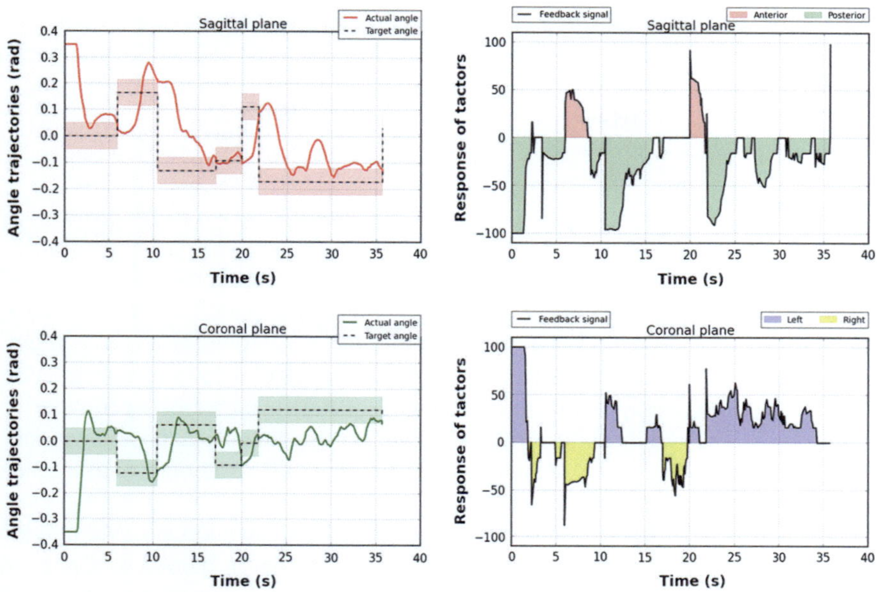

Fig. 7.7 The *first and second row* of diagrams correspond to the sagittal and coronal plane movement respectively. The *first column* shows the respective ankle angles with respect to time as they were recorded during the trials. The *green/red shaded area* around the target angle (*black dashed line*) illustrates the selected tolerance in angle which is dictated by the accuracy of the sensor used (i.e., Microsoft Kinect). The *second column* of plots shows the vibration of the tactors with respect to time, as generated by the optimal COR interface

training session. The metric used for preliminary evaluation of COR-based vibrotactile feedback perception is the mean time-to-target (TTT) normalized over distance between initial CoG and target CoG. Among the four trials, TTT was found to decrease from an initial mean value of 40 s/m over the first few targets, to approximately 20 s/m by average for the last target. This preliminary result is an indication that sensory feedback perception based on COR interfaces improved over time for all four subjects resulting in an overall decreased reaction time to stimulus. In addition, optimal COR-generated signals were found to be sufficiently instructive to always drive the subject towards the target posture.

However, a few exceptions were noted. In particular, in instances where reaching the assigned CoG target required diagonal motion from the subject, TTT exploded (up to approximately 120 s/m) to account for the fact that subjects were incapable of interpreting the tactile signals as a cue for diagonal motion (i.e., on the sagittal and coronal plane simultaneously).

Overall, it was concluded that the subjects were able to interpret the COR-generated vibrotactile signals correctly in order to track the targets, while the mean time-to-target (TTT) normalized over distance was found to decrease over the first subsequent targets and remain fairly constant for the remaining targets. This preliminary result is promising in that it promotes an alternative implementation of COR interfaces as a means for human–robot collaboration through sensory augmentation.

7.5 Conclusion

This chapter presented an algebraic expression to compute optimal actions for nonlinear systems at any time. We demonstrated how a line search can resolve short durations to apply these actions to provide long time horizon improvement in tracking. Compared to standard methods where optimal controls are synthesized based on finite switching durations, these methods completely avoid constrained iterative optimization. They facilitate the rapid exchange between human and computer control required to enable a new shared control paradigm where automated assistance can be provided quickly and on-demand.

To demonstrate the performance of these enhanced COR interfaces, we utilized ROS to apply optimal control-on-request to standing balance therapy and posture reinforcement. In particular, the chapter focused on a shared control example where bursts of optimal assistance are applied to a simulated standing figure to prevent it from falling. This approach presents an opportunity for rehabilitation. Robot-assisted balance therapy is challenging as control authority is shared between the robot and the user, rendering their integration problematic. In an ideal rehabilitation setting, the robot should only provide as much assistance as needed to prevent a fall, requiring active patient participation. However, existing assistive control strategies either rely on sustained control or assume the use of force/pressure sensors which are generally not available in current over-ground lower-limb exoskeletons. To address these issues, we considered employing an optimal COR interface to assist therapy. The short duration of activation ensures the feasibility of our design in a shared control scenario, alleviating the need for force feedback. The controller is *easy to implement*, has *low computational demands* and *runs in real time*. Our real-time simulation studies showed that a therapist equipped with a COR interface can *ensure patient safety* while requiring *active participation*.

Finally, we used the proposed COR design to provide instructive vibrotactile feedback for posture reinforcement during standing. Preliminary results from a short human study indicated that the subjects are able to interpret the sensory cues correctly in order to achieve a target posture with a decreasing normalized mean time-to-target (TTT) over time.

Acknowledgements This work was supported by the National Science Foundation Graduate Research Fellowship program and Grant CBET 1637764. Any opinions, findings and conclusions or recommendations expressed in this material are those of the authors and do not necessarily reflect the views of the National Science Foundation.

References

1. Abdallah M, Goswami A (2005) A biomechanically motivated two-phase strategy for biped upright balance control. In: IEEE international conference on robotics and automation, pp 1996–2001
2. Anderson BDO, Moore JB (1990) Optimal control: linear quadratic methods. Prentice-Hall Inc, New Jersey
3. Ansari A, Murphey T (2015) Control-on-request: short-burst assistive control for long time horizon improvement. In: American control conference, pp 1173–1180
4. Banala SK, Kim SH, Agrawal SK, Scholz JP (2009) Robot assisted gait training with active leg exoskeleton (ALEX). IEEE Trans Neural Syst Rehabil Eng 17(1):2–8
5. Boyd S, Vandenberghe L (2004) Convex optimization. Cambridge University Press, Cambridge
6. Caldwell TM, Murphey TD (2013) Projection-based optimal mode scheduling. In: IEEE conference on decision and control (CDC)
7. Chen Z, Haykin S (2002) On different facets of regularization theory. Neural Comput 14(12):2791–2846
8. Cipriani C, Zaccone F, Micera S, Carrozza MC (2008) On the shared control of an EMG-controlled prosthetic hand: analysis of user-prosthesis interaction. IEEE Trans Robot 24(1):170–184
9. De Vito E, Rosasco L, Caponnetto A, Piana M, Verri A (2004) Some properties of regularized kernel methods. J Mach Learn Res 5:1363–1390
10. Demeester E, Hüntemann A, Vanhooydonck D, Vanacker G, Van Brussel H, Nuttin M (2008) User-adapted plan recognition and user-adapted shared control: a bayesian approach to semi-autonomous wheelchair driving. Auton Robots 24(2):193–211
11. Deng-Peng X, Xu L (2011) Multiple balance strategies for humanoid standing control. Acta Autom Sin 37(2):228–233
12. Ding Z-Q, Chen I-M, Yeo SH (2010) The development of a real-time wearable motion replication platform with spatial sensing and tactile feedback. In: IEEE international conference on intelligent robots and systems, pp 3919–3924
13. Egerstedt M, Wardi Y, Axelsson H (2003) Optimal control of switching times in hybrid systems. In: IEEE international conference on methods and models in automation and robotics
14. Egerstedt M, Wardi Y, Axelsson H (2006) Transition-time optimization for switched-mode dynamical systems. IEEE Trans Autom Control 51(1):110–115
15. Elbert T, Sterr A, Rockstroh B, Pantev C, Matthias MM, Edward T (2002) Expansion of the tonotopic area in the auditory cortex of the blind. J Neurosci 22(22):9941–9944
16. Emken JL, Bobrow JE, Reinkensmeyer DJ (2005) Robotic movement training as an optimization problem: designing a controller that assists only as needed. In: IEEE international conference on rehabilitation robotics, pp 307–312
17. Forster A, Young J (1995) Incidence and consequences offalls due to stroke: a systematic inquiry. Bmj 311(6997):83–86
18. Franklin JN (1978) Minimum principles for ill-posed problems. SIAM J Math Anal 9(4):638–650
19. Greena SA, Billinghurstb M, Chena X, Chasea JG (2008) Human-robot collaboration: a literature review and augmented reality approach in design. Int J Adv Robot Syst 5(1)
20. Griffiths P, Brent Gillespie R (2004) Shared control between human and machine: haptic display of automation during manual control of vehicle heading. In: Proceedings 12th international symposium on haptic interfaces for virtual environment and teleoperator systems. IEEE, pp 358–366
21. Hansen PC (1999) The L-curve and its use in the numerical treatment of inverse problems. Technical Universityof Denmark, IMM, Department of Mathematical Modelling
22. Hogan N (1984) Impedance control: an approach to manipulation. In: American control conference, pp 304–313

23. Hogan N, Krebs HI, Rohrer B, Palazzolo JJ, Dipietro L, Fasoli SE, Stein J, Hughes R, Frontera WR, Lynch D et al (2006) Motions or muscles? Some behavioral factors underlying robotic assistance of motor recovery. J Rehabil Res Dev 43(5):605
24. Jezernik S, Colombo G, Keller T, Frueh H, Morari M (2003) Robotic orthosis lokomat: a rehabilitation and research tool. Neuromodul: Technol Neural. Interface 6(2):108–115
25. Jezernik S, Colombo G, Morari M (2004) Automatic gait-pattern adaptation algorithms for rehabilitation with a 4-DOF robotic orthosis. IEEE Trans Robot Autom 20(3):574–582
26. Krebs HI, Hogan N, Aisen ML, Volpe BT (1998) Robot-aided neurorehabilitation. IEEE Trans Rehabil Eng 6(1):75–87
27. Krebs HI, Palazzolo JJ, Dipietro L, Ferraro M, Krol J, Rannekleiv K, Volpe BT, Hogan N (2003) Rehabilitation robotics: performance-based progressive robot-assisted therapy. Auton Robots 15(1):7–20
28. Kuo AD (1995) An optimal control model for analyzing human postural balance. IEEE Trans Bio-Med Eng 42(1):87–101
29. Lieberman J, Breazeal C (2007) TIKL: development of a wearable vibrotactile feedback suit for improved human motor learning. IEEE Trans Robot 23(5):919–926
30. Liu C, Atkeson CG (2009) Standing balance control using a trajectory library. In: IEEE international conference on intelligent robots and systems, pp 3031–3036
31. Loram ID, Gollee H, Lakie M, Gawthrop PJ (2011) Human control of an inverted pendulum: Is continuous control necessary? Is intermittent control effective? Is intermittent control physiological? J Physiol 589(2):307–324
32. Loram ID, Lakie M (2002) Human balancing of an inverted pendulum: position control by small, ballistic-like, throw and catch movements. J Physiol 540(3):1111–1124
33. Mavrommati A, Ansari A, Murphey TD (2015) Optimal control-on-request: an application in real-time assistive balance control. In: IEEE international conference on robotics and automation (ICRA), pp 5928–5934
34. Milton J, Cabrera JL, Ohira T, Tajima S, Tonosaki Y, Eurich CW, Campbell SA (2009) The time-delayed inverted pendulum: implications for human balance control. Chaos: an interdisciplinary. J Nonlinear Sci 19(2):026110–026110
35. Ott C, Roa MA, Hirzinger G (2011) Posture and balance control for biped robots based on contact force optimization. In: IEEE international conference on humanoid robots, pp 26–33
36. Perry J, Garrett M, Gronley JK, Mulroy SJ (1995) Classification of walking handicap in the stroke population. Stroke 26(6):982–989
37. Philips J, Millan JdR, Vanacker G, Lew E, Galan F, Ferrez PW, Van Brussel H, Nuttin M (2007) Adaptive shared control of a brain-actuated simulated wheelchair. In: IEEE 10th international conference on rehabilitation robotics (ICORR), pp 408–414
38. Rauschecker JP (1995) Compensatory plasticity and sensory substitution in the cerebral cortex. Trends Neurosci 18(1):36–43
39. Robot Operating System (2014)
40. Sekhon LHS, Fehlings MG (2001) Epidemiology, demographics, and pathophysiology of acute spinal cord injury. Spine 26(24S):S2–S12
41. Sklar AE, Sarter NB (1999) Good vibrations: tactile feedback in support of attention allocation and human-automation coordination in event-driven domains. Hum Factors: J Hum Factors Ergon Soc 41(4):543–552
42. Stephens B (2007) Integral control of humanoid balance. In: IEEE international conference on intelligent robots and systems, pp 4020–4027
43. Sun M, Ren X, Cao X (2010) Effects of multimodal error feedback on human performance in steering tasks. J Inf Process 18:284–292
44. `trep` software package development (2014)
45. Tzorakoleftherakis E, Mussa-Ivaldi F, Scheidt R, Murphey TD (2014) Effects of optimal tactile feedback in balancing tasks: a pilot study. In: IEEE American control conference, pp 261–270
46. Tzorakoleftherakis E, Bengtson MC, Mussa-Ivaldi FA, Scheidt RA, Murphey TD (2015) Tactile proprioceptive input in robotic rehabilitation after stroke. In: IEEE international conference on robotics and automation

47. Veneman JF, Kruidhof R, Hekman EEG, Ekkelenkamp R, Van Asseldonk EHF, Van Der Kooij H (2007) Design and evaluation of the lopes exoskeleton robot for interactive gait rehabilitation. IEEE Trans Neural Syst Rehabil Eng 15(3):379–386
48. Wardi Y, Egerstedt M (2012) Algorithm for optimal mode scheduling in switched systems. In: American control conference, pp 4546–4551
49. West RG (2004) Powered gait orthosis and method of utilizing same. US Patent 6,689,075, 10 Feb 2004
50. Winstein CJ, Gardner ER, McNeal DR, Barto PS, Nicholson DE (1989) Standing balance training: effect on balance and locomotion in hemiparetic adults. Arch Phys Med Rehabil 70(10):755–762
51. Winter DA (2009) Biomechanics and motor control of human movement. Wiley, New York
52. Wolbrecht ET, Chan V, Reinkensmeyer DJ, Bobrow JE (2008) Optimizing compliant, model-based robotic assistance to promote neurorehabilitation. IEEE Trans Neural Syst Rehabil Eng 16(3):286–297
53. Wright SJ, Nocedal J (1999) Numerical optimization, vol 2. Springer New York
54. Yu H, Matthew S, Steven D (2003) An adaptive shared control system for an intelligent mobility aid for the elderly. Auton Robots 15(1):53–66
55. Zabczyk J (2009) Mathematical control theory: an introduction. Springer, Berlin

Chapter 8
Intelligent Human–Robot Interaction Systems Using Reinforcement Learning and Neural Networks

Hamidreza Modares, Isura Ranatunga, Bakur AlQaudi, Frank L. Lewis and Dan O. Popa

8.1 Introduction

Human–robot interaction (HRI) systems have received considerable attention in robotics and control society because of their potential applications in industry, entertainment, teleoperation, household, and healthcare, to name just a few. Unlike ordinary industrial robotics where the environment is structured and known, in HRI systems, the robots interact with humans who may have very different skills and capabilities. Therefore, in HRI systems, robots should adjust themselves to the level of the skills and capability of the human and compensate for possible human mistakes due to fatigue, stress, etc. This increases the performance of the overall HRI system in terms of safety, human force, and precision. In order to design such an intelligent HRI system, one major issue is that of how to best design a learning-based controller for the HRI system so that the assistance and safety are provided to the human using minimum knowledge about the human and robot dynamics, while avoiding compromise on performance.

Industrial robots are usually programmed to follow desired trajectories. Adaptive robot controllers using neural networks (NNs) have been widely used to provide highly effective controllers in yielding guaranteed position control for robot manipulators with unknown nonlinear dynamics, and modeling inaccuracies [7, 8, 17, 20, 21]. However, when the robot manipulator is in contact with an object or a human, it must be able to control not only positions, but also forces. Impedance control [10]

This chapter was developed and enhanced from an earlier paper published as [25] © 2016 IEEE.

H. Modares (✉)
Missouri University of Science and Technology,
Rolla, MO, USA
e-mail: modaresh@mst.edu

I. Ranatunga · B. AlQaudi · F.L. Lewis · D.O. Popa
The University of Texas at Arlington UTA Research Institute (UTARI),
Arlington, TX, USA

Y. Wang and F. Zhang (eds.), *Trends in Control and Decision-Making for Human–Robot Collaboration Systems*, DOI 10.1007/978-3-319-40533-9_8

provides an effective method for control of both position and force simultaneously in trajectory-following tasks. Adaptive impedance control techniques using NNs have been developed to tune the impedance model to be followed using the tracking error dynamics based on various considerations [6, 9, 11, 12, 14, 38].

For modern interactive HRI systems to be capable of performing a wide range of tasks successfully, it is required to include the effects of both the robot dynamics and the human dynamics. Human factors studies have shown that in coordinated motion with a robot, human learning has two components [1, 4, 15, 30]. The operator learns a robot-specific inverse dynamics model to compensate for the nonlinearities of the robot [31, 33] and simultaneously he learns a feedback control component that is specific to the successful performance of the task. Recently, adaptive impedance methods have been extended to the control of HRI system to improve its performance and safety. The impedance parameters are tuned based on various criteria such as the stability [3], the input torque [13], minimizing a cost function [26], and incorporating the human intention [22]. These methods, however, have not taken into account the skill differences of different operators. Impedance control methods based on human factor studies have been recently developed for HRI systems [5, 15, 16, 30]. These methods take into account the skill differences of the operators by adjusting the impedance of the robot according to the identified operator's model dynamics. Moreover, the impedance parameters are tuned based on a Lyapunov function to assure stability. But, stability is a bare minimum requirement for a controlled system, and it is desired to tune the impedance parameters to optimize the long-term performance of the system. References [24, 32] performed the adjustment of the impedance based on estimation of the operator's impedance parameters. These techniques did not concern to optimize the overall long-term performance of the human–robot systems. Li et al. [23] and Wang et al. [37] developed an adaptive impedance method for HRI systems to find the optimal parameters of the robot impedance model.

In this chapter, an intelligent HRI system with adjustable robot behavior is developed. Unlike [5, 15, 16], the proposed method does not require the knowledge of the human model and does not need to estimate or identify the human impedance characteristics. A robot-specific inner loop is first designed to make the robot with unknown dynamics behave like a simple prescribed robot impedance model as perceived by a human operator. In contrast to most previous work, this is not a trajectory-following objective and no information of the task performance is used in the inner loop. Next, a task-specific outer loop is developed, taking into account the human transfer characteristics, to find the optimal parameters of the prescribed robot impedance model depending on the task. The problem of finding the optimal parameters of the prescribed robot impedance model is formulated as an optimal control problem and reinforcement learning (RL) [2, 18, 19, 21, 27, 29, 34–36] is used to learn its solution to obviate the requirement of the knowledge of the human model.

8.2 HRI Control: Motivation and Structure Overview of the Proposed Approach

Robots interacting with humans should be able to adjust themselves to the level of the skill of the human and compensate for possible human mistakes due to fatigue, stress, etc. To clarify this, consider a simple mechanical system with a joystick. A skillful human operator is able to handle faster joystick movements and thus achieve better performance using a small damping coefficient. On the other hand, an inexperienced human operator cannot handle fast movements and may harm the system or himself if the damping coefficient is small. Therefore, the damping coefficient of the joystick should be tuned so that the response of the system is adapted to the level of the skill of the human operator. To achieve this goal, one can design an intelligent HRI system using a learning-based controller for the HRI system so that the assistance and safety are provided to the human using minimum knowledge about the human and robot dynamics.

An intelligent HRI system is designed here. The proposed HRI system has two loops. The first loop is a robot-specific inner loop which does not require any information of the task. The second loop is a task-specific outer loop which includes the human operator dynamics, the robot, and the task performance details. The objective in the robot-specific loop is to make the unknown robot manipulator dynamics behave like a prescribed robot impedance model as perceived by a human operator. Therefore, the human only needs to interact with the simplified impedance model. An adaptive NN controller is used to compensate for the unknown robot nonlinearities. This is not the same as the bulk of the work in robot impedance control and NN control, which is directed towards making a robot follow a prescribed trajectory, and/or causing the trajectory error dynamics to follow a prescribed impedance model. No trajectory information and knowledge of the prescribed robot impedance model is needed for the inner loop design. This leaves the freedom to incorporate task information in an outer loop design. The robot-specific inner loop controller is shown in Fig. 8.1.

Given the robot-specific inner loop design, the optimal parameters of the prescribed robot impedance model are determined in the task-specific outer loop to guarantee motion tracking and also assist the human to perform the task with minimum effort. To this end, the problem of determining the optimal parameters of the prescribed robot impedance model is transformed into a linear quadratic regulation (LQR) problem. This design must take into account the unknown human dynamics as well as the desired overall performance of the human–robot system. Reinforcement learning is used to solve the LQR problem without any knowledge of the human operator dynamics. Figure 8.2 shows the task-specific outer loop control design. Figure 8.3 shows the schematic of the proposed two-loop control design method for the HRI system. The details of the inner loop design and the outer loop design are given in the subsequent sections.

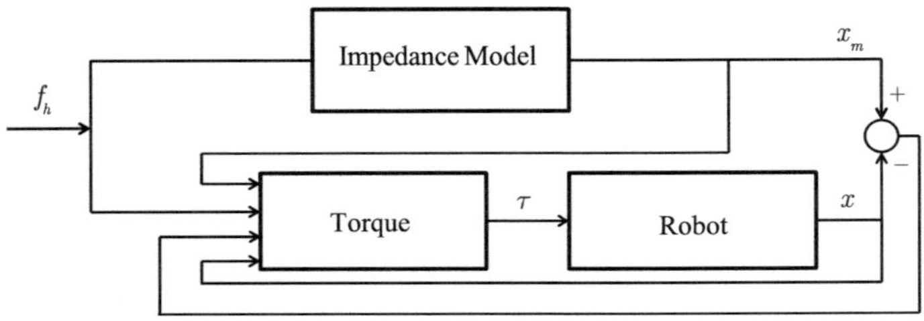

Fig. 8.1 The inner loop robot-specific layer

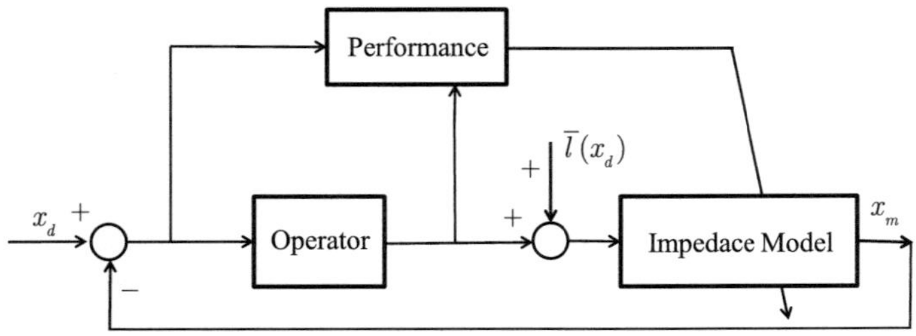

Fig. 8.2 The outer loop task-specific layer

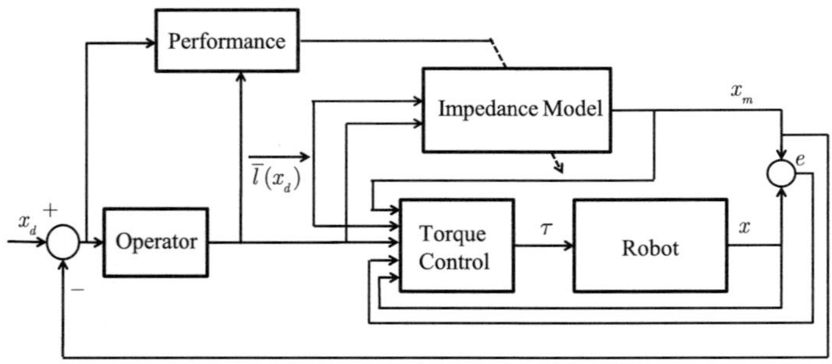

Fig. 8.3 The overall structure of the proposed HRI systems

8.3 Inner Robot-Specific Loop

In this section, the design of the inner loop controller is given. The aim of the inner loop controller is to make the robot manipulator behave like a prescribed robot impedance model. In contrast to trajectory-following based methods, the proposed method does not require a reference motion trajectory. Moreover, in the proposed method, the designed control torque does not require any knowledge of the prescribed robot impedance model. Consider the dynamical model of robot manipulator in Cartesian space [20]

$$M(q)\,\ddot{x} + C(q,\dot{q})\,\dot{x} + F_c(\dot{q}) + G(q) + \tau_d = \tau + f_h, \tag{8.1}$$

with $M = J^{-T} M^* J^{-1}$, $C = J^{-T}(C^* - M^* J^{-1} \dot{J}) J^{-1}$, $M = J^{-T} M^* J^{-1}$, $F_c = J^{-T} F^*$, $G = J^{-T} G^*$, $\tau = J^{-T} \tau^*$, where $q \in R^n$ is the vector of generalized joint coordinates, n is the number of joints, $x \in R^n$ is the end effector Cartesian position, the control input force is $\tau = J^{-T} \tau^*$ with τ^* the vector of generalized torques acting at the joints, $M^* \in R^{n \times n}$ is the symmetric positive definite mass (inertia) matrix, $C^*(q, \dot{q}) \dot{q} \in R^{n \times 1}$ is the vector of Coriolis and centripetal forces, $F_c{}^*(\dot{q}) \in R^{n \times 1}$ is the Coulomb friction term, $G^*(q) \in R^{n \times 1}$ is the vector of gravitational torques, τ_d is a general nonlinear disturbance, f_h is the human control effort, and J is the Jacobian matrix.

Consider the prescribed robot impedance model

$$\bar{M}\, \ddot{x}_m + \bar{B}\, \dot{x}_m + \bar{K}\, x_m = f_h + \bar{l}(x_d) \equiv l(f_h, x_d), \tag{8.2}$$

in Cartesian space, where x_m is the output of the prescribed robot impedance model, $\bar{M}$, $\bar{B}$, and $\bar{K}$ are the desired inertia, damping, and stiffness parameter matrices, respectively. These parameters are specified in the task-specific outer control loop design in Sect. 8.4. The auxiliary input $\bar{l}(x_d)$ is a trajectory dependent input and is also designed in Sect. 8.4.

The objective of the inner loop is to design the force τ in (8.1) to make the unknown robot dynamics (8.1) from the human force f_h to the Cartesian coordinates x behave like the prescribed robot impedance model (8.2). The knowledge of the prescribed impedance model parameters is not needed. This allows us to decouple the inner loop robot-specific design from the outer loop task-specific design. That is, it is desired to make the following model-following error go to zero regardless of the values of, $\bar{M}$, $\bar{B}$, and $\bar{K}$.

$$e = x_m - x \tag{8.3}$$

Note this is not a trajectory-following error. Therefore, in contrast to most work on robot torque control, this is a model-following design, and not a trajectory-following design. No task information is required in this section. All task-specific details are taken into account in the next section.

Write the control torque as

$$\tau = \tau_1 - f_h \tag{8.4}$$

Then, the robot dynamics (8.1) becomes

$$M(q)\, \ddot{x} + C(q, \dot{q})\, \dot{x} + F_c(\dot{q}) + G(\dot{q}) + \tau_d = \tau_1 \tag{8.5}$$

It is now required to design a control torque τ_1 to make the robot behave like the prescribed robot impedance model (8.2). Consider the control torque

$$\tau_1 = \hat{W}^T \phi(\hat{V}^T z) + K_v r - v(t) \tag{8.6}$$

where $v(t)$ is a robustifying signal to be specified, K_v is the control gain, and

$$r = \dot{e} + \Lambda_1 e + \Lambda_2 \varepsilon \tag{8.7}$$

is the sliding mode error with

$$\varepsilon = \int_0^t e(\tau)\, d\tau \tag{8.8}$$

Finally,

$$\hat{h}(z) = \hat{W}^T \phi(\hat{V}^T z) \tag{8.9}$$

is a neural network (NN) with $z = [q, \dot{q}, \dot{x}_m, \ddot{x}_m, e, \dot{e}, \varepsilon]^T$ the input to the NN, $\hat{W}$ and $\hat{V}$ the NN weights and $\phi(z)$ the vector of activation functions. As is shown in the proof of Theorem 8.1, the NN controller in (8.6) is used to compensate for the unknown robot function h defined as

$$\begin{aligned} h(z) = {} & M(q)(\ddot{x}_m + \Lambda_1 \dot{e} + \Lambda_2 e) + C(q, \dot{q})(\dot{x}_m + \Lambda_1 e + \Lambda_2 \varepsilon) \\ & + F_c(\dot{q}) + G(q) \end{aligned} \tag{8.10}$$

The two-layer NN which is an approximation of the continuous function $h(z)$ on a compact set $z \in \Omega$ has the following form

$$h(z) = W^T \phi(V^T z) + \varepsilon(z) \tag{8.11}$$

where V is a matrix of first-layer weights, W is a matrix of second-layer weights, and ε is the NN functional approximation error. The ideal weight vectors W and V are unknown and is approximated online. Therefore, $h(z)$ is approximated as (8.9) with $\hat{W}$ and $\hat{V}$ the estimations of W and V, respectively. Define

$$Z = \begin{bmatrix} W & 0 \\ 0 & V \end{bmatrix} \tag{8.12}$$

and $\hat{Z}$ equivalently with $\hat{W}$ and $\hat{V}$ in (8.12), instead of W and V. It is assumed that the ideal NN weights are bounded. The following theorem shows that the proposed control input (8.4) with τ_1 given by (8.6) guarantees the boundness of the model-following error e and the NN weights.

Theorem 8.1 *Consider the robot manipulator dynamics (8.1) and the prescribed robot impedance model (8.2). Let the control input be chosen as (8.4) and (8.6). Let* $\|Z\| \leqslant Z_B$. *Let the update rule for the NN weights be given by*

$$\dot{\hat{W}} = F \hat{\phi} r^T - F \hat{\phi}' \hat{V}^T z r^T - k F \parallel r \parallel \hat{W} \tag{8.13}$$

$$\dot{\hat{V}} = G\,z\,(\hat{\phi}'\,\hat{W}r)^T\,r^T - k\,G\,\|\,r\,\|\,\hat{V} \tag{8.14}$$

where $\hat{\phi} = \phi(\hat{V}^T z)$, $\hat{\phi}' = d\phi(y)/dy|_{y=\hat{V}^T z}$, $F = F^T > 0$, $G = G^T > 0$ *and* $k > 0$ *is a small design parameter. Let the robustifying term be*

$$v(t) = -K_z(\left\|\hat{Z}\right\| + Z_B) \tag{8.15}$$

where $K_z > 0$. *Then,* $e(t)$ *in (8.3) and the NN estimated weights are uniformly ultimately bounded (UUB).*

Proof Differentiating (8.3) with respect to time gives

$$\dot{e} = \dot{x}_m - \dot{x} \tag{8.16}$$

or equivalently

$$\dot{x} = \dot{x}_m - \dot{e} \tag{8.17}$$

Differentiating this equation gives

$$\ddot{x} = \ddot{x}_m - \ddot{e} \tag{8.18}$$

Considering the sliding mode tracking error defined in (8.7), one has

$$\dot{e} = r - \Lambda_1 e - \Lambda_2 \varepsilon \tag{8.19}$$

Differentiating this equation gives

$$\ddot{e} = \dot{r} - \Lambda_1 \dot{e} - \Lambda_2 e \tag{8.20}$$

Using (8.16)–(8.20) in (8.1) yields

$$\begin{aligned} &M(q)(\ddot{x}_m - (\dot{r} - \Lambda_1\dot{e} - \Lambda_2 e)) + C(q,\dot{q})(\dot{x}_m - \\ &(r - \Lambda_1 e - \Lambda_2\varepsilon)) + F_c(\dot{q}) + G(q) + \tau_d = \tau + f_h \end{aligned} \tag{8.21}$$

This gives the sliding mode error dynamics

$$\begin{aligned} M(q)\dot{r} = &-C(q,\dot{q})r + h(q,\dot{q},\dot{x}_m,\ddot{x}_m,e,\dot{e},\varepsilon) + \\ &\tau_d - \tau - f_h \end{aligned} \tag{8.22}$$

with h defined in (8.10). The robot manipulator dynamics (8.1) is assumed to be unknown and therefore h in (8.22) is unknown and approximated online by (8.9). Then, the closed-loop filtered error dynamics (8.22) becomes

$$\begin{aligned} M(q)\dot{r} = &-C(q,\dot{q})r + \hat{W}^T\phi(\hat{V}^T z) + \tau_d - \tau \\ &- f_h + \tilde{h} \end{aligned} \tag{8.23}$$

where $\tilde{h} = h - \hat{h}$ is the estimation error for approximating the function h. Substituting τ form (8.4) and (8.6) in (8.23) gives

$$M(q)\dot{r} = -C(q,\dot{q})r - K_v r + \tau_d + \tilde{h} + v(t) \tag{8.24}$$

A Lyapunov function is now defined as

$$L = \frac{1}{2} r^T M(q) r + tr(\tilde{W}^T F^{-1} \tilde{W}) + tr(\tilde{V}^T F^{-1} \tilde{V}) \tag{8.25}$$

where the weight estimation errors are $\tilde{W} = W - \hat{W}$, $\tilde{V} = V - \hat{V}$. Taking the derivatives of (8.25) and doing some manipulations yields [17].

$$\begin{aligned} &\dot{L} = -r^T K_v r + \tfrac{1}{2} r^T (\dot{M}(q) - 2\,C(q,\dot{q}))\,r + \\ &tr\{\tilde{W}^T (F^{-1}\dot{\tilde{W}} + \hat{\phi}\,r^T - \hat{\phi}'\,\hat{V}^T z r^T)\} \\ &+ tr\{\tilde{V}^T (G^{-1}\dot{\tilde{V}} + z\,r^T\,\hat{W}^T\hat{\phi}')\} + r^T w \end{aligned} \tag{8.26}$$

where $w = \varepsilon + \tau_d$. Using $\dot{M}(q) - 2\,C(q,\dot{q}) = 0$ and the tuning rules, one has

$$\begin{aligned} &\dot{L} = -r^T K_v r + k\,\|r\|\,tr\{\tilde{W}^T (W - \tilde{W})\} + \\ &k\,\|r\|\,tr\{\tilde{V}^T (V - \tilde{V})\} + r^T (w + v) \\ &= -r^T K_v r + k\,\|r\|\,tr\{\tilde{Z}^T (Z - \tilde{Z})\} + r^T (w + v) \end{aligned} \tag{8.27}$$

Since

$$tr\{\tilde{Z}^T (Z - \tilde{Z})\} = < \tilde{Z}, Z > - \left\|\tilde{Z}\right\|_F^2 \le \left\|\tilde{Z}\right\|_F \|Z\|_F - \left\|\tilde{Z}\right\|_F^2 \tag{8.28}$$

One has

$$\begin{aligned} &\dot{L} \le -r^T K_v r + k\,\|r\| \cdot \left\|\tilde{Z}\right\|_F (Z_B - \left\|\tilde{Z}\right\|_F) - K_Z(\left\|\hat{Z}\right\|_F + Z_B)\|r\|^2 \\ &+ \|r\| \cdot \|w\| \le K_{v_{\min}} \|r\|^2 + k\,\|r\| \cdot \left\|\tilde{Z}\right\|_F (Z_B - \left\|\tilde{Z}\right\|_F) - \\ &K_Z(\left\|\hat{Z}\right\|_F + Z_B)\|r\|^2 + \|r\| \left[C_0 + C_1 \left\|\tilde{Z}\right\|_F + C_2\,\|r\| \cdot \left\|\tilde{Z}\right\|_F\right] \\ &\le -\|r\| \left\{K_{v_{\min}} \|r\| - k \cdot \left\|\tilde{Z}\right\|_F (Z_B - \left\|\tilde{Z}\right\|_F) - C_0 - C_1 \left\|\tilde{Z}\right\|_F\right\} \end{aligned} \tag{8.29}$$

where $K_{v_{min}}$ is minimum singular value of K_v and C_0, C_1, and C_2 are known constants (see [21] and [17] for more details). The last inequality $\dot{L}$ is negative as long as the term in braces is positive. Defining $C_3 = Z_B + C_1/k$ and completing the square yields

$$\begin{aligned} &K_{v_{\min}}\|r\| - k\cdot\left\|\tilde{Z}\right\|_F(Z_B - \left\|\tilde{Z}\right\|_F) - C_0 - C_1\left\|\tilde{Z}\right\|_F = \\ &k(\left\|\tilde{Z}\right\|_F - C_3/2)^2 + K_{v_{\min}}\|r\| - C_0 - kC_3{}^2/4 \end{aligned} \tag{8.30}$$

which is guaranteed positive as long as either

$$\|r\| > \frac{C_0 + kC_3{}^2/4}{K_{v_{\min}}} \tag{8.31}$$

or

$$\left\|\tilde{Z}\right\| > \frac{C_3}{2} + \sqrt{\frac{C_0}{k} + \frac{C_3{}^2}{4}} \tag{8.32}$$

This completes the proof. □

Figure 8.4 shows the schematic of the proposed inner loop controller. We call it model reference neuro-adaptive control. This is because a prescribed robot impedance model is used as a reference model for the robot to follow its behavior and an adaptive NN is used to compensate for the unknown nonlinear robot dynamics.

Remark 8.1 Function $h(z)$ in (8.10) does not contain the parameters $\bar{M}$, $\bar{D}$ and $\bar{K}$ of the prescribed robot impedance model (8.2) and any task information. All task-specific information is considered in the outer loop design in Sect. 8.4. This means that the NN does not need to estimate the impedance model, which is already known. This is in contrast to methods that design the torque τ in (8.1) to guarantee following

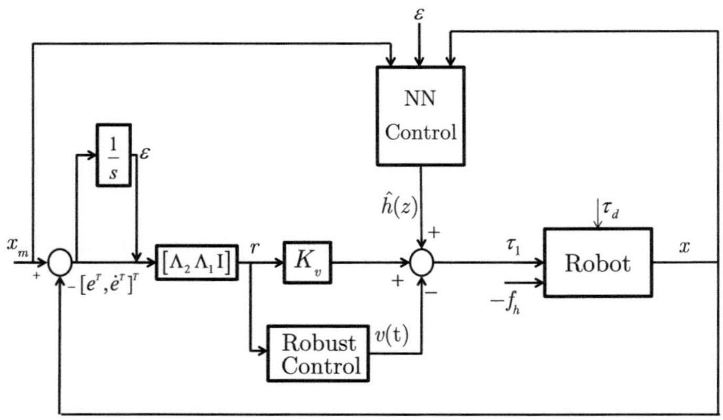

Fig. 8.4 The inner loop neuro-adaptive control design

a desired trajectory, and demanding that the trajectory tracking error dynamics follow a prescribed impedance model.

8.4 Outer Task-Specific Loop Control

In this section the design of the outer task loop controller is detailed. The aim of the task-specific outer loop controller is to find the optimal values of the prescribed impedance parameters $\bar{B}$, $\bar{K}$ and $\bar{M}$ in (8.2) to minimize the human control effort f_h and optimize the tracking performance depending on the task. The human transfer function from e_d to f_h is considered as [30]

$$G(s) = \frac{K_d\, s + K_p}{Ts + 1} \tag{8.33}$$

where K_d, K_p and T are unknown matrix gains. These parameters vary from one individual to another and depend on the specific task.

8.4.1 Task-Specific Outer Loop Control Method: An LQR Approach

In addition to the adaptive impedance loop that specifies the optimal impedance parameters, an assistive feedforward input $\bar{l}(x_d)$ is employed to make the steady-state tracking error go to zero. The optimal prescribed impedance parameters in (8.2) are determined so that the human effort and the tracking error are minimized for a given task and human operator. Figure 8.5 shows the block diagram of the outer loop after the transforming the optimal model parameters into an optimal control problem.

In the following, the problem of finding optimal values of the impedance parameters $\bar{B}$, $\bar{K}$ and $\bar{M}$ is transformed into an LQR problem, and these parameters are obtained by solving an algebraic Riccati equation (ARE).

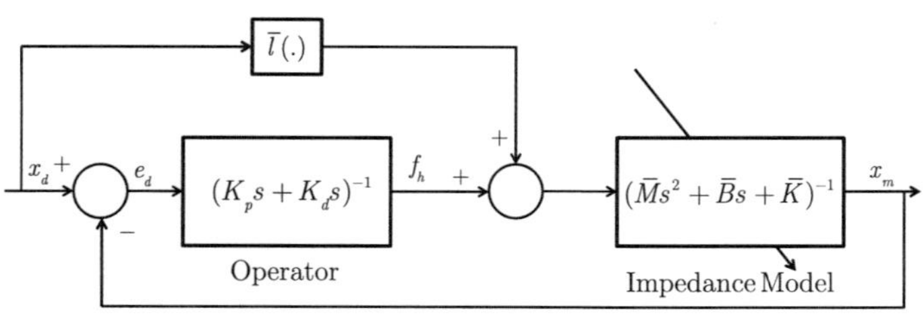

Fig. 8.5 The outer loop design for tuning impedance model parameters

To this end, define the tracking error

$$e_d = x_d - x_m \in R^n \tag{8.34}$$

and

$$\bar{e}_d = [e_d^T \; \dot{e}_d^T]^T = \bar{x}_d - \bar{x} \in R^{2n} \tag{8.35}$$

with $\bar{x} = [x_m^T \; \dot{x}_m^T]^T \in R^{2n}$ and $\bar{x}_d = [x_d^T \; \dot{x}_d^T]^T \in R^{2n}$.

Based on this tracking error, define the performance index

$$J = \int_t^\infty (\bar{e}_d^T Q_d \, \bar{e}_d + {f_h}^T Q_h \, f_h + {u_e}^T R \, u_e) \, d\tau \tag{8.36}$$

where $Q_d = Q_d^T > 0$, $Q_h = Q_h^T > 0$, $R = R^T > 0$, and u_e is a feedback control input, as shown later, which depends linearly on the tracking error $\bar{e}_d$ and the human effort f_h. That is,

$$u_e = K_1 \, \bar{e}_d + K_2 \, f_h \tag{8.37}$$

It is shown in Theorem 8.2 that the control input (8.37) has two components. The first component, i.e., K_1, tunes the prescribed impedance parameters $\bar{B}$, $\bar{K}$ and the second component, i.e., K_2, tunes $\bar{M}$.

By defining the augmented state

$$X = \begin{bmatrix} \bar{e}_d \\ f_h \end{bmatrix} \in R^{3n} \tag{8.38}$$

the performance index (8.36) can be written as

$$J = \int_t^\infty (X^T Q \, X + {u_e}^T R \, u_e) \, d\tau \tag{8.39}$$

where $Q = \text{diag}(Q_d, Q_h)$, and $u_e = K \, X$ with $K = [\, K_1 \;\; K_2 \,]$.

The dynamics of the system with augmented state (8.38) are now given. Using (8.2) one has

$$\bar{x} = \begin{bmatrix} 0 & I_{n\times n} \\ 0 & 0 \end{bmatrix} \bar{x} + \begin{bmatrix} 0 \\ I_{n\times n} \end{bmatrix} u \equiv A_q \bar{x} + B_q u \tag{8.40}$$

where

$$u = \bar{M}^{-1}(-K_q \bar{x} + f_h) + \bar{M}^{-1} \, \bar{l}(x_d) \tag{8.41}$$

with

$$K_q = [\,\bar{K}\ \bar{B}\,] \tag{8.42}$$

On the other hand, the human model (8.33) can be written as

$$\dot{f}_h = -T^{-1} f_h + T^{-1} [K_p\ K_d]\, \bar{e}_d \tag{8.43}$$

where $\bar{e}_d$ is defined in (8.35). The following theorem shows how the problem of finding the optimal parameters of the prescribed impedance model and the human gain are obtained by solving an LQR problem.

Theorem 8.2 *Consider the prescribed robot impedance model (8.2). Based on dynamics in (8.40) and (8.43), define augmented matrices A and B by*

$$A = \begin{bmatrix} A_q & 0 \\ -T^{-1} & T^{-1}[K_p\ K_d] \end{bmatrix}, \quad B = \begin{bmatrix} B_q \\ 0 \end{bmatrix} \tag{8.44}$$

Define

$$K = [\bar{M}^{-1} K_q\ \ \bar{M}^{-1}] \tag{8.45}$$

as the matrix of the impedance parameters and the human gain. Then, the optimal value of K which minimizes the performance index (8.36) is given by

$$K = \ R^{-1} B^T P \tag{8.46}$$

where P is the solution to the algebraic Riccati equation (ARE)

$$0 = A^T P + P\ A + P\ B\ R^{-1} B^T P + Q \tag{8.47}$$

Then, the optimal feedback control is given by

$$u_e = \bar{M}^{-1} \bar{K}\ e_d + \bar{M}^{-1} \bar{B}\ \dot{e}_d + \bar{M}^{-1} f_h \tag{8.48}$$

Proof Manipulating (8.41) gives

$$\begin{aligned} u &= \bar{M}^{-1}(K_q \bar{e}_d + f_h) + \bar{M}^{-1}\,(\bar{l}(x_d) - K_q \bar{x}_d) \\ &\equiv u_e + u_d \end{aligned} \tag{8.49}$$

where $\bar{e}_d$ and $\bar{x}_d$ are defined in (8.35), and

$$u_e = \bar{M}^{-1} K_q \bar{e}_d + \bar{M}^{-1} f_h \tag{8.50}$$

is a feedback control input, and

$$u_d = \bar{M}^{-1} \left(\bar{l}(x_d) - K_q \bar{x}_d \right) \tag{8.51}$$

is a feedforward control input. The steady state or feedforward term is used to guarantee perfect tracking. That is, in the steady state one has

$$\dot{\bar{x}}_d = A_q \bar{x}_d + B_q u_d \tag{8.52}$$

Therefore,

$$\begin{aligned} \bar{l}(q_d) = \bar{M}\, u_d + K_q \bar{x}_d = \bar{M}\, B_q{}^{-1} (\dot{\bar{x}}_d - A_q \bar{x}_q) \\ + K_q \bar{x}_d \end{aligned} \tag{8.53}$$

Taking derivative of $\bar{e}_d$ and using (8.40) and (8.52), and some manipulations give

$$\dot{\bar{e}}_d = A_q \bar{e}_d + B_q u_e \tag{8.54}$$

Using the augmented state (8.38), and using (8.43) and (8.54) one has

$$\begin{aligned} \dot{X} = \begin{bmatrix} \dot{\bar{e}}_d \\ \dot{f}_h \end{bmatrix} = \begin{bmatrix} A_q & 0 \\ -T^{-1} & T^{-1}[K_p \;\; K_d] \end{bmatrix} \begin{bmatrix} \bar{e}_d \\ f_h \end{bmatrix} + \begin{bmatrix} \bar{B}_q \\ 0 \end{bmatrix} u_e \\ \equiv AX + Bu_e \end{aligned} \tag{8.55}$$

The control input u_e in terms of the augmented state can be written as

$$u_e = \bar{M}^{-1} K_q \bar{e}_d + \bar{M}^{-1} f_h = K\, X \tag{8.56}$$

Finding the optimal feedback control (8.56) to minimize the performance index (8.36) subject to the augmented system (8.55) is an LQR problem and its solution is given by [25]

$$u_e{}^* = -R^{-1} B^T P X \tag{8.57}$$

where P is the solution to the Riccati equation (8.47). Equating the right-hand sides of (8.56) and (8.57) yields

$$K = [\bar{M}^{-1} K_q \;\; \bar{M}^{-1}] = -R^{-1} B^T P \tag{8.58}$$

This completes the proof. □

Remark 8.2 The outer loop control design consists of two components: An adaptive impedance component which finds the optimal values of the parameters (8.42) of the prescribed impedance model, and an assistive component including the feedforward term $\bar{l}(q_d)$ to help the human to minimize the tracking error.

8.4.2 *Learning Optimal Parameters of the Prescribed Impedance Model Using Integral Reinforcement Learning*

Reinforcement learning (RL) technique, inspired by learning mechanisms observed in mammals, is a computational approach to learning from interactions with the surrounding environment and concerned with how an agent or actor ought to take actions so as to optimize a cost of its long-term interactions with the environment. In the context of control, the environment is the dynamic system, the agent corresponds to the controller, and actions correspond to control signals. The RL objective is to find a strategy that minimizes an expected long-term cost. Unlike traditional optimal control solutions, RL does not require the exact knowledge of the system dynamics. Instead, RL largely relies upon experience gathered from tracking actions and directly interacting with the system dynamics.

Solving (8.47) requires the knowledge of the matrix A in (8.44) and consequently the knowledge of the human model. In order to obviate the requirement of knowledge of the human model, the integral reinforcement learning (IRL) algorithm [34, 36] is used to solve the given LQR problem.

Note that for time interval $\Delta t > 0$, the value function (8.36) satisfies

$$V(X(t)) = \int_{t}^{t+T} (X^T QX + u_e{}^T Ru_e)\, d\tau + V(X(t+T)) \tag{8.59}$$

It is well known that the value function for the LQR problem is quadratic in terms of the state of the system state. That is, $V(X) = X^T P X$. Using this quadratic form in (8.59) yields the IRL Bellman equation

$$\begin{aligned} X(t)^T P\, X(t) = \int_{t}^{t+\Delta t} \left[X(t)^T Q\, X(t) + u_e{}^T R\, u_e\right] d\tau + \\ X(t+\Delta t)^T P\, X(t+\Delta t) \end{aligned} \tag{8.60}$$

Using (8.60) for policy evaluation step and an update law in form of (8.57) to find an improved policy, the following IRL-based algorithm is obtained for solving (8.47).

Algorithm 8.1 Online IRL algorithm for the outer loop control design Initialization: Start with an admissible control input $u^0 = K_1{}^0 X$ Policy evaluation: Given a control policy u^i, find P^i using the Bellman equation

$$\begin{aligned} X(t)^T P^i X(t) = \frac{1}{2} \int_{t}^{t+\Delta t} [X(t)^T Q\, X(t) + (u_e{}^i)^T R\, (u_e{}^i)]\, d\tau \\ + X(t+\Delta t)^T P^i\, X(t+\Delta t) \end{aligned} \tag{8.61}$$

Policy improvement: update the control input using

$$u_e^{i+1} = -R^{-1} B_1^T P^i X \tag{8.62}$$

The policy evaluation and improvement steps (8.61) and (8.62) are repeated until the policy improvement step no longer changes the present policy, thus convergence to the optimal controller is achieved [36]. The solution for P^i in the policy evaluation step (8.62) is generally carried out in a least squares (LS) sense.

Note that Algorithm 8.1 solves ARE (8.47) and does not require knowledge of the A matrix which contains knowledge of the human dynamics. In fact, the information of A is embedded in the online measurement of system data. The details about the implementation of this algorithm can be found in [36].

8.5 Simulation Results

In this section, simulation and experimental results are provided to verify the efficiency of the proposed method.

Example 1: two degree-of-freedom planer xy-stage

In order to compare the propose method to the method presented in [15], a haptic interface system consisting of a two degree-of-freedom planer xy-stage is considered. A point-to-point task is considered by setting a target point in the work space [30]. The dynamics of the stage is as follows.

$$m_p \ddot{x}_p + d_p \dot{x}_p = \tau + f_h \tag{8.63}$$

The subscripts of x and y are omitted because controllers of the x- and y-axes can be designed separately [30]. Similar to [30], the human transfer function is considered as (8.33) with $K_p = 779$, $K_d = 288$, and $T = 0.18$. The goal is to move x_p to a desired point x_d which is assumed to be origin here. That is, $x_d = 0$.

Based on (8.2), consider the prescribed impedance model as

$$m \ddot{x}_m + d \dot{x}_m + k x_m = f_h \tag{8.64}$$

The method of [15, 30] gives the final parameters as $m = 10, d = 88$, and $k = 0$ after 400 s. The optimal parameters of the prescribed model are now found using the IRL Algorithm 8.1 and the results of two methods are compared. The integral reinforcement is chosen as $T = 0.02$. To do a fair comparison, similar to [15], the initial values of the impedance model are chosen as $m = 50$, $d = 50$, and $k = 0$. Figure 8.6 shows the convergence of the parameters of the impedance model to their optimal values. In fact, the optimal values are $m = 3.21, d = 113.91$, and $k = 44.32$. From this figure, it is obvious that the proposed method finds the set of optimal parameters for the impedance model after 6 iterations, i.e., 1.2 s. This confirms that the proposed method is considerably faster than the method of [15, 30]. To compare

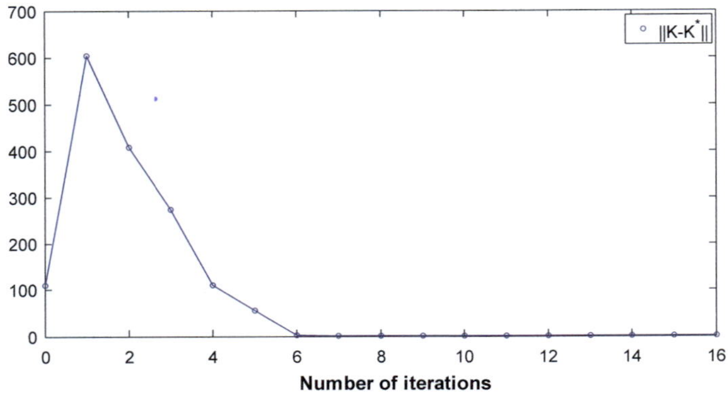

Fig. 8.6 Convergence of the prescribed impedance parameters using the proposed method

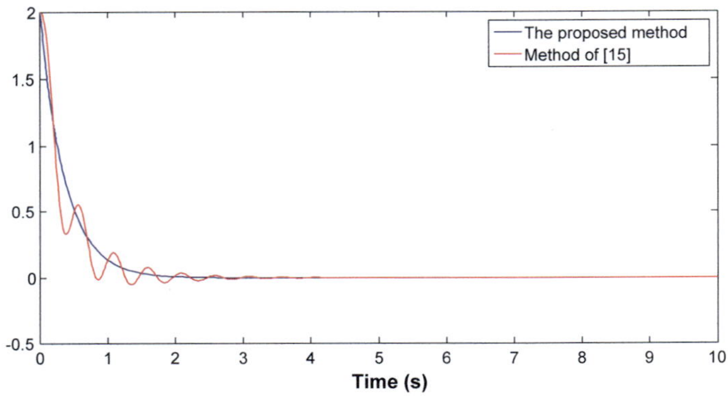

Fig. 8.7 The tracking error for the proposed method and the method of [15]

the performance, two resulting HRI systems are simulated starting from the same initial point. Figures 8.7, 8.8, and 8.9 show the performance of both methods in terms of the tracking error, its derivative, and the required human effort. It is obvious that although the proposed method does not require any knowledge of the system dynamics, it has a better transient performance in terms of the tracking error and requires less human effort to perform the task.

Example 2: two-link planar robot arm

The proposed method is now applied to a two-link robot arm. The links length are assumed $L_1 = 1$ m and $L_2 = 1$ m. The rigid links masses are assumed $m_1 = 0.8$ kg and $m_2 = 2.3$ kg. The gravitational acceleration is $g = 9.8\ \text{m/s}^2$. It is now first shown that the proposed inner loop control design method makes the robot two-link planar robot arm behave like an impedance model. Then, it is shown that the proposed approach successfully finds optimal parameters of the impedance parameters to make the tracking error go to zero.

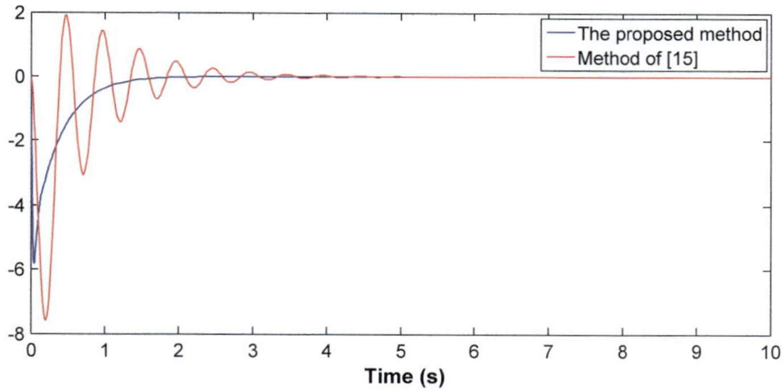

Fig. 8.8 The derivative of tracking error for the proposed method and the method of [15]

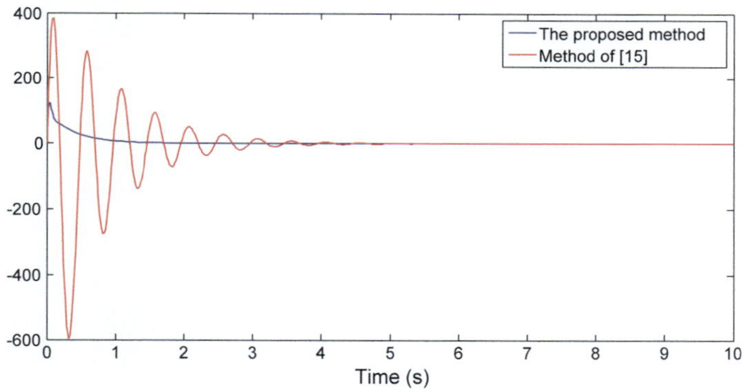

Fig. 8.9 The required human effort for the proposed method and the method of [15]

—Inner Loop

In this simulation, the human force is assumed sinusoidal in both directions. Two different sets of impedance gains are considered for the prescribed impedance model to show that the robot two-link planar robot arm behaves like the impedance model regardless of the impedance model parameters.

Case (1) The first set of matrix gains for the prescribed impedance model in (8.2) are given as

$$\bar{M} = \begin{bmatrix} 10 \\ 01 \end{bmatrix}, \bar{B} = \begin{bmatrix} 50 \\ 05 \end{bmatrix}, \bar{K} = \begin{bmatrix} 20 \\ 02 \end{bmatrix}.$$

Case (2) The second set of matrix gains for the prescribed impedance model in (8.2) are given as

$$\bar{M} = \begin{bmatrix} 20 \\ 02 \end{bmatrix}, \bar{B} = \begin{bmatrix} 150 \\ 015 \end{bmatrix}, \bar{K} = \begin{bmatrix} 200 \\ 020 \end{bmatrix}.$$

Figures 8.10 and 8.11 show the error between trajectories of the prescribed impedance model and the robot arm for both cases. This confirms that the robot arm behaves like the prescribed impedance model, regardless of its parameters. This gives us freedom to tune the impedance parameters in the outer loop such that the tracking error and the required human workload is minimized.

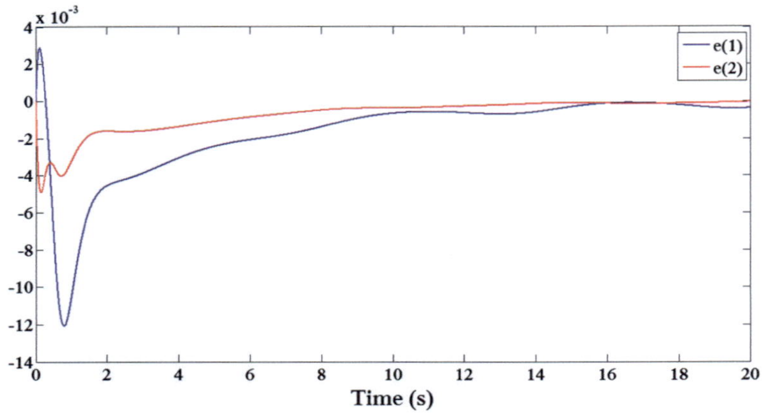

Fig. 8.10 The error between the trajectory of the robot arm and the prescribed impedance model for case 1

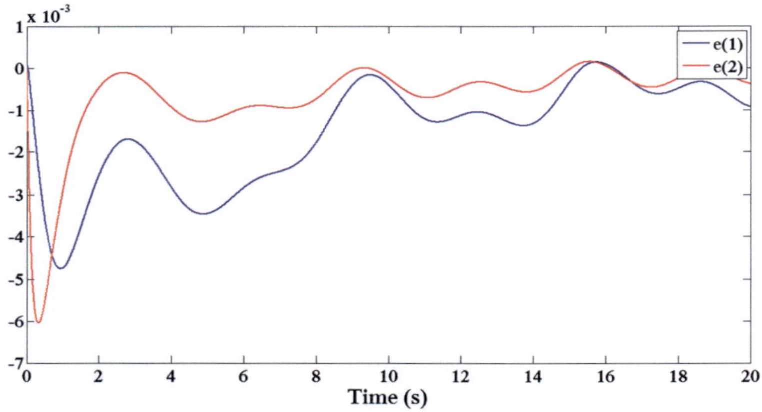

Fig. 8.11 The error between the trajectory of the robot arm and the prescribed impedance model for case 2

—Outer Loop

In this simulation, the mass matrix for the prescribed impedance model is initialized as identity of appropriate dimension and the desired trajectory to be followed by the robot arm is considered as $x_d = [\sin(t), \cos(t)]$. It is also assumed that the human admittance parameters are $K_d = 10$ and $K_p = 20$ and $h = 1$. The matrices A and B in (8.44) then become

$$A = \begin{bmatrix} 0 & 0 & 1 & 0 & 0 & 0 \\ 0 & 0 & 0 & 1 & 1 & 0 \\ 0 & 0 & 0 & 0 & 0 & 0 \\ 0 & 0 & 0 & 0 & 0 & 0 \\ 0.1 & 0 & 0 & 0 & -2 & 0 \\ 0 & 0.1 & 0 & 0 & 0 & -2 \end{bmatrix}, \quad B = \begin{bmatrix} 0 & 0 \\ 0 & 0 \\ 1 & 0 \\ 0 & 1 \\ 0 & 0 \\ 0 & 0 \end{bmatrix} \tag{8.65}$$

The solution to the ARE (8.47) and consequently the control gain (8.46) are then given by

$$P^* = \begin{bmatrix} 507.3640.0007.076 - 0.0003.378 - 0.000 \\ 0.000507.3640.0007.076 - 0.0003.378 \\ 7.0750.0007.170 - 0.0000.0460.000 \\ -0.0007.076 - 0.0007.170 - 0.0000.046 \\ 3.378 - 0.0000.046 - 0.00099.995 - 0.000 \\ -0.0003.3780.0000.046 - 0.00099.995 \end{bmatrix} \tag{8.66}$$

$$K^* = \begin{bmatrix} -70.7580.000 - 71.7040.000\,0.4580.000 \\ 0.000 - 70.7580.000 - 71.704 - 0.0000.458 \end{bmatrix}$$

The optimal gain matrices for the prescribed impedance model yield $\bar{B} = \begin{bmatrix} 71.704 - 0.000 \\ -0.00071.704 \end{bmatrix}$, $\bar{K} = \begin{bmatrix} 70.7580.000 \\ -0.00070.758 \end{bmatrix}$,

It is shown in [38] how the proposed Algorithm 8.1 successfully finds these optimal gains online without requiring the knowledge of the human dynamics. The final gain and kernel matrix found by Algorithm 8.1 are

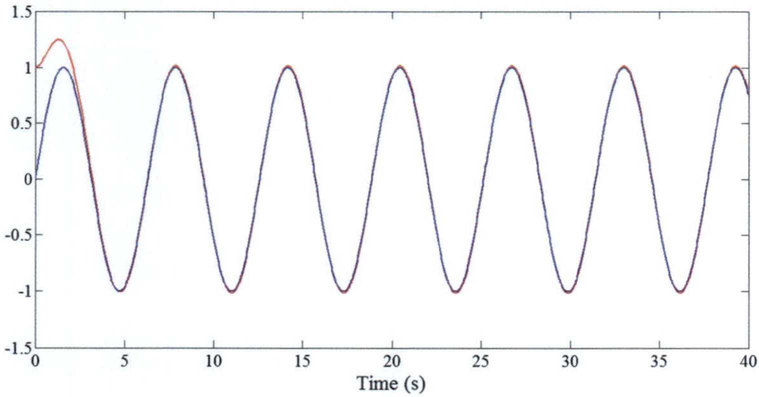

Fig. 8.12 The trajectory of the robot arm and the desired trajectory in x direction

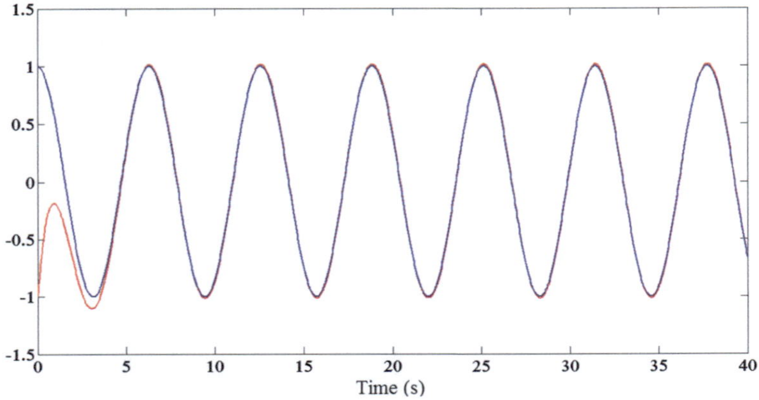

Fig. 8.13 The trajectory of the robot arm and the desired trajectory in y direction

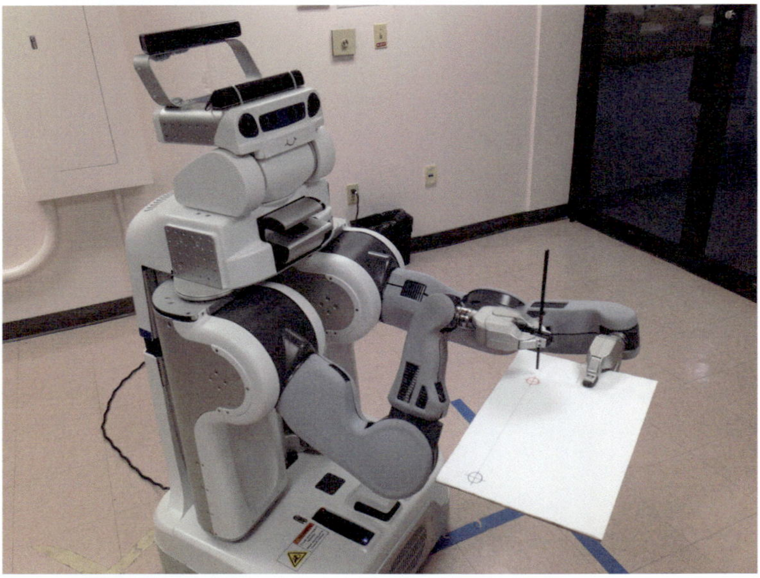

Fig. 8.14 The trajectory of the robot arm and the desired trajectory in y direction

$$P_{16} = \begin{bmatrix} 507.3640.0007.076 - 0.0003.378 - 0.003 \\ 0.000507.3640.0007.076 - 0.0033.378 \\ 7.0750.0007.170 - 0.0000.0460.000 \\ -0.0007.076 - 0.0007.170 - 0.0000.046 \\ 3.378 - 0.0000.046 - 0.00099.984 - 0.000 \\ -0.0003.3780.0000.046 - 0.00099.984 \end{bmatrix} \tag{8.67}$$

$$K_{16} = \begin{bmatrix} -70.7580.000 - 71.7040.000\,0.4610.000 \\ 0.000 - 70.7580.000 - 71.704 - 0.0000.461 \end{bmatrix}$$

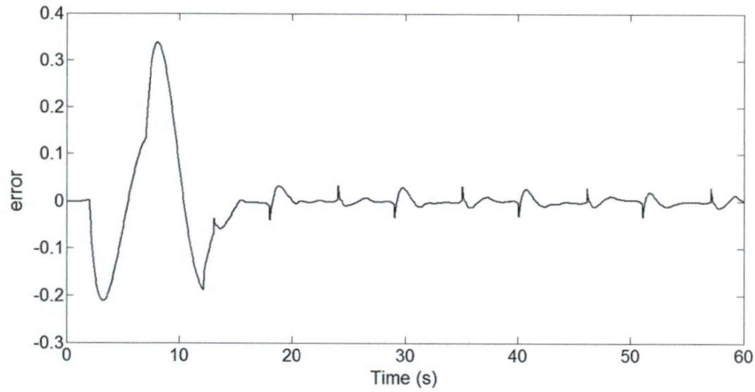

Fig. 8.15 The trajectory of the robot arm and the desired trajectory in y direction

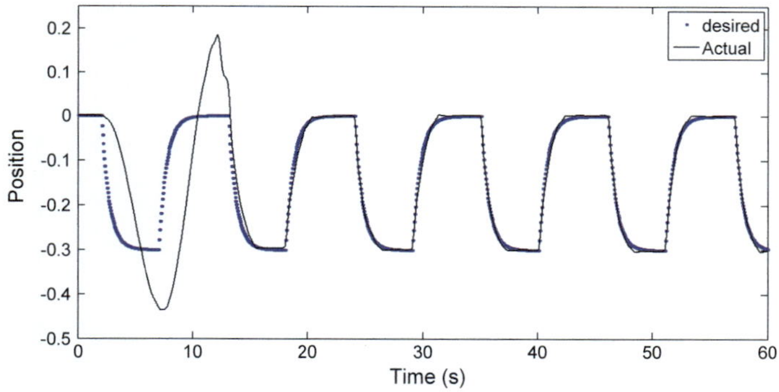

Fig. 8.16 The trajectory of the robot arm and the desired trajectory in y direction

By comparing (8.66)–(8.67), one can see that the solution found by Algorithm 8.1 is the same as the one found by solving the ARE (8.47). Figures 8.12, 8.13 show the results of the proposed method during and after learning. It can be seen that the system states follow the desired trajectories very fast after the learning is finished.

Experimental results: a PR2 robot

A practical experiment is now performed on a PR2 robot shown in Fig. 8.14. In this experiment, the operator holds the PR2 gripper to perform point-to-point motion between red and blue points along the y-axis. Human force is measured using an ATI Mini40 FT sensor attached between the gripper and forearm of the PR2. The controller is implemented using the real-time controller manager framework of the PR2 in ROS Groovy. The real-time loop on the PR2 runs at 1000 Hz and communicates with the sensors and actuators on an EtherCAT network.

Figures 8.15, 8.16 show the results of this experiment. Figure 8.15 shows the error between the trajectory of the robot and the trajectory of the prescribed impedance model in the inner loop. Figure 8.16 shows the results of the outer loop controller. The prescribed impedance model is initialized with a set of nonoptimal parameters and thus the performance of the overall systems is not satisfactory at the beginning. After a short time of interaction, the outer loop controller learns the optimal impedance parameters and therefore the HRI system tracks the desired trajectory successfully.

8.6 Conclusion

A novel control approach is presented to design an optimized assistive human–robot interaction system. The proposed control structure has two loops. In the first loop, i.e., the inner loop, a neuro-adaptive controller is employed to make the unknown nonlinear robot behave like a prescribed impedance model. The proposed inner loop does not require the knowledge of the task or the prescribed impedance parameters. This decomposes the robot-specific control design from the task specific design. In the second loop, i.e., the task-specific loop, which includes the human, the robot and their interaction, the optimal parameters of the prescribed impedance parameters are found based on the human skill and the given task to assist the human to perform the task with less effort and optimal performance.

References

1. Baron S, Kleinman DL, Levison WH (1970) An optimal control model of human response. part ii: prediction of human performance in a complex task. Automaica 6:371–383
2. Bertsekas DP (2012) Dynamic programming and optimal control: approximate dynamic programming, 4th edn. Athena Scientific, Massachusetts
3. Duchaine V, Gosselin C (2009) Safe, stable and intuitive control for physical human-robot interaction. In: Proceedings of the IEEE international conference on robotics and automation. Kobe, Japan, pp 3383–3388
4. Frankin S, Wolpert DM, Franklin DM (2012) Visuomotor feedback gains upregulate during the learning of novel dynamics. J Neurophysiol 108:467–478
5. Furuta K, Kado Y, Shiratori S (2006) Assisting control in human adaptive mechatronics-single ball juggling. In: Proceedings of the IEEE international conference on control applications. Munich, Germany, pp 545–550
6. Ge SS, Hang CC, Woon LC, Chen XQ (1998) Impedance control of robot manipulators using adaptive neural networks. Int J Intell Control Syst 2:433–452
7. Ge SS, Harris CJ (1998) Adaptive neural network control of robotic manipulators. World Scientific, Singapore
8. Ge SS, Lee TH, Wang ZP (2001) Adaptive neural network control for smart materials robots using singular perturbation technique. Asian J Control 3:143–155
9. Gribovskaya E, Kheddar A, Billard A (2011) Motion learning and adaptive impedance for robot control during physical interaction with humans. In: IEEE international conference on robotics and automation. Shanghai, China, pp 4326–4333
10. Hogan N (1985) Impedance control: an approach to manipulation. i: theory. ii: implementation. iii: applications. ASME Trans J Dyn Syst Meas Control 107:1–24
11. Huang L, Ge SS, Lee TH (2002) Neural network based adaptive impedance control of constrained robots. In: Proceedings of the IEEE international symposium on intelligent control. Vancouver, Canada, pp 615–619
12. Hussain S, Xie SQ, Jamwal PK (2013) Adaptive impedance control of a robotic orthosis for gait rehabilitation. IEEE Trans Cybern 43:1025–1034

13. Ikeura R, Moriguchi T, Mizutani K (2002) Optimal variable impedance control for a robot and its application to lifting an object with a human. In: Proceedings of the 11th IEEE international workshop robot-human interactive communication, Berlin, Germany, pp 500–505
14. Jung S, Hsia TC (1998) Neural network impedance force control of robot manipulator. IEEE Trans Ind Electron 45:451–461
15. Kosuge K, Furuta K, Yokoyama T (1987) Virtual internal model following control of robot arms. In: Proceedings of the IEEE international conference on robotics and automation, pp 1549–1554
16. Kurihara K, Suzuki S, Harashima F, Furuta K (2004) Human adaptive mechatronics (HAM) for haptic system. In: Proceedings of the 30th IEEE Annual Conference of the Industrial Electronics, Busan, Korea, pp 647–652
17. Lewis FL, Dawson DM, Abdallah CT (2003) Robot manipulator control: theory and practice, 2nd edn. CRC Press, Florida
18. Lewis FL, Vraibe D, Syrmos V (2012) Optimal control, 3rd edn. Wiley, New Jersey
19. Lewis FL, Vraibe D, Vamvoudakis KG (2014) Reinforcement learning and feedback control: using natural decision methods to design optimal adaptive controllers. IEEE Control Syst Mag 32:76–105
20. Lewis FL, Yesildirek A (1995) Neural net robot controller with guaranteed tracking performance. IEEE Trans Neural Netw 6:703–715
21. Lewis FL, Yesildirek A, Liu K (1996) Multilayer neural net robot controller with guaranteed tracking performance. IEEE Trans Neural Netw 7:388–399
22. Li Y, Ge SS (2014) Human? robot collaboration based on motion intention estimation. IEEE/ASME Trans Mechatron 19:1007–1014
23. Li Y, Ge SS, Yang C (2011) Impedance control for multi-point human-robot interaction. In: Proceedings of the 8th Asian Control Conference, Kaohsiung, Taiwan, pp 1187–1192
24. Mitsantisuk C, Ohishi K, Katsura S (2011) Variable mechanical stiffness control based on human stiffness estimation. In: Proceedings of the IEEE international conference on mechatronics. Istanbul, Turkey, pp 731–736
25. Modares H et al (2015) Optimized assistive human? robot interaction using reinforcement learning. IEEE Trans Cybern. doi:10.1109/TCYB.2015.2412554
26. Oh S, Woo H, Kong K (2014) Frequency-shaped impedance control for safe human-robot interaction in reference tracking application. IEEE/ASME Trans Mechatron 19:1907–1916
27. Powell WB (2007) Approximate dynamic programming: solving the curses of dimensionality. Wiley-Interscience, New York
28. Stulp F et al (2012) Model-free reinforcement learning of impedance control in stochastic environments. IEEE Trans Auton Mental Develop 4:330–341
29. Sutton RS, Barto AG (1998) Reinforcement learning: an introduction. Cambridge University Press, Cambridge
30. Suzuki S, Furuta K (2012) Adaptive impedance control to enhance human skill on a haptic interface system. J Control Sci Eng 2012:1–10
31. Tsuji T, Tanaka Y (2005) Tracking control properties of human-robotic systems based on impedance control. IEEE Trans Syst Man Cybern Part A 35:523–535
32. Tsumugiwa T, Yokogawa R, Hara K (2002) Variable impedance control based on estimation of human arm stiffness for human-robot cooperative calligraphic task. In: Proceedings of the IEEE international conference on robotics and automation, pp 644–650
33. Tustin A (1947) The nature of the operator's response in manual control and its implications for controller design. J Inst Electric Eng 94:190–202
34. Vrabie D et al (2009) Adaptive optimal control for continuous-time linear systems based on policy iteration. Automatica 45:447–484
35. Vrabie D, Lewis FL (2009) Neural network approach to continuous-time direct adaptive optimal control for partially unknown nonlinear systems. Neural Netw 22:237–246
36. Vraibe D, Vamvoudakis KG, Lewis FL (2012) Optimal adaptive control and differential games by reinforcement learning principles, control engineering series. IET Press

37. Wang C et al (2013) Continuous critic learning for robot control in physical human-robot interaction. In: Proceedings of the 13th international conference on control, automation and system. Gwangju, Korea, pp 833–838
38. Xu G, Song A (2009) Adaptive impedance control based on dynamic recurrent fuzzy neural network for upper-limb rehabilitation robot. In: IEEE international conference on control, automation. Christchurch, New Zealand, pp 1376–1381

Chapter 9
Regret-Based Allocation of Autonomy in Shared Visual Detection for Human–Robot Collaborative Assembly in Manufacturing

S.M. Mizanoor Rahman, Zhanrui Liao, Longsheng Jiang and Yue Wang

9.1 Introduction

Assembly is one of the important steps in manufacturing due to extensive utilization of manpower and utilities in this step, and the overall manufacturing performance largely depends on the assembly. However, manual assembly is usually tedious, less efficient, and harmful to musculoskeletal systems of human workers. On the other hand, fully automated assembly is usually expensive and less flexible. In this situation, human–robot collaboration (HRC) in assembly exploiting the complementary advantages of human and robot will significantly enhance the assembly performance [1, 2]. Recent advancements in light-weight and low-cost industrial manipulators such as Baxter [3], Kinova [4], Kuka [5], Universal [6], ABB [7], Motoman [8], Fanuc [9], Epson [10], and others have raised the possibility of effectiveness of such collaboration.

In human–robot collaborative assembly in manufacturing, various sensing and detection phenomena are necessary to maintain quality and safety, e.g., detection of correct assembly parts and correct assembly style is necessary to ensure assembly quality [11, 12], and detection of obstacles is necessary to ensure operation

This chapter was developed and enhanced from an earlier paper published as [12] © 2016 IEEE.

S.M. Mizanoor Rahman · Z. Liao · L. Jiang · Y. Wang (✉)
Department of Mechanical Engineering, Clemson University, Clemson, SC, USA
e-mail: rrahman@clemson.edu

Z. Liao
e-mail: zhanrul@g.clemson.edu

L. Jiang
e-mail: longshj@g.clemson.edu

Y. Wang
e-mail: yue6@g.clemson.edu

Y. Wang and F. Zhang (eds.), *Trends in Control and Decision-Making for Human–Robot Collaboration Systems*, DOI 10.1007/978-3-319-40533-9_9

safety [13]. These detections can be performed by the human co-worker manually using human vision ("manual detection"), however it needs additional concentration of the human that may impose physical and mental workload and thus may increase fatigue and reduce efficiency. Visual sensor-based autonomous sensing and detection by the robot is an alternative approach ("autonomous detection"). However, robot's visual sensing is usually prone to environmental disturbances (e.g., sudden shadow of human body and piled materials) and perturbation (e.g., wrong input parts, arrival of input parts at wrong time and position), and thus the sensing is uncertain and erroneous. The robot's visual sensing may not be reliable even in ideal environment (without disturbances and perturbation) due to limitations in sensing technologies and calibration. All these errors and uncertainties in sensing will lead to false alarms and missed detections by the robot. Thus, it is assumed that autonomous detection with human's intervention is more useful, practical, and flexible in some situations especially in human–robot collaborative assembly in manufacturing ("shared sensing/detection") [14]. The allocation of autonomy between human and robot in shared sensing, i.e., 'what the human and the robot will share and how they will share' is hence critical.

Optimal allocation of autonomy in shared detections under uncertainty is challenging, which can be performed using a Bayesian sequential decision-making strategy [15]. However, such optimal strategy may not be similar as the style of human decision-making under uncertainty [16]. It has been shown that teams with members sharing similar mental models perform better than teams with members having more accurate but less similar mental models in decision-making [17]. Hence, it is necessary to have a similar decision-making strategy between the human and the robot (sensors) regarding the shared detection phenomena under uncertainty for collaborative assembly. Since it is difficult to change how humans think, we propose to develop a humanlike decision-making strategy for the robot in order to maintain such similarity [18, 19].

Regret has been shown to play a critical role in rational human decision-making under uncertainty. Humans experience regret when they perceive that they could achieve better if they could adopt another option in decision-making [20–22]. Humans always anticipate potential regret during decision-making and tend to make choices to avoid such regretful experiences, i.e., humans prefer to make suboptimal decisions to avoid potential regret in making a wrong decision [18], [23]. Again, humans tend to be risk-averse when posed with potential gains, and risk-seeking when posed with potential losses, which will also influence the intensity of human's regret in the intervention in the shared detection. Regret-based decision-making is usually studied in economics [22], business [21], transportation management [20], etc. We believe that a regret-based, humanlike decision strategy, will benefit the decision-making regarding allocation of autonomy between human and robot in shared detection phenomena in human–robot collaborative assembly as well. The regret-based decision-making is suboptimal in term of detection accuracy, but such suboptimal approach fits with human's natural decision-making style, which will positively impact the collaboration and hence the overall assembly performance.

We in this chapter include regret analysis in Bayesian sequential optimal decision-making strategy of allocation of autonomy regarding the shared detection phenomena during human–robot collaborative assembly to provide a risk-based and humanlike decision-making framework. By the term "allocation of autonomy" in shared detection/sensing, we mean whether (i) only the human will make additional observations ("manual mode"), (ii) only the robot will make additional observations ("autonomous mode"), and (iii) both the human and the robot will make additional observations ("mixed mode") in the shared detection/sensing during the collaborative assembly. The strategy dynamically switches among manual, autonomous, and mixed modes of autonomy allocation based on modified Bayes risk functions and detection capability of both the human and the robot co-worker at the current time step or at the current run time. As an illustration, we consider a case regarding the detection of correctness of assembly parts by the robot before the parts are manipulated by the robot to the human for assembly. The decision of correctness of the parts is performed by the robot based on its visual detection facility. The robot will take more observations to make a final decision regarding the correctness of the parts based on its current detection capability. However, the human may join in the detection phenomenon through making one or more manual observations and sharing the information to the robot. This illustration serves as a representative shared detection scenario in human–robot collaborative assembly in manufacturing in particular, and in all types of HRC in general. We then evaluate the effectiveness of the framework in a human–robot collaborative assembly task in the laboratory setting.

The remaining of the chapter is organized as follows: Sect. 9.2 introduces the hybrid cell developed for in-lab pilot study of the regret-based Bayesian decision-making framework in human–robot collaborative assembly in manufacturing. Section 9.3 explains the usual detection phenomena in the human–robot collaborative assembly, and formulates a problem on the detection phenomena. Section 9.4 introduces in details the Bayesian sequential optimal decision-making algorithm for allocation of autonomy through switching among manual, autonomous, and mixed modes regarding the detection phenomena. Section 9.5 introduces a method of inclusion of regret in Bayesian optimal decision-making algorithm which results in a suboptimal decision approach as well as explains the effects of human's risk-seeking and risk-averse attitude in decision-making. Section 9.6 illustrates the decision-making approaches. Section 9.7 describes the implementation plan of the regret-based Bayesian decision-making framework in the developed hybrid cell for human–robot collaborative assembly. Section 9.8 describes the experimental evaluation method of the regret-based Bayesian decision-making framework, and Sect. 9.9 presents the evaluation results with analyses. Section 9.10 contains the concluding remarks based on the evaluation results and future works.

9.2 The Hybrid Cell for Human–Robot Collaborative Assembly

Based on our case study experience in an automotive manufacturing company, we identified the center console assembly where three different automotive parts namely faceplate, switch rows, and I-drive as shown in Fig. 9.1 are assembled together to

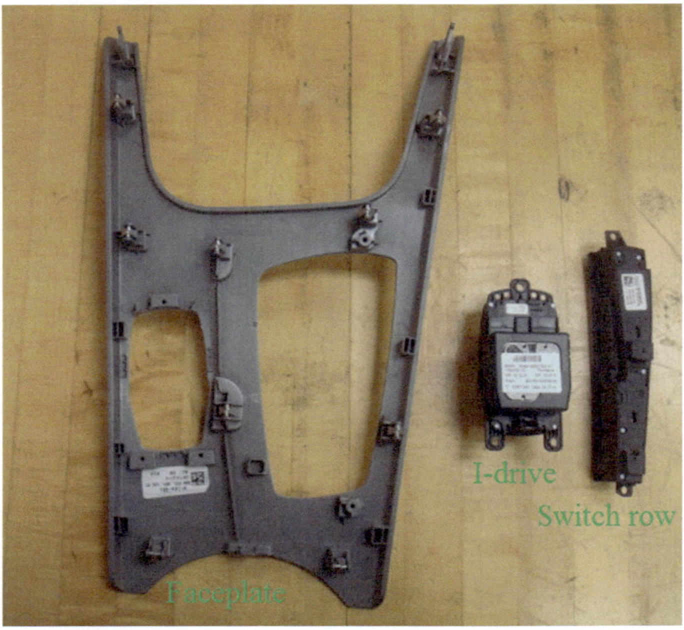

Fig. 9.1 The automotive parts that are assembled to produce a center console product [courtesy: BMW Manufacturing Co., Spartanburg, SC, USA]

form a center console product. Currently, the assembly is performed manually. In the factory environment as shown in Fig. 9.2, the human co-worker collects the parts from different shelves, puts the parts on a table, and then assembles the parts together using a powered screwdriver. This assembly task can be performed in collaboration between a human co-worker and a robot where the entire task is divided into multiple subtasks and the subtasks are assigned to the human co-worker and the robot. In this chapter, we present how this assembly task can be performed in collaboration between a human and a robot. However, this assembly task may be considered as a representative of all human–robot collaborative assembly tasks in general.

To facilitate the proposed collaborative assembly task in an in-lab pilot study, we developed a human–robot hybrid cell as shown in Fig. 9.3, where a human and a robot can collaborate to assemble the automotive parts into a finished (assembled) center console product [1, 11, 13, 24]. A hybrid cell is considered as a dedicated assembly environment where a human and a robot perform assembly simultaneously side by side without any physical separation. The hybrid cell is usually preferred for collaborative assembly due to its various advantages such as convenience in task allocation and scheduling, ease in communication, resource mobilization, and supervision [1, 11, 13, 24].

As shown in the figure, the hybrid cell we considered in this work includes two agents-a humanoid robot (Baxter robot made by Rethink Robotics) and a human co-worker. The automotive parts are put on different locations ("A" and "B") on a

Fig. 9.2 The factory environment where the center console is assembled in current practice [courtesy: BMW Manufacturing Co., Spartanburg, SC, USA]

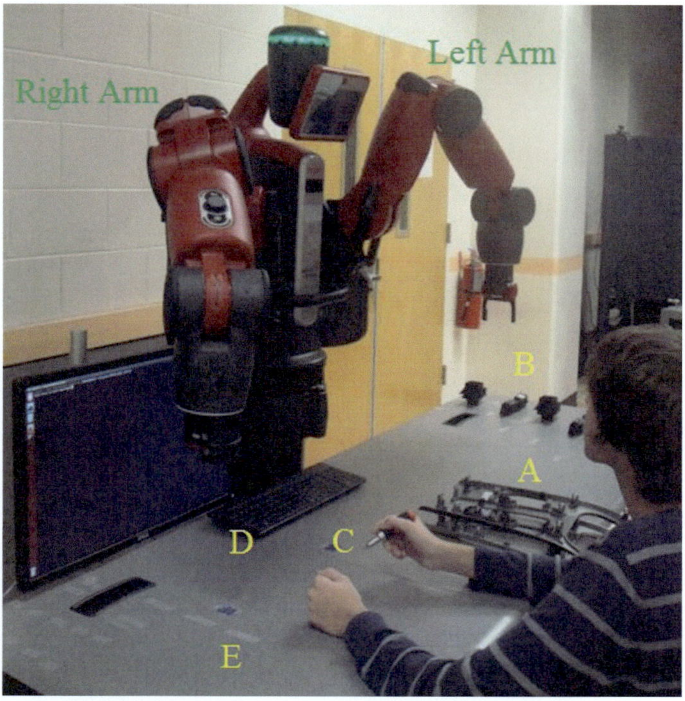

Fig. 9.3 The hybrid cell developed for in-lab study of human–robot collaborative assembly (© 2016 IEEE. Reprinted, with permission, from [12])

table. The human co-worker sits in front of the robot to collaborate with the robot for the assembly (the human can also perform the assembly with standing posture depending on the nature and requirements of the assembly tasks). The faceplate parts are piled on the table at the right side of the human (at "A"). The small parts (I-drive and switch rows) are kept at "B." At first, the human picks a faceplate part from "A" using his/her dominant hand and places it in front of him/her (at "C"). The robot then sequentially picks the I-drive and switch rows parts from "B," and places the parts in front of the human co-worker (near "C") using its left arm. Before picking a part, the robot uses its vision-based detection system embedded in its left wrist to identify whether the part is correct. Once a correct part is manipulated to the human by the robot, the human attaches the part with the large part (faceplate) manually at "C" using screws and a screwdriver. Once the attachment of the I-drive and switch rows parts with the faceplate part is finished, the human exposes the finished part at "D" so that the robot can use its vision-based detection system embedded in its right wrist to determine whether the assembly style is correct. Then, the human places the assembled and inspected (by the robot) center console product in his/her left side on the table (at "E"). The face screen of the robot displays the assembly related instructions (e.g., assembly sequence, style) in real time. A computer monitor along with a mouse and a keyboard help the human co-worker and the experimenter add inputs to the robot programming codes based on situations. An appropriate level of illumination is maintained in the hybrid cell area.

9.3 Detection Problem Formulation with Focus on the Selected Assembly Task

9.3.1 Description of the Problem

As shown in Fig. 9.3, the robot needs to make decisions on various detection phenomena during the collaborative assembly. Based on our experiences, three types of detections are most useful during the assembly: (i) detecting whether the input assembly parts are correct, (ii) detecting whether the parts are assembled in correct style, and (iii) detecting whether there is any obstacle or unsafe event along the path of the robot manipulator. The first two detections are related to assembly quality, and the third detection phenomenon is related to the safety of the robot, human co-worker, materials, and the environment. In all three detection cases, the robot (sensor) at first makes a visual observation on the part or on the event and then makes its decision. In this chapter, we want to confine to the first two types of detection phenomena.

Figure 9.4a shows a method that the robot uses to detect whether an input assembly part is correct. We first consider the exclusively autonomous sensing mode. As illustrated, before picking the part by the robot for manipulating it to the human co-worker for assembly, the robot uses the vision sensor embedded in its left wrist to take the image of the barcode attached to the part and then processes the image

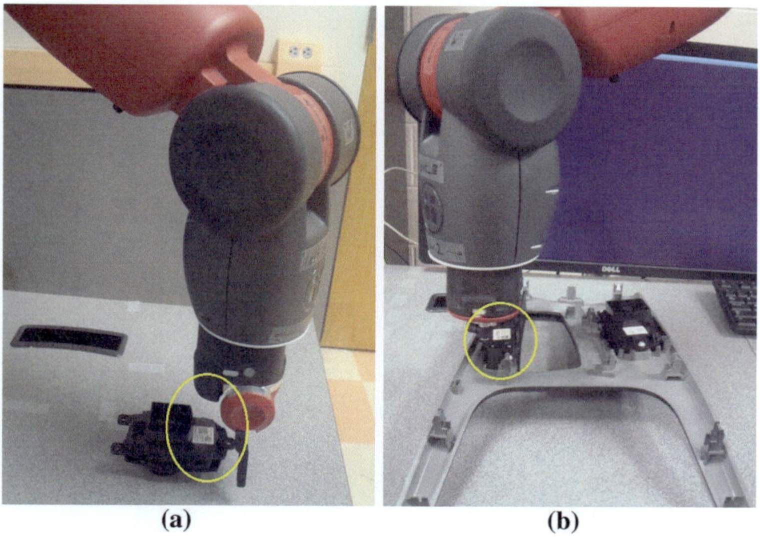

Fig. 9.4 a Detection of correctness of an input assembly part by the robot using the camera embedded in its left wrist, and b Detection of correctness of the assembly style by the robot in an assembled product using the camera embedded in its right wrist (© 2016 IEEE. Reprinted, with permission, from [12])

using OpenCV library to extract the unique information for the part archived in the barcode, and then decides about the correctness of the input part comparing the barcode information with the reference information. If the part is decided as the correct one, the robot will pick the part using the left arm and manipulate it to the human co-worker for assembling with other parts, otherwise the robot will not pick the part. Instead, the robot will issue a warning message displayed in its face screen to inform the human co-worker of the status of the input part. In this case, the experimenter replaces the wrong part by a new part, the robot repeats the detection process, and continues the part manipulation to the human co-worker for assembly. Once the parts are manipulated to the human, the human assembles the parts together following the instructions displayed in the robot face screen. Once the assembly is finished, the human places the assembled product in front of the robot. As illustrated in Fig. 9.4b, the robot uses the vision sensor embedded in its right wrist to take an image of the barcode attached onto each assembled part separately and then processes the image using OpenCV library to extract the orientation of the barcode. Then, the OpenCV compares the orientation of the barcode of each of the attached parts with the reference orientation, and decides whether the style of the assembly is correct. Figure 9.5a, b illustrate the detection cases regarding correct and wrong assembly style respectively. If the style is decided as the correct one, the robot will pick the finished product using the right arm and manipulate it to store in a shelve (currently, the human manually stores the finished product in a place on his/her left side on the table). Otherwise, the robot will not pick the finished product to dispatch it. Instead,

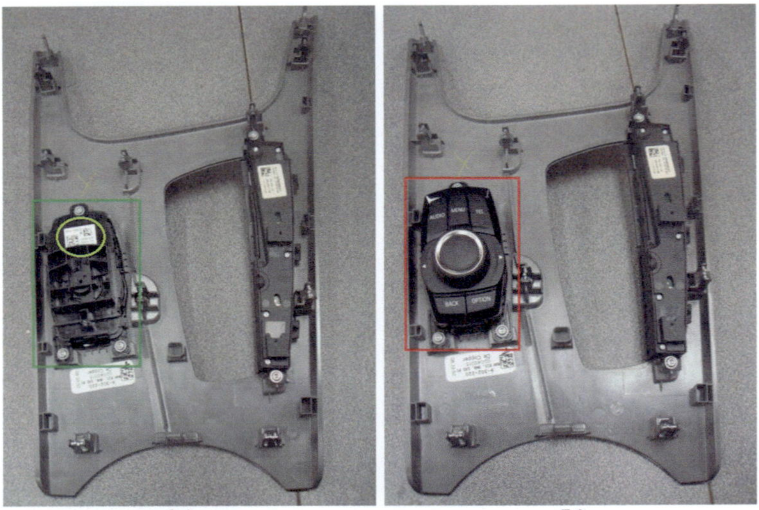

(a) (b)

Fig. 9.5 **a** Example of a correct assembly style. **b** Example of a wrong assembly style

the robot will issue a warning message displayed in its face screen to inform the human co-worker of the status of the correctness of the assembled product.

In both sensing/detection cases as illustrated above, there are uncertainties in the observations of the robot. The major causes of the uncertainties are: (i) disturbances such as shadow of human or robot body and of piled materials on the target part or on the assembled product, (ii) perturbation such as critically wrong or similar input parts, (iii) poor illumination of the factory room, (iv) limitations in sensing technologies and calibration, etc. All these errors and uncertainties in sensing will lead to wrong decision-making by the robot, which indicates the necessity of the human co-worker's intervention in the decision-making. The human co-worker can intervene the decision-making process by providing more information to the robot through making additional manual observations on the targeted part or product. The human co-worker's interventions in the observations for decision-making result in the shared-detection phenomena [14]. The human is usually intelligent and robust in observations and decision-making. Hence, human's intelligence and robustness will be complementary to the robot's capability and thus the human's interventions (additional observations) will help the robot make correct decisions for the detection phenomena. We believe that these are the benefits of the shared-detection approach.

In this chapter, for illustrative purpose, we will formulate the problem with the first type of detection phenomenon (detecting whether the input assembly parts are correct) as shown in Fig. 9.4a only. There are three modes in the decision-making regarding the shared detection/sensing phenomena: (i) manual sensing mode-the robot makes a decision on the detection of the correctness of the assembly part and seeks help from the human for manual observation. In other words, the human makes observations on the target part, supplies the observation information to the robot and thus helps the robot make the decision about the correctness of the part, (ii) autonomous sensing mode–the robot makes an observation on the target part using its vision system and makes the decision on the detection of the correctness of the assembly part, and (iii) mixed sensing mode–the robot makes an observation on the target part using its vision system and makes the decision on the detection of the correctness of the assembly part, but the robot can also ask the human to make additional observations so that the robot can use the observation information to make the decision about the correctness of the part. The selection of the mode will be real time which depends on various dynamic parameters such as (i) probability of correctness in the observation (sensing) by the human, (ii) probability of correctness in the observation (sensing) by the robot, (iii) costs and risk involved in making an additional observation by the human or the robot, (iv) human worker's risk-seeking or risk-averse tendency, etc.

It is clear that in all three modes, the decision on the detection of the correctness of the assembly part is actually made by the robot, but the robot sometimes itself makes additional observations or asks the human co-worker to make additional observation so that the robot can make correct decisions. Again, the change in the modes will not occur randomly. Instead, the changes will occur when it is deemed necessary based on the status of one or more of the dynamic parameters as mentioned above.

9.3.2 Problem Formulation

We will derive the problem for our particular case where a human collaborates with a robot that detects an assembly part. Denote o as the targeted object. Let the binary random variable $X(o)$ represent the state of the object o,

$$X(o) = \begin{cases} 1 & \text{if the target object is } o \\ 0 & \text{if the target object is not } o. \end{cases} \tag{9.1}$$

We assume that the sensing capacity of the robot is denoted by β_r and the robot can sense only one object at a given time step (run). Now, we define a binary observation variable $Y(o, run)$ for the robot, when $Y(o, run) = 1$, $Y(o, run) = 0$ indicate that the robot observes the object being o and not o, respectively. Let β_H be the sensing capability of the human (probability of correctness in manual sensing). Similarly, $Y_H(o, run) = 1$, $Y_H(o, run) = 0$ indicate that the human observes the object being o and not o, respectively.

Modes of autonomous and manual sensing will be switched. Let us consider $W_h(run)$ as the model for the human workload (human utilization rate) as the following,

$$W_h(run) = W_h(previous\,run) + \frac{m(run) - W_h(previous\,run)}{\tau}. \tag{9.2}$$

$$m(run) = \begin{cases} 1 & \text{if manual mode of sensing} \\ 0 & \text{if autonomous mode of sensing.} \end{cases} \tag{9.3}$$

$m(run)$ denotes the sensing mode of the robot at the current run and the time constant $\tau > 0$ reflects the extent to which previous utilization affects the current utilization. $W_h(run)$ is bounded between 0 and 1.

Let $\alpha(run)$ be the difficulty or easiness of the sensing task in a run where $\alpha(run) = 1$ means that the sensing task is easy and $\alpha(run) = 0$ means that the sensing task is difficult.

9.4 Bayesian Sequential Decision-Making Algorithm for Allocation of Autonomy

A probabilistic Bayesian sequential decision-making strategy is proposed in this section to obtain the optimal choice among the three available options regarding the autonomy allocation in sensing. We denote the decision rule as Δ, the number of observations as L, and the prior probability of object being correct or wrong as $P(X(o) = x(o); run), x(o) = 0, 1$. Then, the expected Bayes risk function is defined as follows:

$$\begin{aligned} risk(\Delta, L, P(X(o) = x(o); run), run) = R(\Delta, L|X(o) = 0; run)P(X(o) = 0; run) \\ +R(\Delta, L|X(o) = 1; run)P(X(o) = 1; run). \end{aligned} \tag{9.4}$$

where Δ is a $2 \times (L + 1)$ matrix, for general $L \geq 1$ observations. The definition of $R(\Delta, L|X(o) = x(o); run)$ is as follows, if in the autonomous mode,

$$\begin{aligned} R(\Delta, L|X(o) = x(o); run) &\triangleq R^A(\Delta, L|X(o) = x(o); run) \\ &= \text{Prob(decide } X(o) \neq x(o))|X(o) = x(o)) \\ &\quad + c_r(o, run)\mathbf{E}[N(\phi)|X(o) = x(o)] \\ &= COST\Delta B_r^{x(o)} + c_r(o, run)\mathbf{E}[N(\phi)|X(o) = x(o)]. \end{aligned} \tag{9.5}$$

If in the manual mode,

$$\begin{aligned} R(\Delta, L|X(o) = x(o); run) &\triangleq R^M(\Delta, L|X(o) = x(o); run) \\ &= \text{Prob(decide } X(o) \neq x(o))|X(o) = x(o)) \\ &\quad + c_H(o, run)\mathbf{E}[N(\phi)|X(o) = x(o)] \\ &= COST\Delta B_H^{x(o)} + c_H(o, run)\mathbf{E}[N(\phi)|X(o) = x(o)]. \end{aligned} \tag{9.6}$$

The function $R(\Delta, L|X(o) = x(o); run)$ is the conditional Bayes risk providing state $X(o) = x(o)$. For this work, the sensor on the agent (i.e., robot camera, human eye) is always assumed to be good, which restricts the amount of "reasonable" decision rules to a small number. In the conditional Bayes risk functions (9.5) and (9.6), $c_r(o, run), c_H(o, run)$ is the cost induced by taking more observations under the autonomous and manual mode, respectively. It is assumed that human observation cost is heavier than that in the autonomous mode, i.e., $c_H(o, run) > c_r(o, run)$. $B^{x(o)}$, $B_H^{x(o)}$ are the columns in the conditional probability matrix of the autonomous and manual mode, respectively [18, 19]. $N(\phi)$ is the time of stoppage for taking observation and arriving at a decision regarding whether the object is correct or wrong. Because the observation is random, $N(\phi)$ is a random variable. $COST$ is the penalty of making a decision on the object status/state given the actual state. To simplify the problem, currently a uniform cost assignment (UCA) is assumed such that,

$$COST = \begin{cases} 0 & \text{if decision is consistent with actual state} \\ 1 & \text{otherwise} \end{cases}$$

The initial default mode can be chosen as the autonomous mode without loss of generality. The minimum expected Bayes risk $risk^*$ for an arbitrary run is found by search for a pair of decision rule and observation length (Δ^*, L^*), as follows.

$$\begin{aligned} & risk^*(\Delta^*, L^*, P(X(o) = x(o); run), run) \\ = {} & R(\Delta^*, L^*|X(o) = 0, run)P(X(o) = 0; run) \\ & + R(\Delta^*, L^*|X(o) = 1, run)P(X(o) = 1; run). \end{aligned} \tag{9.7}$$

In the special case when no observation is taken, the robot must directly make a decision that object is wrong or correct. The decision is made according to the expected Bayes risk functions,

$$risk(\Delta = \text{decide object wrong}, L = 0, P(X(o) = 1; run)) = P(X(o) = 1; run). \tag{9.8}$$

$$risk(\Delta = \text{decide object correct}, L = 0, P(X(o) = 1; run)) = P(X(o) = 0; run). \tag{9.9}$$

9.5 Inclusion of Regret in Bayesian Decision-Making Algorithm for Allocation of Autonomy

The expected Bayesian risk function can be translated as expected utility in expected utility theory. In behavior economics literature [16, 21, 22], it has been suggested that expected utility theory is not obeyed by a human decision-maker. People react differently to different decision-making circumstances. For instance, when humans face a choice problem of two mutually alternative options, each has the same expected utility as the other, systematically, they change their decisions from one to another depending on different probability distributions to the consequences in the options. This change is recorded as the tendency towards a gain with higher certainty (low risk, risk-aversion) or a loss with more uncertainty (high risk, risk-seeking). Other violations of expected utility theory like the Allais paradox are evidenced through experiments [25]. In a word, humans do not make decisions based on the optimization of the expected utility function. It is inferred that the expected utility functions must have ignored something that is important in human decision-making. Regret theory is an alternative for expected utility theory that can explain human's risk-seeking/aversion behaviors. Regret theory derives from the psychological idea that people feel regret when they think that they could achieve better if they had chosen the alternative options. Thus the theory asserts, when making a decision, people systematically predict the regret and bias certain options more in order to avoid it. Evidence indicates that most people indeed act as predicted by regret theory [16, 22].

What is more, in the study of teamwork, the overall performance of a team has been shown to be better if all the team members share similar mental models [17]. In an HRC team, if the robot makes decisions exactly like the human worker does, the collaboration is more predictable for the human. The human, thus, is more satisfied and the overall performance will increase. Adopting regret in the standard Bayesian sequential strategy for a robot is a natural means to make the robot decision-making more humanlike. We propose a method to integrate regret analysis into the Bayesian strategy introduced in Sect. 9.4.

If the manual (autonomous) mode is chosen when the actual state is $X(o) = 1$, see Eqs. (9.5) and (9.6), then the conditional Bayes risk is given by $R^A(\Delta, L|X(o) = 1; run)$ and $R^M(\Delta, L|X(o) = 1; run)$, respectively. When a robot takes more observations, the mutually alternative options are the manual mode and the autonomous mode. Regret is predicted when actually choosing the manual (autonomous) mode and giving up the autonomous (manual) mode. In other words, the regret associated with choosing the manual (autonomous) mode depends on both the manual mode and the autonomous mode. Therefore, the modified conditional Bayes risks are,

$$\tilde{R}^A(\Delta, L|X(o) = 0, run) = R_0^A + RGT(R_0^A - R_0^M). \tag{9.10}$$

$$\tilde{R}^M(\Delta, L|X(o) = 0, run) = R_0^M + RGT(R_0^M - R_0^A). \tag{9.11}$$

$$\tilde{R}^A(\Delta, L|X(o) = 1, run) = R_1^A + RGT(R_1^A - R_1^M). \tag{9.12}$$

$$\tilde{R}^M(\Delta, L|X(o) = 1, run) = R_1^M + RGT(R_1^M - R_1^A). \tag{9.13}$$

where $RGT(\cdot)$ is the regret function, R_0^A (R_1^A) and R_0^M (R_1^M) are the short notations for $R^A(\Delta, L|X(o) = 0, run)$ ($R^A(\Delta, L|X(o) = 1, run)$) and $R^M(\Delta, L|X(o) = 0, run)$

$(R^M(\Delta, L|X(o) = 1, run))$, respectively, with the understanding that the discussion is restricted to the same Δ, L and run.

Due to the symmetry of the structure, without loss of generality, we restrict the discussion to $\tilde{R}_1^A$, the short notation of $\tilde{R}^A(\Delta, L|X(o) = 1, run)$, when exploring the properties of the regret function. As indicated in [22], the regret function candidates must enable several properties of the modified conditional Bayes risk function $\tilde{R}_1^A$. First, $\partial \tilde{R}_1^A / \partial R_1^A > 0$, indicating that $\tilde{R}^A$ is increasingly biased if R^A increases. Second, $\partial \tilde{R}_1^A / \partial R_1^M < 0$, meaning the increase of the rival conditional Bayes risk R_1^M leads to the decrease of the bias toward $\tilde{R}_1^A$. What is more, $RGT(\cdot)$ is nondecreasing and $RGT(0) = 0$, i.e., the regret is 0 when the disparity of R_1^A and R_1^M is 0, and not decreasing when R_1^A gets larger than R_1^M. Based on the knowledge, the exponential function in Eq. (9.14) is a good candidate for the regret function [20].

$$RGT(\xi) = \frac{g_2}{2}(1 - e^{-g_1 \xi}). \tag{9.14}$$

where ξ is the independent variable of the function, $g_1 > 0$ and $g_2 > 0$ are constant parameters. The error function of $\tilde{R}^A(\Delta, L|X(o) = 1, run)$ and $\tilde{R}^M(\Delta, L|X(o) = 1, run)$ is defined as follows,

$$\Psi(R_1^A - R_1^M) = \tilde{R}^A(\Delta, L|X(o) = 1, run) - \tilde{R}^M(\Delta, L|X(o) = 1, run). \tag{9.15}$$

Substitute Eqs. (9.12)–(9.14) into Eq. (9.15) to get,

$$\Psi(R_1^A - R_1^M) = R_1^A - R_1^M + g_2 \sinh(g_1(R_1^A - R_1^M)). \tag{9.16}$$

Let $\xi_A^M = R_1^A - R_1^M$, Eq. (9.16) becomes,

$$\Psi(\xi_A^M) = \xi_A^M + g_2 \sinh(g_1 \xi_A^M). \tag{9.17}$$

According to regret theory, $\Psi(\xi_A^M)$ has to be monotonically increasing and convex when $\xi_A^M \geq 0$. Also, it has to be skew-symmetric, $\Psi(-\xi_A^M) = -\Psi(\xi_A^M)$. Figure 9.6 contains the plots of Eq. (9.17) for different valued parameters g_1 and g_2. We can observe from the plots that the proposed function $\Psi(\xi_A^M)$ possesses all the properties required by the regret theory. Furthermore, the value of parameter g_1 and g_2 affects the convexity of function $\Psi(\xi_A^M)$, implying different risk-seeking/risk-aversion attitudes of humans. This function also matches with the quantitative measurements of regret theory [26], which indicates that the proposed regret function can be used.

We adopt this specific regret function. The explicit form of the modified conditional Bayes risk function is,

$$\tilde{R}^A(\Delta, L|X(o) = 1; run) = R_1^A + \frac{g_2}{2}(1 - e^{-g_1(R_1^A - R_1^M)}). \tag{9.18}$$

Likewise,

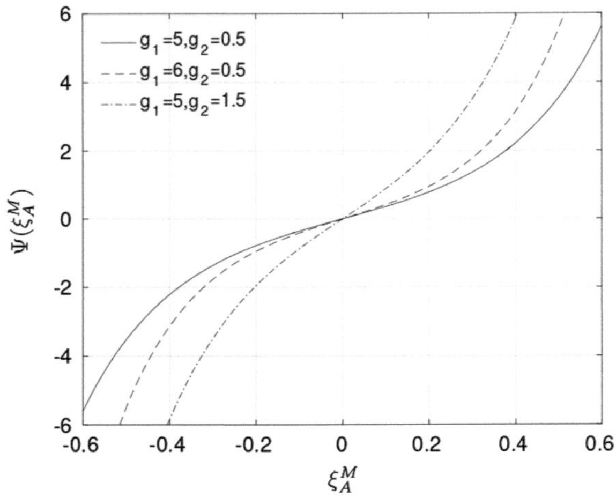

Fig. 9.6 The proposed error function $\Psi(\xi_A^M)$ for different values of parameters in the function

$$\tilde{R}^M(\Delta, L|X(o) = 1; run) = R_1^M + \frac{g_2}{2}(1 - e^{-g_1(R_1^M - R_1^A)}). \tag{9.19}$$

Similarly, given that $X(o) = 0$, we have the modified conditional Bayes risks $\tilde{R}^A(\Delta, L|X(o) = 0; run)$ and $\tilde{R}^M(\Delta, L|X(o) = 0; run)$. The values of g_1 and g_2 are tuned through experiments. In this work, for simplification, g_2 is fixed as a constant, g_1 is regarded as the regret intensity factor. In Sect. 9.7.3, qualitative measurement of human worker's regret is conducted during the experiment to tune the value of g_1.

In the special case when no observation is taken, the modified expected Bayes risk function, like the original expected Bayes risk function, is given by Eqs. (9.8) and (9.9). Otherwise, the modified Bayes risk function is,

$$\begin{aligned} \widetilde{risk}(\Delta, L, P(X(o) = x(o); run), run) = \tilde{R}(\Delta, L|X(o) = 0; run)P(X(o) = 0; run) \\ + \tilde{R}(\Delta, L|X(o) = 1; run)P(X(o) = 1; run). \end{aligned} \tag{9.20}$$

where $\tilde{R} = \tilde{R}^M$ for the manual mode and $\tilde{R} = \tilde{R}^A$ for the autonomous mode.

9.6 Illustration of the Decision-Making Approach

In this section, we illustrate the optimal Bayesian decision-making approach as well as the inclusion of regret in the Bayesian approach.

9.6.1 *Illustration of the Optimal Bayesian Decision-Making Approach*

Three modes without regret are considered as the optimal decision-making scheme as shown in Figs. 9.7, 9.8 and 9.9. We choose different values of β_r, β_H, c_r, c_H,

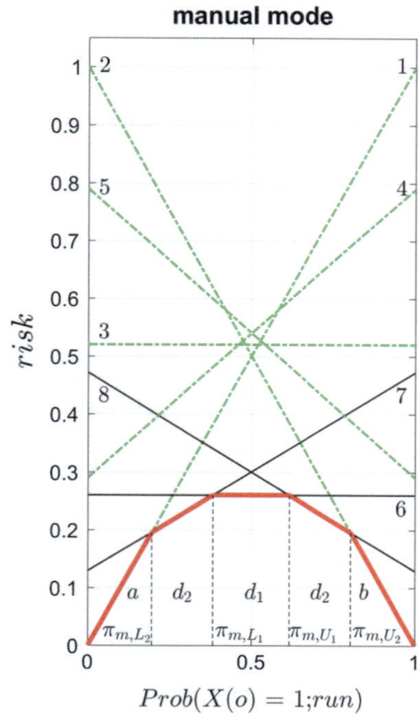

Fig. 9.7 The manual mode (© 2016 IEEE. Reprinted, with permission, from [12])

which are listed in Table 9.1, to obtain several relative positions of the Bayes risk curves. Certain intersections of different decision curves represent the threshold probabilities $\pi_{m,L}, \pi_{m,U}, \pi_{a,L}, \pi_{a,U}$ respectively that divide the neighboring regions. The minimum Bayes risk curve is marked with bold red lines. Different decision regions are determined by the minimum Bayes risk curve and different threshold probabilities. The manual mode is shown in Fig. 9.7, in which the robot makes the final decision regarding the object as wrong or correct, or asks the human to take more observations before making the final decision. The autonomous mode is shown in Fig. 9.8, in which the robot makes the final decision regarding the object as wrong or correct, or the robot itself takes more observations to make the final decision. The mixed mode is shown in Fig. 9.9, in which the robot makes the final decision regarding the object as wrong or correct, or the robot itself takes more observations, or it asks the human to take more observations to make the final decision. An optimal decision is made based on which decision region the probability value of the object being right determined by the robot lies in.

In Figs. 9.7, 9.8, and 9.9, Line 1 is the risk function for the robot deciding the object being wrong. Line 2 is the risk function for the robot deciding the object being correct. Line 3 is the risk function for the robot taking one more observation. Lines 4 and 5 are the risk functions for the robot taking two more observations. Line 6 is the risk function for the human taking one more observation. Lines 7 and 8 are the risk functions for the human taking two more observations. Based on these risk function curves, we find the minimum risk function represented by the red bold line. If the

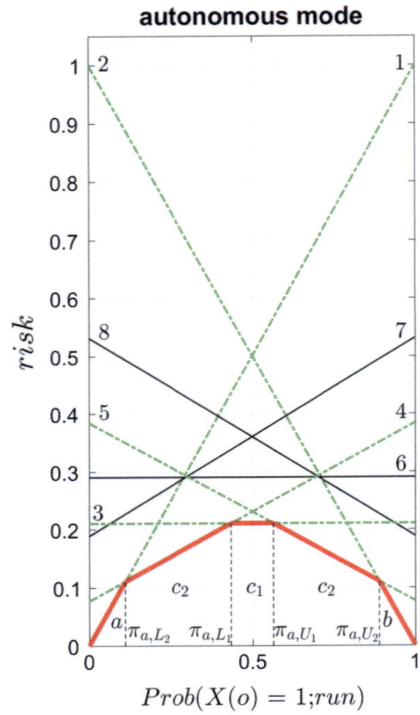

Fig. 9.8 The autonomous mode (© 2016 IEEE. Reprinted, with permission, from [12])

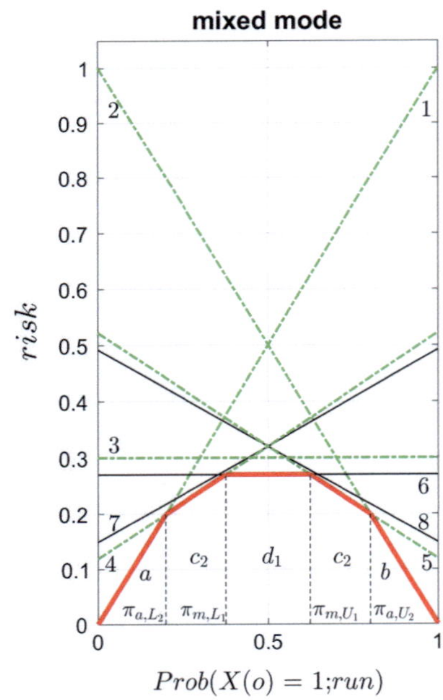

Fig. 9.9 The mixed mode (© 2016 IEEE. Reprinted, with permission, from [12])

Table 9.1 Chosen values of different parameters for different modes for simulation

	β_r	β_H	c_r	c_H	π_{m,L_2}	π_{m,L_1}	π_{m,U_1}	π_{m,U_2}	π_{a,L_2}	π_{a,L_1}	π_{a,U_1}	π_{a,U_2}
Manual	0.50	0.78	0.02	0.04	0.195	0.383	0.616	0.804	–	–	–	–
Autonomous	0.81	0.78	0.02	0.07	–	–	–	–	0.110	0.435	0.565	0.890
Mixed	0.72	0.78	0.02	0.05	–	0.376	0.624	–	0.198	–	–	0.801

probability value of object being correct ($Prob(X(o) = 1; run)$) determined by the robot lies in region (a), the final decision is that the object is wrong. If the probability lies in region (b), the final decision is that the object is correct. If the probability lies in region (c_1 or c_2), the decision is that the robot takes one or two more observations respectively to make the final decision. If the probability lies in region (d_1 or d_2), the decision is that the robot will ask the human to take one or two more observations to help it make the final decision.

9.6.2 Illustration of the Regret-Based Modified Decision-Making Approach

We here illustrate the inclusion of regret analysis in the mixed mode only though such illustrations are possible for autonomous and manual modes. Inclusion of regret in Bayes risk curve for the mixed mode may also result in the mixed mode (the modified mixed mode), but with different probability regions. In extreme, the mixed mode will also result in a manual or autonomous mode depending on the sensing parameters. As illustrated in Fig. 9.10, for the case 1 when the bias is to have more manual observations, the region (c_2) in the modified mixed mode is smaller and the region (d_1) is larger compared to those in the Bayesian optimal mixed mode. Similarly, we can illustrate the case 2 when the bias is to have more autonomous observation as shown in Fig. 9.11.

9.7 Implementation Scheme of the Regret-Based Bayesian Decision-Making Approach for the Assembly Task

9.7.1 The Overall Scheme in a Flowchart

The overall regret-based Bayesian decision-making scheme for the assembly task is shown in Fig. 9.12 as a flowchart. As shown in the flowchart, the robot sensor at first detects whether the type of the input assembly part is correct or not, and displays the robot's observation in its face screen. The probability of correctness in the robot's sensing in the current run is also displayed in the robot face screen. Then, the system constructs the Bayes risk curve using the current values of β_r, β_H, c_r, and c_H. Then,

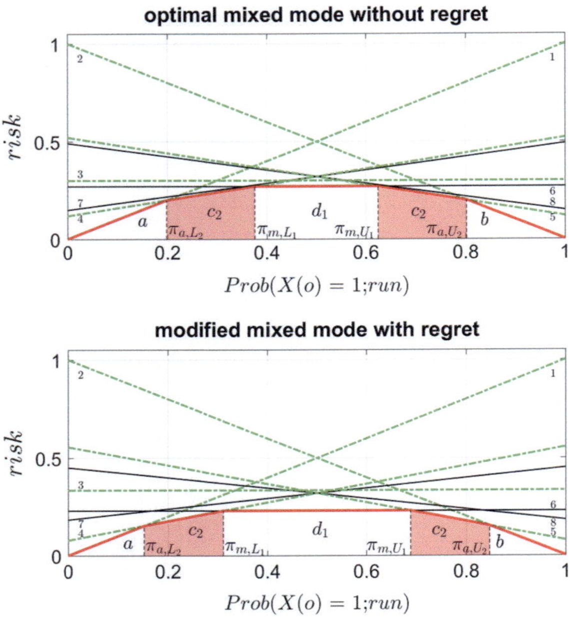

Fig. 9.10 Regret-based modified mixed mode when there is a bias towards manual observation (© 2016 IEEE. Reprinted, with permission, from [12])

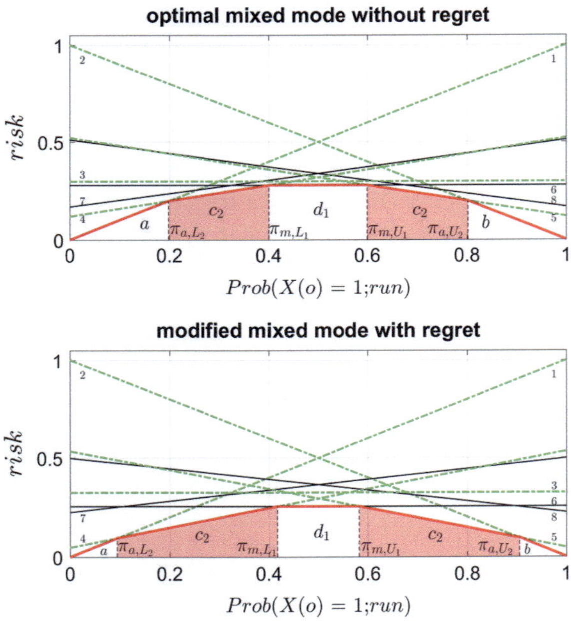

Fig. 9.11 Regret-based modified mixed mode when there is a bias toward autonomous observation

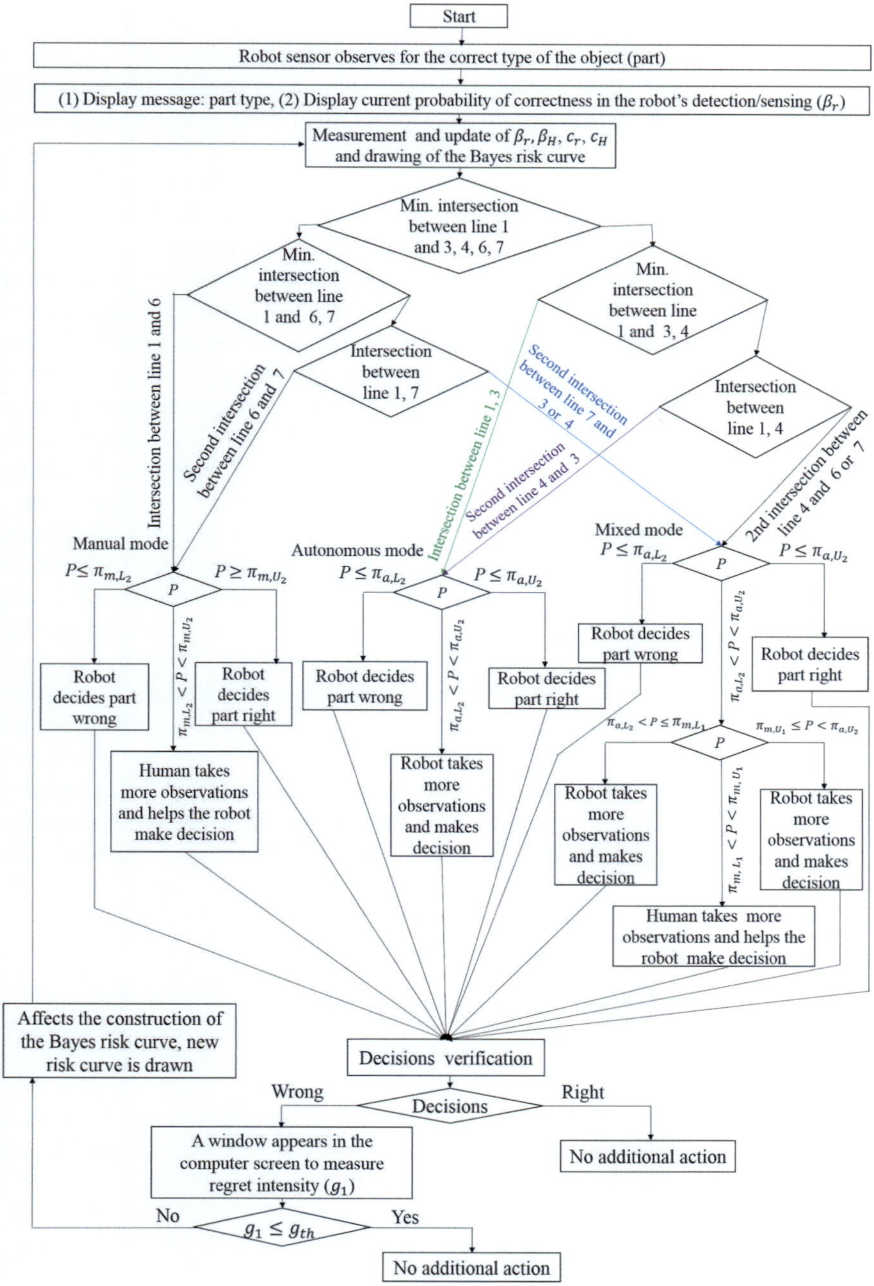

Fig. 9.12 The overall implementation scheme in a flowchart. Here, P indicates $Prob(X(o) = 1; run)$

depending on the minimum intersection points (Bayes risk, probability) between line 1 and other lines (3, 4, 5, 6), one of the three modes (manual, autonomous, and mixed mode) of allocation of autonomy in the detection for the assembly part is selected. If the input part is finally decided as correct, the robot picks it and manipulates it to the human, and the assembly task moves forward. However, if the input part is finally decided as wrong, the robot does not manipulate it to the human. Instead, it issues a warning message to display on its face screen to make the human and the experimenter aware of the status of the input part. If the experimenter replaces the wrong part by another part, the robot observes it again and if found correct then picks it and manipulates it to the human. While the assembly is going on, the experimenter verifies the decisions about the correctness of the input part. If the decisions are found correct, there is no action to take. However, if any decision is identified as wrong, the experimenter inputs this information to the computer programming that runs the robot system. Then the assembly task stops temporarily, a window appears in the computer screen asking the human co-worker to measure his/her regret intensity for the wrong decisions. If the regret intensity is below a threshold (g_{th}), no additional action is required. However, if the regret intensity is above a threshold, then based on the algorithms, Bayes risk is adjusted with the regret, new Bayes risk curve is constructed, and a different mode of allocation of autonomy in the detection will be selected based on the newly constructed Bayes curve characteristics (Bayes risk, probability).

In fact, all the decisions about the correctness of the parts are made by the robot, and the human sometimes helps the robot through taking observations. Our approach assesses human co-worker's regret for the wrong decisions and the regret value is used as the robot's regret to generate a humanlike decision-making strategy developed in Sect. 9.5. The subject repeats the trial with the new mode that is decided considering the regret intensity so that the subject can experience the robot's humanlike decision-making behavior.

9.7.2 Measurement of Sensing Probability and Observation Cost

The probability of correctness in the robot's sensing (β_r) can be measured following (9.21), where n_w is the number of wrong decisions and n_t is the number of total decisions made by the robot up to the previous run.

$$\beta_r(run) = (1 - \frac{n_w}{n_t}) \times 100\%. \tag{9.21}$$

Similarly, the probability of correctness in the human sensing (β_H) can be measured following (9.22), where n_{w0} is the number of wrong observations and n_{t0} is the number of total observations made by the human up to the previous run.

$$\beta_H(run) = (1 - \frac{n_{w0}}{n_{t0}}) \times 100\%. \tag{9.22}$$

The observation cost c_r can be measured following (9.23), where T_r is the total time (in seconds) needed by the robot to make observations in the recent past run.

$$c_r(\text{run}) = \begin{cases} 0, & T_r(\text{previous run}) = 0\,\text{s} \\ 0.1, & 0\,\text{s} < T_r(\text{previous run}) \leq 10\,\text{s} \\ 0.2, & 10\,\text{s} < T_r(\text{previous run}) \leq 20\,\text{s} \\ \ldots & \ldots \\ 1.0, & T_r(\text{previous run}) \geq 100\,\text{s} \end{cases} \tag{9.23}$$

Similarly, the observation cost c_H can be measured following (9.24), where T_H is the total time (in seconds) needed by the human to make observations in the recent past run.

$$c_H(\text{run}) = \begin{cases} 0, & T_H(\text{previous run}) = 0\,\text{s} \\ 0.1, & 0\,\text{s} < T_H(\text{previous run}) \leq 10\,\text{s} \\ 0.2, & 10\,\text{s} < T_H(\text{previous run}) \leq 20\,\text{s} \\ \ldots & \ldots \\ 1.0, & T_H(\text{previous run}) \geq 100\,\text{s} \end{cases} \tag{9.24}$$

Usually, the observation time for the human and the robot is small. In this small time period, all other cost elements such as utilities, facilities, etc. usually do not change much. Indirect costs due to human workload and fatigue also do not change much. This is why, we cannot directly estimate the observation costs based on the utility and facility costs used for the observation. Alternatively, we here consider the observation time as a sole measure of the observation costs because the actual observation costs increase as the observation time becomes longer.

9.7.3 *Measurement Method for Regret Intensity*

In this subsection, we present a software-based approach of regret measurement [26, 27], as follows. In our scenario, regret measurement is event-based because regret measurement is needless when no mistake in the decision-making is identified. Once a mistake is identified (by the experimenter), the regret measurement window appears in the computer screen. The wrong decisions may occur due to wrong observations made by the human or the robot itself. If the wrong decision was due to the human's wrong observations, then the regret assessment window looks like as in Fig. 9.13. If the wrong decision was due to the robot's wrong observations, then the regret assessment window looks like as in Fig. 9.14. Then, the human responds the questionnaire in the window, and the human's regret intensity (measure of g_1) is automatically calculated and normalized between 0 and 1. This human regret is used as the robot's regret.

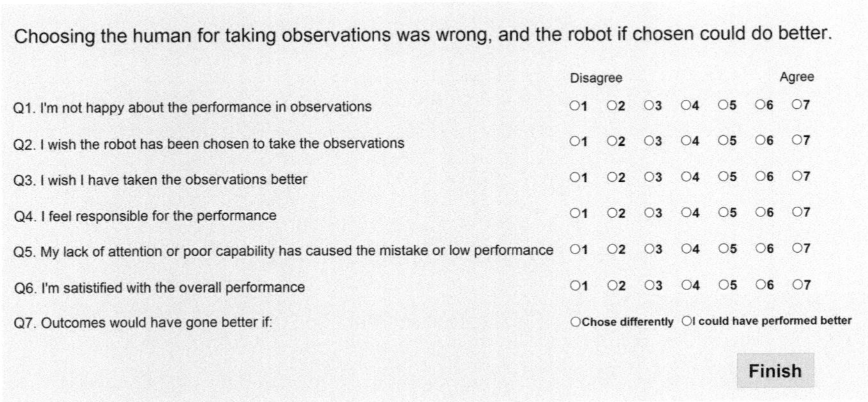

Choosing the human for taking observations was wrong, and the robot if chosen could do better.

	Disagree						Agree
Q1. I'm not happy about the performance in observations	O1	O2	O3	O4	O5	O6	O7
Q2. I wish the robot has been chosen to take the observations	O1	O2	O3	O4	O5	O6	O7
Q3. I wish I have taken the observations better	O1	O2	O3	O4	O5	O6	O7
Q4. I feel responsible for the performance	O1	O2	O3	O4	O5	O6	O7
Q5. My lack of attention or poor capability has caused the mistake or low performance	O1	O2	O3	O4	O5	O6	O7
Q6. I'm satistified with the overall performance	O1	O2	O3	O4	O5	O6	O7

Q7. Outcomes would have gone better if: OChose differently OI could have performed better

Finish

Fig. 9.13 Questionnaire for regret assessment when wrong decision was due to the human's wrong observations

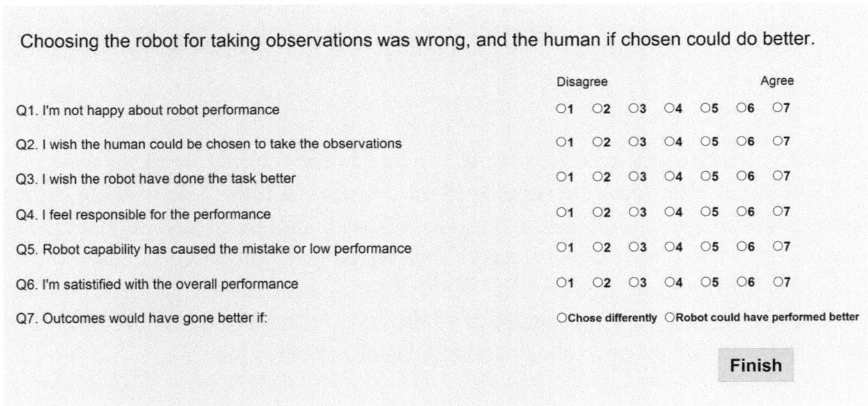

Choosing the robot for taking observations was wrong, and the human if chosen could do better.

	Disagree						Agree
Q1. I'm not happy about robot performance	O1	O2	O3	O4	O5	O6	O7
Q2. I wish the human could be chosen to take the observations	O1	O2	O3	O4	O5	O6	O7
Q3. I wish the robot have done the task better	O1	O2	O3	O4	O5	O6	O7
Q4. I feel responsible for the performance	O1	O2	O3	O4	O5	O6	O7
Q5. Robot capability has caused the mistake or low performance	O1	O2	O3	O4	O5	O6	O7
Q6. I'm satistified with the overall performance	O1	O2	O3	O4	O5	O6	O7

Q7. Outcomes would have gone better if: OChose differently ORobot could have performed better

Finish

Fig. 9.14 Questionnaire for regret assessment when wrong decision was due to the robot's wrong observations

Actually, the questionnaires are "Likert-Type Scales" [11, 13, 28] with 7 questions to measure the regret. In this decision-making approach regarding who will take additional observations if needed, we analyze regret along with disappointment–the feeling of sadness caused by the nonfulfillment of expectations in the decision-making [27]. The human (robot) will experience disappointment when the human (robot) is overloaded and the robot decides about the detection with a low performance (makes a wrong decision). In this case, the human (robot) does not feel regret about the choice due to the workload but is not satisfied with the results, which will affect the trust level of the human in the robot and the vice versa [11, 29]. In each case, the questionnaire contains 7 items. The items are explained below as an example when the wrong observations are made by the human during manual observations in manual or mixed modes:

- Item 1 (I am not happy about the performance in observations) is an "Affective reaction."
- Item 2 (I wish the robot has been chosen to take the observations) is a "Regret counterfactual." The human might have higher value on this item when the human feels like that the robot could make a better performance if it was chosen. In other words, the human is not satisfied with the outcomes of the detection, and he/she will experience regret.
- Item 3 (I wish I have taken the observations better) is a "Disappointment counterfactual." The human might have higher value on this item due to low trust in the robot. In other words, although the human is not satisfied with the outcomes, the human does not feel regret about "choosing the human" as the human has a relatively low trust in the robot.
- Item 4 (I feel responsible for the performance) is an "Internal attribution."
- Item 5 (My lack of attention or poor capability has caused the mistake or low performance) is an "External attribution"
- Item 6 (I am satisfied with the overall performance) is a "Control item." If the result had higher values on this control item than on the affective reaction item (item 1, "I am not happy about the performance in observations"), this questionnaire needs to be excluded from the subsequent analysis.
- Item 7 (Outcomes would have gone better if) is a "Choice between counterfactuals."

Similar approach is followed to measure the regret intensity when the wrong observations are made by the robot during observations in autonomous or mixed modes.

9.8 Experimental Evaluation of the Regret-Based Bayesian Decision-Making Approach

9.8.1 Objective

The objective is to evaluate the effectiveness of the regret-based Bayesian decision-making approach as introduced above.

9.8.2 Hypothesis

The regret-based Bayesian suboptimal decision-making approach is more humanlike than the Bayesian sequential optimal decision-making approach as it will generate similar mental models between the robot and its human co-worker, and thus the regret-based approach will result in better human–robot (cognitive) interaction, which will enhance the overall performance in the human–robot collaborative assembly in manufacturing.

9.8.3 The Evaluation Criteria

We categorize the evaluation criteria as follows: (i) category-1: system parameters, (ii) category-2: human–robot interaction (HRI), and (iii) category-3: assembly performance. The system parameters are those that are used to model the algorithms. These parameters are $W_h(run)$, β_r, β_H, c_r, c_H, risk function value, regret value, and sensing task difficulty or easiness (α). These parameters are measured/assessed following the methodologies as presented above. We will measure human utilization ratio (human workload) and sensing task difficulty (or easiness) following the analytical methods introduced at the problem formulation in Sect. 9.3.2. These parameters help derive the algorithms, and also may help evaluate the effectiveness of the algorithms. It is to be noted that the human workload (utilization ratio) as used here is not the human's cognitive workload. Instead, the human workload (utilization ratio) is the extent of human's intervention required in the sensing during the collaborative assembly. Of course, higher human utilization ratio will result in higher cognitive workload. The HRI parameters are the human–robot team fluency, human's cognitive workload, human's trust in the robot, and other cognitive interactions. The assembly performance is expressed through efficiency and quality. The quality is also measured in terms of percentage of correct detections during the assembly.

9.8.4 The Experiment Design

The independent variable is the decision-making approach, i.e., Bayesian decision-making approach versus regret-based Bayesian decision-making approach. The dependent variables are the resulted system parameter values, HRI, and the collaborative assembly performance. However, in this chapter, we confine to the evaluation of the effectiveness of the proposed algorithms based on the system parameters only, i.e., the dependent variables are the system parameter values only.

9.8.5 Subjects

Twenty (20) Mechanical Engineering students are recruited to participate in the experiments voluntarily. Ten (10) subjects form Group I and participate in the collaborative assembly under the regret-based Bayesian decision-making approach. The remaining ten (10) subjects form Group II and participate in the collaborative assembly under the Bayesian decision-making approach. The subjects have no prior experiences in the similar assembly tasks. The subjects are separately instructed about the experimental procedures before the start of the experiments. The study involving human subjects is approved by the Institutional Review Board (IRB). All the required IRB principles regarding human subject tests are obeyed.

9.8.6 *The Experimental Procedures*

In the first phase, each subject conducts some mock practices of the assembly with the robot, which will make the subject familiar with the task and to achieve a steady-state learning status. Then, in the second phase, the formal collaborative assembly is conducted by the subject as shown in Fig. 9.3. The collaboration is performed following the collaborative assembly scheme introduced in Fig. 9.12. Each subject in Group II separately performs the collaborative assembly for the Bayesian decision-making approach, and each subject in Group I separately performs the assembly for regret-based Bayesian decision-making approach. The values of the system parameters are measured and recorded for each trial of each subject under each of the two approaches.

9.9 Evaluation Results and Analyses

Figure 9.15 shows the human and robot's detection probabilities and costs for more observations made by the robot and the human for the trials for the Bayesian decision-making approach. The values for trial 1 are the initial values that we determine based on the initial conditions, i.e., initially we consider that both the human and the robot have 100 percent detection probabilities, and the costs for more observations made by the robot and the human are zero as no such observations are actually made at the initial trial. The results for the trials from 2 to 11 are based on the actual data that are measured and recorded during the collaboration. The results show that the detection probabilities for both the human and the robot are dynamic in nature, i.e., the probabilities vary with trials (with time, situations). We also note that the detection probability for the human is higher than that for the robot. This may happen due to vulnerability and limitation of the robot sensing capability compared to the human's natural sensing capability. We also note that the costs for more observations for the human are more than that for the robot. We believe that it happens as the human took longer time to perform the observations than the time taken by the robot. In general, the results show that the human takes longer time, but the sensing observation is more accurate. On the contrary, the robot takes less time to finish an observation, but the observation outcome is less accurate.

Figure 9.16 compares the risk values for the trials between Bayesian decision-making with and without regret based on the system parameter values shown in Fig. 9.15. The results show that the risks are dynamic in nature, i.e., the risks vary with trials (with time, situations). Such changes in Bayes risks should also change the sensing modes based on changes in the configuration of the Bayes risk curves.

Figure 9.17 compares the human workload for the trials between Bayesian decision-making with and without regret. The results show that the human workload reduces due to consideration of regret with the Bayesian decision-making. The consideration of regret also reduces the risks as shown in Fig. 9.16, which changes

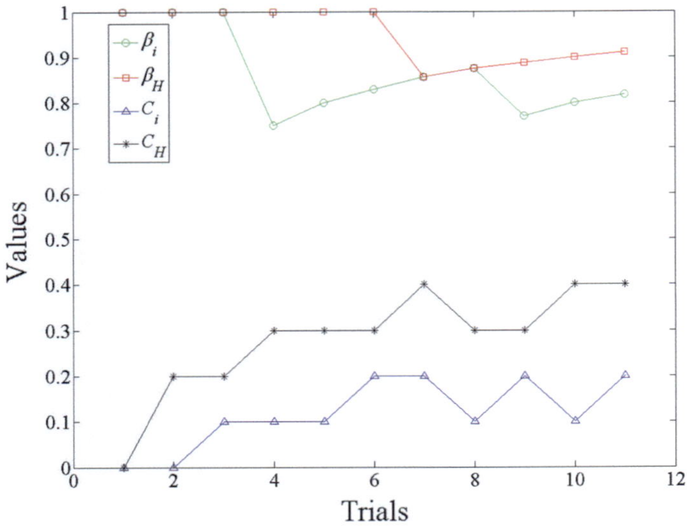

Fig. 9.15 Human and robot's detection probabilities and costs for more observations for the trials under the Bayesian decision-making approach during the collaborative assembly

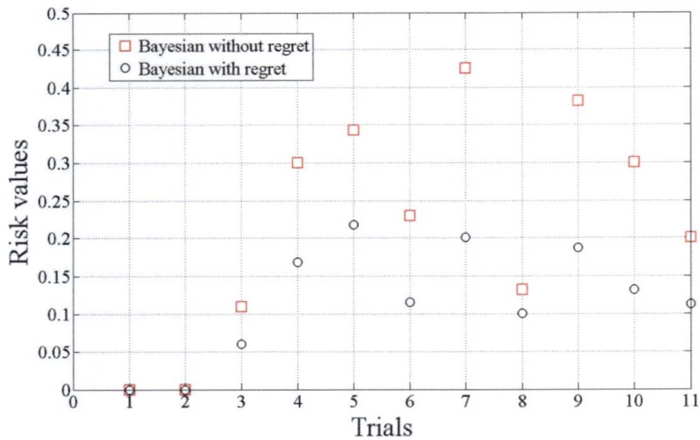

Fig. 9.16 The risk values for the trials between Bayesian decision-making with and without regret for different trials under the (two) subject groups

the mode of allocation of autonomy in the sensing and enhances the incidents with the autonomous mode that then reduces the workload of the human. Figure 9.18 compares the task easiness (or the level of task difficulty) for the trials between Bayesian decision-making with and without regret. The results show that the task easiness increases due to inclusion of regret in the decision-making. We believe that it happens because the incidents with the autonomous mode increases and the robot

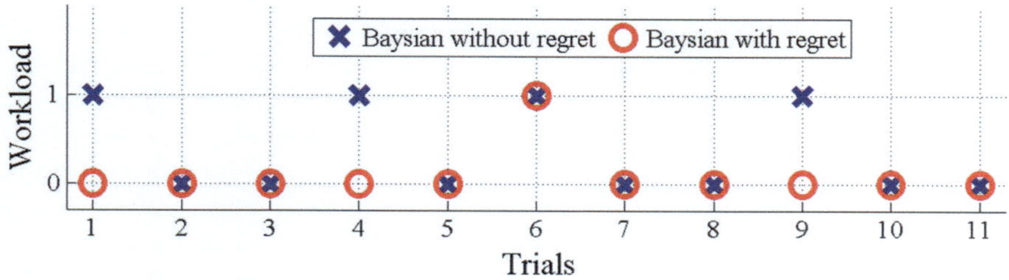

Fig. 9.17 Comparison of human workload for the trials between Bayesian decision-making with and without regret

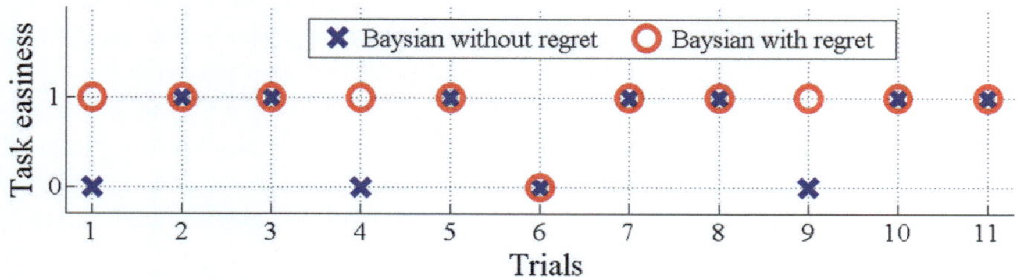

Fig. 9.18 Comparison of task easiness for the trials between Bayesian decision-making with and without regret (lower task easiness means higher task difficulty and vice versa)

sends fewer requests to the human for observation, which contributes to increased task easiness based on the task easiness formulation method explained in Sect. 9.3.2.

In general, the experimental results show the effectiveness of both the Bayesian decision-making of allocation of sensing autonomy for the collaborative assembly task with and without regret in terms of the system parameters. However, the impact of inclusion of regret produces better results than the results produced by the Bayesian decision-making without regret. The effectiveness of the regret-based approach may also be proven in terms of HRI and assembly performance, which will be addressed in future.

9.10 Conclusions and Future Innovations

We proposed an HRC system for assembly in manufacturing where a human collaborated with a robot to assemble a few input parts into final assembled products. The robot manipulated the input parts to the human and the human manually assembled the parts. To ensure the correctness in the input assembly parts, the robot vision system was used to detect the correctness of the parts before the robot manipulated the parts to the human for assembly. Assuming that the robot's detection/sensing process would be influenced by disturbances and uncertainties, either of the three additional steps was associated with the robot's detection procedures: (i) the robot directly decided the part as correct or wrong based on the current probability of correctness in its detection process, (ii) the robot took more observations to make the decision, (iii) the robot seeked help to the human to take more observations for the robot. Such allocation of autonomy between the human and the robot was determined based on a Bayesian sequential decision-making approach in terms of Bayes risk and probability of correctness in robot's decision. In order to make the decision-making more humanlike and to maintain similar mental models between the human and the robot during the assembly, we then included human's regret in the decision-making in the case when any mistake was identified in the detection decision.

We developed the real human–robot collaborative system for assembly and formulated the detection problem. We then derived the Bayesian decision-making approach in terms of Bayes risk. We then incorporated the human regret in the Bayes risk function. We explained the algorithms using examples. We then validated the effectiveness of the proposed algorithms through experiments using a hybrid cell in lab settings. The results showed that the Bayesian optimal decision-making produced good performance in terms of human workload and task easiness. The results also showed that inclusion of regret further enhanced the system performance through reducing the workload and increasing the task easiness.

In future, we will conduct experiments to investigate the effectiveness of the approaches in terms of HRI such as team fluency, symbiosis, human's cognitive workload, human's trust in the robot, etc. We will also evaluate the approaches for assembly performance in terms of efficiency and quality.

Acknowledgements This work was supported in part by the National Science Foundation (NSF) under the grant No. CMMI-1454139. The authors also thank BMW Manufacturing Co., Spartanburg, SC, USA for supporting with the Baxter robot and the assembly parts used in the experiment.

References

1. Kaipa KN, Morato C, Liu J, Gupta SK (2014) Human-robot collaboration for bin-picking tasks to support low-volume assemblies. In: Proceedings of RSS workshop on human-robot collaboration for industrial manufacturing
2. Wilcox R, Nikolaidis S, Shah J (2012) Optimization of temporal dynamics for adaptive human-robot interaction in assembly manufacturing. In: Proceedings of robotics: science and systems conference (RSS)
3. http://www.rethinkrobotics.com/baxter/. Accessed 11 Feb 2016
4. http://www.kinovarobotics.com/. Accessed 11 Feb 2016
5. http://www.kuka-robotics.com/. Accessed 11 Feb 2016
6. http://www.universal-robots.com/. Accessed 11 Feb 2016
7. http://www.abb.us/robotics. Accessed 11 Feb 2016
8. http://www.motoman.com/. Accessed 11 Feb 2016
9. http://www.fanucamerica.com/. Accessed 11 Feb 2016
10. http://www.robots.epson.com/, Accessed 11 Feb 2016
11. Rahman SMM, Sadr B, Wang Y (2015) Trust-based optimal subtask allocation and model predictive control for human-robot collaborative assembly in manufacturing. In: Proceedings of ASME dynamic systems and controls conference, Columbus, Ohio, October 28–30, 2015, Paper No. DSCC2015-9850, pp V002T32A004, 10 pages

12. Mizanoor Rahman SMM, Liao Z, Jiang L, Wang Y (2016) Regret-based allocation of autonomy in human-robot shared visual detection of faulty parts in collaborative assembly in manufacturing. In: IEEE international conference on automation science and engineering (CASE), 2016, Fort Worth, Texas
13. Rahman SMM, Wang Y (2015) Dynamic affection-based motion control of a humanoid robot to collaborate with human in flexible assembly in manufacturing. In: Proceedings of ASME dynamic systems and controls conference, Columbus, Ohio, October 28–30, 2015, Paper No. DSCC2015-9841, pp V003T40A005, 10 pages
14. Truong X, Yoong V, Ngo T (2014) Dynamic social zone for human safety in human-robot shared workspaces. In: Proceedings of 2014 11th international conference on ubiquitous robots and ambient intelligence (URAI), pp 391–396, 12–15 Nov 2014
15. Schlueter R, Nussbaum-Thom M, Ney H (2012) Does the cost function matter in Bayes decision rule? IEEE Trans Pattern Anal Mach Intell 34(2):292–301
16. Kahneman D, Tversky A (1979) Prospect theory: an analysis of decision under risk. Econometrica 47(2):263292
17. Mathieu JE, Heffner TS, Goodwin GF, Salas E, Cannon-Bowers JA (2000) The influence of shared mental models on team process and performance. J Appl Psychol 85(2):273
18. Wang Y (2015) Regret-based Bayesian sequential decision-making for human-agent collaborative search tasks. In: Proceedings of American control conference (ACC), pp 5738–5743, 1–3 July 2015
19. Wang Y (2016) Regret-based automated decision-making aids for domain search tasks using human-agent collaborative teams. IEEE Trans Control Syst Technol pp 1–16
20. Chorus CG (2012) Regret theory-based route choices and traffic equilibria. Transportmetrica 8(4):291–305
21. Bell D (1982) Regret in decision making under uncertainty. Oper Res 30:961–981
22. Loomes G, Sugden R (1982) Regret theory: an alternative theory of rational choice under uncertainty. Econ J 92:805–824
23. Coricelli G, Dolan RJ, Sirigu A (2007) Brain, emotion and decision making: the paradigmatic example of regret. Trends Cogn. Sci 11(6):258–265
24. Tan J, Duan F, Zhang Y, Watanabe K, Kato R, Arai T (2009) Human-robot collaboration in cellular manufacturing: design and development. In: Proceeding of IEEE/RSJ international conference on intelligent robots and systems, p 2934
25. Allais M (1953) Le comportement de l'homme rationnel devant le risque: critique des postulats et axiomes de l'cole amricaine. Econom J Econom Soc 21(4):503–546
26. Bleichrodt H, Cillo A, Diecidue E (2010) A quantitative measurement of regret theory. Manag Sci 56(1):161–175
27. Marcatto F, Ferrante D (2008) The regret and disappointment scale: an instrument for assessing regret and disappointment in decision making. Judgm Decis Mak 3:87–99
28. Carifio J, Rocco JP (2007) Ten common misunderstandings, misconceptions, persistent myths and urban legends about Likert scales and Likert response formats and their antidotes. J Soc Sci 3(3):106–116
29. Hoff KA, Bashir M (2014) Trust in automation: integrating empirical evidence on factors that influence trust. Hum Factors: J Hum Factors Ergon Soc 57(3):407–434

Chapter 10
Considering Human Behavior Uncertainty and Disagreements in Human–Robot Cooperative Manipulation

José Ramón Medina, Tamara Lorenz and Sandra Hirche

10.1 Introduction

Physical cooperation between humans and robots has high impact in many critical application areas. In future manufacturing scenarios, the close cooperation between humans and robots is envisioned, where robots take over the physically exhausting tasks. The impact of physical robotic partners in medical settings cannot be overstated. Applications such as rehabilitation or physical training are greatly enhanced by robotic partners regulating the appropriate human effort. Service robots for domestic use also require intuitive physical interaction with human users. Similarly, elderly care requires efficient and comfortable mobility aids from robot companions. The high impact of physical human–robot cooperation in such a broad spectrum of application domains highlights the need for intuitive and effective physically cooperating robotic partners.

Among all aspects involved in the realization of physical helpers, robot control is a crucial component; it determines robot behavior while physically coupled to a human

This chapter was developed and enhanced from an earlier paper published as [33] © 2015 IEEE.

J.R. Medina (✉)
École Polytechnique Fédérale de Lausanne (EPFL), 1015 Lausanne, Switzerland
e-mail: jose.medina@epfl.ch

T. Lorenz
Department of Psychology, Department of Mechanical Engineering and Materials Science & Department of Electrical Engineering and Computing Systems, University of Cincinnati, ML 0376 Edwars 1, Cincinnati, OH 45221-0376, USA
e-mail: tamara.lorenz@uc.edu

S. Hirche
Chair of Information-oriented Control, Technische Universität München, Arcisstrasse 21, 80333 Munich, Germany
e-mail: hirche@tum.de

Y. Wang and F. Zhang (eds.), *Trends in Control and Decision-Making for Human–Robot Collaboration Systems*, DOI 10.1007/978-3-319-40533-9_10

partner. Although *passive* robots naturally react to human applied wrenches, they do not efficiently reduce human effort. In contrast, a robot partner pursuing a common movement in a *proactive* manner is more general, desirable, and sometimes even the only alternative [27]. As an example consider the cooperative transport of a rigid object. A passive robot behaves complying to sensed wrenches through the object, thereby rendering an additional mass in the overall system. In contrast, a proactive partner does not only react to the human input but also actively progresses toward the common movement goal, a priori accounting for all required effort. It also provides increased cooperative manipulability, a requirement for fulfilling tasks involving navigation through narrow passages. In fact, these are the only control strategies capable of achieving successful cooperation, which requires working *jointly* toward a common known goal [42]. Within proactive approaches, robot assistance based on human behavior prediction has demonstrated superior performance in terms of human effort minimization [24]. By anticipating human intentions, robot actions ideally apply all necessary redundant effort. Such an approach strongly relies on human *behavior models*, providing inherently uncertain predictions of human actions. However, when the robot's expectations mismatch human intentions, unintended interaction wrenches appear incurring safety risks and discomfort. Identifying these internal wrench components and understanding their relation with prediction errors/disagreements are instrumental steps in the design of proactive robotic partners. In addition, the role of prediction uncertainty is also crucial: it is a decisive indicator of possible prediction errors that should be considered in robot control in order to avoid potential disagreements.

This chapter presents an optimal control scheme adapting to both human behavior uncertainty and disagreements. First, we present a novel characterization of effective (motion-inducing) and internal (squeezing) force/torque components and their relation with load distribution and disagreements. Second, we introduce a risk-sensitive optimal control scheme that anticipates human actions while adapting to both uncertainty and internal force/torque components. Results demonstrate superior performance in terms of both implicit and subjective measures in an experiment with human users. The main objective of this work is to identify suitable control strategies for pHRI and to evaluate the benefits of including human behavior uncertainty and disagreements into the control design—also from a human user point of view. In consequence, we focus on user evaluation of the proposed control approaches rather than providing an in-depth stability analysis.

The remainder of this chapter is structured as follows: Sect. 10.2 presents the prototypical problem of human–robot cooperative manipulation. A decomposition of interaction wrenches into effective and internal components that characterize the load distribution and disagreements is presented in Sect. 10.3. Section 10.4 exposes the proposed optimal robot control scheme and its evaluation in an experiment with human users.

10.2 Human–Robot Cooperative Manipulation

As a prototypical physical human–robot interaction task, we consider the cooperative transport of an object, as shown in Fig. 10.1a. Other scenarios, such as kinesthetic teaching Fig. 10.1b or physical rehabilitation are straightforwardly transformed into this setting by considering their respective kinematic constraints. After defining the task for multiple cooperative agents, in this section we expose the control challenges that the presence of a human partner implies followed by an overview on different reactive and proactive approaches found in the literature.

10.2.1 Cooperative Manipulation

Consider $N \in \mathbb{Z}^+$ manipulators tightly grasping a rigid object. Let $\boldsymbol{x} \in SE(3)$ be the state of the manipulated object composed by its Cartesian position and orientation in the inertial frame I as depicted in Fig. 10.1a. The object's twist is given by

$$\dot{\boldsymbol{x}} = \begin{bmatrix} \dot{\boldsymbol{p}} \\ \boldsymbol{\omega} \end{bmatrix} , \tag{10.1}$$

where $\dot{\boldsymbol{p}} \in \mathbb{R}^3$ is the translational velocity and $\boldsymbol{\omega} \in \mathbb{R}^3$ the rotational velocity. The dynamics of the object are given by

$$M_o\ddot{\boldsymbol{x}}(t) + \boldsymbol{h_o}(\boldsymbol{x}(t), \dot{\boldsymbol{x}}(t)) = \boldsymbol{u}(t) , \tag{10.2}$$

where $M_o \in \mathbb{R}^{6\times 6}$ represent the object mass and inertia and $\boldsymbol{h_o}$ is the sum of friction and gravitation,

$$\boldsymbol{u} = \begin{bmatrix} \boldsymbol{f} \\ \boldsymbol{\tau} \end{bmatrix} , \tag{10.3}$$

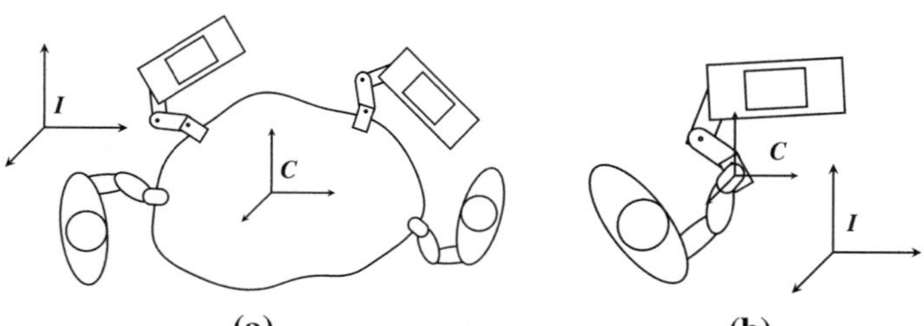

(a) (b)

Fig. 10.1 Physical multiagent interaction in a cooperative transportation (**a**) and kinesthetic teaching (**b**) setting is the applied wrench, where $\boldsymbol{f} \in \mathbb{R}^3$ and $\boldsymbol{\tau} \in \mathbb{R}^3$ denote the force and torque components in frame C, respectively, see Fig. 10.1.

The wrench $\boldsymbol{u}$ results from the addition of each agent's individual wrench contributions. Let $\widetilde{\boldsymbol{u}}_i$ be the wrench applied by agent i in frame C and let $\boldsymbol{p}_i \in \mathbb{R}^3$ be the vector from the i-th grasp position to the object frame. The resulting wrench is given by

$$\boldsymbol{u}(t) = \sum_{i=1}^{N} G_i \widetilde{\boldsymbol{u}}_i(t), \tag{10.4}$$

where G_i is the Jacobian of the kinematics constraints of agent i, usually referred to as partial grasp matrix in this context [39]. The matrix G_i takes the form

$$G_i = \begin{bmatrix} I_3 & 0_3 \\ P_i & I_3 \end{bmatrix}, \tag{10.5}$$

where $P_i = [\boldsymbol{p}_i]_\times$ is the cross-product matrix for $\boldsymbol{p}_i$ that renders the torques generated by applying forces on the grasping position.

For mathematical convenience, we adopt the stacked form

$$G = \begin{bmatrix} G_1 \ G_2 \ \cdots \ G_L \end{bmatrix} \qquad \widetilde{\boldsymbol{u}} = \begin{bmatrix} \widetilde{\boldsymbol{u}}_1^{\mathrm{T}} \ \widetilde{\boldsymbol{u}}_2^{\mathrm{T}} \ \cdots \ \widetilde{\boldsymbol{u}}_L^{\mathrm{T}} \end{bmatrix}^{\mathrm{T}},$$

such that $\boldsymbol{u}(t) = G\widetilde{\boldsymbol{u}}(t)$.

Given this setting, the control problem of cooperative manipulation consists of the design of the agents' exerted wrenches $\widetilde{\boldsymbol{u}}$ fulfilling the following objective [4]:

Definition 10.1 The objective of a cooperative manipulation task is given by

$$\lim_{t\to\infty} \boldsymbol{x}(t) \to \boldsymbol{x}_d(t) , \tag{10.6}$$

where $\boldsymbol{x}_d(t) \in \mathbb{R}^6$ is the desired object trajectory and

$$\lim_{t\to\infty} \widetilde{\boldsymbol{u}}(t) \to \widetilde{\boldsymbol{u}}_d(t) , \tag{10.7}$$

where $\widetilde{\boldsymbol{u}}_d(t) \in \mathbb{R}^{6N}$ are the desired manipulator wrench trajectories.

Note that objective (10.7) is necessary to avoid unnecessary wrench components between agents that only produce undesired squeeze or internal object stress.

Apart from a position control scheme, in order to fulfill objective (10.7), the regulation of wrenches is instrumental. Among all force control methods, due to its convenient properties, the most recurrent approach in the literature is impedance/admittance control [21]. It regulates the apparent inertia, damping, and stiffness of the system by means of force feedback enabling *compliant* behavior. The simplest example is given by a second-order system rendering a mass-damper system. In this case, the impedance-controlled robot dynamics are

$$M_r\ddot{\boldsymbol{x}}(t) + D_r\dot{\boldsymbol{x}}(t) = \boldsymbol{u}_{\text{imp}}(t) + \boldsymbol{u}_{\text{ext}}(t) , \tag{10.8}$$

where $\boldsymbol{e} = \boldsymbol{x}_d - \boldsymbol{x}$, $\boldsymbol{u}_{\text{ext}} \in \mathbb{R}^6$ is an external wrench input and $M_r, D_r \in \mathbb{R}^{6\times6}$ are the positive-definite matrices representing desired virtual mass, inertia, and damping[1] and $\boldsymbol{u}_{\text{imp}}(t)$ is a virtual force rendering a desired task-dependent behavior, for instance following a desired configuration trajectory $\boldsymbol{x}_d$. A valid example tracks $\boldsymbol{x}_d$ with a proportional derivative controller such that

$$\boldsymbol{u}_{\text{imp}}(t) = K_{\dot{x}}\dot{\boldsymbol{e}}(t) + K_{x}\boldsymbol{e}(t) , \tag{10.9}$$

where $K_{\dot{x}}, K_x \in \mathbb{R}^{6\times6}$ are the positive-definite stiffness and damping matrices, respectively. Following this idea, in [41] two robotic manipulators are controlled such that the resulting object dynamics follow impedance control law (10.8) and (10.9). Synthesizing the desired system behavior (10.8) considering the real object dynamics requires exerting a desired resulting wrench on the object (10.2) given by

$$\boldsymbol{u}_d(t) = \boldsymbol{u}_{\text{dyn}}(t) + \overline{\boldsymbol{u}}_{\text{imp}}(t) - \boldsymbol{u}_{\text{ext}}(t) , \tag{10.10}$$

where

$$\boldsymbol{u}_{\text{dyn}}(t) = -M_o M_r^{-1} D_r\dot{\boldsymbol{x}}(t) + \boldsymbol{h}_o(\boldsymbol{x}(t), \dot{\boldsymbol{x}}(t))$$

compensates for dynamics (10.2) and renders desired behavior (10.8) and

$$\overline{\boldsymbol{u}}_{\text{imp}}(t) = M_o M_r^{-1}\boldsymbol{u}_{\text{imp}}(t) .$$

The computation of the corresponding wrench trajectories for each manipulator $\tilde{\boldsymbol{u}}(t)$ yielding $\boldsymbol{u}_d(t)$ is an ill-posed problem in case of input redundancy. The possible solutions depend on a decomposition matrix $A \in \mathbb{R}^{6N\times6}$ which distributes the load among agents such that

$$\tilde{\boldsymbol{u}}(t) = A\boldsymbol{u}_d(t) \qquad \text{and} \qquad \boldsymbol{u}(t) = GA\boldsymbol{u}_d(t) \tag{10.11}$$

hold. A usual approach in the literature is given by the Moore–Penrose pseudoinverse of G, denoted G^+, which yields the solution minimizing $||\boldsymbol{u}_d - A\tilde{\boldsymbol{u}}||$ with minimum norm $\tilde{\boldsymbol{u}}$ and thereby assigning a uniform load allocation. The wrench of each manipulator is then given by

$$\tilde{\boldsymbol{u}}_d(t) = G^+\boldsymbol{u}_d(t) . \tag{10.12}$$

[1]Note that for orientation degrees of freedom, this expression holds only if $\boldsymbol{\omega} \times \boldsymbol{\omega} J_d$ holds, where J_d represents the desired inertia. For simplicity of illustration, we stick to this particular case. A more rigorous analysis can be found in [41].

Under ideal conditions, the controls resulting from (10.8), (10.9), and (10.12) fulfill objectives (10.6) and (10.7) [46]. However, the deployment of real cooperative manipulation systems is significantly more challenging due to sensorimotor uncertainties, which give rise to unexpected internal stress [10]. High interaction forces and torques do not just hinder tracking performance but also induce discomfort or even compromise safety.

10.2.2 Control Challenges in Physical Human–Robot Interaction

Consider now the setting from Sect. 10.2.1 with an additional human agent. Let $\widetilde{\boldsymbol{u}}_h \in \mathbb{R}^6$ be the applied wrench of the human in frame C and $\boldsymbol{p}_h \in \mathbb{R}^3$ the vector from the human grasp position to the object frame. The corresponding human partial grasp matrix defined as in (10.5) is denoted as G_h. Object dynamics from (10.2) are now given by

$$M_o\ddot{\boldsymbol{x}}(t) + \boldsymbol{h_o}(\boldsymbol{x}(t), \dot{\boldsymbol{x}}(t)) = \boldsymbol{u}(t) + \boldsymbol{u}_h(t) , \tag{10.13}$$

where $\boldsymbol{u}_h = G_h\widetilde{\boldsymbol{u}}_h(t)$. Although task Definition 10.1 holds for the pHRI case, control design based on (10.8) and (10.12) is in general not attainable anymore. The presence of a physically coupled human partner implies two additional challenging issues:

- The desired object configuration trajectory $\boldsymbol{x}_d(t)$ from (10.6) is generally unknown to the robot. Even when a final goal configuration is known to both the human and the robot, any assumption on the human-in-the-loop desired object trajectory at every time instant is, in most settings, unrealistic. In addition, objective (10.6) does not hold anymore as it considers an infinite horizon while a human might want to reach $\boldsymbol{x}_d(t)$ in a finite amount of time.
- The human-desired applied wrench load share is unknown. This fact hinders the design of matrix A which distributes the load among agents as given in (10.11).

In order to cope with these issues there are two options in the literature. On one side, *reactive* assistants consider the human an exogenous environmental agent. The human applied wrench is modeled as an external wrench component and is accommodated by means of a force control scheme. On the other side, *proactive* assistants estimate or assume feedforward models λ of the human-in-the-loop desired object trajectory $\boldsymbol{x}_d(t)$. These approaches yield model-based control schemes that anticipate human actions instead of just compensating for them. The following subsections review in detail related works following these two different approaches.

10.2.3 Reactive Assistants

Reactive assistants consider the human wrench input as an environmental disturbance and compensate for it by means of force control. Following the approach presented in [41] and explicitly modeling the human operator as an exogenous input, interaction of humans with multiple robots is enabled in [25] by rendering a mass-damper system

$$M_r \ddot{\boldsymbol{x}}(t) + D_r \dot{\boldsymbol{x}}(t) = \boldsymbol{u}_{\text{ext}}(t) = \boldsymbol{u}_h(t) \ . \tag{10.14}$$

This approach is further extended in a decentralized manner in [20]. Assuming the human hand behaves as a passive element plus an independent exogenous signal [6], the stability of this system is guaranteed [36].

In terms of Definition 10.1, no desired trajectory is considered, i.e., $\boldsymbol{u}_h(t) = \boldsymbol{0}$ is the only control goal to be fulfilled. The agents' load distribution A from (10.11) is also given by the Moore–Penrose pseudoinverse but the human operator is not considered as an agent of the system and thereby is ignored in the decomposition.

Reactive assistants, although effective and intuitive, do not reduce significantly human effort but rather serve as a physical human–robot interface.

10.2.4 Proactive Assistants

In contrast to reactive assistants, proactive approaches solve the cooperative manipulation problem considering the human as a constitutive part of the multiagent system and estimating its preferences. From a psychological perspective, this approach is necessary for successful *cooperation*. Psychologists emphasize that cooperation allows multiple agents to jointly work toward a shared goal. However, this implies awareness of other interacting agents in terms of reasoning and knowing about their intentions and capabilities. It further imposes the requirement of implicitly or explicitly communicating the goal [37]. In this context, in [53] a reactive control scheme was successfully transferred to a humanoid robot and combined with a predefined movement generation routine based on directions from human speech inputs. However, wherever explicit communication between robots and humans is infeasible, predicting outcomes of the partner's actions is a requirement for successful interaction during joint action [42]. Following this idea, proactive assistants consider *behavior models* λ, comprising motion, force, or both of them, estimating the human-desired trajectory $\boldsymbol{x}_d$ and desired interaction behavior $\widetilde{\boldsymbol{u}}_d$.

Definition 10.2 A *behavior model* λ is given by a function $\boldsymbol{f} : \mathbb{R}^n \rightarrow \mathbb{R}^{6+6N}$ such that

$$\begin{bmatrix} \boldsymbol{x}_d(t) \\ \widetilde{\boldsymbol{u}}_d(t) \end{bmatrix} = \boldsymbol{f}(\boldsymbol{\xi}(t)) \ ,$$

where $\boldsymbol{\xi}(t) = \begin{bmatrix} \boldsymbol{x}(t) & \dot{\boldsymbol{x}}(t) & \widetilde{\boldsymbol{u}}(t) & \boldsymbol{\xi}_{\text{env}}(t) \end{bmatrix}$ is the input space and $\boldsymbol{\xi}_{\text{env}}$ are environment or task-related features.

These models serve as a predictive reference for the robot in order to effectively cooperate with its partner and urge toward the human-desired goal. Proactive control schemes achieve higher transparency than reactive ones, i.e., they are able to follow/anticipate human movements with lower human-perceptible resistive forces [24]. Furthermore, goal-oriented behavior is even necessary to solve a manipulation task if parts of the task state space are not reachable for a single agent or difficult to control by the human, for example, when transporting a voluminous object thorough a cluttered environment [27].

In general, proactive assistants render an overall system behavior given by

$$M_r\ddot{\boldsymbol{x}}(t) + D_r\dot{\boldsymbol{x}}(t) = \boldsymbol{u}_r(t) + \boldsymbol{u}_h(t) , \tag{10.15}$$

where $\boldsymbol{u}_r \in \mathbb{R}^6$ is the resulting proactive control input of the robots. Instead of only acting as a human–robot interface as in (10.14), the presence of the additional input $\boldsymbol{u}_r$ enables anticipation by exploiting the feedforward knowledge provided by λ.

The characterization of behavior models λ for pHRI is very diverse. Some behavior models have discretized latent state spaces yielding hybrid control schemes that switch between different admittance parameters triggered by the recognition of predefined sets of haptic and motion patterns in [45, 50]. Similarly, predicting human intention from the force derivative provides a way to adapt the rendered damping D_r while ensuring stability [9].

A different line of work rely on *analytical models* of human behavior in order to estimate $\boldsymbol{x}_d$. In this case predictions $\boldsymbol{f}(\boldsymbol{\xi}(t))$ from Definition 10.2 are given by an a priori assumed analytical function of the inputs. The well-known minimum jerk principle [15] is exploited either on a position level in [31] or on a velocity level in [7]. However, this principle does not hold for cooperative manipulation tasks [34]. When tasks of higher complexity involving sequences of multiple, previously unknown primitive actions are considered, these works can serve only as guidelines as they just consider local predictive models. As an alternative, planning approaches provide suitable task solving behavior models when both the common goal and the environment are known. Especially compelling are feedback motion planning methods which continuously adapt their predictions when human deviations occur [28]. However, besides requiring accurate environment models, human dynamic preferences are not taken into account.

A recent trend in haptic assistance follows the application of *learning techniques* and more precisely the Programming by Demonstration (PbD) paradigm [2] to generate models of human motion behavior. In this context, the cooperation behavior synthesis presented by [14] represents the first step of strategies involving learning and recognizing human force patterns in a large-scale setup to generate an appropriate object position trajectory. The recognition performance of left-to-right HMMs on Fourier-transformed force data as well as the generation of handcrafted robot actions serve as a starting point for developments directed toward more reliable intention recognition and intuitive robotic actions in the recognized task phases. Acquiring demonstrations through teleoperation, humanlike interactive behavior is reproduced considering learned motions and interaction forces [13, 17]. Interpreting the interaction as a continuous teaching process, the robot may evolve from a passive follower into an active contributor using iterative learning techniques [22, 32]. Although these approaches rely on probabilistic models, their decision-making schemes do not exploit the probabilistic nature of predictions.

Regardless of the method used, the goal-oriented behavior that all above-mentioned proactive approaches synthesize implies the possibility of conflicts with the interacting partner. An insight on the consequences of this problem is acquired by observing haptic interaction between physically coupled human–human dyads during point-to-point tasks. Humans behave in a goal-oriented fashion and, while they do cooperate toward a final common goal, they have to continuously negotiate a trajectory to follow. Studies highlight the importance of haptic feedback for the continuous negotiation process between partners [19] and indicate the existence of different roles [40]. These results motivate the synthesis of interaction roles, i.e., the load share that each partner is responsible for, defined by matrix A in (10.11). Following this idea, a possible framework is presented in [12], where different roles are achieved using a controller blending scheme. This concept is exploited by adjusting the robot role depending on the level of prediction confidence in [47]. Other options smoothly adapt the role of the robot depending on the human force magnitude [35] a measure of disagreement between partners [38], or game theory [30], while following a predefined trajectory. The large-scale human user study presented in [35] shows the advantages of a dynamic role allocation scheme. Similar results are outlined in [1], where the scope of roles is explored using haptic shared control in a driver-supporting scenario. These works rely on the robot role adaptation based on some measure of disagreement. However, there is no consensus in the literature about how to evaluate physical disagreement between N goal-oriented agents or how to adapt the robotic role.

10.3 Interaction Wrench Decomposition

A deep understanding of interaction wrenches is crucial for estimating human intention and assessing the interaction. They communicate intention, disagreements and determine the effort of each agent. Consider the N-agents system from the previous section with object-centered dynamics

$$M_o\ddot{\boldsymbol{x}}(t) + \boldsymbol{h}_o(\boldsymbol{x}(t), \dot{\boldsymbol{x}}(t)) = \boldsymbol{u}(t) = G\tilde{\boldsymbol{u}}(t) = GA\boldsymbol{u}_d(t)\,. \tag{10.16}$$

This section explores two complementary problems:

- Given a desired resulting wrench $\boldsymbol{u}_d$, what are the possible physically consistent decomposition matrices A that produce no counteracting wrenches? (Sect. 10.3.1)
- Given the observed agents' wrenches $\tilde{\boldsymbol{u}}$, what is the corresponding load share that each agent accounts for and which are the effective and counteracting components? (Sect. 10.3.2)

Physically consistent definitions of disagreement and load share are further proposed in Sect. 10.3.3.

10.3.1 Nonuniform Wrench Decomposition Matrices

The design of the matrix A in (10.16) determines the agents' wrenches $\widetilde{\boldsymbol{u}}$ that yield a desired $\boldsymbol{u}_d$. A desirable characteristic is the absence of counteracting components in the resulting $\widetilde{\boldsymbol{u}}$, which produce unnecessary work but have no influence on dynamics (10.16), i.e., they are not motion-inducing. Recall that the Moore–Penrose pseudoinverse G^+ yields the solution minimizing $||\boldsymbol{u}_d - A\widetilde{\boldsymbol{u}}||$ with minimum norm $\widetilde{\boldsymbol{u}}$ and thereby assuming a uniform allocation policy. This approach lacks physical consistency as the norm of a wrench, consisting of both forces [N] and torques [Nm] has no physical meaning. A valid option are generalized inverses [8], which define some predefined weighting in order to define the otherwise physically inconsistent norm [49], similarly yielding uniform allocations. However, humans clearly exhibit nonuniform distributions and even favor asymmetric load distributions interacting in dyads [40]. In this section, following recent results in this field [11], we aim for a physically consistent decomposition that supports nonuniform allocation policies.

To ensure physical consistency, we first abstract the analysis from object geometry and grasping constraints by setting it in the object frame as in [48]. In this basis, set force and torque components are decoupled and treated independently. Let $\overline{\boldsymbol{u}}_i$ be the applied wrench of agent i on the object frame, such that $\overline{\boldsymbol{u}}_i = G_i\widetilde{\boldsymbol{u}}_i$. Its force and torque components and the resulting wrench are denoted

$$\overline{\boldsymbol{u}}_i = \begin{bmatrix} \overline{\boldsymbol{f}}_i \\ \overline{\boldsymbol{\tau}}_i \end{bmatrix} \qquad \boldsymbol{u}_d = \begin{bmatrix} \boldsymbol{f}_d \\ \boldsymbol{\tau}_d \end{bmatrix} .$$

In the object frame the decomposition problem is the design of A such that

$$\overline{\boldsymbol{u}} = A\boldsymbol{u}_d \qquad \boldsymbol{u}_d = \overline{G}A\boldsymbol{u}_d, \tag{10.17}$$

where we defined the stacked forms

$$\overline{G} = \begin{bmatrix} I_1 \; I_2 \cdots I_N \end{bmatrix} \qquad \overline{\boldsymbol{u}} = \begin{bmatrix} \overline{\boldsymbol{u}}_1 \; \overline{\boldsymbol{u}}_2 \cdots \overline{\boldsymbol{u}}_N \end{bmatrix} ,$$

with similar definitions for force and torque components.

The decomposition matrices A representing allocation policies inducing no internal stress are defined in the following proposition.

Proposition 10.1 *The family of solutions satisfying* (10.17) *yielding no counteracting wrenches is given by*

$$A = \begin{bmatrix} \overline{A}_1 \, \overline{A}_2 \, \dots \, \overline{A}_N \end{bmatrix}^{\mathrm{T}} ,$$

where

$$\overline{A}_i = \mathrm{diag}\begin{bmatrix} \alpha_{f,i} \; \alpha_{f,i} \; \alpha_{f,i} \; \alpha_{\tau,i} \; \alpha_{\tau,i} \; \alpha_{\tau,i} \end{bmatrix} , \tag{10.18}$$

and

$$\begin{aligned}
&\alpha_{f,i}, \alpha_{\tau,i} \geq 0 \qquad \forall i = 1 \cdots N \\
&\sum_{i=1}^{N} \alpha_{f,i} = 1 \qquad \sum_{i=1}^{N} \alpha_{\tau,i} = 1 \,.
\end{aligned} \tag{10.19}$$

Proof See Appendix.

The duple $\{\alpha_{f,i}, \alpha_{\tau,i}\}$ represents the load share in terms of the fractions of the overall mass and inertia of the system (10.16) that agent i accounts for. Consider now a dyadic setting where a human interacts with a robot with load shares $\{\alpha_{f,h}, \alpha_{\tau,h}\}$ and $\{\alpha_{f,r}, \alpha_{\tau,r}\}$, respectively. It is interesting to comment extreme cases of load distributions

- $\{\alpha_{f,h} = 1, \alpha_{\tau,h} = 1\}$ and $\{\alpha_{f,r} = 0, \alpha_{\tau,r} = 0\}$, which correspond to a passive robotic partner.
- $\{\alpha_{f,h} = 0, \alpha_{\tau,h} = 0\}$ and $\{\alpha_{f,r} = 1, \alpha_{\tau,r} = 1\}$, which correspond to a proactive robotic partner that accounts for all necessary effort.

All load distributions lying in between correspond to cooperative settings where the load is effectively shared. Note that the generalized Moore–Penrose inverse solution yields $\{\alpha_{f,h} = 0.5, \alpha_{\tau,h} = 0.5\}$ and $\{\alpha_{f,r} = 0.5, \alpha_{\tau,r} = 0.5\}$, uniformly distributing load among agents.

10.3.2 Effective and Internal Wrenches

Proposition 10.1 defines the decomposition matrices A yielding no counteracting wrenches in the object frame and enables control design with arbitrary load distributions. Another relevant issue in pHRI settings is the inverse problem, where, given an observed $\overline{\boldsymbol{u}}$, the respective *effective* and *internal* components are identified. The separation of these two additive components is defined as

$$\begin{aligned}
\overline{\boldsymbol{u}} &= \overline{\boldsymbol{u}}_{\text{eff}} + \overline{\boldsymbol{u}}_{\text{int}} \\
s.t \quad \overline{G}\,\overline{\boldsymbol{u}}_{\text{eff}} &= \overline{G}\,\overline{\boldsymbol{u}} = \boldsymbol{u}_d \\
\overline{G}\,\overline{\boldsymbol{u}}_{\text{int}} &= \boldsymbol{0} \ ,
\end{aligned} \tag{10.20}$$

where

- the effective wrenches $\overline{\boldsymbol{u}}_{\text{eff}}$ are components that constitute the resulting wrench $\boldsymbol{u}_d$ and can be written in terms of an instance of A from Proposition 10.1 as

$$\overline{\boldsymbol{u}}_{\text{eff}} = A\boldsymbol{u}_d \,. \tag{10.21}$$

- the internal wrenches $\widetilde{\boldsymbol{u}}_{\text{int}}$ produce unnecessary internal stress in the object and lie in

$$\overline{\boldsymbol{u}}_{\text{int}} \in \text{Null}(\overline{G}) \,, \tag{10.22}$$

where $\text{Null}(\overline{G})$ denotes the nullspace of $\overline{G}$.

For a given $\overline{\boldsymbol{u}}$, the possible effective wrench components are defined in the following proposition:

Proposition 10.2 *The* effective wrench *components are given by*

$$\overline{\boldsymbol{u}}_{\mathit{eff},i} = \begin{bmatrix} \alpha_{f,i} \boldsymbol{f}_d \\ \alpha_{\tau,i} \boldsymbol{\tau}_d \end{bmatrix}, \tag{10.23}$$

where $\alpha_{f,i} = \frac{m_i}{m_0}, \alpha_{\tau,i} = \frac{j_i}{j_0} \in \mathbb{R}$ *satisfy*

$$\begin{aligned} &\min_{m_i} \sum_{i=1}^{N} m_i || \ddot{\boldsymbol{p}} - \frac{\overline{\boldsymbol{f}}_i}{m_i} || \\ &s.t. \, m_i \geq 0 \\ &\sum_{i=1}^{N} m_i = m_0 \end{aligned} \qquad \begin{aligned} &\min_{j_i} \sum_{i=1}^{N} j_i || \dot{\boldsymbol{\omega}} - \frac{\overline{\boldsymbol{\tau}}_i}{j_i} || \\ &s.t. \, j_i \geq 0 \\ &\sum_{i=1}^{N} j_i = j_0 \end{aligned}$$

and $m_0 \in \mathbb{R}$ *and* $j_0 \in \mathbb{R}$ *are the mass and moment of inertia of the object, respectively.*

Proposition 10.2 follows from ensuring Gauss's principle of least constraint. In this chapter, we only consider dyadic settings. This particular case constraining $N = 2$ admits simple solutions exposed in the following proposition.

Proposition 10.3 *The* effective wrench *components of a 2-agent system are given by*

$$\overline{\boldsymbol{u}}_{\mathit{eff},i} = \begin{bmatrix} \alpha_{f,i} \boldsymbol{f}_d \\ \alpha_{\tau,i} \boldsymbol{\tau}_d \end{bmatrix}, \tag{10.24}$$

where $\alpha_{f,i}, \alpha_{\tau,i} \in \mathbb{R}$ *are given by*

$$\begin{aligned} \alpha_{f,i} &= \min\left(\max\left(\frac{\overline{\boldsymbol{f}}_i^{\mathrm{T}} \boldsymbol{f}_d}{||\boldsymbol{f}_d||^2}, 0 \right), 1 \right) \\ \alpha_{\tau,i} &= \min\left(\max\left(\frac{\overline{\boldsymbol{\tau}}_i^{\mathrm{T}} \boldsymbol{\tau}_d}{||\boldsymbol{\tau}_d||^2}, 0 \right), 1 \right) . \end{aligned} \tag{10.25}$$

Proof See Appendix.

The definition of the internal wrench components follows from (10.20) as

$$\overline{\boldsymbol{u}}_{\text{int}} = \overline{\boldsymbol{u}} - \overline{\boldsymbol{u}}_{\text{eff}} \ . \tag{10.26}$$

The proposed decomposition is depicted in Fig. 10.2 in comparison with the generalized pseudoinverse for a simple two-dimensional example with two agents. In contrast to the generalized pseudoinverse $\overline{G}^+$ from Fig. 10.2a which assumes effective load shares $\alpha_{f,1} = \alpha_{f,2} = 0.5$, the decomposition from Proposition 10.3 depicted in Fig. 10.2b follows from the computation of the effective mass that each agent accounts for applying (??), yielding $\alpha_{f,1} = 0.756$ and $\alpha_{f,2} = 0.244$.

The decomposition of observed wrenches in effective and disagreement components enables the definition of interesting performance measures for physical multi-agent interaction tasks.

Definition 10.3 The *effective work* done by agent i of the multiagent system between time instants t_1 and t_2 is given by

$$\int_{t_1}^{t_2} \overline{\boldsymbol{u}}_{\text{eff},i}^{\mathrm{T}} \dot{\boldsymbol{x}} dt \ .$$

Definition 10.4 The *counteracting work* done by agent i of the multiagent system between time instants t_1 and t_2 is given by

$$\int_{t_1}^{t_2} \overline{\boldsymbol{u}}_{\text{int},i}^{\mathrm{T}} \dot{\boldsymbol{x}} dt \ .$$

Similar measures for the work of the overall system are straightforwardly computed.

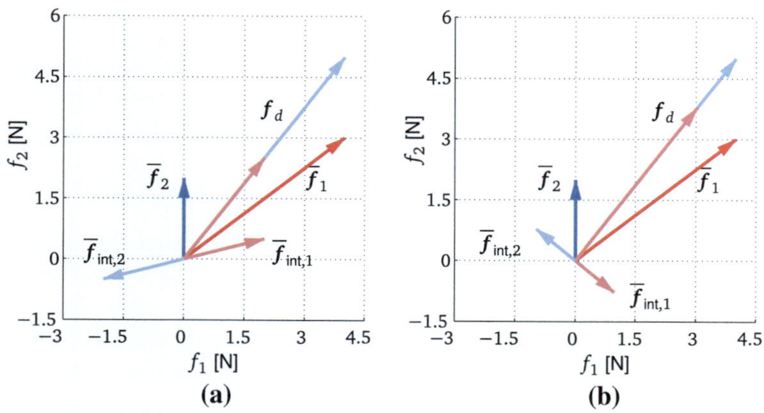

Fig. 10.2 Exemplary force decomposition for two agents using the Moore–Penrose pseudoinverse (**a**) and decomposition from Proposition 10.3 and (10.26) depicted in (**b**)

An analysis of possible causes yielding internal wrench components is also relevant in order to understand their role in the interaction. In the following, we study the decomposition of wrenches arising from the interaction of decentralized systems, a condition more in accordance with human–robot settings. Consider now the *decentralized* case where each agent follows its own desired resulting wrench and decomposition matrix. Assume now that each agent follows

$$\overline{\boldsymbol{u}} = A_i \boldsymbol{u}_{d\,i} \,, \tag{10.27}$$

where $\boldsymbol{u}_{d\,i}$ is the i-th agent desired resulting wrench and A_i an assumed load distribution from Proposition 10.1. If all desired wrenches and load distribution matrices are equal, i.e., $A_i = A_j \;\wedge\; \boldsymbol{u}_{d\,i} = \boldsymbol{u}_{d\,j}\ \forall 1 \le i, j \le N$, then the standard problem considered in Proposition 10.1 is recovered and no internal wrench components arise. Consider the case when desired resulting wrenches match but load distributions are different, i.e., $A_i \ne A_j \wedge \boldsymbol{u}_{d\,i} = \boldsymbol{u}_{d\,j} \forall 1 \le i, j \le N$. It is straightforward to show that although only a scaled desired resulting wrench is effectively exerted, no internal wrenches are produced. In contrast, when desired resulting wrenches differ, if the angle between desired resulting forces or torques is not zero, internal wrenches appear as illustrated in the following proposition.

Proposition 10.4 *For a decentralized system where each agent follows* (10.27), $\overline{\boldsymbol{u}}_{int} \ne \boldsymbol{0}$

$$\textit{iff}\ \ \exists i, j \mid \frac{\boldsymbol{f}_{d\,i}^{\mathrm{T}} \boldsymbol{f}_{d\,j}}{||\boldsymbol{f}_{d\,i}||||\boldsymbol{f}_{d\,j}||} < 1 \quad or \quad \exists i, j \mid \frac{\boldsymbol{\tau}_{d\,i}^{\mathrm{T}} \boldsymbol{\tau}_{d\,j}}{||\boldsymbol{\tau}_{d\,i}||||\boldsymbol{\tau}_{d\,j}||} < 1 \,. \tag{10.28}$$

Proof See Appendix.

Proposition 10.4 together with the previous analysis yields interesting conclusions:

- If all desired resulting wrenches $\boldsymbol{u}_{d\,i}$ are equal, a discrepancy in load shares does not produce any internal wrenches; it yields a resulting wrench scaled by a factor.
- If desired resulting wrenches $\boldsymbol{u}_{d\,i}$ are not equal but have same direction, no internal wrenches are exerted.
- The presence of internal wrenches *implies* $\boldsymbol{u}_{d\,i}$'s with different directions.

As a result, internal wrenches play a key role during haptic interaction. In terms of haptic feedback, *internal wrench components are the sole communication channel available in order to express the desire to change the direction of the object's acceleration*. In other words, internal wrench components can be interpreted as the **haptic negotiation channel** concerning the desired acceleration's direction. In a human–robot setting, assuming that the human partner behaves as in (10.27), observed internal components imply errors in the estimation of human-desired resulting wrench.

10.3.3 Load Share and Disagreement

There is no consensus in the literature about how to measure disagreement between physically interacting agents. In [40], disagreement between human dyads is interpreted as the difference between applied wrenches. However, this measure only provides no disagreement when agents apply the same wrench, classifying any deviations from this uniform distribution as disagreement. In [18], disagreement is related to internal forces in one dimension, i.e., counteracting force components that do not contribute to the resulting force. Here, we follow this idea and relate disagreement to $\overline{\boldsymbol{u}}_{\text{int}}$. We extend this concept to 6D by means of Proposition 10.3, motivated by classical mechanics principles. Note that the observed load share is implicitly identified in the computation of the internal wrenches. The definitions of disagreement, effective load, load share, and load distribution follow.

Definition 10.5 The *translational* and *rotational disagreement* of the i-th agent are given by $||\overline{\boldsymbol{f}}_{\text{int},i}||$ and $||\overline{\boldsymbol{\tau}}_{\text{int},i}||$ and measured in [N] and [N/m], respectively.

Definition 10.6 The *effective load* of agent i is given by $\left\{\alpha_{f,i} m_0, \alpha_{\tau,i} j_0\right\}$ and it represents the effective mass and moment of inertia that agent i accounts for.

Definition 10.7 The *load share* of agent i is given by $\left\{\alpha_{f,i}, \alpha_{\tau,i}\right\}$ and it represents the effective share of mass and moment of inertia that agent i accounts for.

Definition 10.8 The *load distribution* of the multiagent system is given by the set $\left\{\alpha_{f,1}, \alpha_{f,2} \cdots \alpha_{f,N}\ \alpha_{\tau,1}, \alpha_{\tau,2} \cdots \alpha_{\tau,N}\right\}$ comprising all load shares.

These measures, together with Definitions 10.3 and 10.4 are instrumental for the evaluation of cooperative manipulation between goal-oriented agents as they capture the physical properties of the interaction.

10.4 Optimal Robot Assistance Considering Human Behavior Uncertainty and Disagreements

In this section, we approach the problem of control design of a proactive robotic partner for cooperative manipulation. For simplicity, we consider a common interaction contact point between the robot and its human partner at the end effector. In addition, we assume that the nonlinear robot dynamics is feedback-linearized and an impedance control law renders the robot behavior [44]

$$M_r \ddot{\boldsymbol{x}}(t) + D_r \dot{\boldsymbol{x}}(t) = \boldsymbol{u}(t) = \boldsymbol{u}_h(t) + \boldsymbol{u}_r(t), \tag{10.29}$$

where $\boldsymbol{x} \in \mathbb{R}^6$ is the pose of the end effector, $\boldsymbol{u}_h \in \mathbb{R}^6$ is the applied wrench by the human, $\boldsymbol{u}_r \in \mathbb{R}^6$ is the assistive control input of the robot, and $M_r, D_r \in \mathbb{R}^{6\times6}$ are positive-definite matrices representing the rendered inertia and viscous friction, respectively. For later convenience, we discretize the system from (10.29) with a sampling time interval Δ, yielding impedance-controlled dynamics

$$\boldsymbol{\xi}_{k+1} = \boldsymbol{f}(\boldsymbol{\xi}_k, \boldsymbol{u}_{h,k} + \boldsymbol{u}_{r,k}) , \tag{10.30}$$

where k is the time index such that $t = k\Delta$ and $\boldsymbol{\xi}_k = [\ \boldsymbol{x}_k\ \dot{\boldsymbol{x}}_k\ ]^{\mathrm{T}}$ the state of the system.[2]

In addition to the *reactive behavior* represented by the compliance control (10.29), the robot implements an anticipatory *proactive behavior* given by $\boldsymbol{u}_r$. To this end, the robot relies on a behavior model λ, which provides predictions of the next human-desired state and human internal wrench in terms of normal distributions $\hat{\boldsymbol{\xi}}_{d,k} \sim \mathcal{N}(\boldsymbol{\xi}_d | \hat{\boldsymbol{\mu}}_{\xi,k}, \hat{\boldsymbol{\Sigma}}_{\xi,k})$ and $\hat{\boldsymbol{u}}_{\mathrm{int},k} \sim \mathcal{N}(\boldsymbol{u}_{\mathrm{int},k} | \hat{\boldsymbol{\mu}}_{\boldsymbol{u}_{\mathrm{int}},k}, \hat{\boldsymbol{\Sigma}}_{\boldsymbol{u}_{\mathrm{int}},k})$. The behavior model represents expected trajectories based on multiple previously performed trials. The resulting normally distributed state-space predictions represent the trial-to-trial statistical mean and variance of human-desired motion. Internal human wrench components, studied in depth in Sect. 10.3, represent counteracting wrenches not influencing dynamics (10.29) that imply disagreement between human and robot or, similarly, discrepancies between desired accelerations. This additional unnecessary effort is not observable in state-space trajectories. Consider as an example a binary decision problem while approaching an obstacle, where both agents disagree on which way to go, either left or right. We further assume that both partners have a common goal and deviations from the predicted trajectory induced by the human are small and do not express a divergence or a tendency toward a different goal. Given this problem setting, the focus of this work lies on the design of a possibly time-varying assistive control law synthesizing $\boldsymbol{u}_r$ taking uncertainties into account.

10.4.1 Anticipatory Assistance Based on Learned Models

In this work, the objective governing haptic assistance is human force minimization. From an optimality point of view, this concept is expressed by the minimization of a cost function in the form

$$J = \sum_{k=0}^{T-1} \|\boldsymbol{u}_{h,k}\|^2, \tag{10.31}$$

where T is the time horizon for the optimization.[3] The human has an urge toward a goal as predicted by the task model λ but relying on a human force estimation of T steps potentially accumulates high tracking errors in task space. A more suitable alternative considers the minimization problem in task space based on the tracking

---

[2]The time-discretized system dynamics will be specified in Sect. 10.4.2 for a two-dimensional example.

[3]Note that we will later employ a receding horizon control scheme, i.e., without fixed final time T. For simplicity of presentation we consider for now the finite horizon control problem (10.31).

error between task model predictions $\hat{\boldsymbol{\xi}}_d = \{\hat{\boldsymbol{\xi}}_{d,0} \cdots \hat{\boldsymbol{\xi}}_{d,T}\}$ and the current state $\boldsymbol{\xi}$, i.e.,

$$J = \sum_{k=0}^{T} \|(\hat{\boldsymbol{\xi}}_{d,k} - \boldsymbol{\xi}_k)\|^2 \,. \tag{10.32}$$

The solution to this optimization problem leads to a control scheme which does not consider any energy expenditure of the robot nor any torque constraints of the robot actuators. In addition, disturbances to the tracking may result in rather abrupt behavior of the robot, which might not be desirable when targeting intuitive assistance. In order to keep the computational complexity low to achieve real-time computation, we add a soft penalty on the energy expenditure in terms of a quadratic term. As a result the cost function takes the form as in the classical linear quadratic optimal control problem

$$J = \|(\hat{\boldsymbol{\xi}}_{d,T} - \boldsymbol{\xi}_T)\|^2_{Q_T} + \sum_{k=0}^{T-1} \|(\hat{\boldsymbol{\xi}}_{d,k} - \boldsymbol{\xi}_k)\|^2_{Q_k} + \|\boldsymbol{u}_{h,k}\|^2_{R_k} , \tag{10.33}$$

where $\|x\|^2_Q$ stands for the quadratic form $x^{\mathrm{T}} Q x$, Q_T is the final cost weight, and Q_k and R_k are weighting factors that define the trade-off between tracking performance and robot contribution. Observe that two factors challenge the equivalence between (10.31) and human intentions: (i) the prediction error of $\hat{\boldsymbol{\xi}}_d$; (ii) the weighting factors Q_k and R_k may differ from human preferences. As perfect predictions and weighting factors design is usually not feasible when dealing with humans, corrections/disagreements from the human partner are expected. This source of variability, together with the normally distributed nature of predictions define the uncertainty that the robotic assistance needs to cope with. In the following, we analyze the consequences of uncertainty in the robot behavior and synthesize their effect as variations in the cost function to optimize (10.33) and in the predicted dyad dynamics.

Remark 10.1 Following the analysis from Sect. 10.3, the minimization of cost function (10.33) yielding the robot assistive input $\boldsymbol{u}_r$ implicitly assumes an a priori load share distribution $\{\alpha_{f,h} = 0, \alpha_{\tau,h} = 0\}$ and $\{\alpha_{f,r} = 1, \alpha_{\tau,r} = 1\}$, i.e., a priori the robot accounts for all load. Other load sharing policies are straightforwardly computed by multiplying the resulting input by the desired decomposition matrix A from Proposition 10.1.

10.4.1.1 Considering Motion Prediction Uncertainty

Cost function (10.33) depends on a multivariate normally distributed reference $\hat{\boldsymbol{\xi}}_d$. A classical realization of the cost function for optimal tracking control measures the Euclidean distance to the reference's mean, i.e.,

$$J = \|(\hat{\boldsymbol{\mu}}_{\xi,T} - \boldsymbol{\xi}_T)\|^2_{Q_T} + \sum_{k=0}^{T-1} \|(\hat{\boldsymbol{\mu}}_{\xi,k} - \boldsymbol{\xi}_k)\|^2_{Q_k} + \|\boldsymbol{u}_{h,k}\|^2_{R_k} \,. \tag{10.34}$$

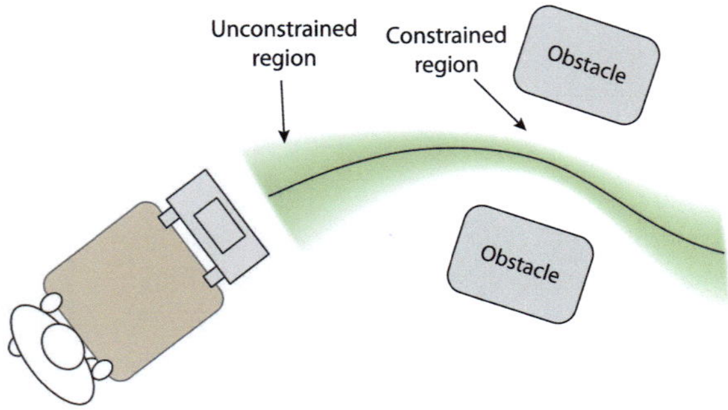

Fig. 10.3 Exemplary representation of the predicted state mean and variance $\hat{\xi} = \{\hat{\mu}_\xi, \hat{\Sigma}_\xi\}$. Low variance directions suggest the possibility of constrained passages, while high variances areas imply unconstrained areas (© 2015 IEEE. Reprinted, with permission, from [33])

However, in order to accordingly measure the distance to a multivariate Gaussian, the weighted *Mahalanobis distance* is a suitable option as it includes the covariance of the prediction into the distance metric. In this case, the cost function becomes

$$J = \|(\hat{\mu}_{\xi,T} - \xi_T)\|^2_{\hat{Q}_T} + \sum_{k=0}^{T-1} \|(\hat{\mu}_{\xi,k} - \xi_k)\|^2_{\hat{Q}_k} + \|u_{h,k}\|^2_{R_k} , \tag{10.35}$$

where $\hat{Q}_k = \hat{\Sigma}_{\xi,k}^{-\frac{1}{2}} Q_k \hat{\Sigma}_{\xi,k}^{-\frac{1}{2}}$ and $\hat{Q}_T = \hat{\Sigma}_{\xi,T}^{-\frac{1}{2}} Q_T \hat{\Sigma}_{\xi,T}^{-\frac{1}{2}}$ and $\hat{\Sigma}_{\xi,i}$ is per definition a symmetric positive-definite matrix $\forall i \;\; 0 \leq i \leq T$. This formulation diminishes the cost of tracking errors in directions with high prediction variance while it increases the penalty of deviations in regions with low variance. This is a desirable behavior in most applications, where a low variance of the learned desired trajectory over repeated trials indicates some importance to keep track of it. As an example consider the situation depicted in Fig. 10.3. Low variance directions might be caused, for example, by environmental constraints while high variance regions imply unconstrained areas. This suggests flexibility in unconstrained directions and strict tracking through narrow passages, a behavior in accordance with the minimization of the Mahalanobis distance as in (10.35). Note that the opposite effect is achieved using the inverse of the covariance matrix, i.e., $\hat{Q}_k = \hat{\Sigma}_{\xi,k}^{\frac{1}{2}} Q_k \hat{\Sigma}_{\xi,k}^{\frac{1}{2}}$ and $\hat{Q}_T = \hat{\Sigma}_{\xi,T}^{\frac{1}{2}} Q_T \hat{\Sigma}_{\xi,T}^{\frac{1}{2}}$. In this case the cost increases inversely with the variance: directions with low predicted variance diminish their cost with respect to high variance regions, which increase their penalty. The idea of employing the measure of variance as a way to inform the system on how systematically it should follow a reference trajectory was also applied in [26] for an impedance control scheme and in [5] for kinematic control.

10.4.1.2 Considering Disagreement

In addition to motion prediction, the interaction behavior model λ also provides human internal wrench predictions $\hat{u}_{\text{int}}$. We assume that there exists a human-desired nominal motion trajectory. In consequence, if the task model λ contains enough trials, the mean of the predicted trajectory $\hat{\boldsymbol{\xi}}_d$ approaches the human-desired nominal motion trajectory, tracked by the robot. From Proposition 10.4, if desired resulting wrenches match in direction, no internal wrenches are expected and therefore $\mathsf{E}[\boldsymbol{u}_{\text{int},k}] = \hat{\boldsymbol{\mu}}_{\boldsymbol{u}_{\text{int}},k} \approx \boldsymbol{0}$. However, the human internal wrench covariance encodes valuable information: regions with high variance are caused by significant or recurrent intertrial corrections, meaning a conflictive area, see Fig. 10.4 for an illustration. To model this unbiased variability, we denote $\boldsymbol{\epsilon}_e \sim \mathcal{N}(\boldsymbol{\epsilon}_e|\boldsymbol{0}, \hat{\boldsymbol{\Sigma}}_{\boldsymbol{u}_{\text{int}}})$ the *expected disagreement* with the human partner. Applying the same idea to current observations of $\boldsymbol{u}_{\text{int}}$, an estimation of the *current disagreement*, $\boldsymbol{\epsilon}_c \sim \mathcal{N}(\boldsymbol{\epsilon}_c|\boldsymbol{0}, \boldsymbol{\Sigma}_{\boldsymbol{u}_{\text{int}}})$ is computed empirically at time k as the sample covariance over a window of W samples

$$\boldsymbol{\Sigma}_{\boldsymbol{u}_{\text{int}}} = \frac{1}{W} \sum_{i=k-W+1}^{k} \boldsymbol{u}_{\text{int},i} \boldsymbol{u}_{\text{int},i}^{\mathrm{T}} . \tag{10.36}$$

To consider both estimations, see Fig. 10.4 for visualization, the maximum of both disagreements

$$\boldsymbol{\epsilon} = \max(\boldsymbol{\epsilon}_e, \boldsymbol{\epsilon}_c) , \tag{10.37}$$

is considered a reasonable solution. Note that keeping the zero mean assumption for the current disagreement is a conservative approach. In fact, the intratrial mean along a window W that the current disagreement considers may differ from zero. Interpreting human wrench corrections as unbiased entails higher uncertainty for the robot instead of (signed) errors. The consequences of this design will be more apparent in the analysis of the optimal control solution.

Disagreements result in variability of the dyad's motion due to conflicts with the human partner. Expressing this idea in mathematical terms, dynamics (10.30) contains an additional additive disagreement vector $\boldsymbol{\epsilon}$, yielding the stochastic dyad dynamics

$$\boldsymbol{\xi}_{k+1} = \boldsymbol{f}(\boldsymbol{\xi}_k, \boldsymbol{u}_{r,k} + \boldsymbol{\epsilon}_k) . \tag{10.38}$$

Given initial state $\boldsymbol{\xi}_0$, the optimal robotic assistance results from minimizing cost function (10.35), i.e.,

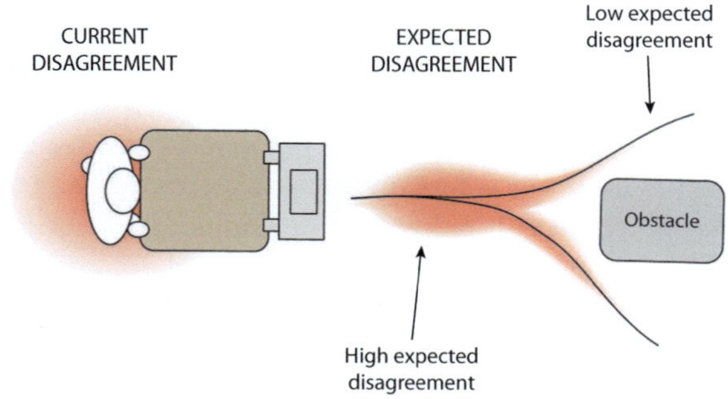

Fig. 10.4 Schematic representation of the expected disagreement $\boldsymbol{\epsilon}_e$ and the current disagreement $\boldsymbol{\epsilon}_c$ (© 2015 IEEE. Reprinted, with permission, from [33])

$$\min_{\boldsymbol{u}_{r,0}\cdots\boldsymbol{u}_{r,T-1}} J(\boldsymbol{\xi}_0, \boldsymbol{u}_{r,0}\cdots\boldsymbol{u}_{r,T-1}) , \tag{10.39}$$

constrained to the dyad dynamics (10.38). As the dynamics are constantly updated with the current disagreement from (10.36), a recalculation of the solution is necessary. To this end, we adopt a Model Predictive Control (MPC) scheme [16], where the dynamics (10.38) are considered stationary for the optimization horizon T and the optimal control problem is solved at every time step. Cost function (10.35) at time step k formulated in a receding horizon fashion becomes

$$J_k = \|(\hat{\boldsymbol{\mu}}_{\xi,k+T} - \boldsymbol{\xi}_{k+T})\|^2_{\hat{Q}_T} + \sum_{i=k}^{k+T-1} \|(\hat{\boldsymbol{\mu}}_{\xi,i} - \boldsymbol{\xi}_i)\|^2_{\hat{Q}_i} + \|\boldsymbol{u}_{r,i}\|^2_{R_k} . \tag{10.40}$$

Although the optimization is solved for a horizon of T steps, only the solution for the first step is applied as assistive robot control.

10.4.1.3 Risk-Sensitive Optimization

Problem (10.39) is a stochastic optimization problem due to the stochastic nature of the dynamics (10.38). Traditionally, it is solved by minimizing the *expected cost*, $\mathsf{E}[J]$. However, the uncertainty induced by the human behavior represented in $\boldsymbol{\epsilon}$ is not characterized by its expected value but its higher order moments. In order to capture the uncertain human influence in the dynamics, we propose to employ a risk-sensitive optimization [23, 51]. The risk-sensitive solution considers all infinite cumulants of the cost distribution [52]. The risk-sensitive optimization criterion considers a cost function in the form

$$\gamma_k(\theta) = -2\theta^{-1} \ln \mathsf{E}[\exp^{-\frac{1}{2}\theta J_k}] , \tag{10.41}$$

where $\theta \in \mathbb{R}$ is the risk-sensitivity parameter and J_k is given in (10.40). The Taylor series expansion of $\gamma(\theta)$ around $\theta = 0$,

$$\gamma(\theta) = \mathsf{E}[J] - \frac{1}{4}\theta \,\mathsf{Var}[J] + \dots ,$$

shows the influence of higher order moments in the solution, parameterized by θ. If $\theta = 0$ the controller is risk-neutral and corresponds to the classical optimization of the expected cost. For $\theta < 0$ and $\theta > 0$ the controller becomes risk-averse and risk-seeking, respectively. In our robot assistance scenario, the variance of the cost is determined by previously experienced and current disagreements with the human partner (10.37), and θ determines its assessment. In the risk-seeking case, disagreement is considered beneficial and decreases the resulting cost. In contrast, a risk-averse controller considers disagreements as a detrimental influence and the cost increases.

10.4.2 The Two-Dimensional Translational Case

The risk-sensitive optimization problem can be solved efficiently through the Riccatti equation for linear dynamics. Real-time computations are fundamental for intuitive haptic assistance; the physical coupling between human and robot requires immediate adaptation. To satisfy this prerequisite, we limit our study to a linear implementation. We consider now dynamics (10.29) in the plane, i.e., $\boldsymbol{x} \in \mathbb{R}^2$. In this case, the discrete-time dynamics from (10.38), written in the form $\boldsymbol{\xi}_{k+1} = A\boldsymbol{\xi}_k + B\boldsymbol{u}_k$ are given by

$$\begin{bmatrix} \boldsymbol{x}_{k+1} \\ \dot{\boldsymbol{x}}_{k+1} \end{bmatrix} = \begin{bmatrix} 1 & \Delta \\ 0 & 1 - M_r^{-1} D_r \Delta \end{bmatrix} \begin{bmatrix} \boldsymbol{x}_k \\ \dot{\boldsymbol{x}}_k \end{bmatrix} + \begin{bmatrix} 0 & 0 \\ 0 & M_r^{-1} \Delta \end{bmatrix} (\boldsymbol{u}_{rk} + \boldsymbol{\epsilon}_k) \,, \tag{10.42}$$

where $M_r, D_r \in \mathbb{R}^{2\times 2}$.

The disagreement $\boldsymbol{\epsilon}_k$ from (10.37), given by the maximum of the expected and current disagreement is approximated by $\mathcal{N}(\boldsymbol{\epsilon}|0, \widetilde{\boldsymbol{\Sigma}}_{\boldsymbol{u}_{\text{int}}})$, where $\widetilde{\boldsymbol{\Sigma}}_{\boldsymbol{u}_{\text{int}}}$ is the Löwner–John hyperellipsoid [3]. This Gaussian approximation calculates the minimum volume hyperellipsoid around the set defined by $\mathcal{N}(\boldsymbol{\epsilon}_e|\boldsymbol{0}, \hat{\boldsymbol{\Sigma}}_{\boldsymbol{u}_{\text{int}}})$ and $\mathcal{N}(\boldsymbol{\epsilon}_c|\boldsymbol{0}, \boldsymbol{\Sigma}_{\boldsymbol{u}_{\text{int}}})$.

For this particular problem setting and for comparison purposes, we present two different solutions: a classical linear quadratic regulator (LQR), which considers the reference prediction uncertainty (10.40) but discards the variability of the disagreement in its optimal solution and, in contrast, a linear exponential quadratic regulator (LEQR) which considers both uncertainty sources in its risk-sensitive optimization (10.41).

10.4.2.1 Linear Quadratic Regulator Solution

The minimization of the expected cost $\mathsf{E}[J_k]$ with J_k as in (10.40) and dynamics (10.42) is an instance of a linear quadratic problem. Considering only the feedback term, the solution to this problem at sample time k for a receding horizon of T steps yields a feedback control law given by

$$\boldsymbol{u}_{ri} = -K_i(\hat{\boldsymbol{\mu}}_{\xi,i} - \boldsymbol{\xi}_i) \,, \tag{10.43}$$

where K_i is the feedback matrix of the Ricatti recursion

$$K_i = -R^{-1}B^{\mathrm{T}}(BR^{-1}B^{\mathrm{T}} + \Pi_{i+1}^{-1})^{-1}A \tag{10.44}$$

and the cost-to-go is given by

$$\Pi_i = \hat{Q}_i + A^{\mathrm{T}}(BR^{-1}B^{\mathrm{T}} + \Pi_{i+1}^{-1})^{-1}A \,, \tag{10.45}$$

being $\Pi_T = \hat{Q}_T$.

Only the feedback matrix for time step $i = k$, K_k, is applied in the system. It is in the form

$$K_k = \begin{bmatrix} 0\,0 \\ K_{x,k}\ K_{\dot{x},k} \end{bmatrix}, \tag{10.46}$$

where $K_{x,k}$ and $K_{\dot{x},k}$ are the position and velocity gains, respectively.

The resulting assistive tracking control synthesizes a variable impedance control law with its gains depending on prediction uncertainty. This becomes more obvious when applying the control law for the limit $\Delta \to 0$ in the system dynamics (10.29)

$$M_r\ddot{\boldsymbol{x}} + D_r\dot{\boldsymbol{x}} = \boldsymbol{u_h} - \left(K_x(t)(\boldsymbol{x_d} - \boldsymbol{x}) + K_{\dot{x}}(t)(\dot{\boldsymbol{x}}_d - \dot{\boldsymbol{x}})\right)\ , \tag{10.47}$$

where $\boldsymbol{x_d}$, $\dot{\boldsymbol{x}}_d$, $K_x(t)$ and $K_v(t)$ are the position and velocity components of $\hat{\boldsymbol{\mu}}_\xi$ and the feedback gains $K_{x,k}$ and $K_{v,k}$ in the limit $\Delta \to 0$, respectively.

This control strategy tracks the predicted trajectory considering its variance by means of the Mahalanobis distance present in $\hat{Q}_i$ and $\hat{Q}_T$, see Fig. 10.5. As a result, the robot gets stiffer in the directions where the motion prediction has low variance and becomes more compliant for high variance. However, the variability induced by disagreements with the human partner is neglected.

10.4.2.2 Linear Exponential Quadratic Regulator

The optimization problem considers now cost function (10.41) with J_k as in (10.40). In this case, the regulation term of this optimization problem follows from a modified form of the standard Ricatti recursion [43]

$$K_i = -R_i^{-1}B^{\mathrm{T}}(BR_i^{-1}B^{\mathrm{T}} + \theta B\widetilde{\boldsymbol{\Sigma}}_{\boldsymbol{u}_{\mathrm{int}},i}B^{\mathrm{T}} + \Pi_{i+1}^{-1})^{-1}A\ , \tag{10.48}$$

and

$$\Pi_i = \hat{Q}_i + A^{\mathrm{T}}(BR_i^{-1}B^{\mathrm{T}} + \theta B\widetilde{\boldsymbol{\Sigma}}_{\boldsymbol{u}_{\mathrm{int}},i}B^{\mathrm{T}} + \Pi_{i+1}^{-1})^{-1}A\ , \tag{10.49}$$

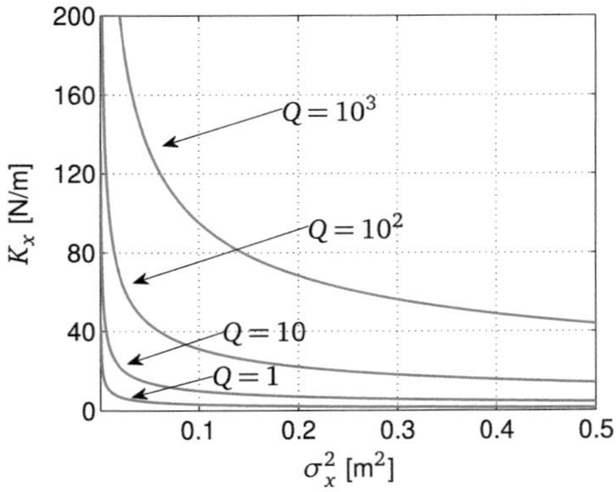

Fig. 10.5 Infinite horizon position gains w.r.t position prediction variance for several precision parameters Q with $R = I$ (© 2015 IEEE. Reprinted, with permission, from [33])

with $\Pi_T = \hat{Q}_T$.

As in the LQR case, motion prediction uncertainty is equivalently taken into account by means of the Mahalanobis considered in $\hat{Q}_i$ and $\hat{Q}_T$. However, the additive term $\theta B \widetilde{\boldsymbol{\Sigma}}_{\boldsymbol{u}_{\text{int}},i} B'$ present in the risk-sensitive solution leads to different results, see Fig. 10.6. The risk-neutral case $\theta = 0$ yields the same solution as for the LQR canceling the additional term. For a risk-seeking optimization $\theta > 0$ the expected variability $\widetilde{\boldsymbol{\Sigma}}_{\boldsymbol{u}_{\text{int}},i}$ is assumed to be collaborating as if it was doing part of the work toward fulfilling the task. Accordingly, the resulting gain from (10.48) decreases adopting a more compliant behavior. Understanding the risk-sensitivity parameter as the robot's attitude, a *recessive* attitude is achieved implementing a risk-seeking controller: the robot becomes compliant under disagreement with its partner. In contrast, the risk-averse solution $\theta < 0$ increases the overall cost as if variability were directing the system toward an undesired state. As a result, gains are increased and the robot becomes stiffer. This case corresponds to a *dominant* attitude by generating an aggressive response to disagreements.

10.4.3 Experiments

To evaluate the proposed controllers and the preferences of humans interacting with it, we designed an experiment in which a human actor has to transport a virtual object from an initial position toward a defined goal position.

Human Behavior Model Acquisition: an interaction behavior model must reflect both the human task execution preferences and the interaction with the robotic partner. Due to complexity, we rely on learning by demonstration techniques. A preliminary model of the task was acquired by initially letting the human lead, i.e., $\boldsymbol{u}_r = 0$. With this first rough representation, the robot actively assists and additionally observes internal wrench patterns during further task trials. Modeling state and human internal wrench input trajectories, $\{\boldsymbol{\xi}, \boldsymbol{u}_{\text{int}}\}$, the robot acquires a task

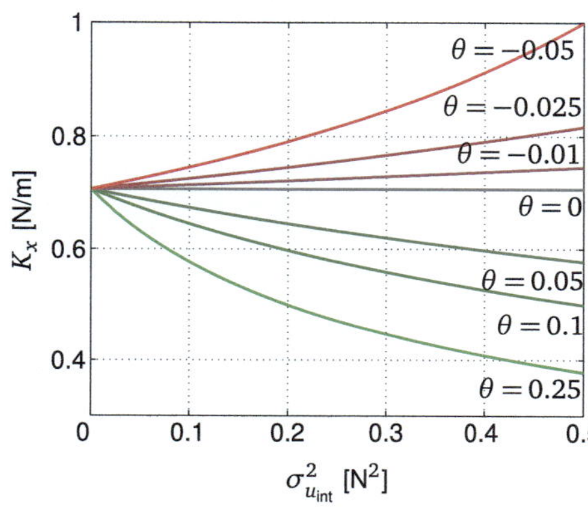

Fig. 10.6 Infinite horizon position gains w.r.t disagreement variance for several risk-sensitivities in an optimization with $Q = R = I$ (© 2015 IEEE. Reprinted, with permission, from [33])

model that represents both the human-desired state trajectory and the expected disagreement levels. Observations are encoded into a time-based Hidden Markov Model (HMM), as detailed in [29]. For simplicity, in our implementation we rely on the complete human wrench $\boldsymbol{u}_h$ instead of internal wrench components $\boldsymbol{u}_{\text{int}}$, yielding higher uncertainty.

Experimental Setup: the human actor applies forces to a haptic interface in order to move the virtual object through a maze. The interface consists of a two degrees-of-freedom (anteroposterior and mediolateral plane of the user standing in front) linear actuated device (*ThrustTube*) with a free-spinning handle (superoinferior direction of the user) at the grasp point, see Fig. 10.7. Virtual obstacles are visually and haptically displayed by means of a visuo-haptic rendering framework. The shared object from dynamics (10.42) is physically rendered with a mass of $M_r = diag\{m, m\}$ with $m = 90$ kg and damped by a viscous friction of $D_r = diag\{d, d\}$ with $d = 200$ Ns/m, emulating a heavy object. All virtual obstacles and walls are haptically rendered in order to provide a haptic feedback of the environment to the user. The HMM used to encode observations has 40 states and predictions are updated with a rate of 50 Hz.

On top of the interface, a virtual maze is presented on a screen, see Fig. 10.8. On the right side the maze includes two obstacles moving horizontally in order to potentially provoke disagreements between the human and the robot. Note that the human behavior model does not consider the obstacle's state, which results in an increased predicted variance in the region perturbed by the obstacle's motion. This condition aims for evaluating the assistive performance when the model is not accurate enough and variability becomes more relevant. On the left side, a navigation task requires high movement precision caused by a narrow and curved path.

Task and Procedure: in total 19 persons (5 female) participated in the experiment. They were between 23 and 31 years old ($M = 26.84$ years). The experimental task was to move the virtual object represented by a red dot from the upper right corner (start position) to the upper left corner of the maze (green dot at target position) without hitting any object or wall, see Fig. 10.8. Participants were instructed to move at their comfortable speed and to finish the task even if they might hit the wall or an object.

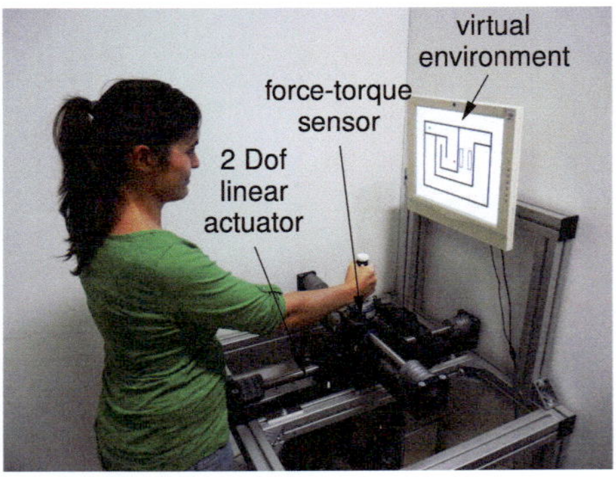

Fig. 10.7 2 DoF linear actuator setup with an attached force/torque sensor at the handle and a virtual scene displayed on the monitor (© 2015 IEEE. Reprinted, with permission, from [33])

Fig. 10.8 Virtual environment (maze). With the handle of the two-dimensional haptic interface participants were able to move the virtual object (*red dot*) to the goal position (*green dot*) (© 2015 IEEE. Reprinted, with permission, from [33])

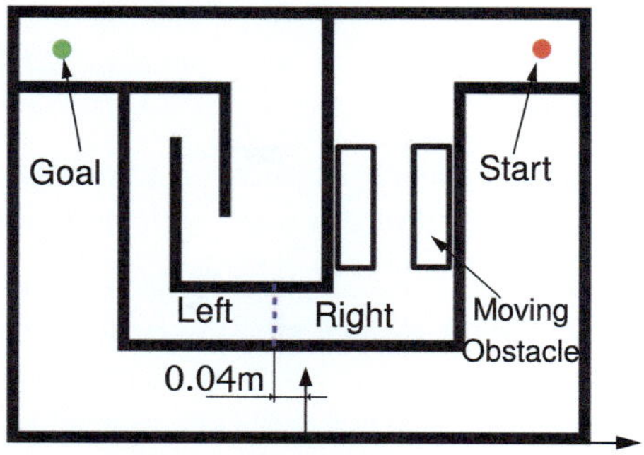

Overall the experiment was divided into seven parts corresponding to seven different control laws of three trials each. In the first trial, the human lead a passive robot, i.e., $\boldsymbol{u_r} = 0$, and the resulting observations were encoded into the task model. This model was used during the second trial, where the robot actively assisted its human partner. This second task execution already captures negotiation forces between the human and the robot. Observations of the first and second trials were used to train the task model for the third trial, providing already an estimation of the expected trajectory, both in state and internal wrench components, $\left\{\hat{\xi}_d, \hat{\boldsymbol{u}}_{\text{int}}\right\}$. The third trial was the only one considered for further evaluations. We tested seven different control laws for each participant:

(a) No active assistance: $\boldsymbol{u_r} = 0$.
(b) LQR using cost function (10.34).
(c) Risk-neutral with Mahalanobis distance: $\theta = 0$ using cost function (10.41).
(d) Risk-averse with Mahalanobis distance and expected process noise: $\theta = -\alpha$, using cost function (10.41) and $\boldsymbol{\epsilon} = \boldsymbol{\epsilon_e}$.
(e) Risk-averse with Mahalanobis distance considering both expected and current process noise: $\theta = -\alpha$, using cost function (10.41) and $\boldsymbol{\epsilon} = \max(\boldsymbol{\epsilon_e}, \boldsymbol{\epsilon_c})$.
(f) Risk-seeking with Mahalanobis distance and expected process noise: $\theta = \beta$, using cost function (10.41) and $\boldsymbol{\epsilon} = \boldsymbol{\epsilon_e}$.
(g) Risk-seeking with Mahalanobis distance considering both expected and current process noise: $\theta = \beta$, using cost function (10.41) and $\boldsymbol{\epsilon} = \max(\boldsymbol{\epsilon_e}, \boldsymbol{\epsilon_c})$.

All implementations rely on the solutions presented in Sect. 10.4.2. In our experiments $\beta = 8.1 \cdot 10^4$ and $\alpha = 8.1 \cdot 10^2$, $R = I$ and $Q_k = Q_T = diag\{\omega_p, \omega_v\}$, with ω_p and ω_v being the position and velocity weightings. For (b), we chose $\omega_p = 10^{10}$ and $\omega_v = 10^7$. For (c)–(g) we set $\omega_p = 10^5$ and $\omega_v = 10$, due to the low variance values, in the order of 10^{-5}. The receding horizon for the optimization was $T = 0.2$ s and to estimate the current noise we used a window W corresponding to 0.05 s.

The experimental procedure was as follows: participants were asked to face the haptic device and grasp the handle. Next, the experimenter initialized the control algorithm and told the participant to start moving. After reaching the green target, participants were asked to free the handle which was moved back to the initial position automatically. Before every third trial participants were verbally informed that this was going to be the trial they had to rate. Subsequently they had to rate the perceived helpfulness when

- passing a *moving obstacle* (subtask MO).
- navigating through a *narrow channel* (subtask NC).

Ratings were done on a 6-point scale from 1 (counter productive) to 6 (helpful) and resulted in the explicit measure *perceived helpfulness*. Every participant performed $3 \times 7 = 21$ trials of which seven trials were rated.

Regarding implicit measures we evaluated:

- the mean power exerted by the human M_{P_h}, defined as $\frac{1}{t}\int_0^t \boldsymbol{u_h}^{\mathrm{T}} \dot{\boldsymbol{x}} d\tau$.
- the mean disagreement $M_{\|\boldsymbol{u_D}\|}$ between both agents, defined as $\frac{1}{t}\int_0^t \|\boldsymbol{u_D}\| d\tau$, where

$$\boldsymbol{u_D} = \begin{cases} \dfrac{-\boldsymbol{u_h}}{\|\boldsymbol{u_h}\|} \cdot \boldsymbol{u_r}, & \text{if } -\boldsymbol{u_h} \cdot \boldsymbol{u_r} > 0 \\ & \wedge\, \boldsymbol{u_h} \neq \boldsymbol{0} \\ 0, & \text{otherwise.} \end{cases}$$

- the mean human applied force $M_{\|\boldsymbol{u_h}\|}$, defined as $\frac{1}{t}\int_0^t \|\boldsymbol{u_h}\| d\tau$.
- the mean contact forces during collisions with the virtual environment $M_{\|\boldsymbol{u_c}\|}$, defined as $\frac{1}{t}\int_0^t \|\boldsymbol{u_c}\| d\tau$, where $\boldsymbol{u_c}$ is the force applied on virtual obstacles.
- the mean execution time M_t.

Means were taken over all participants for the respective controller. To gain knowledge on the human perception and response during different kinds of tasks the presented maze was divided into two subtasks. For measures passing the *moving obstacle*, data was calculated until the participant was passing the turning point indicated by the dashed line shown in Fig. 10.8, placed at 0.04 m to the left of the origin. The rest of the task execution corresponded to subtask *narrow channel*. Data analysis was done in Matlab and statistical analysis was carried out with SPSS (Statistical Package for the Social Sciences).

Results and Discussion: To access the rating of *perceived helpfulness*, see Fig. 10.9(i), a 2×7 repeated measures ANOVA was performed with the between-subject factors subtask (MO vs. NC) and control method (a)–(g). Marginal differences were observed between subtasks, $F(1, 18) = 3.95$, $p = 0.062$, which shows equally perceived helpfulness of the systems response on both sides. Numerically higher ratings for NC ($M = 3.96$) compared to MO ($M = 3.51$) reflect the fact that the task model is not aware of the moving obstacle and is not as accurate as for NC. Regarding control schemes there was a significant main effect, $F(6, 108) = 2.46$, $p < 0.05$. Comparisons show that control laws (a), (e), and (g) resulted in significantly higher ratings compared to controllers (b) and (d), see Tables 10.1 and 10.2. This shows that control laws (e) and (g), both risk-sensitive and considering the current disagreement, were perceived as more helpful than method (b), the classical LQR control. It is also remarkable that numerically all control methods considering the Mahalanobis

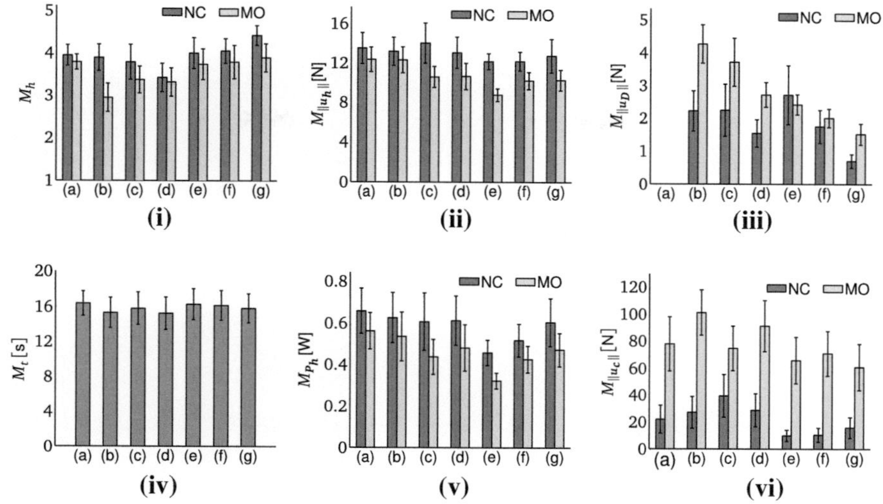

Fig. 10.9 Experimental results. Each of the figures shows the mean and the standard error of the different evaluated measures for the seven control laws (*a–g*). (**i**) shows the mean perceived helpfulness M_h, (**ii**) depicts the human exerted force $M_{\|\boldsymbol{u}_h\|}$, (**iii**) reflects the disagreement $M_{\|\boldsymbol{u}_D\|}$, (**iv**) shows the execution times M_t, (**v**) depicts the human power M_{P_h}, and **vi** shows the collision forces $M_{\|\boldsymbol{u}_c\|}$ (© 2015 IEEE. Reprinted, with permission, from [33])

Table 10.1 Average values for perceived helpfulness M_h, human power M_{P_h}, disagreement $M_{\|\boldsymbol{u}_D\|}$, human force $M_{\|\boldsymbol{u}_h\|}$, and collision forces $M_{\|\boldsymbol{u}_c\|}$ (© 2015 IEEE. Reprinted, with permission, from [33])

Cont.	M_h	M_{P_h} [W]	$M_{\|\boldsymbol{u}_D\|}$ [N]	$M_{\|\boldsymbol{u}_h\|}$ [N]	$M_{\|\boldsymbol{u}_c\|}$ [s]
(a)	3.90	0.61	–	12.92	1.32
(b)	3.42	0.58	3.15	12.73	2.42
(c)	3.58	0.52	2.91	12.30	2.02
(d)	3.47	0.55	2.28	11.85	2.29
(e)	3.92	**0.39**	2.72	**10.46**	1.38
(f)	3.76	0.47	2.17	11.19	**1.03**
(g)	**4.11**	0.54	**1.29**	11.50	1.12

distance (c)–(g) were rated higher than the classical LQR control (b). Furthermore, only control laws considering the current disagreement in the dynamics, (e) and (g), were numerically rated higher than the pure passive follower (a).

For all implicit measures 2×7 repeated measures ANOVAs were carried out with the between-subject factors subtask (NC vs. MO) and control method (a–g). If the sphericity criterion was not met, Greenhouse-Geisser correction was applied.

Regarding subtasks, there was higher human power on NC (M=0.58 W) compared to MO (M=0.46 W), $F(1, 18) = 15.59$, $p < 0.01$, see Fig. 10.9(v). After correction, the differences between applied power for the control methods only mar-

Table 10.2 Overview of results on planned comparisons for perceived helpfulness, human power, and disagreement (© 2015 IEEE. Reprinted, with permission, from [33])

Perceived Helpfulness M_h			Power M_{P_h} [W]			Disagreement $M_{\|u_D\|}$ [N]		
Comparison	$F(1, 18)$	p	Comparison	$F(1, 18)$	p	Comparison	$F(1, 18)$	p
(a) > (b)	7.15	<0.05	(a) > (e)	11.35	<0.01	–	–	–
(e) > (b)	5.90	<0.05	(b) > (e)	4.54	<0.05	(b) > (g)	32.47	<0.001
(g) > (b)	7.08	<0.05	(c) > (e)	2.58	n.s.	(c) > (g)	16.86	<0.01
(a) > (d)	5.00	<0.05	(d) > (e)	2.91	n.s.	(d) > (g)	11.37	<0.01
(e) > (d)	5.50	<0.05	(f) > (e)	2.23	n.s.	(e) > (g)	10.26	<0.01
(g) > (d)	7.21	<0.05	(g) > (e)	4.46	<0.05	(f) > (g)	6.08	<0.05

ginally reached significance, $F(3.0, 54.8) = 2.44$, $p = 0.074$. Concerning control schemes, a tendency toward higher human power applied in control laws (a), (b), and (g) is identified compared to control law (e), see Table 10.2. The risk-averse control law (e) becomes stiffer under the presence of disagreement and, as it also considers the current disagreements with its partner, it is the most agressive partner. The observed differences suggest that this control method reduces the power applied by the human as the robot constantly takes a leading role.

Disagreement, see Fig. 10.9(iii), was marginally higher on MO ($M = 2.77$ N) than on NC ($M = 1.86$ N), $F(1, 18) = 4.22$, $p = 0.055$, in accordance with the results from perceived helpfulness. The behavior models lack obstacle awareness and provide low prediction performance for MO thereby yielding disagreements. Looking at the results for control law (g), a highly significant main effect was observed, $F(5, 90) = 5.97$, $p < 0.001$. Planned comparisons show that all control methods cause a higher disagreement than (g), see Table 10.2. As (g) is risk-seeking, it consequently tends to avoid disagreements adopting a more passive role. The fact that it also considers the current disagreement boosts this effect.

Applied forces, Fig. 10.9(ii), were higher on NC ($M = 12.97$ N) than on MO ($M = 10.74$ N), $F(1, 18) = 15.76$, $p < 0.01$. Although differences between control methods were not significant after correction, $p > 0.1$, the risk-averse with current disagreement, control law (e), numerically required the lowest applied force.

On NC, the force from collisions, see Fig. 10.9(vi), was smaller ($M = 0.65$ N) compared to MO ($M = 2.67$ N), $F(1, 18) = 28.73$, $p < 0.001$.

No significant differences were found between movement times of the different control methods, see Fig. 10.9(iv), $F(6, 108) = 0.46$, $p > 0.8$.

In summary, the consideration of uncertainty in the optimization is shown beneficial for haptic assistance, as exhibited by the superior performance of all proposed control schemes with respect to a naive LQR control ignoring any variability. The inclusion of the Mahalanobis distance in the optimization criterion leads to higher perceived helpfulness as it considers the learned trajectory's uncertainty. The benefits of a risk-sensitive optimization are only significant when an estimation of current disagreement is additionally considered. The risk-seeking control law achieves minimal disagreement as it adopts a recessive attitude during negotiation with its partner. This is especially noticeable passing through the obstacle: as the behavior model lacks any obstacle awareness, counteracting forces arise frequently due to poor prediction performance. In terms of human power, the risk-averse policy is more effective. However, the risk-seeking variant is preferred by human users, favoring disagreement avoidance to effort minimization. If the behavior model is accurate, risk-aversion effectively minimizes human effort but the complexity of human behavior makes fulfilling this prerequisite highly improbable. Nevertheless, other pHRI application domains may take advantage of risk-averse policies. For instance, consider a kinesthetic teaching scenario where the robot has the role of the teacher. In this case the robot aims for minimizing variations in human performance, a condition in accordance with a risk-averse control scheme.

10.5 Conclusions

The synthesis of goal-oriented physical robotic helpers has promising applications but also challenging problems to address. In this chapter, we first identified internal and effective wrenches arising during interaction by previously analyzing possibly nonuniform load decomposition policies. Internal wrenches indicate errors in terms of desired accelerations and are a decisive indicator of disagreements. Based on these findings, we proposed a anticipatory control scheme for navigation tasks adapting to prediction uncertainty and disagreements. Formulated as an optimal control problem minimizing human force, both the uncertainty of the expected desired trajectory and the disagreements with the human become decisive components of robot assistance. The presented scheme demonstrated superior performance in terms of human effort minimization and perceived helpfulness in an experiment with human users. Promising future works are, for instance, the regulation of load shares. A suitable alternative for approaching this problem is the study of this process in human dyads for later application in human–robot settings.

Acknowledgements This research is partly supported by the ERC Starting Grant "Control based on Human Models (con-humo)" under grant agreement 337654.

Appendix

Proof (*Proof of Proposition* 10.1) Consider first only the force components. Gauss's principle of least constraint or least forcing states that the true motion of a mechanical system of N masses minimizes

$$J = \sum_{i=1}^{N} m_i || \ddot{\boldsymbol{p}} - \frac{\overline{\boldsymbol{f}}_i}{m_i} || , \tag{10.50}$$

where m_i is the i-th effective mass and $\boldsymbol{p}$ is the object's position. The multiagent formulation of (10.16) represents a multibody system where each body is constrained to the same position $\boldsymbol{p}$ and share the overall mass of the object m_0, i.e.,

$$\sum_{i=1}^{N} m_i = m_0 \quad m_i \geq 0\,. \tag{10.51}$$

From (10.17) and (10.18), the assigned force component of the i-th agent is given by

$$\overline{\boldsymbol{f}}_i = \alpha_{f,i} \boldsymbol{f}_d \tag{10.52}$$

Applying (10.18) produces a resulting wrench $\sum_{i=1}^{N} \alpha_{f,i} \boldsymbol{f_d} = \boldsymbol{f_d}$ and an acceleration $\ddot{\boldsymbol{p}} = \frac{\boldsymbol{f}_d}{m_0}$.

The effective mass that agent i accounts for must fulfill Gauss's principle and therefore minimizes J. The stationary point fulfills

$$\frac{\partial J}{\partial m_i} = \frac{\partial \left(m_i \left\| \frac{\boldsymbol{f}_d}{m_0} - \frac{\alpha_{f,i} \boldsymbol{f}_d}{m_i} \right\| \right)}{\partial m_i} = 0\,.$$

Solving for m_i yields

$$m_i = \alpha_{f,i} m_0\,,$$

and therefore $\alpha_{f,i} = \frac{m_i}{m_0}$ represents the *normalized effective mass* for agent i. Conditions (10.19) ensure the fulfillment of constraints (10.51).

From (10.51), the resulting acceleration considering the sum of effective masses must be consistent with the overall acceleration, i.e.,

$$\frac{\boldsymbol{f}_d}{\sum_{i=1}^{N} m_i} = \frac{\boldsymbol{f}_d}{m_0}\,.$$

As a result the components accounting for the effective masses fulfill

$$\frac{\sum_{i=1}^{N} m_i}{m_0} \boldsymbol{f}_d = \sum_{i=1}^{N} \alpha_{f,i} \boldsymbol{f}_d = \boldsymbol{f}_d\,,$$

yielding definition (10.52) and ensuring that $\overline{\boldsymbol{f}}_i$ comprises only effective wrenches.

The proof for torque components yielding the normalized effective moment of inertia $\alpha_{\tau,i}$ is derived in an identical manner.

Proof (*Proof of Proposition* 10.3) Consider first only the force components. The stationary point of the unconstrained expression of (10.50) is

$$\frac{\partial J}{\partial m_i} = \frac{\partial \left(m_i \left\| \frac{\boldsymbol{f}_d}{m_0} - \frac{\overline{\boldsymbol{f}}_i}{m_i} \right\| \right)}{\partial m_i} = 0 \, .$$

Let $\alpha_{f,i} = \frac{m_i}{m_0}$. Solving for m_i yields

$$\alpha_{f,i} = \frac{\overline{\boldsymbol{f}}_i^{\mathrm{T}} \boldsymbol{f}_d}{||\boldsymbol{f}_d||^2} \, .$$

The minimum and maximum from expression (??) ensure constraints (10.51) and hold only for $N = 2$.

The normalized effective moment of inertia components $\alpha_{\tau,i}$ are derived in an identical manner.

Proof (*Proof of Proposition* 10.4) By contradiction. Consider only force components. The contradictory case assumes $\overline{\boldsymbol{f}}_{\text{int}} \neq 0$ and

$$\forall i, j \mid \frac{\boldsymbol{f_d}_i^{\mathrm{T}} \boldsymbol{f_d}_j}{||\boldsymbol{f_d}_i|| ||\boldsymbol{f_d}_j||} = 1 \tag{10.53}$$

From (10.27), the applied wrench of the i-th agent is

$$\overline{\boldsymbol{f}}_i = \widetilde{\alpha}_{f,i} \boldsymbol{f_d}_i \, ,$$

where $\widetilde{\alpha}_{f,i}$ is the mass share assumed by agent i in A_i. The overall resulting force is

$$\boldsymbol{f_d} = \sum_{i=1}^{N} \widetilde{\alpha}_{f,i} \boldsymbol{f_d}_i \, .$$

From (10.26), the internal forces are

$$\overline{\boldsymbol{f}}_{\text{int}} = \widetilde{\alpha}_{f,i} \boldsymbol{f_d}_i - \alpha_{f,i} \boldsymbol{f_d} \, . \tag{10.54}$$

From Proposition 10.3, the unconstrained expression for the effective load share $\alpha_{f,i}$ is given by

$$\alpha_{f,i} = \frac{\widetilde{\alpha}_{f,i} \boldsymbol{f_d}_i \boldsymbol{f_d}}{||\boldsymbol{f_d}||^2} = \frac{\widetilde{\alpha}_{f,i} ||\boldsymbol{f_d}_i||}{||\boldsymbol{f_d}||} \, . \tag{10.55}$$

Note that conditions (10.53) ensure constraints (10.51). Substituting (10.55) in (10.54) yields

$$\overline{\boldsymbol{f}}_{\text{int}} = \widetilde{\alpha}_{f,i}\boldsymbol{f}_{d_i} - \frac{\widetilde{\alpha}_{f,i}||\boldsymbol{f}_{d_i}||}{||\boldsymbol{f}_d||}\boldsymbol{f}_d \,. \tag{10.56}$$

From (10.53), all $\boldsymbol{f}_{d_i}$ are in the same direction and therefore $\boldsymbol{f}_d$ too. $\boldsymbol{f}_d/||\boldsymbol{f}_d||$ is the unit vector of $\boldsymbol{f}_{d_i}$ and therefore $\overline{\boldsymbol{f}}_{\text{int}} = \boldsymbol{0}$, concluding the proof.

The proof for the inverse condition is also done by contradiction starting from (10.28) and assuming no internal force components and sketched as follows: condition (10.28) implies that in (10.55) a cosinus appears due to the different directions of desired resulting wrenches. Therefore, the difference between applied wrench and effective wrench from (10.56) is nonzero yielding the contradiction.

The proof for torque components is derived in an identical manner.

References

1. Abbink DA, Mulder M, Boer ER (2012) Haptic shared control: smoothly shifting control authority? Cognit Technol Work 14(1):19–28
2. Billard A, Calinon S, Dillmann R, Schaal S (2008) Robot programming by demonstration. In: Siciliano B, Khatib O (eds) Handbook of robotics. Springer, Berlin, pp 1371–1394
3. Boyd S, Ghaoui L, Feron E, Balakrishnan V (1994) Linear matrix inequalities in system and control theory. Society for industrial and applied mathematics
4. Caccavale F, Uchiyama M (2008) Cooperative manipulators. Springer handbook of robotics, pp 701–718
5. Calinon S, Guenter F, Billard A (2007) On learning, representing, and generalizing a task in a humanoid robot. IEEE Trans Syst Man Cybern B 37(2):286–298
6. Colgate JE, Hogan N (1988) Robust control of dynamically interacting systems. Int J Control 48(1):65–88
7. Corteville B, Aertbelien E, Bruyninckx H, De Schutter J, Van Brussel H (2007) Human-inspired robot assistant for fast point-to-point movements. In: Proceedings of IEEE ICRA, pp 3639–3644
8. Doty KL, Melchiorri C, Bonivento C (1993) A theory of generalized inverses applied to robotics. Int J Robot Res 12(1):1–19
9. Duchaine V, Gosselin C (2009) Safe, stable and intuitive control for physical human-robot interaction. In: Proceedings of IEEE ICRA, pp 3383–3388
10. Erhart S, Hirche S (2013) Adaptive force/velocity control for multi-robot cooperative manipulation under uncertain kinematic parameters. In: Proceedings of IEEE/RSJ IROS, pp 307–314
11. Erhart S, Hirche S (2015) Internal force analysis and load distribution for cooperative multi-robot manipulation. IEEE Trans Robot 31(5):1238–1243
12. Evrard P, Kheddar A (2009) Homotopy switching model for dyad haptic interaction in physical collaborative tasks. In: Proceedings of EHS EuroHaptics, pp 45–50
13. Evrard P, Gribovskaya E, Calinon S, Billard A, Kheddar A (2009) Teaching physical collaborative tasks: object-lifting case study with a humanoid. In: Proceedings of IEEE Humanoids, pp 399–404
14. Fernandez V, Balaguer C, Blanco D, Salichs MA (2001) Active human-mobile manipulator cooperation through intention recognition. In: Proceedings of IEEE ICRA, pp 2668–2673
15. Flash T, Hogan N (1985) The coordination of arm movements: an experimentally confirmed mathematical model. J Neurosci 5:1688–1703

16. Garcia CE, Prett DM, Morari M (1989) Model predictive control: theory and practice-a survey. Automatica 25(3):335–348
17. Gribovskaya E, Kheddar A, Billard A (2011) Motion learning and adaptive impedance for robot control during physical interaction with humans. In: Proceedings of IEEE ICRA, pp 4326–4332
18. Groten RK (2011) Haptic human-robot collaboration: how to learn from human dyads. Dissertation, Technische Universität München, München
19. Groten R, Feth D, Peer A, Buss M (2010) Shared decision making in a collaborative task with reciprocal haptic feedback - an efficiency-analysis. In: Proceedings of IEEE ICRA
20. Hirata Y, Kume Y, Wang Z-D, Kosuge K (2003) Decentralized control of multiple mobile manipulators based on virtual 3-D caster motion for handling an object in cooperation with a human. Proceedings of IEEE ICRA 1:938–943
21. Hogan N (1984) Impedance control: an approach to manipulation. In: Proceedings of ACC, pp 304–313
22. Ikemoto S, Amor HB, Minato T, Jung B, Ishiguro H (2012) IEEE Robot Autom Mag
23. Jacobson D (1973) Optimal stochastic linear systems with exponential performance criteria and their relation to deterministic differential games. IEEE Trans Autom Control 18(2):124–131
24. Jarrassé N, Paik J, Pasqui V, Morel G (2008) How can human motion prediction increase transparency? In: Proceedings of IEEE ICRA, pp 2134–2139
25. Kosuge K, Yoshida H, Fukuda T (1993) Dynamic control for robot-human collaboration. In: Proceedings of IEEE workshop on robot and human, communication, pp 398–401
26. Kronander K, Billard A (2012) Online learning of varying stiffness through physical human-robot interaction. In: Proceedings of IEEE ICRA
27. Lawitzky M, Mörtl A, Hirche S (2010) Load sharing in human-robot cooperative manipulation. In: Proceedings of IEEE Ro-Man, pp 185–191
28. Lawitzky M, Medina JR, Lee D, Hirche S (2012) Feedback motion planning and learning from demonstration in physical robotic assistance: differences and synergies. In: Proceedings of IEEE/RSJ IROS, pp 3646–3652
29. Lee D, Ott C (2011) Incremental kinesthetic teaching of motion primitives using the motion refinement tube. Auton Robot, pp 1–17
30. Li Y, Tee KP, Chan WL, Yan R, Chua Y, Limbu DK (2015) Continuous role adaptation for human-robot shared control. IEEE Trans Robot 31(3):672–681
31. Maeda Y, Hara T, Arai T (2001) Human-robot cooperative manipulation with motion estimation. Proc IEEE/RSJ IROS 4:2240–2245
32. Medina JR, Lawitzky M, Mörtl A, Lee D, Hirche S (2011) An experience-driven robotic assistant acquiring human knowledge to improve haptic cooperation. In: Proceedings of IEEE/RSJ IROS, pp 2416–2422
33. Medina JR, Lorenz T, Hirche S (2015) Synthesizing anticipatory haptic assistance considering human behavior uncertainty. IEEE Trans Robot 31(1):180–190. doi:10.1109/TRO.2014.2387571. IEEE
34. Miossec S, Kheddar A (2008) Human motion in cooperative tasks: moving object case study. In: Proceedings of IEEE ROBIO, pp 1509–1514
35. Mörtl A, Lawitzky M, üçükyılmaz A, Sezgin M, Basdogan C, Hirche S (2012) The role of roles: physical cooperation between humans and robots. Int Robot Res
36. Niemeyer G, Slotine J-JE (1991) Stable adaptive teleoperation. IEEE J Ocean Eng 16(1):152–162
37. Parker L (2008) Distributed intelligence: overview of the field and its application in multi-robot systems. J Phys Agents 2:5–14
38. Passenberg C, Groten R, Peer A, Buss M (2011) Towards real-time haptic assistance adaptation optimizing task performance and human effort. In: Proceedings of IEEE WHC, pp 155–160
39. Prattichizzo D, Trinkle JC (2008) Grasping. Springer handbook of robotics, pp 671–700
40. Reed KB, Peshkin MA (2008) Physical collaboration of human-human and human-robot teams. IEEE Trans Haptics 1(2):108–120

41. Schneider S, Cannon R (1992) Object impedance control for cooperative manipulation: theory and experimental results. IEEE Trans Robot Automat 8(3):383–394
42. Sebanz N, Knoblich G (2009) Prediction in joint action: what, when, and where. Top Cognit Sci 2:353–367
43. Shaiju AJ, Petersen IR (2008) Formulas for discrete time LQR. LQG LEQG and minimax LQG optimal control. In: Proceedings of IFAC
44. Spong MW, Hutchinson S, Vidyasagar M (2006) Robot modeling and control, vol 3. Wiley, New York
45. Stefanov N, Peer A, Buss M (2010) Online intention recognition for computer-assisted teleoperation. In: Proceedings of IEEE ICRA, pp 5334–5339
46. Szewczyk J, Plumet F, Bidaud P (2002) Planning and controlling cooperating robots through distributed impedance. J Robot Syst 19(6):283–297
47. Thobbi A, Gu Y, Sheng W (2011) Using human motion estimation for human-robot cooperative manipulation. In: Proceedings of IEEE/RSJ IROS, pp 2873–2878
48. Uchiyama M, Dauchez P (1988) A symmetric hybrid position/force control scheme for the coordination of two robots. In: Proceedings of IEEE ICRA, pp 350–356
49. Walker ID, Freeman RA, Marcus SI (1991) Analysis of motion and internal loading of objects grasped by multiple cooperating manipulators. Int J Robot Res 10(4):396–409
50. Wang Z, Peer A, Buss M (2009) An HMM approach to realistic haptic human-robot interaction. In: Proceedings of IEEE EuroHaptics, pp 374–379
51. Whittle P (1981) Risk-sensitive linear/quadratic/gaussian control. Adv Appl Probab 13(4):764–777
52. Won C (2004) Cost distribution shaping: the relation between bode integral, entropy, risk-sensitivity, and cost cumulant control. Proc ACC 3:2160–2165
53. Yokoyama K, Handa H, Isozumi T, Fukase Y, Kaneko K, Kanehiro F, Kawai Y, Tomita F, Hirukawa H (2003) Cooperative works by a human and a humanoid robot. In: Proceedings of IEEE ICRA, pp 2985–2991

Chapter 11
Designing the Robot Behavior for Safe Human–Robot Interactions

Changliu Liu and Masayoshi Tomizuka

11.1 Introduction

Recent advances in robotics suggest that human–robot interaction (HRI) is no longer a fantasy, but is happening in various fields. In factories, robots are leaving their cages and starting to work cooperatively with human workers [16]. The manufacturers are interested in combining human's flexibility and robot's productivity in flexible production lines [15, 19]. On the other hand, as automated driving is widely viewed as a promising technology to revolutionize today's transportation system [5], substantial research efforts are directed into the field from research groups and companies [1]. As a consequence, human drivers and autonomous vehicles will interact on the road, which poses new challenges in road safety [25]. Another example is in the field of rehabilitation. In order to rebuild the sensory connection of a patient after a stroke, robots, or exoskeletons are needed to guide and assist the patient in walking. There are close physical interactions between the patient and the robot [14]. Others like nursing robots [27] or robot guide dogs [31] are also in great demand and involve HRI.

This chapter was developed and enhanced from an earlier paper published as [20] ©2015 IEEE.

C. Liu (✉) · M. Tomizuka
University of California at Berkeley, Berkeley, CA 94720, USA
e-mail: changliuliu@berkeley.edu

M. Tomizuka
e-mail: tomizuka@berkeley.edu

Y. Wang and F. Zhang (eds.), *Trends in Control and Decision-Making for Human–Robot Collaboration Systems*, DOI 10.1007/978-3-319-40533-9_11

11.1.1 The Safety Issues and Existing Solutions

Human safety is one of the biggest concerns in HRI [32]. Two different approaches can be used to address the safety issues. One way is to increase the intrinsic safety level of the robot through hardware design, so that even if collision happens, the impact on the human is minimized [11]. The other way is to let the robot behave safely, which is called 'interactive safety' as opposed to the intrinsic safety [33]. In this chapter, interactive safety will be addressed in the context of decision making, motion planning, and control of the robot during HRI.

Conventional approach to address the interactive safety is conservative, which slows down the robot when human is nearby, hence sacrifices productivity for the sake of safety. However, to make the interaction desirable, both safety and efficiency need to be considered in designing the robot behavior.

Less conservative methods in the context of obstacle avoidance in robot motion planning have also been used to address safety in HRI. Some authors use potential field methods, e.g., introducing a virtual force which pushes the robot away from the obstacles [13, 26]. Some use sliding mode methods to locally modify the reference trajectory in the presence of obstacles [10]. These two methods result in closed-form analytical control laws; but they do not emphasize optimality (or efficiency). Some authors formulate the problem as an optimization or optimal control problem with hard constraints to represent the safety requirements [7]. Unfortunately, these non-convex optimization problems are generally hard to solve analytically [12] and different approximations and numerical methods are used [29, 30].

However, as humans will respond to the robot's movement, interactions need to be considered explicitly in the design of the robot behavior, which implies that humans cannot be simply treated as obstacles. Moreover, the algorithms must be designed such that it can be executed fast enough for timely responses in emergency situations, which implies that those numerical methods may not be desirable for online applications.

These requirements are pushing for new perspectives in understanding and dealing with HRI. The robot should be designed intelligent enough to conduct social behavior [34] to interact with humans safely and efficiently even in emergency situations.

11.1.2 Safety Problems in HRI: Conflicts in Multiagent Systems

By designing the robot behavior, the safety issues need to be understood in the human–robot systems. Since the intelligent robots are sharing the space with humans, the system can be viewed as a multiagent system (MAS), where all humans and robots in the environment are regarded as agents [28]. This multiagent situation complicates the interactions as the individual optima usually does not coincide with the system optima [17]. An agent's interest is to be efficient, i.e., finish the task as soon as

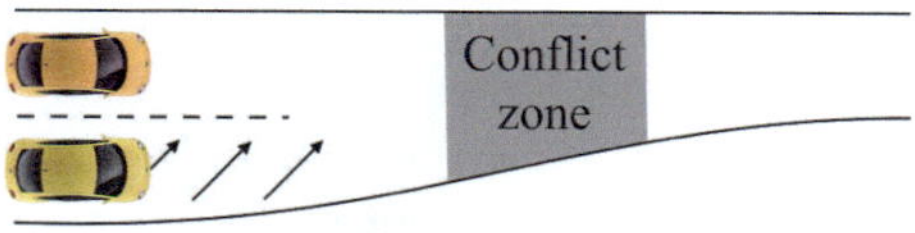

Fig. 11.1 Conflict between two vehicles during lane merging

Fig. 11.2 Conflict between a human worker and an autonomous guided vehicle (AGV) in a factory floor

possible, while staying safe. For example, an automated vehicle's interest is to go to the destination in minimum time without colliding with surrounding vehicles. The safety is the mutual interests for all agents in the system, while the efficiency goals may conflict with one another, as the individual tasks are not identical and the resources are limited. Figure 11.1 shows the conflict between two vehicles during lane merging. The two vehicles cannot pass the conflict zone at the same time (since the road resources are limited), which implies that one of the vehicle must yield (i.e., sacrifice efficiency) to ensure safety. Figure 11.2 shows the conflict between an autonomous guided vehicle (AGV) and a human worker as their future trajectories intersect. The intersection point is a potential conflict zone if neither the AGV or the human detours.

There are plenty of methods to make a robot efficient [6]. However, literature is limited on solving the conflicts among agents when there is no designated controlling agent. In the real world, conflicts of interests can be solved by active communication and negotiation [24], whereas in a human–robot coexisting environment, communications are limited. Classic game theory [3] offers reasoning strategies for cooperation to evolve in limited communications when the behavioral rules of the agents are known. However, in the real world, agents are "strangers" to each other. This situation runs into the bayesian game regime [2]. To ensure safety, the robot should actively learn the human behavior and only seek to maximize its utility within a set of actions that is safe with respect to what human would do and how human would react to the robot motion.

11.1.3 Safe Control and Exploration

Take human's social behavior as an example, to avoid conflicts of interests, a human's behavior is usually constrained. The constraint comes from at least two factors: the social norm and the uncertainties that he perceives for other people. The social norm guides behavior in a certain situation or environment as "mental representations of appropriate behavior". For example, in the AGV case, the social norm will suggest the AGV to keep a safe distance from human and detour if necessary. In practice, the perceived uncertainties also affect human's decision. Similarly, whether the AGV needs to detour and how much the AGV should detour highly depends on how certain the AGV is about its prediction of the human's trajectory. Moreover, the uncertainties can be attenuated through active learning (refer to uncertainty reduction theory [4]). A common experience is: a newcomer tends to behave conservatively in a new environment due to large uncertainties. But through observing and learning his peers, he will gradually behave freely due to the reduction of uncertainties.

Based on these observations, a safety-oriented method is introduced to design the robot controller. The HRI is modeled in the MAS framework. A safe set is defined in the system state space (which contains both the robot state and the human state) to represent the social norm. The safe set algorithm (SSA) [18] enforces invariance in the safe set according to the predicted human behavior. The safe exploration algorithm (SEA) further constrains the robot motion by uncertainties in the predictions [20]. By actively learning human behaviors, the robot will be more "confident" about its prediction of human motion. The robot can access a larger subset of the safe set when the uncertainty is smaller.

The remainder of the chapter is organized as follows: in Sect. 11.2, a multiagent interaction model will be introduced, followed by the discussion of the safety-oriented design methodology in Sect. 11.3. The safe set algorithm (SSA) and the safe exploration algorithm (SEA) will be discussed in Sects. 11.4 and 11.5 respectively. A method to combine SSA and SEA will be discussed in Sect. 11.6. Section 11.7 points out directions for future works. Section 11.8 concludes the chapter.

11.2 Modeling the Human–Robot Interactions

11.2.1 The Agent Model

Suppose there are N agents in the system and are indexed from 1 to N. R denotes the set of all indices for the robots, and H for the humans. Denote agent i's state as x_i, its control input as u_i, its information set as π_i and its goal as G_i for $i = 1, \ldots, N$. For simplicity, write x_R as the union of the states of all robots and x_H as the union of the states of all humans. Denote the system state as $x = [x_R^T, x_H^T]^T \in X$ where X is the state space of the system. The open loop system dynamics can be written as

$$\dot{x} = f(x, u_1, u_2, \ldots, u_N, w) \tag{11.1}$$

where w is a Gaussian noise term.

In the case that there is no physical contact among humans and robots as shown in Figs. 11.1 and 11.2, the system is decomposable,[1] i.e.,

$$\dot{x}_i = f_i\left(x_i, u_i, w_i\right), \forall i = 1, \ldots, N \tag{11.2}$$

Moreover, it is assumed that the dynamics of all robots are affine in the control term, i.e.,

$$\dot{x}_i = f_{ix}^*\left(x_i, w_i\right) + f_{iu}^*\left(x_i\right) u_i, \forall i \in R \tag{11.3}$$

Agent i chooses the control u_i based on the information set π_i and its goal G_i. The information set is a combination of the measured data and the communicated information. In this chapter, it is assumed that there is no direct communication among the defined agents. If a group of agents do communicate with each other, then they can be coordinated and will be treated as one agent. In this way, agent i's information set at time T contains all the measurements up to time T, i.e., $\pi_i\left(T\right) = \left\{y_i\left(t\right)\right\}_{t\in[t_0,T]}$ where

$$y_i = h_i\left(x, v_i\right), \forall i = 1, \ldots, N \tag{11.4}$$

and v_i is the measurement noise. The controller can be written as

$$u_i = g_i\left(\pi_i, G_i\right), \forall i = 1, \ldots, N \tag{11.5}$$

11.2.2 The Closed-Loop System

Using (11.4) and (11.5) in the open loop dynamics (11.1) or (11.2), the closed-loop dynamic equation becomes

$$\dot{x} = f^{cl}\left(x, G, v_1, \ldots, v_N, w_1, \ldots, w_N\right) \tag{11.6}$$

where $G = \left[G_1^T, \ldots, G_N^T\right]^T \in X$ is the system goal. The system block diagram for the general multiagent system is shown in Fig. 11.3 based on (11.1), (11.4) and (11.5). The block diagram for the decomposable system is shown in Fig. 11.4 based on (11.2), (11.4) and (11.5). Due to measurement feedbacks, all agents are coupled together in the closed-loop system no matter it is decomposable or not in the open loop. The difference is that in the decomposable system, interaction happens only in the measurement and controller side instead of in the open loop dynamics. In this chapter, the decomposable system will be studied.

[1] In certain cases, the open loop system may not be decoupled. For example, in the case of rehabilitation, the robot can affect the human's dynamics directly by assisting the human to accomplish special tasks, such as walking. When the robot's input enters the human's dynamic equation, (11.2) does not hold.

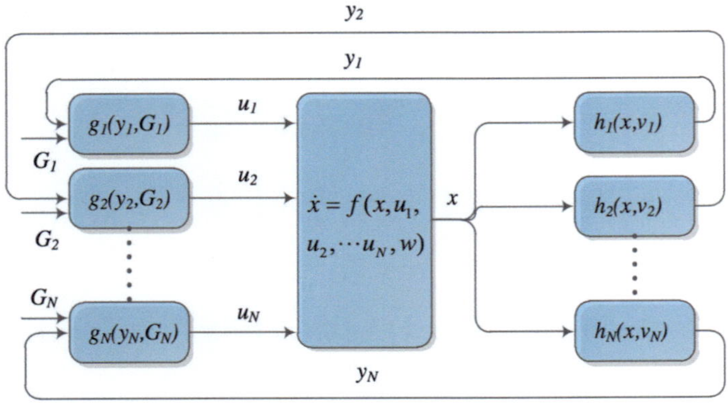

Fig. 11.3 Multiagent system block diagram

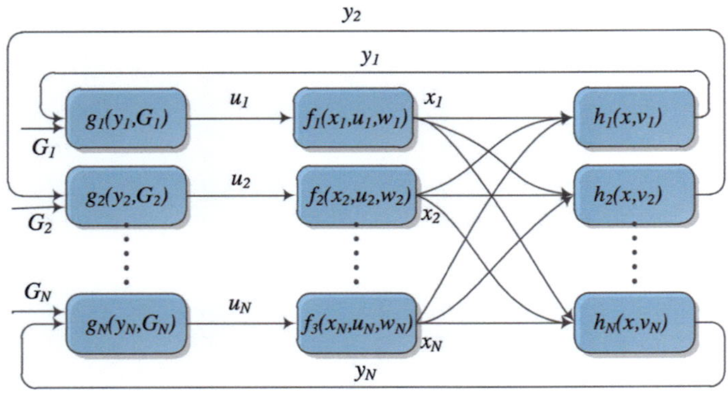

Fig. 11.4 Decomposable multiagent system block diagram (© 2015 IEEE. Reprinted, with permission, from [20])

11.2.3 Information Structure

In a multiagent system, an agent should be considerate to other agents. Before choosing the control u_i, agent i needs to consider the "strategies" that others will play (the control signal u_j's that others would choose).

Information structure [3] specifies what information is available to which agent when the agents are deciding on their control strategies. When all agents are operating on the same information set but independently before they take the next move, this is a simultaneous game and the optimal solution of this system defines a Nash Equilibrium (or a Bayesian Nash Equilibrium if the utilities of the agents are not publicly known). When some agents know the others' u_j (by observing x_j) before deciding on their own controls, this is a sequential game and the optimal solution defines a Stackelberg equilibrium (or a perfect Bayesian equilibrium if the utilities of the agents are not publicly known). The agents being observed are considered as leaders and the agents that take the observation are considered as followers.

It is assumed that safety-oriented HRI can be modeled as a sequential game and the robot being controlled is always a follower [18]. There are several advantages in doing so. First, it is a conservative strategy, thus good from the safety point of view. Second, it is assumed that robots have a smaller reaction time than humans. Thus they can quickly adapt to human motion, which qualifies them as followers. Moreover, to analyze a human's control strategy, the robot needs to know the human's cost function, which, however, is hard to obtain. So it is better to assume all humans' strategies are revealed in their past moves before the robot takes the move. Then the robot must be a follower, which plays a reactive strategy.

11.3 The Safety-Oriented Behavior Design

The robot behavior is determined by the robot controller g_i for $i \in R$. The goal in designing g_i is (1) to find the bounds on robot actions, which are safe with respect to what human would do and how human would react and (2) to maximize the robot's utility or minimize the robot's cost within the bounds.

When there are more than one robot in the system, they can be coordinated by a central controller. In this way, all robots can be treated as one agent. If it is not possible to design a coordinator as in the case of autonomous vehicles, those robots that cannot be coordinated are simply regarded as human agents. Then the number of effective robots in the system is reduced to one, denoted by R.

11.3.1 The Safety Principle

Denote the system's safe set as X_S, which is a closed subset of the state space X that is safe, e.g., collision-free. Then the state space constraint R_S for the robot depends on humans' states, e.g., $R_S(x_H) = \{x_R : [x_R^T, x_H^T]^T \in X_S\}$. If humans are followers and will take care about the safety, then the safe set for the robot is

$$R_S^1 = \{x_R : x_R \in R_S(x_H) \text{ for some } x_H\} \tag{11.7}$$

However, based on the discussion in Sect. 11.2.3, humans are assumed to be the leaders, so the safety problem should be taken care of by the robot. In the case that the robot knows human's next move $\hat{x}_H$, the safety bound for the robot becomes

$$R_S^2 = \{x_R : x_R \in R_S(\hat{x}_H)\} \tag{11.8}$$

However, due to noises and uncertainties, the estimate $\hat{x}_H$ may not be accurate. The human state x_H may lie in a set Γ_H containing $\hat{x}_H$. Then the robot motion should

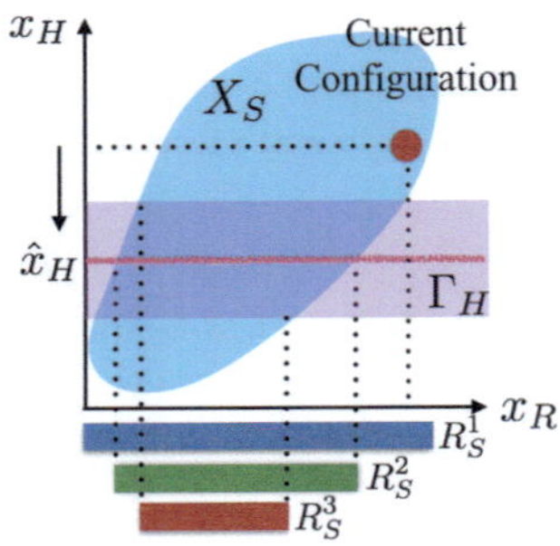

Fig. 11.5 The state space safety constraints X_S, R_S^1, R_S^2 and R_S^3

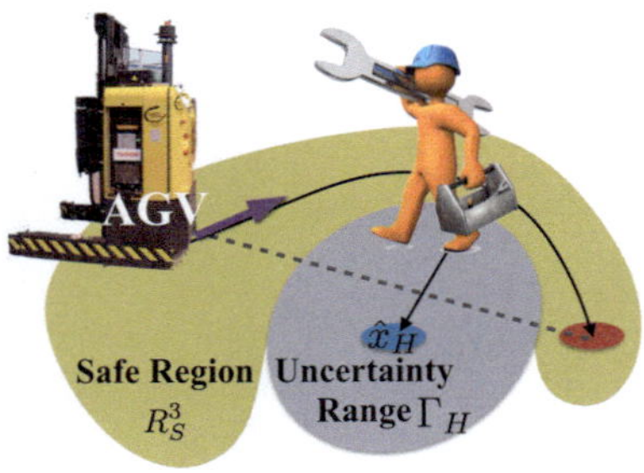

Fig. 11.6 Solving the conflict by re-planning in the safe region R_S^3

be constrained in a smaller set

$$R_S^3 = \{x_R : x_R \in R_S(x_H), \forall x_H \in \Gamma_H\} \tag{11.9}$$

Figure 11.5 illustrates the safe set X_S and the state space constraints R_S^1, R_S^2 and R_S^3. It is clear that $R_S^3 \subset R_S^2 \subset R_S^1$.

The Safety Principle: the function $g_i(.,.)$, $i \in R$ should be chosen such that X_S is invariant, i.e., $x(t) \in X_S$ for all t, or equivalently, $x_R(t) \in R_S^3(t)$ for $\Gamma_H(t)$ which accounts for almost all possible noises $v_1, \ldots, v_N, w_1, \ldots, w_N$ and human decisions $g_i(.,.)$, $i \in H$ (those with negligible probabilities will be ignored).

Figure 11.6 illustrates the expected outcome of the robot behavior under the safety principle. In view of the potential conflict, the robot re-plans a trajectory in the safe region R_S^3.

11.3.2 The Safety Index

The safety principle requires the designed control law to make the safe set invariant with respect to time. In addition to constraining the motion in the safe region R_S^3, the robot should also be able to cope with any unsafe human movement. Given the current configuration in Fig. 11.5, if the human is anticipated to move downwards, the robot should go left in order for the combined trajectory to stay in the safe set.

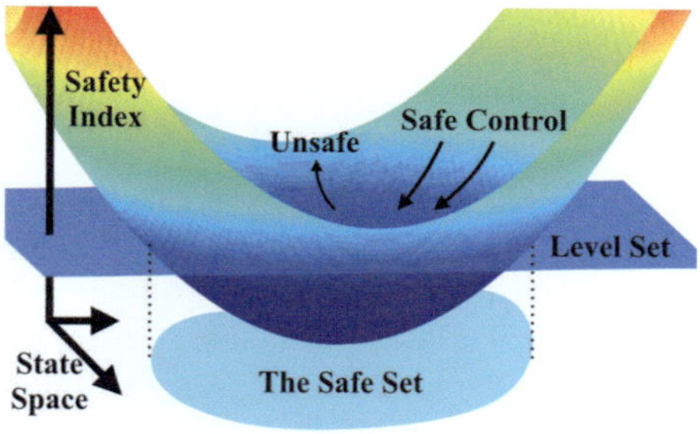

Fig. 11.7 The safety index

To make the robot better understand the dynamics, a safety index is introduced as shown in Fig. 11.7. The safety index $\phi : X \to \mathbb{R}$ is a function on the system state space such that

1. ϕ is differentiable with respect to t, i.e., $\dot{\phi} = (\partial\phi/\partial x)\dot{x}$ exists everywhere;
2. $\partial\dot{\phi}/\partial u_R \neq 0$;
3. The unsafe set $X \setminus X_S$ is not reachable given the control law $\dot{\phi} < 0$ *when* $\phi \geq 0$ and the initial condition $x(t_0) \in X_S$.

The first condition is to ensure that ϕ is smooth. The second condition is to ensure that the robot input can always affect the safety index. The third condition presents a quick criteria to determine whether a control input is safe or not, e.g., all the control inputs that drive the state below the level set 0 are safe and unsafe otherwise.

Lemma: (Existence of the Safety Index) The function ϕ satisfying all three conditions exists for any $X_S = \{x : \phi_0(x) \leq 0\}$, where $\phi_0(x)$ is a smooth function on the state space.[2]

To ensure safety, the robot's control must be chosen from the set of safe control $U_S(t) = \{u_R(t) : \dot{\phi} \leq -\eta_R \; when \; \phi \geq 0\}$ where $\eta_R \in \mathbb{R}^+$ is a safety margin. By (11.3), the derivative of the safety index can be written as

$$\dot{\phi} = \frac{\partial\phi}{\partial x_R} f^*_{Ru} u_R + \frac{\partial\phi}{\partial x_R} f^*_{Rx} + \sum_{j\in H} \frac{\partial\phi}{\partial x_j} \dot{x}_j \tag{11.10}$$

Then the set of safe control is

$$U_S(t) = \{u_R(t) : L(t) u_R(t) \leq S(t, \dot{x}_H)\} \tag{11.11}$$

where

[2]The Lemma is proved in [18]. ϕ can be constructed in the following procedure: first, check the order from ϕ_0 to u_R in the Lie derivative sense, denote it by n; then define ϕ as $\phi_0 + k_1\dot{\phi}_0 + \cdots + k_{n-1}\phi_0^{(n-1)}$. The coefficients $k_1, \ldots, k_n$ are chosen such that the roots of $1 + k_1 s + \cdots + k_{n-1}s^{n-1} = 0$ all lie on the negative real line.

$$L(t) = \frac{\partial \phi}{\partial x_R} f_{Ru}^* \tag{11.12}$$

$$S(t, \dot{x}_H) = \begin{cases} -\eta_R - \sum_{j \in H} \frac{\partial \phi}{\partial x_j} \dot{x}_j - \frac{\partial \phi}{\partial x_R} f_{Rx}^* & \phi \geq 0 \\ \infty & \phi < 0 \end{cases} \tag{11.13}$$

$L(t)$ is a vector at the "safe" direction, while $S(t, \dot{x}_H)$ is a scalar indicating how much control effort is needed to be safe, which can be broken down into three parts: the margin, the term to compensate human motion and the term to compensate the inertia of the robot itself. In the following arguments when there is no ambiguity, $S(t, \dot{x}_H)$ denotes the value in the case $\phi \geq 0$ only. As could be expected, under different assumptions of the human behavior, $S(t)$ varies. The sets of safe control correspond to R_S^1, R_S^2 and R_S^3 are

$$U_S^1(t) = \{u_R(t) : L(t) u_R(t) \leq S(t, \dot{x}_H) \text{ for some } \dot{x}_H\} \tag{11.14}$$

$$U_S^2(t) = \left\{u_R(t) : L(t) u_R(t) \leq S\left(t, \hat{\dot{x}}_H\right)\right\} \tag{11.15}$$

$$U_S^3(t) = \left\{u_R(t) : L(t) u_R(t) \leq S(t, \dot{x}_H) \text{ for all } \dot{x}_H \in \dot{\Gamma}_H\right\} \tag{11.16}$$

where $\hat{\dot{x}}_H$ is the velocity vector that moves the current configuration x_H of human to $\hat{x}_H$ and $\dot{\Gamma}_H$ is the set of velocity vectors that move x_H to Γ_H. Computationally, $\hat{\dot{x}}_H = \frac{\hat{x}_H - x_H}{T_s}$ where T_s is the sampling time. Obviously $U_S^3 \subset U_S^2 \subset U_S^1$. When the uncertainties in the estimation of $\hat{\dot{x}}_H$ reduces, U_S^3 converges to U_S^2.

The difference between R_S and U_S is that R_S is static as it is on the state space, while U_S is dynamic as it concerns with the "movements". Due to the safety index, the non-convex state-space constraints R_S are transformed to convex state space constraints U_S. For example, in Fig. 11.6, the safe region R_S^3 for the AGV is the space outside the uncertainty range. But the set U_S^3 according to (11.16) is a half space.

According to the safety principle, the robot control should be restricted to the set U_S^3. However, the choice of the uncertainty bound $\dot{\Gamma}_H$ differs. One way is to use a constant bound based on the mean prediction error. Another way is to let the uncertainty bound depend on the level of uncertainties in real time. The first way corresponds to the safe set algorithm (SSA) [18] while the second one corresponds to the safe exploration algorithm (SEA) [20], which will be discussed in the following two sections.

11.4 The Safe Set Algorithm (SSA)

The safe set algorithm offers a fast online solution concerning the safety principle. In this section, the control algorithm and the learning algorithm in SSA will be discussed, followed by an application on a robot arm.

11.4.1 The Control Algorithm

In SSA, a constant $\lambda_R^{SSA} \in \mathbb{R}^+$ is introduced to bound the noises and uncertainties in the estimation, i.e.,

$$S^{SSA}\left(t, \hat{x}_H\right) = S(t, \hat{x}_H) - \lambda_R^{SSA} = -\eta_R - \lambda_R^{SSA} - \sum_{j \in H} \frac{\partial \phi}{\partial x_j}\hat{\dot{x}}_j - \frac{\partial \phi}{\partial x_R} f_{Rx}^* \tag{11.17}$$

where $\hat{\dot{x}}_j = \frac{\hat{x}_j(k+1|k) - \hat{x}_j(k|k)}{T_s}$ for all $j \in H$. k represents the time step of the last measurement before t, while $k+1$ is the time step for the next anticipated measurement. T_s is the sampling time. $\hat{x}_j(p|k)$ is the estimate of $x_j(p)$ given information up to time step k. $L(t)$ and $S(t)$ will be written as $L(k)$ and $S(k)$ to denote that the last measurement is taken at k-th time step. The control algorithm can also be designed in discrete time as discussed in [18] which is more conservative.

The optimal control problem for robot motion planning can be posed as [19]:

$$\min \quad E(J(x_R, u_R, G_R)) \tag{11.18}$$

$$s.t. \quad u_R \in U_S^{SSA}, u_R \in \Omega, x_R \in \Gamma_R \tag{11.19}$$

where $U_S^{SSA}(t) = \{u_R : L(t)u_R \leq S^{SSA}(t, \hat{x}_H)\}$. J is the cost function that evaluates the efficiency of the robot motion. Ω is the control saturation and Γ_R is the state space constraint such as joint limits and speed limits. Ω and Γ_R is usually convex, while J can be designed to be convex. Then (11.18)–(11.19) is a convex optimization problem, which is easy to solve online.

An analytic control law that solves the optimization approximately can be obtained by breaking down the problem (11.18)–(11.19). Assume the state space constraint Γ_R is considered in the safe set X_S. Suppose $u_R^o(t)$ is the solution of the optimal control problem without the safety constraint $u_R \in U_S^{SSA}$ (which can be solved offline). Then the safe control law $u_R^*(t)$ can be found by mapping $u_R^o(t)$ to the set of safe control $U_S^{SSA}(t)$ according to the following cost function

$$u_R^* = \min_{u_R \in U_S^{SSA} \cap \Omega} \frac{1}{2}\left(u_R - u_R^o\right)^T Q\left(u_R - u_R^o\right) \tag{11.20}$$

where Q is a positive definite matrix. Let $c = \min_{u \in U_S^{SSA}(t)} \left|L(t)\left(u - u_R^o(t)\right)\right|$. By (11.20), the safe control law $u_R^*(t)$ when Ω is not tight is [18]:

$$u_R^*(t) = u_R^o(t) - c\frac{Q^{-1}L(t)^T}{L(t)Q^{-1}L(t)^T} \tag{11.21}$$

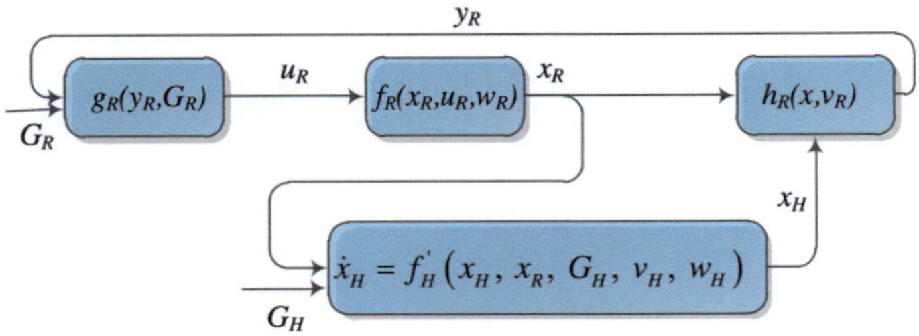

Fig. 11.8 Simplified system model (© 2015 IEEE. Reprinted, with permission, from [20])

11.4.2 Online Learning and Prediction of Humans' Dynamics

Since only one effective robot is considered, the system in Fig. 11.4 can be reduced to the system in Fig. 11.8 by combining all the blocks for human agents to the closed-loop block $\dot{x}_H = f_H'(x_H, x_R, G_H, v_H, w_H)$. The objective is to find the estimate $\hat{x}_H(k+i|k)$ to minimize the expected prediction error, e.g.,

$$\hat{x}_H(k+i|k) = \arg\min_a E_{x_H(k+i)}(\|x_H(k+i) - a\|^2) \tag{11.22}$$

The human's closed-loop dynamics f_H' need to be estimated for state prediction. The nonlinear continuous time dynamic function $\dot{x}_H = f_H'(.)$ can be linearized and discretized as

$$x_H(k+1) = A_H(k)x_H(k) + B_H(k)u_H^c(k) + w_H^*(k) \tag{11.23}$$

where $u_H^c(k) = \left[\hat{x}_R(k|k)^T, G_H^T(k)\right]^T$.[3] $\hat{x}_R$ is the estimate of the robot state from a state estimator (e.g., Kalman Filter). $A_H(k)$ and $B_H(k)$ are time varying parameters. $w_H^*(k)$ is a noise term assumed to be zero-mean Gaussian and white. Assume the robot's measurement of the human is:

$$y_R^H(k) = x_H(k) + v_R^H(k) \tag{11.24}$$

Equations (11.23) and (11.24) form a linear time varying (LTV) Gaussian system with unknown parameters. A recursive least square parameter adaptation algorithm (RLS-PAA) [9] is developed to identify the system online such that the prediction $\hat{x}_H(k+1|k)$ minimizes the expected prediction error (11.22). Define $\hat{A}_H(k)$ and $\hat{B}_H(k)$ to be the estimates of the matrices given the information up to the k-th time step.

[3] Methods for inferring $G_H(k)$ are discussed in [19]. In this chapter, it is assumed to be known.

- State Estimation
  At $k + 1$-th time step, $\hat{x}_H$ is first updated according to the closed-loop dynamics in (11.25). Then the measurement information is incorporated in the a posteriori estimate in (11.26). A constant update gain $\alpha \in (0, 1)$ is chosen to ensure that the measurement information is always incorporated.

$$\hat{x}_H(k+1|k) = \hat{A}_H(k)\,\hat{x}_H(k|k) + \hat{B}_H(k)\,u_H^c(k) \tag{11.25}$$

$$\hat{x}_H(k+1|k+1) = (1-\alpha)\,\hat{x}_H(k+1|k) + \alpha y_R^H(k+1) \tag{11.26}$$

- Parameter Estimation
  The closed-loop matrices are estimated using RLS-PAA:

$$\begin{aligned}\left[\hat{A}_H(k+1), \hat{B}_H(k+1)\right] &= \left[\hat{A}_H(k), \hat{B}_H(k)\right] \\ &+ \left(\hat{x}_H(k+1|k+1) - \hat{x}_H(k+1|k)\right)\varphi(k)^T F(k+1)\end{aligned} \tag{11.27}$$

where $\varphi(k) = \left[\hat{x}_H(k|k)^T \; u_H^c(k)^T\right]^T$. F is the learning gain such that

$$F(k+1) = \frac{1}{\lambda}\left[F(k) - \frac{F(k)\,\varphi(k)\varphi^T(k)F(k)}{\lambda + \varphi^T(k)\,F(k)\varphi(k)}\right] \tag{11.28}$$

where $\lambda \in (0, 1)$ is a forgetting factor.

11.4.3 Applications

In the factories of the future, there will be close interactions among human workers and robot arms [21]. In this section, the SSA is applied to a planar robot arm to demonstrate the effectiveness of the algorithm.

The environment is shown in Fig. 11.9 where both the robot arm and the human need to approach their respective goals in minimum time without colliding with each other. The robot arm has two links with joint positions θ_1 and θ_2 and the joint velocities $\dot{\theta}_1$ and $\dot{\theta}_2$. Let $\theta = [\theta_1, \theta_2]^T$. Considering the kinematics of the robot arm, the control input is defined to be the joint accelerations, i.e., $u_R = \left[\ddot{\theta}_1, \ddot{\theta}_2\right]^T$. The end

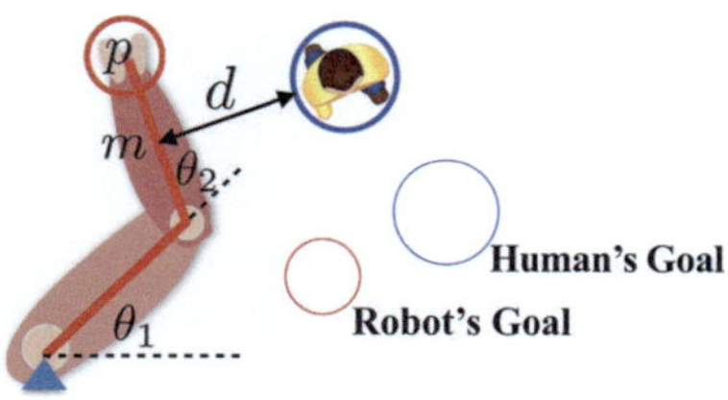

Fig. 11.9 The interaction between a robot arm and a human

point position of the robot is denoted as $p = (p_x, p_y)$. The optimal control law is designed to be

$$u_R^o = J_p^{-1}\left\{K_p\begin{bmatrix} p_x - g_x \\ p_y - g_y \end{bmatrix} + K_v\begin{bmatrix} \dot{p}_x \\ \dot{p}_y \end{bmatrix} - H_p\right\} \tag{11.29}$$

where J_p is the Jacobian matrix at p and $H_p = \dot{J}_p\dot{\theta}$. (g_x, g_y) is the goal point in the work space. $K_p \in \mathbb{R}^{2\times 2}$ and $K_v \in \mathbb{R}^{2\times 2}$ are the control gain.

The closest point to the human is $m = (m_x, m_y)$. Define the robot state as $x_R = [m_x, m_y, \dot{m}_x, \dot{m}_y]^T$. The state space equation is

$$\dot{x}_R = A_R x_R + B_R J_m u_R + B_R H_m \tag{11.30}$$

where

$$A_R = \begin{bmatrix} 0 & I_2 \\ 0 & 0 \end{bmatrix}, B_R = \begin{bmatrix} 0 \\ I_2 \end{bmatrix}$$

and J_m is the Jacobian matrix at m with $H_m = \dot{J}_m\dot{\theta}$.

The human is simplified as a circle, whose state is taken as $x_H = [h_x, h_y, \dot{h}_x, \dot{h}_y]$ where h_x and h_y is the position and $\dot{h}_x$ and $\dot{h}_y$ is the velocity. Define the safe set as $X_S = \{x : d \geq d_{min}\}$ where d measures the smallest distance between the human and the robot arm and d_{min} is a constant. Based on the discussion in Sect. 11.3.2, the safety index is designed as $\phi = D - d^2 - k_\phi\dot{d}$ where $D = d_{min}^2 + \eta_R T_s + \lambda_R^{SSA} T_s$ and $k_\phi > 0$ are constants [18]. Let the relative distance, velocity and acceleration vectors be $\mathbf{d} = \begin{bmatrix} I_2 & 0 \end{bmatrix}(x_R - x_H)$, $\mathbf{v} = \begin{bmatrix} 0 & I_2 \end{bmatrix}(x_R - x_H)$ and $\mathbf{a} = \begin{bmatrix} 0 & I_2 \end{bmatrix}(\dot{x}_R - \dot{x}_H)$. Then $d = |\mathbf{d}|$ and

$$\begin{aligned}\dot{\phi} &= -2d\dot{d} - k_\phi\ddot{d} = -2\mathbf{d}^T\mathbf{v} - k_\phi\frac{\mathbf{d}^T\mathbf{a} + \mathbf{v}^T\mathbf{v} - \dot{d}^2}{d} \\ &= -2\mathbf{d}^T\mathbf{v} - k_\phi\frac{\mathbf{d}^T(J_m u_R + H_m) - \mathbf{d}^T\begin{bmatrix} 0 & I_2 \end{bmatrix}\dot{x}_H + \mathbf{v}^T\mathbf{v}}{d} + k_\phi\frac{(\mathbf{d}^T\mathbf{v})^2}{d^3}\end{aligned} \tag{11.31}$$

Hence

$$L(t) = -k_\phi\frac{\mathbf{d}^T}{d}J_m \tag{11.32}$$

$$S(t, \dot{x}_H) = -\eta_R + 2\mathbf{d}^T\mathbf{v} + k_\phi\frac{\mathbf{d}^T H_m - \mathbf{d}^T\begin{bmatrix} 0 & I_2 \end{bmatrix}\dot{x}_H + \mathbf{v}\cdot\mathbf{v}}{d} - k_\phi\frac{(\mathbf{d}^T\mathbf{v})^2}{d^3} \tag{11.33}$$

In the simulation, several goals were assigned for the human. Before parameter adaptation, the robot inferred on the human's current goal first [19]. Sampling time $T_s = 0.1s$. Figure 11.10a shows the human avoidance behavior of the robot. In (1), the human and the robot were both near their respective goals. However, since the human was heading toward the robot in high speed, the robot went backward in (2)

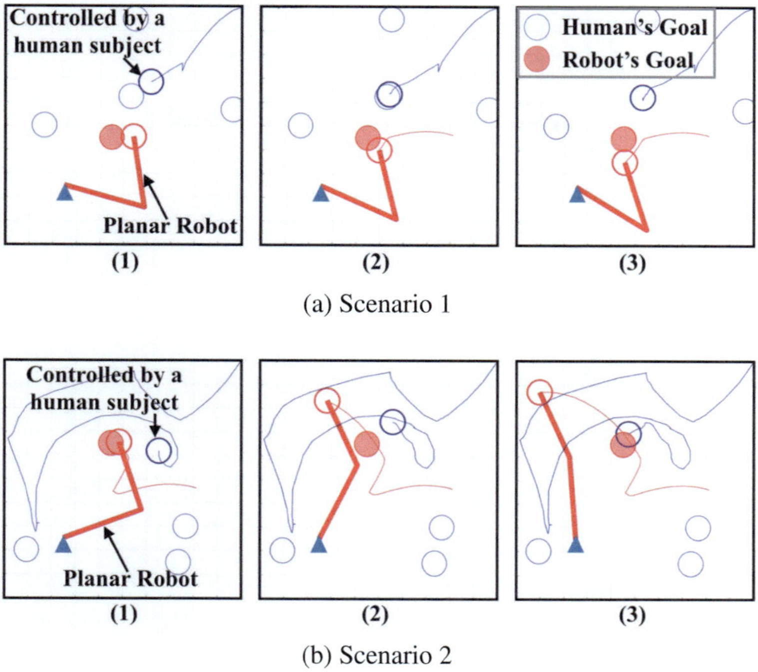

Fig. 11.10 Two scenarios in the simulation of SSA on a planar robot arm. The objective: the human and the robot arm should approach their respective goals in minimum time. Safety requirement: the minimum distance between the human and the robot arm should be greater than a threshold

and (3). Figure 11.10b shows the robot behavior under unexpected human behavior. In (1), the human suddenly changed his course. Although all of his goal points were in the lower part of the graph, the human started to go up. By observing that, the robot went away from the human in (2) and (3). The simulation results confirms the effectiveness of the algorithm.

11.5 The Safe Exploration Algorithm (SEA)

One of the limitation of SSA is that the bound for the uncertainties (i.e., λ_R^{SSA}) is a constant. However, the mean squared estimation error (MSEE) of the human's state is changing from time to time. A larger bound is needed if the MSEE is larger. To capture this property, the safe exploration algorithm (SEA) is introduced, where the control strategy changes for different levels of uncertainties.

Table 11.1 Definitions of notations in state estimation

	State estimate	Estimation error	MSEE
a posteriori	$\hat{x}_j(k\|k)$	$\tilde{x}_j(k\|k)$	$X_j(k\|k)$
a priori	$\hat{x}_j(k+1\|k)$	$\tilde{x}_j(k+1\|k)$	$X_j(k+1\|k)$

11.5.1 The Safe Set in the Belief Space

In a belief space [7], the state estimate of x_j for $j \in H$ is no longer a point but a distribution, i.e., $\mathcal{N}\left(\hat{x}_j, X_j\right)$ where X_j is the covariance which represents the level of uncertainties in the estimation. All the distributions are assumed to be Gaussian. Since $x_j \sim \mathcal{N}\left(\hat{x}_j, X_j\right)$, the covariance can be written as $X_j = E\left[\left(x_j - \hat{x}_j\right)\left(x_j - \hat{x}_j\right)^T\right]$, which is the mean squared estimation error (MSEE).

The definition of the a priori and a posteriori estimates, estimation errors and MSEEs are shown in Table 11.1, where $\tilde{x}_j(k|k) = x_j(k) - \hat{x}_j(k|k)$ and $\tilde{x}_j(k+1|k) = x_j(k+1) - \hat{x}_j(k+1|k)$. At the k-th time step, the best prediction for $x_j(k+1)$ has the following distribution

$$\mathcal{N}\left(\hat{x}_j(k+1|k), X_j(k+1|k)\right) \tag{11.34}$$

In the belief space, since the distribution of $x_j(k+1)$ is unbounded, the inequality in (11.16) is ill-defined. Indeed, u_R needs to satisfy a probability constraint

$$P\left(\left\{x_j(k+1) : L(k)u_R \leq S(k, x_H)\right\}\right) \geq 1 - \epsilon, \forall j \in H \tag{11.35}$$

where $\epsilon > 0$ is a small number. A bounded set $\Gamma_j(k)$ can be defined for $j \in H$ such that the probability density of $x_j \notin \Gamma_j(k)$ is small and $P\left(x_j \in \Gamma_j(k)\right) \geq 1 - \epsilon$. For a Gaussian distribution, the probability mass lying within the 3σ deviation is 0.997. Set $\epsilon = 0.003$ and let $\Delta x_j = x_j - \hat{x}_j(k+1|k)$, then the set Γ_j can be defined as

$$\Gamma_j(k) = \left\{x_j : \Delta x_j^T X_j(k+1|k)^{-1} \Delta x_j \leq 9\right\} \tag{11.36}$$

By (11.16) and (11.35), the constraint in U_R^3 is equivalent to the following optimization problem,

$$L(k)u_R \leq S^{SEA}(k) = \min_{x_j(k+1)\in\Gamma_j(k), \forall j \in H} \{S(k)\} \tag{11.37}$$

By (11.13), the RHS of (11.37) can be decoupled as a sequence of optimization problems, i.e., for all $j \in H$,

$$\min_{x_j(k+1)\in\Gamma_j(k)} \frac{\partial\phi}{\partial x_j} x_j(k+1) \tag{11.38}$$

By Lagrangian method,[4] the optimal solution $x_j^*(k+1)$ for all $j \in H$ is

$$x_j^*(k+1) = \hat{x}_j(k+1|k) + \frac{3X_j(k+1|k)\left(\frac{\partial\phi}{\partial x_j}\right)^T}{\left[\left(\frac{\partial\phi}{\partial x_j}\right)X_j(k+1|k)\left(\frac{\partial\phi}{\partial x_j}\right)^T\right]^{\frac{1}{2}}} \tag{11.40}$$

Using (11.40), $S^{SEA}(k)$ can be expressed as

$$S^{SEA}(k) = S(k, \hat{x}_H) - \lambda_R^{SEA}(k) = -\eta_R - \lambda_R^{SEA}(k) - \sum_{j\in H}\frac{\partial\phi}{\partial x_j}\hat{\dot{x}}_j - \frac{\partial\phi}{\partial x_R}f_{Rx}^* \tag{11.41}$$

where

$$\lambda_R^{SEA}(k) = \frac{3}{T_s}\sum_{j\in H}\left[\left(\frac{\partial\phi}{\partial x_j}\right)X_j(k+1|k)\left(\frac{\partial\phi}{\partial x_j}\right)^T\right]^{\frac{1}{2}} + \lambda_R^o \tag{11.42}$$

and $\lambda_R^o \in \mathbb{R}^+$ is the bound for other uncertainties. All other equations follow from the safe set algorithm except for the learning and prediction part, where new methods are needed to estimate $X_j(k+1|k)$ online.

11.5.2 Learning in the Belief Space

In this section, the MSEE propagation algorithm in parameter adaptation will be discussed, followed by the discussion of its application to human motion prediction. To simplify the notation, the LTV system is assumed to be

$$x(k+1) = A(k)x(k) + B(k)u(k) + w(k) \tag{11.43}$$

where $x(k) \in \mathbb{R}^n$, $u(k) \in \mathbb{R}^m$ and $w(k)$ is the dynamic noise assumed to be zero-mean, Gaussian and white with covariance W. It is assumed that the state $x(k)$ is fully known. $A(k)$ and $B(k)$ are the unknown parameters that need to be estimated

[4]The objective function is linear while the constraint function defines an ellipsoid as shown in Fig. 11.11. The optimal solution must lie on the boundary of the ellipsoid. Let γ be a Lagrange multiplier. Define the new cost function as:

$$J_j^* = \frac{\partial\phi}{\partial x_j}x_j(k+1) + \gamma\left[9 - \Delta x_j^T X_j(k+1|k)^{-1}\Delta x_j\right] \tag{11.39}$$

The optimal solution satisfies $\frac{\partial J_j^*}{\partial x_j(k+1)} = \frac{\partial J_j^*}{\partial\gamma} = 0$, i.e., $(\frac{\partial\phi}{\partial x_j})^T - 2\gamma X_j(k+1|k)^{-1}\Delta x_j = 0$ and $9 - \Delta x_j^T X_j(k+1|k)^{-1}\Delta x_j = 0$. Then (11.40) follows.

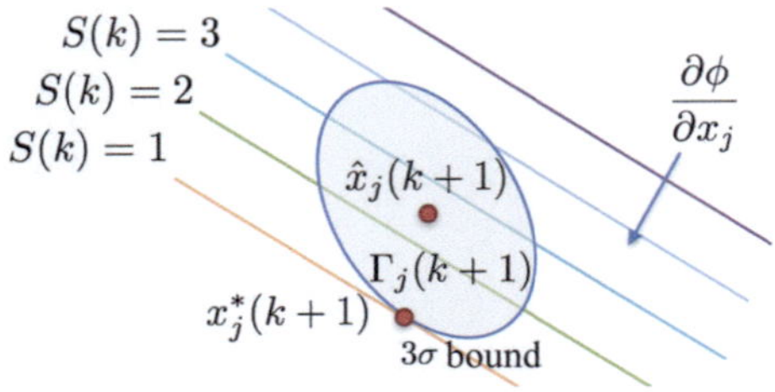

Fig. 11.11 The constraint in the belief space

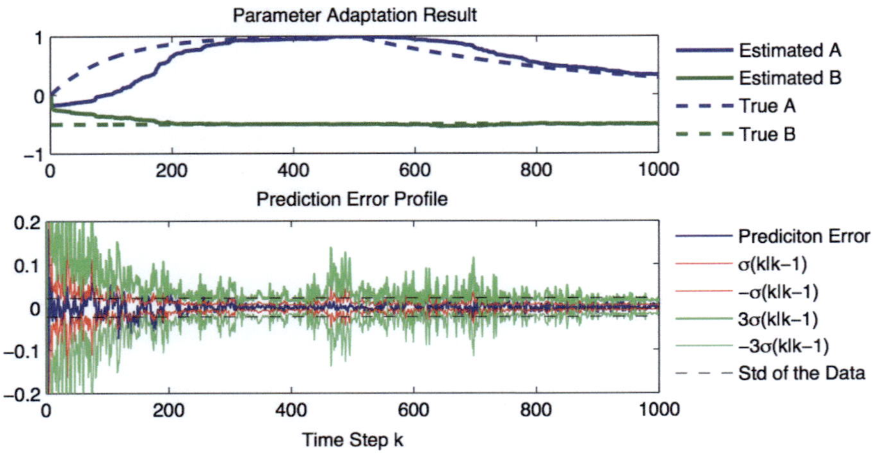

Fig. 11.12 Simulation result of the MSEE propagation algorithm

online. Define the parameter matrix $C(k) = [A(k), B(k)] \in \mathbb{R}^{n\times(n+m)}$ and the data vector $\varphi(k) = \left[x^T(k), u^T(k)\right]^T \in \mathbb{R}^{n+m}$. Since random matrices are hard to deal with, transform matrix $C(k)$ to a vector $\vartheta(k)$. Suppose the row vectors in $C(k)$ are $C_1(k), C_2(k), \ldots, C_n(k) \in \mathbb{R}^{1\times(n+m)}$. Define

$$\vartheta(k) = [C_1(k), C_2(k), \ldots, C_d(k)]^T \in \mathbb{R}^{n(n+m)\times 1} \tag{11.44}$$

Define a new data matrix $\Phi(k)$ as

$$\Phi(k) = \begin{bmatrix} \varphi^T(k) & 0 & \cdots & 0 \\ 0 & \varphi^T(k) & \cdots & 0 \\ \vdots & \vdots & \ddots & \vdots \\ 0 & 0 & \ldots & \varphi^T(k) \end{bmatrix} \in \mathbb{R}^{n\times n(n+m)} \tag{11.45}$$

Using $\Phi(k)$, $\vartheta(k)$, the system dynamics can be written as

$$x(k+1) = \Phi(k)\,\vartheta(k) + w(k) \tag{11.46}$$

Let $\hat{\vartheta}(k)$ be the estimate of $\vartheta(k)$ and $\tilde{\vartheta}(k) = \vartheta(k) - \hat{\vartheta}(k)$ be the estimation error.

- State estimation
  The a priori estimate of the state and the estimation error is

$$\hat{x}(k+1|k) = \Phi(k)\,\hat{\vartheta}(k) \tag{11.47}$$

$$\tilde{x}(k+1|k) = \Phi(k)\,\tilde{\vartheta}(k) + w(k) \tag{11.48}$$

  Since $\hat{\vartheta}(k)$ only contains information up to the $(k-1)$-th time step, $\tilde{\vartheta}(k)$ is independent of $w(k)$. Thus the a priori MSEE is

$$X_{\tilde{x}\tilde{x}}(k+1|k) = E\left[\tilde{x}(k+1|k)\,\tilde{x}(k+1|k)^T\right] = \Phi(k)\,X_{\tilde{\vartheta}\tilde{\vartheta}}(k)\,\Phi^T(k) + W \tag{11.49}$$

  where $X_{\tilde{\vartheta}\tilde{\vartheta}}(k) = E\left[\tilde{\vartheta}(k)\,\tilde{\vartheta}(k)^T\right]$ is the mean squared error of the parameter estimation.

- Parameter estimation
  In the standard PAA, the parameter is estimated as

$$\hat{\vartheta}(k+1) = \hat{\vartheta}(k) + F(k+1)\,\Phi^T(k)\,\tilde{x}(k+1|k) \tag{11.50}$$

  where $F(k+1)$ is the learning gain in (11.28) with $\varphi(k)$ replaced by $\Phi^T(k)$. The parameter estimation error is

$$\tilde{\vartheta}(k+1) = \tilde{\vartheta}(k) - F(k+1)\,\Phi^T(k)\,\tilde{x}(k+1|k) + \Delta\vartheta(k) \tag{11.51}$$

  where $\Delta\vartheta(k) = \vartheta(k+1) - \vartheta(k)$. Since the system is time varying, the estimated parameter is biased and the expectation of the error can be expressed as

$$\begin{aligned} &E\left(\tilde{\vartheta}(k+1)\right) \\ &= \left[I - F(k+1)\,\Phi^T(k)\,\Phi(k)\right] E\left(\tilde{\vartheta}(k)\right) + \Delta\vartheta(k) \\ &= \sum_{n=0}^{k} \prod_{i=n+1}^{k} \left[I - F(i+1)\,\Phi^T(i)\,\Phi(i)\right] \Delta\vartheta(n) \end{aligned} \tag{11.52}$$

  The mean squared error of parameter estimation follows from (11.51) and (11.52):

$$\begin{aligned} &X_{\tilde{\vartheta}\tilde{\vartheta}}(k+1) \\ &= F(k+1)\,\Phi^T(k)\,X_{\tilde{x}\tilde{x}}(k+1|k)\,\Phi(k)\,F(k+1) \\ &\quad - X_{\tilde{\vartheta}\tilde{\vartheta}}(k)\,\Phi^T(k)\,\Phi(k)\,F(k+1) - F(k+1)\,\Phi^T(k)\,\Phi(k)\,X_{\tilde{\vartheta}\tilde{\vartheta}}(k) \\ &\quad + E\left[\tilde{\vartheta}(k+1)\right]\Delta\vartheta^T(k) + \Delta\vartheta(k)\,E\left[\tilde{\vartheta}(k+1)\right]^T - \Delta\vartheta(k)\,\Delta\vartheta(k)^T + X_{\tilde{\vartheta}\tilde{\vartheta}}(k) \end{aligned} \tag{11.53}$$

Since $\Delta\vartheta\ (k)$ is unknown in (11.52) and (11.53), it is set to an average time varying rate $d\vartheta$ in the implementation.

Figure 11.12 shows the simulation result of the proposed learning algorithm on a first order system with a noise covariance $W = 0.005^2$. A forgetting factor $\lambda = 0.98$ is used. The solid and dashed blue lines in the upper figure are $\hat{A}\ (k)$ and $A\ (k)$, while the solid and dashed green lines are $\hat{B}\ (k)$ and the constant parameter B respectively. As shown in the figure, the time varying parameter $A(k)$ is well approximated by $\hat{A}(k)$, while $\hat{B}(k)$ converges to B. In the lower figure, the blue curve is the one step prediction error $\tilde{x}\ (k|k-1)$. The green curves are the 3σ bound ($\sigma = \sqrt{X_{\tilde{x}\tilde{x}}(k|k-1)}$). The black dashed line is the statistical standard deviation (Std) of the data $\tilde{x}\ (k|k-1)$ from $k = 1$ to $k = 1000$. As shown in the figure, the 3σ value offers a good bound for the prediction errors as all measured errors lie between the green curves. Moreover, the MSEE is larger when the parameter is changing faster, which captures the time varying property of the system. On the other hand, the statistical standard deviation does not give a good description of the data in real time.

In Sect. 11.4.2, a RLS-PAA algorithm is adopted in identifying the closed-loop behavior of the humans. In the safe exploration algorithm, the MSEE of $\hat{x}_j\ (k+1|k)$ for all $j \in H$ also needs to be estimated. The system follows from (11.23) and (11.24), which is different from (11.43) in that the state is not exactly known. But it is assumed that the measurement noise is small, thus can be neglected. So the system is approximated by

$$y_H^R(k+1) = A_H(k)y_H^R(k) + B_H(k)u_H^c\ (k) + w_H^*(k) \tag{11.54}$$

This is equivalent to setting $\alpha = 1$ in (11.26). The prediction algorithm then follows from (11.47)–(11.53). In the implementation, the covariance of the noise W, the time varying rate $d\vartheta$ and the initial values are hand-tuned.

11.5.3 A Comparative Study Between SSA and SEA

In this section, a comparative study between SSA and SEA is performed on an autonomous vehicle model shown in Fig. 11.13. The vehicle's state is denoted by $x_R = [R_x, R_y, v_R, \theta_R]^T$ where R_x is the x-position of the vehicle, R_y the y-position, v_R the speed and θ_R the direction. The control input of the vehicle is $u_R = [\dot{v}_R, \dot{\theta}_R]^T$ (saturations apply: $|\dot{v}_R| \leq a_{max}$ and $|\dot{\theta}_R| \leq \omega_{max}$, where a_{max}, ω_{max} are positive constants). The state equation is

$$\dot{x}_R = f_{Rx}^*(x_R) + Bu_R \tag{11.55}$$

where $f_{Rx}^*(x_R) = \left[v_R \cos\theta_R\ v_R \sin\theta_R\ 0\ 0\right]^T$, $B = \left[0\ I_2\right]^T$.

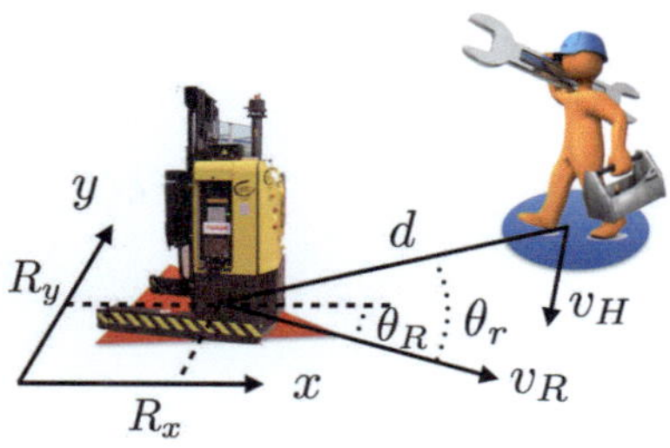

Fig. 11.13 The interaction between an AGV and a human worker

The vehicle can measure its own state directly. It can also measure the relative distance d and the relative direction θ_r toward the nearby human as illustrated in Fig. 11.13. The human's state is $x_H = [h_x, h_y, \dot{h}_x, \dot{h}_y]^T$, which is calculated based on the measurements and the robot state. Suppose that the goal point of the robot is $[G_x, G_y]$. The baseline control law is designed as [20]:

$$\dot{v}_R = -\left[(R_x - G_x)\cos\theta_R + (R_y - G_y)\sin\theta_R\right] - k_v v_R \tag{11.56}$$

$$\dot{\theta}_R = k_\theta \left[\arctan\frac{R_y - G_y}{R_x - G_x} - \theta_R\right] \tag{11.57}$$

where $k_v, k_\theta \in \mathbb{R}^+$ are constants.

The safety index $\phi = D - d^2 - k_\phi \dot{d}$ is chosen as the same as in Sect. 11.4.3. In SSA, D is set to be $d_{min}^2 + \eta_R T_s + \lambda_R^{SSA} T_s$. In SEA, D is set to be $d_{min}^2 + \eta_R T_s + \lambda_R^{SEA}(k) T_s$. The relative distance, velocity and acceleration vectors are

$$\mathbf{d} = [d\cos(\theta_r + \theta_R), d\sin(\theta_r + \theta_R)]^T$$
$$\mathbf{v} = [v\cos(\theta_R) - \dot{h}_x, v\sin(\theta_R) - \dot{h}_y]^T$$
$$\mathbf{a} = \begin{bmatrix} \cos\theta_R & -v_R\sin\theta_R \\ \sin\theta_R & v_R\cos\theta_R \end{bmatrix} u_R - \begin{bmatrix} 0 & I_2 \end{bmatrix} \dot{x}_H$$

Similar to (11.31), the time derivative of the safety index is

$$\begin{aligned}\dot{\phi} &= -2\mathbf{d}^T\mathbf{v} - k_\phi \frac{\mathbf{d}^T\mathbf{a} + \mathbf{v}^T\mathbf{v} - \dot{d}^2}{d} \\ &= -2\mathbf{d}^T\mathbf{v} - k_\phi \frac{[d\cos\theta_r, -dv_R\sin\theta_r]\, u_R - \mathbf{d}^T\begin{bmatrix} 0 & I_2 \end{bmatrix}\dot{x}_H + \mathbf{v}^T\mathbf{v}}{d} + k_\phi \frac{(\mathbf{d}^T\mathbf{v})^2}{d^3}\end{aligned} \tag{11.58}$$

which implies

$$L(t) = k_\phi\, [\cos\theta_r, -v_R\sin\theta_r] \tag{11.59}$$

$$S(t, \dot{x}_H) = -\eta_R + 2\mathbf{d}^T\mathbf{v} + k_\phi \frac{\mathbf{v}^T\mathbf{v} - \mathbf{d}^T\begin{bmatrix} 0 & I_2 \end{bmatrix}\dot{x}_H}{d} - k_\phi \frac{(\mathbf{d}^T\mathbf{v})^2}{d^3} \tag{11.60}$$

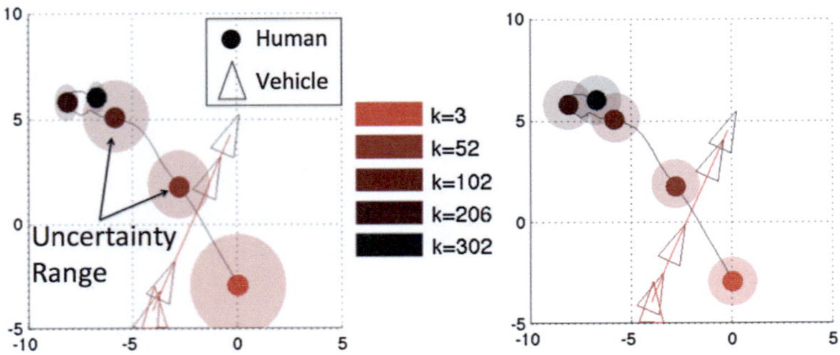

(a) Simulation with SEA (b) Simulation with SSA

Fig. 11.14 Simulation of human vehicle interaction under SEA and SSA (© 2015 IEEE. Reprinted, with permission, from [20])

Then S^{SSA} and S^{SEA} follow from (11.17) and (11.41) respectively, and the final control follows from (11.21).

Figure 11.14 shows the vehicle trajectories under SSA and SEA. The vehicle needed to approach (0, 5) from (−5, −5) while the human went from (0, −3) to (−5, 5). Five time steps are shown in the plots: $k = 3, 52, 102, 206, 302$ from the lightest to the darkest. The solid circles represent the human, which was controlled by a human subject through a multi-touch trackpad in real time (notice there was overshoot as the control was not perfect). The triangles represent the vehicle. The transparent circles in Fig. 11.14a represent the 2D projection of the set $\Gamma_H(k)$ in (11.36), which is shrinking gradually due to the reduction of uncertainties as an effect of learning. In Fig. 11.14b, the transparent circles represent the equivalent uncertainty levels introduced by λ_R^{SSA}, thus the radius remain constant throughout the time.

Figure 11.15 shows the distance profiles and the vehicle velocity profiles under SSA and SEA. Due to large initial uncertainties, the vehicle only started to accelerate after $k = 50$ (when the relative distance is large) in SEA. However, in SSA, the vehicle tried to accelerate in the very beginning, then decelerated when the relative distance to the human decreased. The velocity profile in SSA was serrated, while the one in SEA was much smoother. Meanwhile, in both algorithms, the relative distance was always greater than $d_{min} = 3$. However, before $k = 150$, the relative distance was kept larger in SEA than in SSA, since the vehicle was more conservative in SEA due to large uncertainty. Figure 11.16 shows that the a priori MSEE provides a perfect bound for the prediction error, while the prediction error reduces gradually, which validates the learning algorithms.

In conclusion, the behavior in SSA is: move and modify; while in SEA, it is: move only if confident. The behavior under SEA is better for a newcomer, while the behavior under SSA is better if the robot is already very familiar with the environment, i.e., with low uncertainty levels.

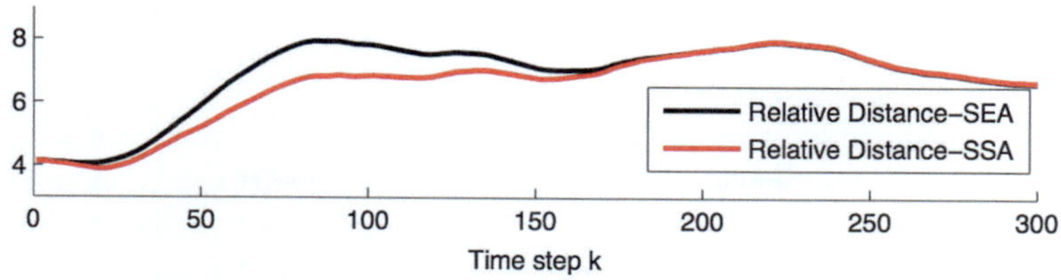

(a) The relative distance profiles in SEA and SSA

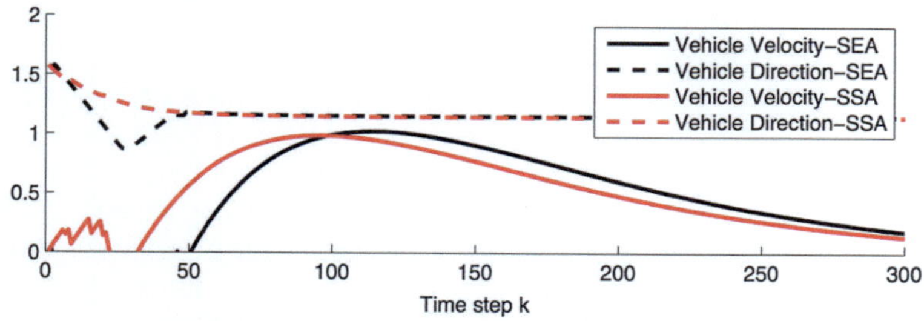

(b) The velocity profiles of the vehicle in SEA and SSA

Fig. 11.15 Comparison between SEA and SSA (© 2015 IEEE. Reprinted, with permission, from [20])

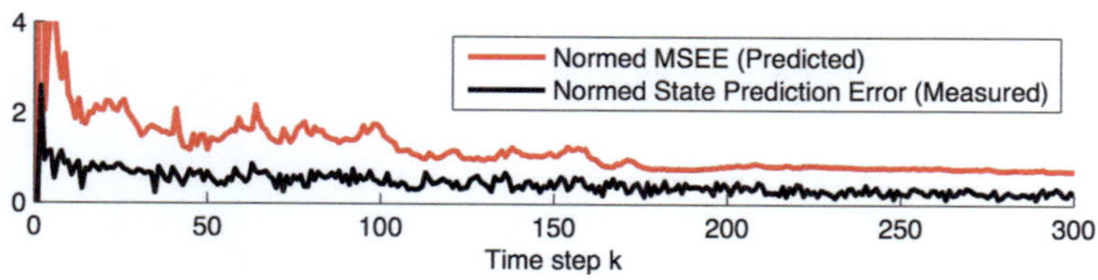

Fig. 11.16 Performance of the a priori MSEE as a bound of the state prediction error (© 2015 IEEE. Reprinted, with permission, from [20])

11.6 Combining SSA and SEA in Time Varying MAS Topology

In real world applications, the system topology is usually time varying, e.g., the robot will encounter different agents at different time in different locations [8]. Mathematically, that means some agents will be decoupled from the system block diagram in Fig. 11.4 and others will join from time to time. The robot is not faced with the same system throughout the time. This scenario is common for mobile robots and automated vehicles [22].

As the robot needs to deal with new agents, SEA is more appropriate than SSA. However, when the number of agents increase, the computation complexity with SEA increases dramatically. In this section, a method to combine SSA and SEA is discussed in order to balance the performance and the computation complexity.

11.6.1 The Control Algorithm

Due to limited sensing capability, the robot can only track humans that are within certain distance. Every agent within this range will be assigned a special identification number. Let $H(k)$ denotes the collection of those identification numbers at time step k. A safety index ϕ_j is designed for each agent $j \in H(k)$, as it is hard to design one analytical safety index that satisfies all the requirements for time varying $H(k)$. In this way, u_R needs to be constrained by

$$U_R^3 = \bigcap_{j \in H(k)} U_{R,j}^3 = \{u_R : L_j(k) u_R(k) \leq S_j(t, \dot{x}_j) \text{ for all } \dot{x}_j \in \dot{\Gamma}_j\} \quad (11.61)$$

where L_j and S_j are calculated with respect to ϕ_j. The uncertainty bound Γ_j is chosen according to SEA only for new agents. Once the MSEE converges, the algorithm is switched to SSA for that agent. The idea is illustrated in the following algorithm, where Π is a set that records the identification numbers of the agents who are no longer considered as new agents.

Algorithm 11.1: The Algorithm Combining SSA and SEA

```
Initialize Π = ∅, k = 0;
while Controller is Active do
    k = k + 1;
    Read current H(k) and y_R^j(k) for j ∈ H(k);
    for j ∈ H(k) do
        calculate the estimate x̂_j(k+1|k) based on measurements y_R^j(k);
        if j ∉ Π then
            calculate the MSEE X_j(k+1|k);
            if X_j converges then
                Π = Π ∪ {j};
            end
        end
    end
    for j ∈ H(k) do
        if j ∈ Π then
            U_{R,j}^3 = U_{R,j}^{SSA} = {u_R : L_j(k) ≤ S_j^{SSA}(k)} (Apply SSA to φ_j);
        else
            U_{R,j}^3 = U_{R,j}^{SEA} = {u_R : L_j(k) ≤ S_j^{SEA}(k)} (Apply SEA to φ_j);
        end
    end
    U_R^3 = ∩_{j∈H(k)} U_{R,j}^3;
    Choose control u_R^* by optimizing over U_R^3;
end
```

11.6.2 The Learning Algorithm

When the system topology is time varying, it is hard to learn humans' closed-loop dynamics in (11.23). When the number of agents increases, the computation complexity regarding (11.23) increases exponentially. That is because the correlation among agents are over estimated, e.g., an agent's motion may only be affect by several surrounding agents instead of by all agents. When the system topology is time varying, it is more appropriate to learn agents' dynamics separately and use low dimension features to represent the correlations among agents. It is assumed that an agent j's motion will be affected by several features f_j^p for $p = 1, 2, \ldots$, e.g., the distance to the nearest agent and distance to the goal point. Then the linearized closed-loop dynamics of agent j can be written as

$$x_j(k+1) = A_j(k)x_j(k) + \sum_p B_j^p(k) f_j^p(k) + w_j^*(k) \tag{11.62}$$

where $w_j^*(k)$ is the noise. Then the parameters $A_j(k)$ and $B_j^p(k)$ can be identified as discussed in Sects. 11.4.2 and 11.5.2.

11.6.3 Performance

The strategy is tested on robot navigation in a crowded environment with multiple humans as shown in Fig. 11.17. The robot is modeled as a double integrator whose input is the acceleration in x and y directions. The position of the humans are controlled in real time by several human subjects who observe the virtual environment though a screen. The humans do not have specific tasks and are just "wandering" in the environment. The robot is required to approach its goal while avoiding humans. The safety index is the same as in Sect. 11.5.3. The features in (11.62) are chosen to be the distance to the closest human and the distance to the robot.

Fig. 11.17 The simulation environment for robot navigation

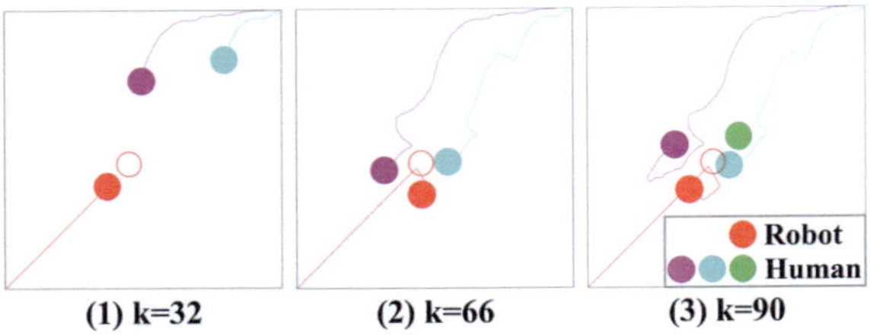

Fig. 11.18 Multi-human simulation results. Objective: the robot needs to navigate to its goal point in a crowded environment. Safety requirement: all relative distances are greater than d_{min}

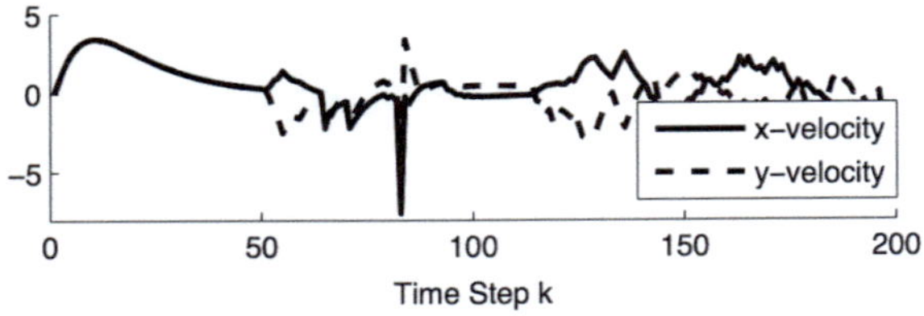

Fig. 11.19 The robot velocity profile in the multi-human simulation. After $k = 50$, the curves are no longer smooth due to the enforcement of the safety constraints

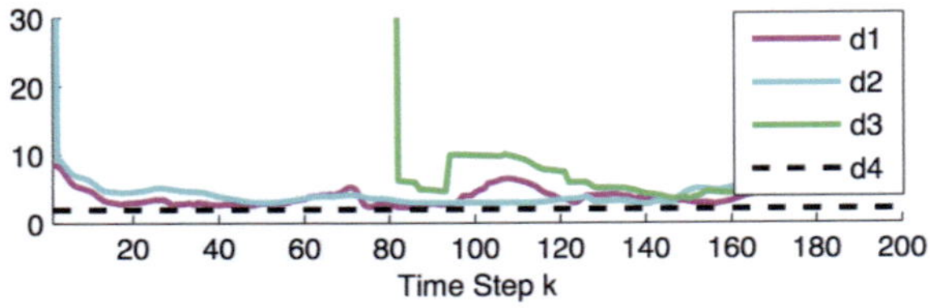

Fig. 11.20 The relative distance profile in the multi-human simulation. $d1$, $d2$, $d3$ are the distance between the robot and the respective human. $d4$ is the threshold d_{min}

The simulation result is shown in Figs. 11.18, 11.19 and 11.20. Before the 50-th time step, the robot was trying to approach its goal. It detoured when the blue agent came close. At the same time, the green agent which was previously hidden by the blue agent showed up in the robot's view. The new agent surprised the robot, as there was a large peak in the robot velocity profile in Fig. 11.19. Algorithm-wise, it was the large uncertainty of the green agent that "pushed" the robot away. The constraint U_R^3 was effective after the 50-th time step as evidenced in Fig. 11.19. The relative distance between the robot and every human agent was always maintained greater than d_{min} as shown in Fig. 11.20.

11.7 Discussions

Although derived from multiagent interaction point of view, the safety-oriented behavior design method can be viewed as an energy based method, which is closely related to other energy based methods such as the potential field method and the Lyapunov method. In Sect. 11.7.1, these methods will be compared. Moreover, although the proposed algorithms can handle a range of typical interactions such as robot navigation in a crowded environment and space sharing interactions among robot arms and human workers, there are certain limitations which will be discussed in Sect. 11.7.2.

11.7.1 The Energy Based Methods

In an energy based method, a scalar energy function is usually defined such that the control objective (e.g., safety) is with low energy. So the desired control should drive the energy function in the negative direction. The potential function, the Lyapunov function and the safety index discussed in Sect. 11.3.2 can all be regarded as energy functions. For a rough comparison, the emphasis in the potential field is the virtual repulsive force; the emphasis in the Lyapunov theory is asymptotic convergence to the control objective; while the emphasis in the safe set algorithm is the time invariance regarding the control objective. To some extent, the potential field method can be regarded as a "force" control, while the safe set method is a "position" control. The safety index is even closely related to the Lyapunov function, as the index can be transformed to an Lyapunov function by setting all negative values to be zero. The novel part of the proposed method is the introduction of multi objectives, e.g., the efficiency objective versus the safety objective. There are various ways to be safe, the robot needs to find the best one.

The multiple objectives is handled in the framework of constrained optimal control. Nonetheless, due to the introduction of the energy method (e.g., the safety index), the non-convex state space safety constraint is transformed to a convex control space constraint, which reduces the computation complexity comparing to other optimization methods. Thus the safe set method takes the advantage of both the energy method and the optimal control method.

11.7.2 Limitations and Future Work

Long Term Behavior Versus Short Term Behavior

Both SSA and SEA are concerned with short term reactive behavior. It is possible that the robot get stuck in local optima. Long term planning is needed for the robot to get out of local optima. However, the uncertainties in the predictions of other agents will accumulate in the long term, which makes such predictions unreliable. In order to make the interactions smoother, the design method of the long term behavior will be studied in the future.

Structured Environment Versus Unstructured Environment

Unstructured environments are also known as free spaces, while certain rules apply for structure environments. A typical structure environment is the traffic lanes shown in Fig. 11.1. Both SSA and SEA work well in unstructured environment. In structured environment, since the environment is discretized, it is important to include discrete choices in the robot control and planning, such as the choice for lane changing or lane following for an automated vehicle [22].

Cooperation Versus Coexistence

Cooperation usually involves physical contacts among agents, which is fundamentally different from coexistence. HRI which involves contacts can only be modeled by the indecomposable model in Fig. 11.3. Then the human dynamics can not be separated from the robot dynamics as in Fig. 11.8, which makes the learning and prediction even harder. Moreover, the safety issues in cooperation no longer come from conflicts as the objectives of the agents are identical, but from mis-interpretation or mis-understandings instead [23]. For example, it is dangerous when the robot motion is unexpected by the human. Cooperation will be analyzed in the future for a better understanding of interactions.

11.8 Conclusion

This chapter discussed the general methodology in designing the robot controller for safe HRI. By modeling the system in a multiagent framework, the safety issues were understood as conflicts in the multiagent system. To solve the conflicts, the robot's behavior was constrained according to the "social norm" and the uncertainties it perceived for the other agents (including humans and other robots). Two algorithms were discussed under the framework: the safe set algorithm (SSA) and the safe exploration algorithm (SEA). In both algorithms, the robot calculated the optimal action to finish the task while staying safe with respect to the predicted human motion. The difference was that SEA actively tracked the uncertainty levels in the prediction and incorporated that information in robot control, while SSA did not. As shown in the human-involved simulations, SEA was better when the uncertainty levels change from time to time, especially in the early stages of human–robot interactions. On the other hand, SSA was better when the predictions were more accurate, e.g., when the robot was "familiar" with the human, as SSA was more computationally efficient than SEA. Finally, a method to combine both algorithms was proposed to take the advantage of both algorithms. Several case studies were presented and demonstrated the effectiveness of the methods.

In the future, the long term behavior, the safety issues in structured environments and the safety issues in human–robot cooperation will be studied.

Acknowledgements This work was supported in part by a Berkeley Fellowship awarded to Changliu Liu.

References

1. (2015) Driverless car market watch. http://www.driverless-future.com
2. Armbruster W, Böge W (1979) Bayesian game theory. In: Game theory and related topics, vol 17, p 28

3. Basar T, Olsder GJ (1995) Dynamic noncooperative game theory, vol 200. Academic press, London
4. Berger CR, Calabrese RJ (1975) Some explorations in initial interaction and beyond: toward a developmental theory of interpersonal communication. Hum Commun Res 1(2):99–112
5. Burns LD (2013) Sustainable mobility: a vision of our transport future. Nature 497(7448):181–182
6. Craig JJ (2005) Introduction to robotics: mechanics and control. Pearson/Prentice Hall, Upper Saddle River
7. Du Toit NE, Burdick JW (2012) Robot motion planning in dynamic, uncertain environments. IEEE Trans Robot 28(1):101–115
8. Franchi A, Secchi C, Son HI, Bülthoff HH, Giordano PR (2012) Bilateral teleoperation of groups of mobile robots with time-varying topology. IEEE Trans Robot 28(5):1019–1033
9. Goodwin GC, Sin KS (2013) Adaptive filtering prediction and control. Courier Dover Publications, New York
10. Gracia L, Garelli F, Sala A (2013) Reactive sliding-mode algorithm for collision avoidance in robotic systems. IEEE Trans Control Syst Technol 21(6):2391–2399
11. Haddadin S, Albu-Schaffer A, Eiberger O, Hirzinger G (2010) New insights concerning intrinsic joint elasticity for safety. In: 2010 IEEE/RSJ international conference on proceedings of intelligent robots and systems (IROS). IEEE, pp 2181–2187
12. Hartl RF, Sethi SP, Vickson RG (1995) A survey of the maximum principles for optimal control problems with state constraints. SIAM Rev 37(2):181–218
13. Khatib O (1986) Real-time obstacle avoidance for manipulators and mobile robots. Int J Robot Res 5(1):90–98
14. Kong K, Bae J, Tomizuka M (2009) Control of rotary series elastic actuator for ideal force-mode actuation in human-robot interaction applications. IEEE/ASME Trans Mechatron 14(1):105–118
15. Krüger J, Lien T, Verl A (2009) Cooperation of human and machines in assembly lines. CIRP Ann.-Manuf. Technol. 58(2):628–646
16. Leber J (2013) At volkswagen, robots are coming out of their cages. http://www.fastcoexist.com/3016848/at-volkswagen-robots-are-coming-out-of-their-cages
17. Liu J (2001) Autonomous agents and multi-agent systems: explorations in learning, self-organization, and adaptive computation. World Scientific, Singapore
18. Liu C, Tomizuka M (2014a) Control in a safe set: addressing safety in human robot interactions. In: ASME 2014 dynamic systems and control conference, ASME, p V003T42A003
19. Liu C, Tomizuka M (2014b) Modeling and controller design of cooperative robots in workspace sharing human-robot assembly teams. In: 2014 IEEE/RSJ international conference on intelligent robots and systems (IROS). IEEE, pp 1386–1391
20. Liu C, Tomizuka M (2015) Safe exploration: addressing various uncertainty levels in human robot interactions. In: American control conference (ACC), pp 465–470
21. Liu C, Tomizuka M (2016a) Algorithmic safety measures for intelligent industrial co-robots. In: 2016 IEEE international conference on proceedings of robotics and automation (ICRA). IEEE, pp 3095–3102
22. Liu C, Tomizuka M (2016b) Enabling safe freeway driving for automated vehicles. In: 2016 American control conference. IEEE, pp 3461–3467
23. Liu C, Tomizuka M (2016c) Who to blame? Learning and control strategies with information asymmetry. In: 2016 American control conference. IEEE, pp 4859–4864
24. Mohr J, Spekman R (1994) Characteristics of partnership success: partnership attributes, communication behavior, and conflict resolution techniques. Strateg Manag J 15(2):135–152
25. Ozguner U, Stiller C, Redmill K (2007) Systems for safety and autonomous behavior in cars: the darpa grand challenge experience. Proc IEEE 95(2):397–412
26. Park DH, Hoffmann H, Pastor P, Schaal S (2008) Movement reproduction and obstacle avoidance with dynamic movement primitives and potential fields. In: 2008 IEEE-RAS international conference on humanoid robots. IEEE, pp 91–98

27. Park HK, Hong HS, Kwon HJ, Chung MJ (2001) A nursing robot system for the elderly and the disabled. International J Hum.-friendly Welfare Robot Syst (HWRS) 2(4):11–16
28. Russell S, Norvig P (1995) Artificial Intelligence: A Modern Approach. Prentice Hall, Upper Saddle River
29. Schulman J, Ho J, Lee A, Awwal I, Bradlow H, Abbeel P (2013) Finding locally optimal, collision-free trajectories with sequential convex optimization. In: Robotics: Science and Systems (RSS), vol 9, pp 1–10
30. Sisbot EA, Marin-Urias LF, Broquere X, Sidobre D, Alami R (2010) Synthesizing robot motions adapted to human presence. Int J Soc Robot 2(3):329–343
31. Tachi S, Komoriya K (1984) Guide dog robot. Control, Planning, and Architecture, Autonomous Mobile Robots, pp 360–367
32. Tadele TS, Vries TJd, Stramigioli S (2014) The safety of domestic robots: a survey of various safety-related publications. Robotics and Automation Magazine, IEEE pp 134–142
33. Tsai CS, Hu JS, Tomizuka M (2014) Ensuring safety in human-robot coexistence environment. In: 2014 IEEE/RSJ international conference on proceedings of intelligent robots and systems (IROS). IEEE, pp 4191–4196
34. Young JE, Hawkins R, Sharlin E, Igarashi T (2009) Toward acceptable domestic robots: applying insights from social psychology. Int J Soc Robot 1(1):95–108

Chapter 12
When Human Visual Performance Is Imperfect—How to Optimize the Collaboration Between One Human Operator and Multiple Field Robots

Hong Cai and Yasamin Mostofi

12.1 Introduction

In recent years, there have been great technological developments in robotics, in areas such as navigation, motion planning, and group coordination. However, while robots are becoming capable of more complicated tasks, there still exist many tasks which robots simply cannot perform to a satisfactory level when compared to humans. A complex visual task, such as recognition and classification in the presence of uncertainty, is one example of such tasks [2]. Thus, proper incorporation of human assistance will be very important to robotic missions.

More recently, the research community has been looking into the role of humans and different aspects of human–robot collaboration. In control and robotics, for instance, the Drift Diffusion Model from cognitive psychology [7, 14] has been heavily utilized in modeling human decision-making and the overall collaboration. Chipalkatty [8] shows how to incorporate human factors into a Model Predictive Control framework, in which human commands are predicted ahead of time. Utilizing machine learning, researchers have also looked into transferring human skills to robots [15] and incorporating human feedback to robot learning [12]. Several human–machine interfaces have been studied. Srivastava [13] has designed a Decision Support System considering the ergonomic factors of the human operator to optimize how the machine should provide information to the human operator. Branson et al. [2] propose a collaboration interface that resembles the 20-question game for bird classification. Experimental studies have been conducted on how humans and robots interact and cooperate in simulated scenarios, such as urban

H. Cai (✉) · Y. Mostofi
Department of Electrical and Computer Engineering, University of California, Santa Barbara, CA, USA
e-mail: hcai@ece.ucsb.edu

Y. Mostofi
e-mail: ymostofi@ece.ucsb.edu

Y. Wang and F. Zhang (eds.), *Trends in Control and Decision-Making for Human–Robot Collaboration Systems*, DOI 10.1007/978-3-319-40533-9_12

search and rescue operations [3, 10]. In [4–6], the fact that human visual performance is not perfect is taken into account in the collaboration between one human operator and a single field robot, emphasizing the importance of properly asking for human's help. More specifically, in [4, 6] we showed how to predict human visual performance for the case where additive noise is the only source of uncertainty. In [5], we proposed an automated machine learning-based approach that allows the robot to probabilistically predict human visual performance for a visual input, with any source of uncertainty, and experimentally validated our approach.

In this chapter, we are interested in the optimization of the human–robot collaboration in visual tasks such that the strengths of both are properly combined in task planning and execution. We know that humans can do complex visual tasks, such as recognition and classification, in the presence of a high level of uncertainty, while robots can traverse harsh and potentially dangerous terrains. *Human visual performance, however, is not perfect as we established in* [4, 5]. We thus incorporate a new paradigm, i.e., *when to ask humans for help* [4, 5], into the optimization of the collaboration between a human operator and *multiple robots*. In this approach, the collaboration properly takes advantage of the human's superior visual performance and the robot's exploration capability, while considering the fact that human visual performance is not perfect, allowing the robots to ask for help in an optimized manner. More specifically, consider a robotic field exploration and target classification task where the robots have limited onboard energy budgets and share a limited number of queries to the human operator. Due to these restrictions, the robots cannot query the human operator all the time for help with classification. On the other hand, they may not have sufficient resources or capabilities to explore the field (and reduce the sensing uncertainty) to the level that their own classification over the whole field becomes acceptable. In this chapter, we then show when the robots should ask the human for help, when they should rely on their own classification, and when they should further explore the environment by co-optimizing their motion, sensing, and communication with the human operator.

In order to solve such co-optimization problems, the robots only need to understand the extent of human visual capabilities and their own performance. For instance, a robot may collect data with a high level of uncertainty. Yet, the human may be able to make sense out of this data and perform an accurate classification of the target of interest. If a robot can properly understand this, it can then judge if it should stop sensing and present the data to the human, or if it should gather more sensing data. In Sect. 12.2, we summarize our previous work [4] on how to probabilistically predict human and robot visual performances when additive noise is the only source of uncertainty. In Sect. 12.3, we then show how to optimize the collaboration between one human operator and multiple field robots when a probabilistic metric of human visual performance is given. We mathematically characterize the optimal decisions that arise from our optimization framework. Based on numerical evaluations, we then verify the efficacy of our design in Sect. 12.4 and show that significant resource savings can be achieved.

The work presented in this chapter is an extension of our previous work [4] to a multi-robot setting. More specifically, in [4], we considered the fact that human visual performance is not perfect in the collaboration of one robot and one human operator. We showed how to predict human visual performance for the case where additive noise is the main source of uncertainty. In this chapter, we extend [4] to a multiple robots setting, with an emphasis on mathematical analysis. More specifically, in this multi-robot setting, interesting new properties arise which we study both mathematically and numerically. We note that while this chapter uses the prediction of human visual performance from [4], a more realistic prediction of human visual performance from [5] can be incorporated in the numerical results as part of future work.

12.2 Human and Robot Performance in Target Classification [4]

In this section, we briefly summarize human and robot classification capabilities in the presence of additive noise based on our previous work [4]. Consider the case where the robot has discovered a target via taking a picture and needs to classify it based on a given set of target possibilities. For example, Fig. 12.1 (left) shows four possible images that are shown to the robot prior to the task. The sensing of the robot in the field, however, is in general subject to noise, low resolution, occlusion, and other uncertainties, which will degrade its classification accuracy. Figure 12.1 (right) shows a sample case where an image is corrupted by an additive Gaussian noise with variance of 2. If the robot could accurately model all the uncertainties and use the best detector accordingly, it would outperform the humans. This, however, is not possible in realistic scenarios as it is impossible for the robot to know/model every source of uncertainty or derive the optimal classifier due to the complexity of a real life visual task. This is why the robot can benefit from the collaboration with the human tremendously by properly taking advantage of human visual abilities. Human performance, however, is not perfect, which requires proper modeling.

In our previous work [4], human and robot performance curves were obtained for the following scenario. The robot takes an image in the field, which is corrupted by an additive Gaussian noise with a known variance but an unknown mean, and then undergoes a truncation process that is unknown to the robot. Figure 12.2 shows the performance curves of the human and the robot using noise variance as the metric. The solid line shows the true probability of correct classification of the robot using the minimum distance detector, which would have been the optimal detector under zero-mean additive Gaussian noise. The dashed line shows the human performance obtained from the data collected utilizing Amazon Mechanical Turk (MTurk). For instance, in Fig. 12.1 (right), humans can achieve an average probability of correct classification of 0.744, which is considerably higher than robot performance (0.5).

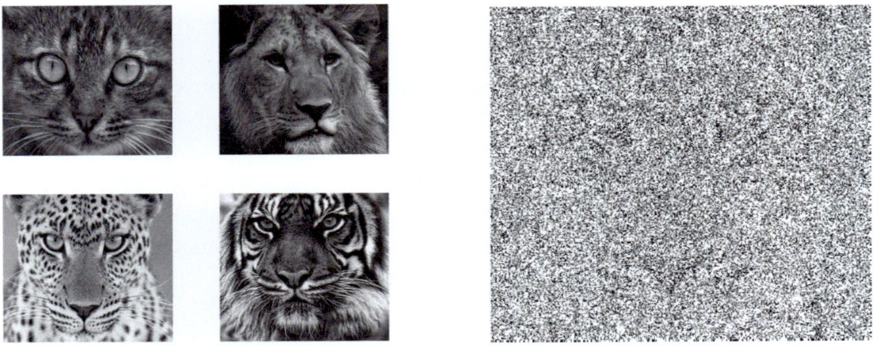

Fig. 12.1 (*left*) Gray-scale test images of cat, leopard, lion, and tiger used in our study [4]. (*right*) A sample corrupted image (leopard) with noise variance of 2 (© 2015 IEEE. Reprinted, with permission, from [4])

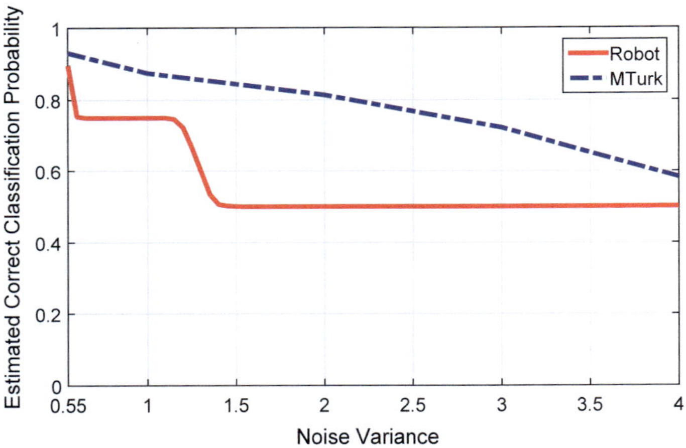

Fig. 12.2 Human and robot performance curves in target classification when additive noise is the main source of performance degradation [4]. The human data is acquired using Amazon MTurk. For more details, see [4] (© 2015 IEEE. Reprinted, with permission, from [4])

While this is a toy example, it captures a possible realistic scenario if additive noise is the main source of performance degradation. For instance, the robot may be able to assess its noise variance based on its distance to the target in the field but may not know the mean of the added noise or the nonlinear truncation that has happened at the pixel level. Our proposed approach of the next section will then utilize these performance curves for the optimization of the overall performance. We refer the readers to [5] for a more comprehensive prediction of human visual performance for any input with any source of uncertainty.

12.3 Optimizing Human–Robot Collaboration for Target Classification

We consider a setup in which the robots have an initial assessment (in the form of acquired images) of N given sites. Each robot is given an individual motion energy budget and they share a limited number of questions to ask the human operator. Two multi-robot scenarios are considered in this section. In the first scenario, it is assumed that each robot is assigned to a predetermined set of sites to classify. For each site that belongs to a robot's assigned set, the robot has three choices: (1) rely on its own classification (based on the initial sensing), (2) use a question and present the data of the site to the human, or (3) spend motion energy to go to the site and sense it better. The robot's second decision of asking the human for help is affected by the other robots' decisions since they share a common number of allowed queries to the remote operator. By studying this case, we capture a realistic situation in which the robots explore the environment and perform their own tasks in geographically separated locations while being monitored by the same remote human operator. In the second scenario, we incorporate site allocation into the optimization framework. Based on the initial sensing, each robot's motion energy cost to visit the sites, and the total number of allowed questions, the collaboration framework determines the sites the robots should query the human about, the sites for which they should rely on the initial sensing and the sites that should be visited. If a site is to be visited, the collaboration approach also determines which robot should visit that site.

12.3.1 Predetermined Site Allocation

In this section, we first discuss the case with a predetermined site allocation. Consider the case where we have a total number of K robots and each robot is assigned a priori to a set of N_k sites. There is a total of $N = \sum_{k=1}^{K} N_k$ sites. The sensing model of the robots is the same as explained in the previous section. In summary, each site contains one of T a priori known targets (see Fig. 12.1 (left) for an example with $T = 4$ targets). The sensing is then corrupted by an additive Gaussian noise with an unknown mean but a known variance, and is then truncated. The probabilities of correct target classification of the ith site assigned to robot k, for $k \in \{1, \ldots, K\}$ and $i \in \{1, \ldots, N_k\}$, are denoted by $p_{r,k,i}$ and $p_{h,k,i}$ for the robot and the human, respectively. These probabilities are obtained from Fig. 12.2, based on the variance assessed during the initial sensing. Note that although we assume a specific form of sensing uncertainty (additive Gaussian noise) here, our proposed optimization framework is general in that it only requires estimates of the human's and the robot's correct classification probabilities given a sensing input. The robots share a total of M allowed questions to the remote human operator and each robot has an individual motion energy budget of $\mathcal{E}_{max,k}$, where k is the index of the robot. Let $\mathcal{E}_{k,i}$ denote the motion energy cost to visit the ith site for the kth robot, which can be numerically evaluated by the robot. If a robot chooses to visit a site, the probability of correct classification increases to a high value of $\tilde{p} \geq p_{r,k,i}, \forall k = 1, \ldots, K, i = 1, \ldots, N_k$. The objective of this collaboration is then to decide which sites to present to the human, which sites to visit and which sites to rely on the robots' own classification based on the initial sensing, in order to maximize the overall average probability of correct classification under resource constraints. Let p_c denote the average probability of correct classification of a site. We have

$$p_c = \frac{1}{N}\sum_{k=1}^{K}\left(\sum_{i=1}^{N_k}\gamma_{k,i}p_{h,k,i} + \sum_{i=1}^{N_k}\eta_{k,i}\tilde{p} + \sum_{i=1}^{N_k}(1-\gamma_{k,i})(1-\eta_{k,i})p_{r,k,i}\right),$$
$$= \frac{1}{N}\sum_{k=1}^{K}\left(\sum_{i=1}^{N_k}\gamma_{k,i}(p_{h,k,i} - p_{r,k,i}) + \sum_{i=1}^{N_k}\eta_{k,i}(\tilde{p} - p_{r,k,i}) + \sum_{i=1}^{N_k}p_{r,k,i}\right),$$

where $\gamma_{k,i}$ is 1 if robot k seeks human's help for its ith site and is 0 otherwise. Similarly, $\eta_{k,i} = 1$ indicates that robot k will visit its ith site and $\eta_{k,i} = 0$ denotes otherwise. We then have the following optimization problem:

$$\begin{aligned}
\max_{\gamma,\eta}.\quad & \sum_{k=1}^{K}\gamma_k^T(p_{h,k} - p_{r,k}) + \eta_k^T(\tilde{p}\mathbf{1} - p_{r,k}) \\
\text{s.t.}\quad & \eta_k^T\mathscr{E}_k \leq \mathscr{E}_{max,k}, \quad \forall k = 1, \ldots, K, \\
& \sum_{k=1}^{K}\gamma_k^T\mathbf{1} \leq M, \\
& \gamma_k + \eta_k \preceq \mathbf{1}, \quad \forall k = 1, \ldots, K, \\
& \gamma, \eta \in \{0, 1\}^N,
\end{aligned} \tag{12.1}$$

where K is the total number of robots, N_k is the total number of sites that robot k needs to classify, $\mathscr{E}_{max,k}$ is the motion energy budget for robot k, M is the number of allowed questions for all the robots, $p_{h,k} = [p_{h,k,1}, \ldots, p_{h,k,N_k}]^T$, $p_{r,k} = [p_{r,k,1}, \ldots, p_{r,k,N_k}]^T$, $\gamma_k = [\gamma_{k,1}, \ldots, \gamma_{k,N_k}]^T$, $\eta_k = [\eta_{k,1}, \ldots, \eta_{k,N_k}]^T$, $\mathscr{E}_k = [\mathscr{E}_{k,1}, \ldots, \mathscr{E}_{k,N_k}]^T$, $\gamma = [\gamma_1^T, \ldots, \gamma_K^T]^T$, $\eta = [\eta_1^T, \ldots, \eta_K^T]^T$ and $N = \sum_{k=1}^{K} N_k$. The second constraint shows the coupling among the robots since they are all querying the same human operator, without which the optimization would be separable.

It can be seen that $(p_{h,k} - p_{r,k})$ and $(\tilde{p}\mathbf{1} - p_{r,k})$ are important parameters as they represent the performance gains by asking the human and visiting the sites, respectively. Note that we do not allow the robots to both query the human and make a visit for the same site. This is because we have already assumed a high probability of correct classification when a robot visits a site. Thus, allowing the robots to both visit and ask about the same site will be a waste of resources in this case.

12.3.1.1 Zero Motion Energy

If $\mathscr{E}_{max,k} = 0, \forall k = 1, \ldots, K$, problem (12.1) reduces to a 0–1 Knapsack Problem [11], which is a combinatorial optimization problem that often arises in resource allocation scenarios. In this case, the robots only need to decide between asking the human and relying on the initial classification, which is shown below.

$$\begin{aligned} \max_{\bar{\gamma}}.\ & \bar{\gamma}^T(\bar{p}_h - \bar{p}_r) \\ \text{s.t.}\ & \bar{\gamma}^T\mathbf{1} \leq M, \\ & \bar{\gamma} \in \{0, 1\}^N, \end{aligned} \tag{12.2}$$

where $\bar{p}_h = [\bar{p}_{h,1}, \ldots, \bar{p}_{h,N}]$, $\bar{p}_r = [\bar{p}_{r,1}, \ldots, \bar{p}_{r,N}]$, $\bar{p}_{h,i}$ and $\bar{p}_{r,i}$ denote the human's and the robot's correct classification probabilities of a site $i \in \{1, \ldots, N\}$, $\bar{\gamma} = [\bar{\gamma}_1, \ldots, \bar{\gamma}_N]$ and $\bar{\gamma}_i$ indicates whether the robots seek human help for site i. The optimal solution to this simplified problem can be obtained easily, which is summarized in the following lemma.

Lemma 12.1 *Suppose that all the N sites are sorted in a descending order according to $\bar{p}_{h,i} - \bar{p}_{r,i}$ such that $\bar{p}_{h,i} - \bar{p}_{r,i} \geq \bar{p}_{h,j} - \bar{p}_{r,j}$ for $i \leq j$. The optimal solution to problem (12.2) is given by*

$$\begin{aligned} \bar{\gamma}_i &= 1, \quad \text{for } i = 1, \ldots, n, \\ \bar{\gamma}_i &= 0, \quad \text{for } i = n+1, \ldots, N, \end{aligned} \tag{12.3}$$

where $\sum_{i=1}^{n} \bar{\gamma}_i = M$.

Proof The results can be easily verified. □

12.3.1.2 Zero Number of Allowed Queries

If $M = 0$, problem (12.1) reduces to K separable 0–1 Knapsack Problems. The optimization problem for the kth robot is shown as follows.

$$\begin{aligned} \max_{\eta_k}.\ & \eta_k^T(\tilde{p}\mathbf{1} - p_{r,k}) \\ \text{s.t.}\ & \eta_k^T \mathscr{E}_k \leq \mathscr{E}_{max,k}, \\ & \eta_k \in \{0, 1\}^{N_k}. \end{aligned} \tag{12.4}$$

Although the optimal solution to optimization problem (12.4) cannot be written directly in this case, its Linear Program (LP) relaxation provides a very close approximation. The LP relaxation of problem (12.4) is obtained by replacing the last binary constraint with $\eta_k \in [0, 1]^{N_k}$.

Lemma 12.2 *Suppose that the sites are sorted in a descending order according to $(\tilde{p} - p_{r,k,i})/\mathscr{E}_{k,i}$ such that $(\tilde{p} - p_{r,k,i})/\mathscr{E}_{k,i} \geq (\tilde{p} - p_{r,k,j})/\mathscr{E}_{k,j}$ for $i \leq j$. The optimal solution to the LP relaxation of problem (12.4) is given by*

$$\begin{aligned}
\eta_{k,i} &= 1, \quad \text{for } i = 1, \ldots, n-1, \\
\eta_{k,i} &= 0, \quad \text{for } i = n+1, \ldots, N_k, \\
\eta_{k,n} &= \frac{\bar{\mathscr{E}}}{\mathscr{E}_n},
\end{aligned}$$

where $\bar{\mathscr{E}} = \mathscr{E}_{max,k} - \sum_{i=1}^{n-1} \mathscr{E}_{k,i}$ and $n = min\{j : \sum_{i=1}^{j} \mathscr{E}_{k,i} > \mathscr{E}_{max,k}\}$.

Proof A graphical proof can be found in [9] and a more formal proof can be found in [11]. □

12.3.1.3 Considering the General Case

Problem (12.1) is in general a Mixed Integer Linear Program (MILP), which makes theoretical analysis difficult. In order to bring a more analytical understanding to this problem, we consider the following LP relaxation of problem (12.1), which is a close approximation to the problem. The LP relaxation allows the decision variables γ and η to take continuous values between 0 and 1.

$$\begin{aligned}
\max_{\gamma,\eta}. \quad & \sum_{k=1}^{K} \gamma_k^T (p_{h,k} - p_{r,k}) + \eta_k^T (\tilde{p}\mathbf{1} - p_{r,k}) \\
\text{s.t.} \quad & \eta_k^T \mathscr{E}_k \leq \mathscr{E}_{max,k}, \quad \forall k = 1, \ldots, K, \\
& \sum_{k=1}^{K} \gamma_k^T \mathbf{1} \leq M, \\
& \gamma_k + \eta_k \preceq \mathbf{1}, \quad \forall k = 1, \ldots, K, \\
& \gamma, \eta \in [0, 1]^N.
\end{aligned} \tag{12.5}$$

We can analyze this LP by applying Karush–Kuhn–Tucker (KKT) conditions [1]. We then have the following expression for the Lagrangian:

$$\begin{aligned}
&\mathscr{L}(\gamma, \eta, \omega, \lambda_1, \lambda_2, \theta, \zeta, \kappa, \tau, \xi, \psi) = -\left(\sum_{k=1}^{K} \gamma_k^T (p_{h,k} - p_{r,k}) + \eta_k^T (\tilde{p}\mathbf{1} - p_{r,k})\right) \\
&+ \mu\left(\sum_{k=1}^{K} \mathbf{1}^T \gamma_k - M\right) + \sum_{k=1}^{K} \lambda_k(\eta_k^T \mathscr{E}_k - \mathscr{E}_{max,k}) + \theta^T(\gamma + \eta - \mathbf{1}) + \psi^T(\gamma - \mathbf{1}) \\
&- \phi^T \gamma + \kappa^T(\eta - \mathbf{1}) - \omega^T \eta,
\end{aligned}$$

where $\mu, \lambda, \theta, \psi, \phi, \kappa, \omega$ are nonnegative Lagrange multipliers, and $\lambda = [\lambda_1, \ldots, \lambda_K]$.

The optimal solution (marked by $\star$) then satisfies the following KKT conditions, in addition to the primal/dual feasibility conditions:

(1) Gradient condition, for $k \in \{1, \ldots, K\}$ and $i \in \{1, \ldots, N_k\}$:

$$\nabla_{\gamma^\star_{k,i}} \mathscr{L} = p_{r,k,i} - p_{h,k,i} + \mu^\star + \theta^\star_{k,i} + \psi^\star_{k,i} - \phi^\star_{k,i} = 0, \tag{12.6}$$

$$\nabla_{\eta^\star_{k,i}} \mathscr{L} = p_{r,k,i} - \tilde{p} + \lambda^\star_k \mathscr{E}_{k,i} + \theta^\star_{k,i} + \kappa^\star_{k,i} - \omega^\star_{k,i} = 0. \tag{12.7}$$

(2) Complementary slackness: $\theta^\star \circ (\gamma^\star + \eta^\star - 1) = \mathbf{0}$, $\psi^\star \circ (\gamma^\star - 1) = \mathbf{0}$, $\phi^\star \circ \gamma^\star = \mathbf{0}$, $\kappa^\star \circ (\eta^\star - 1) = \mathbf{0}$, $\omega^\star \circ \eta^\star = \mathbf{0}$, $\mu(\sum_{k=1}^K \mathbf{1}^T \gamma_k - M) = 0$, $\lambda \circ (\eta^T \mathscr{E} - \mathscr{E}_{max}) = \mathbf{0}$, where $\mathbf{0}$ denotes the vector with all entries equal to 0, $\circ$ denotes the Hadamard product, $\mathscr{E} = [\mathscr{E}_1^T, \ldots, \mathscr{E}_K^T]^T$ and $\mathscr{E}_{max} = [\mathscr{E}_{max,1}, \ldots, \mathscr{E}_{max,k}]^T$.

The following lemmas characterize the optimal solution to the LP relaxation in terms of the optimization parameters.

Lemma 12.3 *Consider two sites i and j that belong to the preassigned sets of robot k_1 and robot k_2, respectively.*[1] *Let $\gamma^\star$ and $\eta^\star$ denote the optimal decision vectors. If $\gamma^\star_{k_1,i} = 1$, $\eta^\star_{k_1,i} = 0$, $\gamma^\star_{k_2,j} = 0$ and $\eta^\star_{k_2,j} = 0$, then $p_{h,k_1,i} - p_{r,k_1,i} \geq p_{h,k_2,j} - p_{r,k_2,j}$.*

Proof Suppose that we have two sites i and j preassigned to robot k_1 and robot k_2 respectively such that $\gamma^\star_{k_1,i} = 1$, $\eta^\star_{k_1,i} = 0$, $\gamma^\star_{k_2,j} = 0$ and $\eta^\star_{k_2,j} = 0$. Applying the complementary slackness conditions results in $\phi^\star_{k_1,i} = \theta^\star_{k_2,j} = \phi^\star_{k_2,j} = 0$. Then the gradient condition gives $p_{r,k_1,i} - p_{h,k_1,i} + \theta^\star_{k_1,i} + \phi^\star_{k_1,i} = p_{r,k_2,j} - p_{h,k_2,j} - \phi^\star_{k_2,j}$. Since $\theta^\star_{k_1,i}$, $\phi^\star_{k_1,i}$ and $\psi^\star_{k_2,j}$ are all nonnegative, it is necessary to have $p_{h,k_1,i} - p_{r,k_1,i} \geq p_{h,k_2,j} - p_{r,k_2,j}$. □

Lemma 12.3 says that if we have any two sites i and j, for which the robots will ask the human and rely on the initial sensing respectively, then the performance gain obtained from querying the human operator for site i should be higher than or equal to that of site j.

Remark 12.1 Lemma 12.3 also holds for the original integer problem (12.1).

Lemma 12.4 *Consider two sites i and j that have been assigned to robot k. Let $\gamma^\star$ and $\eta^\star$ denote the optimal decision vectors. If $\gamma^\star_{k,i} = 0$, $\eta^\star_{k,i} = 1$, $\gamma^\star_{k,j} = 0$ and $\eta^\star_{k,j} = 0$, then $(\tilde{p} - p_{r,k,i})/\mathscr{E}_{k,i} \geq (\tilde{p} - p_{r,k,j})/\mathscr{E}_{k,j}$.*

Proof Suppose that we have two sites i and j assigned to robot k such that $\gamma^\star_{k,i} = 0$, $\eta^\star_{k,i} = 1$, $\gamma^\star_{k,j} = 0$ and $\eta^\star_{k,j} = 0$. We have $\omega^\star_{k,i} = \theta^\star_{k,j} = \kappa^\star_{k,j} = 0$ from the complementary slackness conditions. Equation 12.7 for $\eta^\star_{k,i}$ then becomes: $(p_{r,k,i} - \tilde{p})/\mathscr{E}_{k,i} + \lambda^\star_k + \theta'_{k,i} + \kappa'_{k,i} = 0$, where $\theta'_{k,i} = \theta^\star_{k,i}/\mathscr{E}_{k,i}$ and $\kappa'_{k,i} = \kappa^\star_{k,i}/\mathscr{E}_{k,i}$. Similarly, we have $(p_{r,k,j} - \tilde{p})/\mathscr{E}_{k,j} + \lambda^\star_k - \omega'_{k,j} = 0$ when applying $\nabla_{\eta^\star_{k,j}} \mathscr{L} = 0$. This results in $(p_{r,k,i} - \tilde{p})/\mathscr{E}_{k,i} + \lambda^\star_k + \theta'_{k,i} + \kappa'_{k,i} = (p_{r,k,j} - \tilde{p})/\mathscr{E}_{k,j} + \lambda^\star_k - \omega'_{k,j}$. Since $\theta'_{k,i}$, $\kappa'_{k,i}$ and $\omega'_{k,j}$ are all nonnegative, we have $(\tilde{p} - p_{r,k,i})/\mathscr{E}_{k,i} \geq (\tilde{p} - p_{r,k,j})/\mathscr{E}_{k,j}$. □

[1] Note that robot k_1 and robot k_2 can be the same robot or two different robots.

Lemma 12.4 says that within the set of sites assigned to a robot, if there are two sites i and j, for which the robot will explore and rely on the initial sensing respectively, then the visited site should have a higher performance gain normalized by the energy cost.[2]

Lemma 12.5 *Consider two sites i and j that have been assigned to robot k. Let* $\gamma^\star$ *and* $\eta^\star$ *denote the optimal decision vectors. If* $\gamma^\star_{k,i} = 1$, $\eta^\star_{k,i} = 0$, $\gamma^\star_{k,j} = 0$, $\eta^\star_{k,j} = 1$ *and* $\mathscr{E}_{k,i} \leq \mathscr{E}_{k,j}$, *then* $p_{h,k,i} - p_{h,k,j} \geq 0$.

Proof Consider an optimal solution where we have $\gamma^\star_{k,i} = 1$, $\eta^\star_{k,i} = 0$, $\gamma^\star_{k,j} = 0$, $\eta^\star_{k,j} = 1$ and $\mathscr{E}_{k,i} \leq \mathscr{E}_{k,j}$. We modify the current optimal solution to obtain a new feasible solution in the following way: $\gamma'_{k,i} = \gamma^\star_{k,i} - \delta$, $\eta'_{k,i} = \eta^\star_{k,i} + \delta$, $\gamma'_{k,j} = \gamma^\star_{k,j} + \delta$, $\eta'_{k,j} = \eta^\star_{k,j} - \delta$, where $\delta > 0$ is a small number such that $\gamma'_{k,i}, \eta'_{k,i}, \gamma'_{k,j}, \eta'_{k,j} \in [0, 1]$. The new objective function value becomes $f' = f^\star + \Delta$, where $f^\star$ is the optimum and $\Delta = \delta(\tilde{p} - p_{r,k,i} - (p_{h,k,i} - p_{r,k,i}) + p_{h,k,j} - p_{r,k,j} - (\tilde{p} - p_{r,k,j}))$. Since the current solution is optimal, we should have $\Delta \leq 0$, from which we have $p_{h,k,i} - p_{h,k,j} \geq 0$. □

Consider the case where sites i and j are assigned to robot k. The robot asks for human help for site i and visits site j in the optimal solution. Lemma 12.5 says that in this case, if the motion energy cost of the queried site is less than or equal to that of the visited site, then the human performance of the queried site should be greater than or equal to that of the visited site.

12.3.2 Optimized Site Allocation

In this section, we consider the second collaborative scenario with a human operator and multiple field robots described earlier, where the optimization of site allocation to the robots is also taken into account. Consider the case where there is a total of N sites and K robots. The sensing model is the same as discussed in the previous section. The probabilities of correct target classification of the ith site are denoted by $\tilde{p}_{r,i}$ and $\tilde{p}_{h,i}$ for the robot and the human, respectively. These probabilities are obtained from Fig. 12.2, based on the variance assessed during the initial sensing. The robots share a total of M allowed questions to the remote human operator and each robot has an individual motion energy budget of $\mathscr{E}_{max,k}$, where k is the index of the robot. Let $\tilde{\mathscr{E}}_{k,i}$ denote the motion energy cost to visit the ith site for the kth robot. If a robot chooses to visit a site, the probability of correct classification increases to a high value of $\tilde{p}$. The objective of this collaboration is for the robots to decide on which sites to present to the human, which sites to rely on the initial sensing and which sites to visit. If a site is to be visited, this collaboration also determines which robot should visit the site. Let $\tilde{p}_c$ denote the average probability of correct classification of a site, which we would like to maximize. We have

[2]This lemma is similar to the second condition of Lemma 12.1 of our previous work [4] as it concerns only one robot.

$$\tilde{p}_c = \frac{1}{N}\left(\sum_{k=1}^{K}\sum_{i=1}^{N}\tilde{\eta}_{k,i}\tilde{p} + \sum_{i=1}^{N}\tilde{\gamma}_i\tilde{p}_{h,i} + \sum_{i=1}^{N}(1-\tilde{\gamma}_i)\left(1-\sum_{k=1}^{K}\tilde{\eta}_{k,i}\right)\tilde{p}_{r,i}\right),$$
$$= \frac{1}{N}\left(\sum_{i=1}^{N}\tilde{\gamma}_i(\tilde{p}_{h,i}-\tilde{p}_{r,i}) + \sum_{k=1}^{K}\sum_{i=1}^{N}\tilde{\eta}_{k,i}(\tilde{p}-\tilde{p}_{r,i}) + \sum_{i=1}^{N}\tilde{p}_{r,i}\right),$$

where $\tilde{\gamma}_i$ is 1 if the robots seek human's help for the ith site and is 0 otherwise. Similarly, $\tilde{\eta}_{k,i} = 1$ indicates that robot k will visit the ith site and $\tilde{\eta}_{k,i} = 0$ denotes otherwise. The optimization problem is then given by

$$\begin{aligned}
\max_{\tilde{\gamma},\tilde{\eta}}.\quad & \sum_{k=1}^{K}\tilde{\eta}_k^T(\tilde{p}\mathbf{1}-\tilde{p}_r) + \tilde{\gamma}^T(\tilde{p}_h-\tilde{p}_r) \\
\text{s.t.}\quad & \tilde{\eta}_k^T\tilde{\mathscr{E}}_k \le \mathscr{E}_{\text{max,k}}, \quad \forall k = 1,\ldots,K, \\
& \tilde{\gamma}^T\mathbf{1} \le M, \\
& \tilde{\gamma} + \sum_{k=1}^{K}\tilde{\eta}_k \preceq \mathbf{1}, \\
& \tilde{\gamma}, \tilde{\eta}_k \in \{0,1\}^N, \quad \forall k = 1,\ldots,K,
\end{aligned} \tag{12.8}$$

where K is the total number of robots, N is the total number of sites to classify, $\mathscr{E}_{max,k}$ is the motion energy budget for robot k, M is the total number of allowed questions, $\tilde{p}_h = [\tilde{p}_{h,1},\ldots,\tilde{p}_{h,N}]^T$, $\tilde{p}_r = [\tilde{p}_{r,1},\ldots,\tilde{p}_{r,N}]^T$, $\tilde{\gamma} = [\tilde{\gamma}_1,\ldots,\tilde{\gamma}_N]^T$, $\tilde{\eta}_k = [\tilde{\eta}_{k,1},\ldots,\tilde{\eta}_{k,N}]^T$, $\tilde{\eta} = [\tilde{\eta}_1^T,\ldots,\tilde{\eta}_K^T]^T$ and $\tilde{\mathscr{E}}_k = [\tilde{\mathscr{E}}_{k,1},\ldots,\tilde{\mathscr{E}}_{k,N}]^T$. $\tilde{\gamma}$ and $\tilde{\eta}$ determine whether the robots should ask for human help and visit the sites, respectively.

Problem (12.8) is in the form of a Multiple Knapsack Problems (MKP) [11], which is a natural extension to the basic 0–1 Knapsack Problem discussed in the previous section. This problem arises commonly in optimal decision-making and resource allocation settings.

12.3.2.1 Zero Motion Energy

If $\mathscr{E}_{max,k} = 0, \forall k = 1,\ldots,K$, problem (12.8) reduces to a 0–1 Knapsack Problem.

$$\begin{aligned}
\max_{\tilde{\gamma}}.\quad & \tilde{\gamma}^T(\tilde{p}_h-\tilde{p}_r) \\
\text{s.t.}\quad & \tilde{\gamma}^T\mathbf{1} \le M, \\
& \tilde{\gamma} \in \{0,1\}^N.
\end{aligned} \tag{12.9}$$

The above-reduced problem is very similar to problem (12.2) discussed previously. The optimal solution to this special case can be obtained via the same procedure outlined in Lemma 12.1.

12.3.2.2 Considering the General Case

In order to bring a more theoretical understanding to this setting, we consider the LP relaxation of problem (12.8), which is given as follows.

$$\begin{aligned} \max_{\tilde{\gamma},\tilde{\eta}}. \quad & \sum_{k=1}^{K} \tilde{\eta}_k^T(\tilde{p}\mathbf{1} - \tilde{p}_r) + \tilde{\gamma}^T(\tilde{p}_h - \tilde{p}_r) \\ \text{s.t.} \quad & \tilde{\eta}_k^T \tilde{\mathscr{E}}_k \leq \mathscr{E}_{max,k}, \quad \forall k = 1, \ldots, K, \\ & \tilde{\gamma}^T \mathbf{1} \leq M, \\ & \tilde{\gamma} + \sum_{k=1}^{K} \tilde{\eta}_k \preceq \mathbf{1}, \\ & \tilde{\gamma}, \tilde{\eta}_k \in [0, 1]^N, \quad \forall k = 1, \ldots, K. \end{aligned} \tag{12.10}$$

By allowing the decision variables $\tilde{\gamma}$ and $\tilde{\eta}$ to take continuous values in the interval [0, 1], we can analyze this problem utilizing the KKT conditions, which leads to the following two lemmas.

Lemma 12.6 *Consider two sites i and j. Let* $\tilde{\gamma}^\star$ *and* $\tilde{\eta}^\star$ *denote the optimal decision vectors. If* $\tilde{\gamma}_i^\star = 1$, $\sum_{k=1}^{K} \tilde{\eta}_{k,i}^\star = 0$, $\tilde{\gamma}_j^\star = 0$ *and* $\sum_{k=1}^{K} \tilde{\eta}_{k,j}^\star = 0$, *then* $\tilde{p}_{h,i} - \tilde{p}_{r,i} \geq \tilde{p}_{h,j} - \tilde{p}_{r,j}$.

Proof The proof is similar to that of Lemma 12.3. □

Lemma 12.6 says that if we have two sites i and j, for which the robots will ask the human and rely on the initial sensing respectively, then the performance gain obtained from asking the human should be greater for site i.

Remark 12.2 Lemma 12.6 also holds for the original integer problem (12.8).

Lemma 12.7 *Consider two sites i and j. Let* $\tilde{\gamma}^\star$ *and* $\tilde{\eta}^\star$ *denote the optimal decision vectors. If* $\tilde{\gamma}_i^\star = 0$, $\tilde{\eta}_{k_1,i}^\star = 1$, $\tilde{\gamma}_j^\star = 0$ *and* $\sum_{k=1}^{K} \tilde{\eta}_{k,j}^\star = 0$, *then* $(\tilde{p} - \tilde{p}_{r,i})/\tilde{\mathscr{E}}_{k_1,i} \geq (\tilde{p} - \tilde{p}_{r,j})/\tilde{\mathscr{E}}_{k_1,j}$, *where* k_1 *is the index of the robot that visits site i.*

Proof The proof is similar to that of Lemma 12.4. □

Consider two sites i and j. Suppose that in an optimal solution, site i is visited by robot k_1 and the classification of site j is based on the initial sensing. Lemma 12.7 says that the performance gain obtained from further sensing normalized by robot k_1's motion energy cost should be higher for site i as compared to site j.

Lemma 12.8 *Consider two sites i and j. Let* $\tilde{\gamma}^\star$ *and* $\tilde{\eta}^\star$ *denote the optimal decision vectors. If* $\tilde{\gamma}_i^\star = 1$, $\tilde{\eta}_{k,j}^\star = 1$ *and* $\tilde{\mathscr{E}}_{k,i} \leq \tilde{\mathscr{E}}_{k,j}$, *then* $\tilde{p}_{h,i} - \tilde{p}_{h,j} \geq 0$.

Proof The proof is similar to that of Lemma 12.5. □

Consider two sites i and j. Suppose that in an optimal solution, the robots query the human about site i and have robot k visit site j. Lemma 12.8 says that in this case, if robot k's motion energy cost of the queried site is less than or equal to that of the visited site, then the human performance of the queried site should be greater than or equal to that of the visited site.

Lemma 12.9 *Consider two sites i and j and two robots k_1 and k_2. Let $\tilde{\gamma}^\star$ and $\tilde{\eta}^\star$ denote the optimal motion decision vectors. Suppose that $\tilde{\eta}^\star_{k_1,i} = 1$, $\tilde{\eta}^\star_{k_2,j} = 1$ and $\exists\ m \in \{1, \ldots, N\}$ such that $\tilde{\gamma}^\star_m = 0$ and $\sum_{k=1}^{K} \tilde{\eta}^\star_{k,m} = 0$. Then the following conditions must hold.*

(1) $\tilde{\mathscr{E}}_{k_1,i} \le \tilde{\mathscr{E}}_{k_1,j}$ or $\tilde{\mathscr{E}}_{k_2,i} \ge \tilde{\mathscr{E}}_{k_2,j}$;

(2) If $\tilde{\mathscr{E}}_{k_1,i} \le \tilde{\mathscr{E}}_{k_1,j}$, then $\tilde{\mathscr{E}}_{k_2,j} - \tilde{\mathscr{E}}_{k_2,i} \le \tilde{\mathscr{E}}_{k_1,j} - \tilde{\mathscr{E}}_{k_1,i}$;

(3) If $\tilde{\mathscr{E}}_{k_2,i} \ge \tilde{\mathscr{E}}_{k_2,j}$, then $\tilde{\mathscr{E}}_{k_1,i} - \tilde{\mathscr{E}}_{k_1,j} \le \tilde{\mathscr{E}}_{k_2,i} - \tilde{\mathscr{E}}_{k_2,j}$.

Proof (1) Suppose that $\tilde{\mathscr{E}}_{k_1,i} \ge \tilde{\mathscr{E}}_{k_1,j}$ and $\tilde{\mathscr{E}}_{k_2,i} \le \tilde{\mathscr{E}}_{k_2,j}$. We can let $\tilde{\eta}'_{k_1,i} = 0$, $\tilde{\eta}'_{k_1,j} = 1$, $\tilde{\eta}'_{k_2,i} = 1$ and $\tilde{\eta}'_{k_2,j} = 0$, which will give us the same objective function value but with a less motion energy consumption. The residual energy can be utilized such that $\tilde{\eta}'_{k_1,m} = \delta_{k_1}$ and $\tilde{\eta}'_{k_2,m} = \delta_{k_2}$, where δ_{k_1} and δ_{k_2} are small positive numbers. This constructed solution will be strictly better than the current optimal solution, which is a contradiction. Thus we must have $\tilde{\mathscr{E}}_{k_1,i} \le \tilde{\mathscr{E}}_{k_1,j}$ or $\tilde{\mathscr{E}}_{k_2,i} \ge \tilde{\mathscr{E}}_{k_2,j}$.

(2) and (3) If Condition (2) or (3) fails, we can construct a new feasible solution in a similar way as in the proof of Condition (1), which will be strictly better than the current optimal solution, resulting in a contradiction. □

Consider the case where there exists at least one site, for which the robots will rely on the initial sensing. The first part of Lemma 12.9 says that in this case, if two sites i and j are visited by two robots k_1 and k_2 respectively, then either it should be less costly for robot k_1 to visit site i as compared to site j or it should be less costly for robot k_2 to visit site j as compared to site i. Furthermore, the second part of the lemma says that if it is less costly for robot k_1 to visit site i as compared to site j ($\tilde{\mathscr{E}}_{k_1,i} \le \tilde{\mathscr{E}}_{k_1,j}$), then robot k_2's motion energy cost of visiting site j should not exceed that of site i by $\tilde{\mathscr{E}}_{k_1,j} - \tilde{\mathscr{E}}_{k_1,i}$, which can be thought of as the motion energy saving of robot k_1. The third part can be interpreted in a similar manner. This lemma basically shows that the robots' decisions should be efficient in terms of motion energy usage.

12.4 Numerical Results

In this section, we show the performance of our collaboration design for field exploration and target classification. We first summarize the results for a case where there is only one robot [4] to gain a more intuitive understanding of the optimal behavior that arises from our optimization framework. We then show the numerical results for the case of multiple robots. The optimization problems are solved with the MILP solver of MATLAB by using the collected MTurk data of Fig. 12.2.

12.4.1 Collaboration Between the Human Operator and One Robot [4]

Consider the case where there is only one robot in the field. In this case, both multi-robot formulations (problems (12.1) and (12.8)) reduce to the same optimization problem, which is shown as follows.

$$\begin{aligned} \max_{\gamma,\eta}.\quad & \gamma^T(p_h - p_r) + \eta^T(\tilde{p}\mathbf{1} - p_r) \\ \text{s.t.}\quad & \eta^T \mathscr{E} \leq \mathscr{E}_{\max}, \quad \mathbf{1}^T\gamma \leq M, \\ & \gamma + \eta \preceq \mathbf{1}, \\ & \gamma, \eta \in \{0, 1\}^N. \end{aligned} \tag{12.11}$$

In order to better understand the optimal solution, Fig. 12.3 shows an example of the optimal decisions for the case of 2000 sites, with 500 allowed questions and an energy budget equal to 25% of the total energy needed to visit all the sites. The optimal decision for each site is marked based on solving problem (12.11). Interesting behavior emerges as can be seen. For instance, we can observe that there are clear separations between different decisions. The clearest patterns are two transition points that mark when the robot asks the human operator for help, as shown with

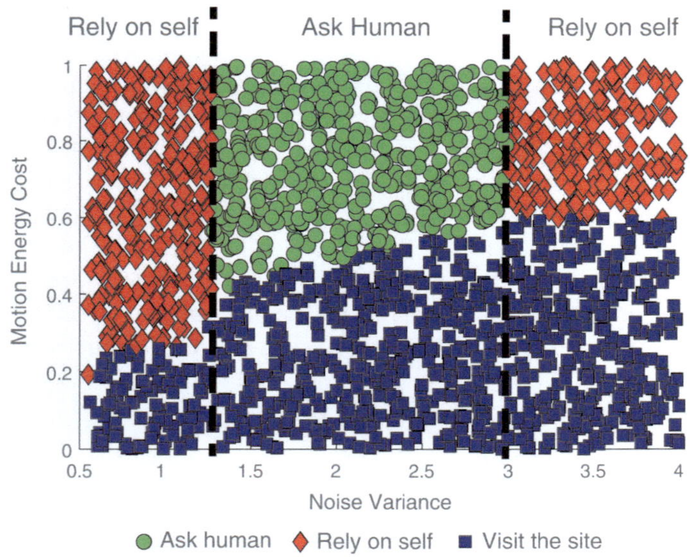

Fig. 12.3 An example of the optimal decisions with 2000 sites, 500 questions, and an energy budget of 25% of the total energy needed to visit all the sites. In this example, the collaboration is between one operator and one robot. This result is from our previous work [4] (© 2015 IEEE. Reprinted, with permission, from [4])

the dashed vertical lines in Fig. 12.3. Basically, the figure suggests that the robot should not bug the human if the variance is smaller than a threshold or bigger than another threshold, independent of the motion cost of a site. This makes sense as the robot itself will perform well for low variances and humans do not perform well for high variances, suggesting an optimal query range. Furthermore, it shows that the robot is more willing to spend motion energy if the sensing of a site has higher noise variance. However, the robot in general only visits the sites where the energy cost is not too high and relies more on itself for the sites with both high variance and high energy cost.

In the following part, we show the energy and bandwidth savings of our proposed approach as compared to a benchmark methodology where human collaboration is not fully optimized. In the benchmark approach, the robot optimizes its given energy budget to best explore the field based on site variances, i.e., it chooses the sites that maximize the sum of noise variances. It then randomly chooses from the remaining sites to ask the human operator, given the total number of questions. In other words, the robot optimizes its energy usage without any knowledge of the human's performance.

12.4.1.1 Energy Saving

Table 12.1 shows the amount of motion energy the robot saves for achieving a desired probability of correct classification by using our approach as compared to the benchmark. The first column shows the desired average probability of correct classification and the second column shows the percentage reduction of the needed energy by using our proposed approach as compared to the benchmark method. In this case, there is a total of $N = 10$ sites and $M = 4$ given queries. The noise variance of each site is randomly assigned from the interval $[0.55, 4]$. $\tilde{p}$ is set to 0.896, which is the best achievable robot performance based on Fig. 12.2. The motion energy cost to visit each site is also assigned randomly and the total given energy budget is taken to be a percentage of the total energy required to visit all the sites. It can be seen that the robot can reduce its energy consumption considerably by properly taking advantage of its collaboration. For instance, it can achieve an average probability of correct

Table 12.1 Energy saving as compared to the benchmark in the one-operator-one-robot case (© 2015 IEEE. Reprinted, with permission, from [4])

Desired ave. correct classification prob.	Energy saving (in %)
0.7	66.67 %
0.75	44.30 %
0.8	27.83 %
0.85	6.3 %
0.9	0.71 %
0.915	Inf

classification of 0.7 with 66.67% less energy consumption. The term "Inf" denotes the cases where the benchmark cannot simply achieve the given target probability of correct classification.

12.4.1.2 Bandwidth Saving

Next, we show explicitly how our proposed approach can also result in a considerable communication bandwidth saving by reducing the number of questions. More specifically, consider the cases with "large bandwidth" and "zero bandwidth". In the first case, the robot has no communication limitation and can probe the human with as many questions as it wants to (10 in this case). In the latter, no access to a human operator is available and thus the robot has to rely on itself to classify the gathered data after it surveys the field. Figure 12.4 compares the performance of our proposed approach with these two cases. The robot is given an energy budget of 30% of the total energy needed to visit all the sites.

As expected, the case of "no bandwidth" performs considerably poorly as the robot could not seek human help in classification. On the other hand, the case of "large bandwidth" performs considerably well as the robot can ask for the human operator's help as many times as it wants. This, however, comes at a cost of excessive communication and thus a high bandwidth usage.[3] It can be seen that our proposed approach can achieve a performance very close to this upper bound with a much less bandwidth usage. For instance, we can see that by asking only 6 questions (40% bandwidth reduction), the robot can achieve an average probability of correct classification of 0.888, which is only 4.3% less than the case of large bandwidth (0.928 in this case).

Table 12.2 shows the amount of bandwidth the robot can save by using our approach, when trying to achieve a desired average probability of correct classification. The first column shows the desired average probability of correct classification while the second column shows the percentage reduction of the needed bandwidth by using our proposed approach as compared to the benchmark. In this case, the robot is given an energy budget of 30% of the total energy needed to visit all the sites. It can be seen that the robot can reduce its bandwidth consumption considerably. For instance, it can achieve an average probability of correct classification of 0.75 with 48.61% less bandwidth usage.

12.4.2 Predetermined Site Allocation

In this section, we numerically demonstrate the efficacy of our approach for the one-operator-multi-robot collaborative scenario when the allocation of the sites to the robots is predetermined. We first show an interesting pattern that characterizes

[3]Bandwidth usage is taken proportional to the number of questions.

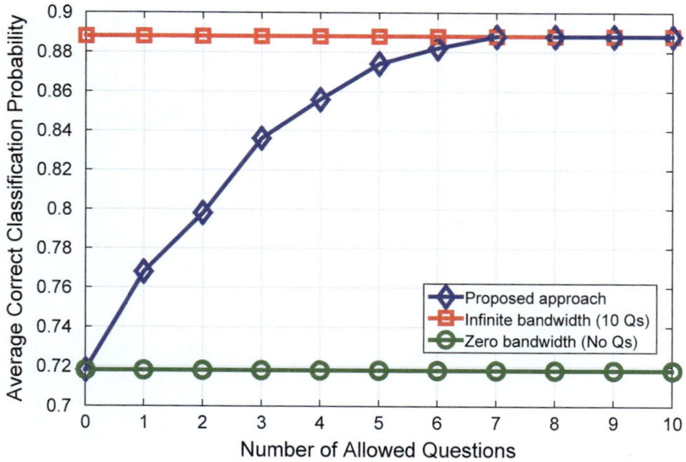

Fig. 12.4 Average probability of correct classification in the one-operator-one-robot collaboration as a function of the total number of given queries. In this example, there is a total of 10 sites and the given motion energy budget is 30% of what is needed to visit all the sites (© 2015 IEEE. Reprinted, with permission, from [4])

Table 12.2 Bandwidth saving as compared to the benchmark in a one-operator-one-robot case (© 2015 IEEE. Reprinted, with permission, from [4])

Desired ave. correct classification prob.	Bandwidth saving (in %)
0.7	37.04 %
0.75	48.61 %
0.8	33.18 %
0.85	7.33 %
0.875	Inf

the conditions under which the robots will visit the sites and ask for human's help respectively. We then illustrate how our approach plans the collaborative operation by showing an example solution to problem (12.1), after which we conduct numerical evaluations to demonstrate how our proposed approach can save resources significantly.

12.4.2.1 Patterns of Optimal Decisions

We solve problem (12.1) with two robots, where each robot is assigned to 1000 sites. There is a total of 500 given queries and the energy budget is taken as 25% of what is needed to visit all the sites in the preassigned set for each robot. The noise variance of each site is randomly assigned from the interval [0.55, 4]. $\tilde{p}$ is set to 0.896, which is the best achievable robot performance based on Fig. 12.2. The motion energy cost to visit each site is also assigned randomly.

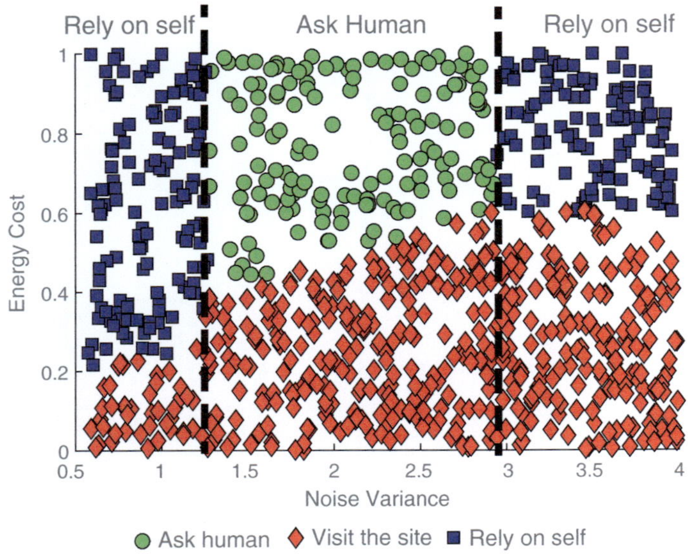

Fig. 12.5 An example of the optimal decisions with two robots. Each robot is assigned to 1000 sites and given an energy budget of 25% of the total energy needed to visit all its preassigned sites. The two robots share a total number of 500 questions. The figure shows the decisions of robot 1

Figure 12.5 shows the optimal decisions of the first robot with the above parameters. Green disks represent the decision of asking for human help, red diamonds represent the decision of visiting the site, and blue squares represent the decision of relying on the initial sensing. It can be seen that the optimal behavior of a robot in the one-operator-multi-robot setting is very similar to that of the one-operator-one-robot case. More specifically, the robot will only query the human operator about sites where the sites' sensing variance is not too low or too high. The robot is more willing to spend motion energy to move to sites with high noise variance for further sensing as long as the energy cost is not too high. The optimal decisions of the second robot have a similar pattern.

To better understand the impact of noise variance and motion energy cost on the optimal decisions, we conduct the following analysis. From Fig. 12.2, we can see that there is a noise variance range within which it is very beneficial to query the human operator ([1.5, 2.5]). Thus the distribution of the values of the noise variance will have a considerable impact on the optimal decisions. For instance, suppose that the noise variance of the sites is drawn from a Gaussian distribution that is mainly concentrated in the interval [1.5, 2.5]. Then, the robot can have a good gain from asking for help if its motion budget is not too large. To further understand these impacts, we perform simulations with two robots, each assigned to 200 sites. We vary the distribution of the noise variance and the given motion energy budgets for the two robots. Figure 12.6 shows the probability density functions (PDFs) of the two noise variance distributions that we will use in the simulations. The first (left)

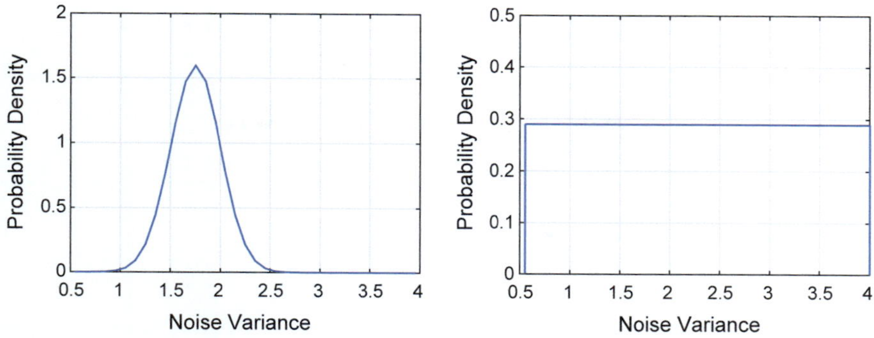

Fig. 12.6 (*left*) The PDF of the truncated Gaussian distribution. (*right*) The PDF of a uniform distribution. Both PDFs have the support of [0.55, 4] and are used to generate noise variances in the simulations

is a truncated Gaussian distribution with mean 1.75 and variance 0.25. The values of the noise variance are truncated so that they stay inside the interval [0.55, 4]. The noise variance produced from this distribution will be mainly within the range where it is most beneficial to query the human operator based on Fig. 12.2. The second distribution is a uniform distribution over the interval [0.55, 4].

Table 12.3 shows the average number of sites asked and visited by each robot. The noise variance of the sites of robot 1 is drawn from the uniform distribution while the noise variance of the sites of robot 2 is drawn from the truncated Gaussian distribution. There is a total of 100 allowed queries. Both robots are given 25% of what is needed to visit all sites from their respective sets. The motion energy cost to visit each site is assigned randomly. The results are averaged over multiple simulations so that the analysis is less dependent on the specific realizations of the two distributions. It can be seen that the average number of sites asked by robot 2 is significantly greater than that by robot 1. This is because the noise variance of the sites of robot 2 mainly lie within the range where it is more beneficial to ask for help. The average number of visited sites is almost the same for both robots as they are given the same energy budget in terms of the percentage of the total energy required to visit all the sites in their respective sets. Thus robot 1 has to rely more on the initial sensing for classification. As we increase the total number of allowed questions, we expect the difference between the number of questions used by the two robots to decrease.

Next, we fix the noise variance distribution and study how different motion energy budgets affect the optimal decisions. Table 12.4 shows the average number of sites asked and visited by each robot. The noise variance of the sites of both robots are drawn from the uniform distribution. There is a total of 100 allowed queries. In terms of energy budget, robot 1 and robot 2 are given 20 and 40% of what is needed to visit all the sites in their respective sets. It can be seen that the average number of queried sites by robot 1 is greater than that of robot 2. This makes sense since the number of visited sites by robot 1 is smaller due to the smaller energy budget.

Table 12.3 Average number of sites asked and visited by each robot. The noise variances for robot 1 and robot 2 are drawn from the uniform distribution and the truncated Gaussian distribution, respectively (see Fig. 12.6). Each robot is assigned to 200 sites and there is a total of 100 allowed queries. Each robot is given an energy budget of 25% of what is needed to visit all the sites in its respective set

	Ave. # of sites visited	Ave. # of sites asked
Robot 1	95.65	34.4
Robot 2	96.3	65.6

Table 12.4 Average number of sites asked and visited by each robot. The noise variances for both robots are drawn from the uniform distribution. Each robot is assigned to 200 sites and there is a total of 100 allowed queries. Robot 1 is given an energy budget of 20% of what is needed to visit all the sites in its set while robot 2 is given an energy budget of 40% of what is needed to visit all the sites in its set

	Ave. # of sites visited	Ave. # of sites asked
Robot 1	85.2	59.5
Robot 2	119.6	40.5

12.4.2.2 Example Solution

In this section, we study a sample solution to problem (12.1). We consider the case with two robots, each assigned to five sites. The noise variance of each site is randomly assigned from the interval [0.55, 4]. $\tilde{p}$ is set to 0.896. The motion energy cost to visit each site is assigned randomly. There is a total of 3 allowed questions and each robot is given an energy budget of 25% of what is needed to visit all the sites in their respectively predetermined sets. The planning results are summarized in Table 12.5. The upper half and lower half of the table show the results for the two robots, respectively. The first column shows the indices of the sites. The second column indicates whether a site is visited. The third column indicates whether a site is selected to query the human operator. The fourth and fifth columns show the performance gains associated with asking for help and visiting the sites respectively ($(p_{h,i} - p_{r,i})$ and $(\tilde{p} - p_{r,i})$). The sixth column shows the motion energy costs for the sites.

For each robot's respective set of sites, it can be seen that for the sites selected for visit, their corresponding performance gains normalized by energy cost are the highest among all unqueried sites, which is consistent with Lemma 12.4. As for sites selected to query the human operator, we can see that the performance gain (5th column) of these sites obtained from asking the human are the highest among all the sites not selected for further sensing (marked by a gray color), which is consistent with Lemma 12.3.

Table 12.5 Example solution to problem (12.1) with predetermined site allocation. There are two robots, each assigned to five sites. The robots share a total of three allowed questions. Each robot is given an energy budget of 25% of what is required to visit all the sites in its own set

	Site Index	Selected for Visit	Selected for Query	Energy Cost	Performance Gain of Query	Performance Gain of Visit
Robot 1	1	0	**1**	0.6474	**0.3218**	0.3365
	2	**1**	0	**0.1434**	0.2917	**0.3960**
	3	**1**	0	**0.0227**	0.1728	**0.3960**
	4	**1**	0	**0.1887**	0.3511	**0.3960**
	5	0	**1**	0.5020	**0.3402**	0.3960
Robot 2	6	**1**	0	**0.2067**	0.2138	**0.3960**
	7	0	**1**	0.8360	**0.3043**	0.3960
	8	0	0	0.6730	0.1712	0.1460
	9	**1**	0	**0.0168**	0.3497	**0.3960**
	10	0	0	0.4823	0.1795	0.1460

12.4.2.3 Energy Saving

Table 12.6 shows the average amount of motion energy the robots save by using our approach when aiming to achieve a given target probability of correct classification. More specifically, the first column shows the target average probability of correct classification while the second column shows the percentage reduction in the average needed energy when using our approach as compared to the benchmark method. In the benchmark method, each robot selects the sites to visit by maximizing the total sum of variances at the sites, after which random sites are selected from the aggregated unvisited ones to query the human operator. In other words, the robots do not have the knowledge of human visual performance but know how their own performance is related to the sensing variance. In the example of Table 12.6, there is a total of four robots, each assigned to 10 sites. The robots share a total number of 10 given queries. The robots' energy budgets are the same as each other in terms of the percentage of the total energy needed to visit all the sites in their respective sets. The noise variance of each site is randomly assigned from the interval $[0.55, 4]$. $\tilde{p}$ is set to 0.896. The motion energy cost to visit each site is also assigned randomly. It can be seen that the robots can reduce the energy consumption considerably by taking advantage of the knowledge of human performance and properly optimizing the collaboration accordingly. For instance, an average probability of correct classification of 0.65 is achieved with 57.14% less energy consumption.

12.4.2.4 Bandwidth Saving

We next show how our approach can also result in a considerable communication bandwidth saving by reducing the number of questions while still achieving the

Table 12.6 Energy saving as compared to the benchmark in the one-operator-multi-robot setting with preassigned sites. In this case, there are four robots, each assigned 10 sites and the robots share a total of 10 questions

Desired ave. correct classification prob.	Energy saving (in %)
0.65	57.14 %
0.7	27.78 %
0.75	27.03 %
0.8	18.75 %
0.85	10.20 %
0.9	Inf

desired performance. We consider the cases with "large bandwidth" and "zero bandwidth" as described in Sect. 12.4.1.2. Figure 12.7 compares the performance of our proposed approach with these two cases. As expected, the case of "no bandwidth" performs considerably poorly as the robots could not seek human help in classification. On the other hand, the case of "large bandwidth" performs considerably well as the robots can ask the human operator as many questions as they want to. It can be seen that our proposed approach can achieve a performance very close to this upper bound with a much less bandwidth usage. For instance, we can see that by asking only 25 questions (37.5% bandwidth reduction), the robot can achieve an average probability of correct classification of 0.817, which is only 2.4% less than the case of large bandwidth (0.835 in this case).

Table 12.7 shows the amount of bandwidth usage the robots can save by using our approach, when trying to achieve a desired average probability of correct clas-

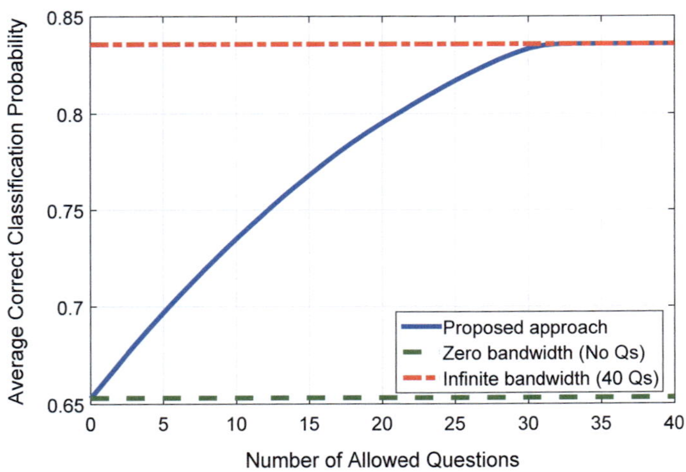

Fig. 12.7 Average probability of correct classification in a human–robot collaboration as a function of the total number of given queries. In this example, there are four robots, each assigned to 10 sites. Each robot is given a motion energy budget equal to 10% of what is needed to visit all the sites in its assigned set

Table 12.7 Bandwidth saving as compared to the benchmark in the one-operator-multi-robot setting with preassigned sites. In this case, there are four robots. Each robot is assigned to 10 sites and given 10% of what is needed to visit all the sites in its preassigned set

Desired ave. correct classification prob.	Bandwidth saving (in %)
0.65	100
0.7	33.75
0.75	22.93
0.8	14.29

sification. More specifically, the first column shows the target average probability of correct classification and the second column shows the percentage reduction of the needed bandwidth by using our approach as compared to the benchmark. In this case, each robot is given an energy budget of 10% of the total energy needed to visit all the sites in its set. It can be seen that the robot can reduce its bandwidth consumption considerably. For instance, it can achieve an average probability of correct classification of 0.7 with 33.75% less bandwidth usage.

12.4.3 Optimized Site Allocation

In this section, we conduct numerical evaluations when our approach also optimizes site allocation, as shown in problem (12.8). We first show patterns of the optimal decisions and illustrate how our approach plans the collaborative operation by showing a sample solution to problem (12.8). We then numerically demonstrate that our proposed approach can save resources significantly.

12.4.3.1 Patterns of Optimal Decisions

We solve problem (12.8) with two robots and a total of 2000 sites. There is a total of 500 given queries. The energy budget is taken as 12.5% of what is needed to visit all the sites for each robot. The noise variance of each site is randomly generated from the interval [0.55, 4]. $\tilde{p}$ is set to 0.896. The motion energy cost to visit each site is also assigned randomly for each robot. The pattern of optimal decisions in terms of asking for human help, visiting a site and relying on the initial sensing for any one of the two robots in this scenario is very similar to those shown in Figs. 12.3 and 12.5.

Here, we show how the optimal decisions are related to the motion energy costs of visiting the sites for the two robots. Figures 12.8 and 12.9 show when the robots will visit a site, when they will query the human operator and when they will rely on the initial sensing. Green disks represent asking for human help, red diamonds indicate that the site is visited by robot 1, yellow triangles indicate that the site is visited by robot 2, and blue squares represent the decision of relying on the initial

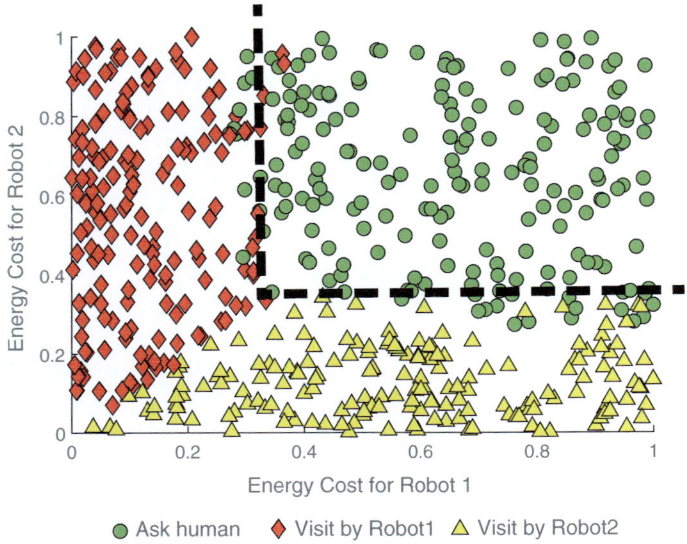

Fig. 12.8 An example of the optimal decisions with two robots with 2000 sites, 500 questions. Each robot's energy budget is 12.5% of the total energy needed for it to visit all the sites

sensing. It can be seen that the two robots select sites that do not require too much motion energy to visit. Note that it may cost the robots different amount of motion energy to visit the same site. For a site that is costly to visit for both robots (sites that reside in the top-right region of Figs. 12.8 and 12.9), the robots will either query the human operator or rely on the initial sensing depending on the noise variance of the site. We note that in Fig. 12.9, there is a number of sites for which the robots will rely on the initial sensing even though the costs of visiting them is not high for at least one of the robots. This is because the noise variances associated with these sites are already low, eliminating the need for further sensing.

12.4.3.2 Example Solution

In this section, we show a sample numerical solution to problem (12.8). We consider the case with two robots and 10 sites. The noise variance of each site is randomly assigned from the interval [0.55, 4]. $\tilde{p}$ is set to 0.896. The motion energy cost to visit each site is also assigned randomly for each robot. There is a total of two allowed questions and each robot is given 12.5% of what is needed for it to visit all the sites. The planning results are summarized in Table 12.8. The first column shows the indices of the sites. The second column indicates whether a site is visited by robot 1 and the third column indicates whether a site is visited by robot 2. The fourth column indicates whether a site is selected to query the human operator. The fifth and sixth columns show the motion energy costs for visiting the sites for robot 1 and

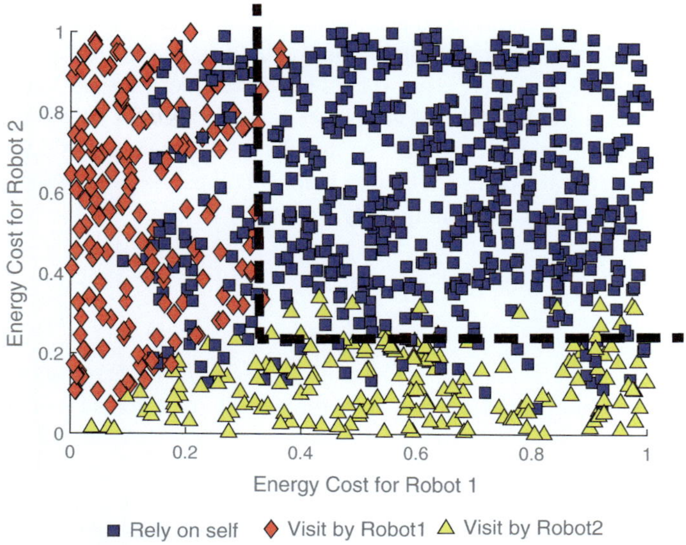

Fig. 12.9 An example of the optimal decisions with two robots with 2000 sites, 500 questions and each robot's energy budget is 12.5% of the total energy needed for it to visit all the sites

Table 12.8 An example solution to problem (12.8) with an optimized site allocation. There are two robots and 10 sites. There is a total of two allowed questions and each robot is given 12.5% of what is required for it to visit all the sites

Site index	Visited by robot 1	Visited by robot 2	Selected for query	Energy cost for robot 1	Energy cost for robot 2	Performance gain of query	Performance gain of visit
1	0	0	0	0.9469	0.9705	0.2879	0.3960
2	0	0	**1**	0.8077	0.8802	**0.3113**	0.3960
3	**1**	0	0	**0.1245**	0.6473	0.1963	**0.2028**
4	0	**1**	0	0.9662	**0.0892**	0.1885	**0.3960**
5	0	0	0	0.6030	0.6419	0.2852	0.3960
6	0	**1**	0	0.6766	**0.0732**	0.3390	**0.3960**
7	0	0	0	0.2427	0.2997	0.1695	0.1460
8	0	0	**1**	0.9348	0.6253	**0.3381**	0.3960
9	0	**1**	0	0.5331	**0.0288**	0.1755	**0.1460**
10	**1**	0	0	**0.0088**	0.1586	0.1807	**0.1860**

robot 2, respectively. The seventh and eighth columns show the performance gains associated with asking for human help and visiting the sites respectively ($(\tilde{p}_{h,i} - \tilde{p}_{r,i})$ and $(\tilde{p} - \tilde{p}_{r,i})$).

We can see that if a robot visits a particular site, the performance gain from visit normalized by this robot's energy cost associated with this site is greater than or equal to that of any other unvisited site, which is consistent with Lemma 12.7. We note that although the performance gain achievable from visiting site 1 is larger than those of sites 3, 9, and 10, site 1 is not visited while the latter three are visited. This is because the motion energy required to visit site 1 is too high for both robots. As for the sites selected for query, it can be seen that these sites have a significantly larger performance gain from asking for human help as compared to any other unvisited site, which is consistent with Lemma 12.6.

12.4.3.3 Energy Saving

Table 12.9 shows the amount of average motion energy the robots save by using our approach when aiming to achieve a desired probability of correct classification. The first column shows the desired average probability of correct classification while the second column shows the percentage reduction of the needed energy by using our proposed approach as compared to a benchmark method similar to the one described in the previous section. In the benchmark method, the robots select the sites to visit by maximizing the total sum of sensing variances under the given energy budgets, after which random sites are selected from the remaining ones to query the human operator. In this case, there is a total of four robots with a total of 40 sites. The robots are given 10 allowed queries. The robots' energy budgets are the same as each other in terms of the percentage of the total energy needed to visit all the sites. The noise variance of each site is randomly assigned from the interval [0.55, 4]. $\tilde{p}$ is set to 0.896. The motion energy cost to visit each site is also assigned randomly for each robot. It can be seen that the robots can reduce their average energy consumption considerably by properly taking advantage of the knowledge of human visual performance. For instance, an average probability of correct classification of 0.7 can be achieved with a 40.00% less energy consumption.

Table 12.9 Energy saving as compared to the case of no proper collaboration in the one-operator-multi-robot setting with an optimized site allocation. In this case, there are four robots with 40 sites and there is a total of 10 questions

Desired ave. correct classification prob.	Energy saving (in %)
0.65	75.00 %
0.7	40.00 %
0.75	25.00 %
0.8	18.18 %
0.85	10.53 %
0.9	Inf

12.4.3.4 Bandwidth Saving

We next show how our approach can also result in a considerable communication bandwidth saving by reducing the number of questions needed while still providing a good performance. We consider the cases of "large bandwidth" and "zero bandwidth" as described in Sect. 12.4.1.2. Figure 12.10 compares the performance of our proposed approach with these two cases. We can see that by asking only 20 questions (50% bandwidth reduction), the robot can achieve an average probability of correct classification of 0.857, which is only 0.8% less than the case of large bandwidth (0.864 in this case).

Table 12.10 shows the amount of bandwidth that the robots can save by using our approach, when trying to achieve a desired average probability of correct classification. The first column shows the target average probability of correct classification and the second column shows the percentage reduction of the needed bandwidth by using our proposed approach as compared to the benchmark. In this case, each robot is given an energy budget of 10% of the total energy needed for it to visit all the sites. It can be seen that by properly designing the collaboration, we can reduce the bandwidth consumption considerably. For instance, an average probability of correct classification of 0.75 can be achieved with a 48.57% less bandwidth usage.

Overall, we can see that by using our proposed collaboration approach, we can reduce the motion energy and bandwidth consumptions considerably.

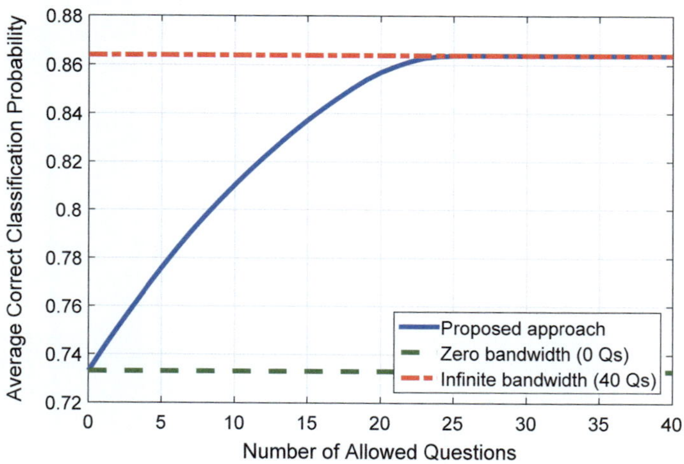

Fig. 12.10 Average probability of correct classification in the one-operator-multi-robot collaboration as a function of the total number of given queries. In this example, there are four robots with 40 sites and each robot's motion energy budget is 10% of what is needed for it to visit all the sites

Table 12.10 Bandwidth saving as compared to the benchmark in the one-operator-multi-robot setting with an optimized site allocation. In this case, there are four robots with 40 sites and each robot's motion energy budget is 10% of what is needed for it to visit all the sites

Desired ave. correct classification prob.	Bandwidth saving (in %)
0.75	48.57 %
0.8	25.23 %
0.85	Inf

12.5 Conclusions

In this chapter, we extended our previously proposed paradigm for human–robot collaboration, namely, "when to ask for human's help", to the case of multiple robots. More specifically, we considered a robotic field exploration and target classification task where a number of robots have a limited communication with a remote human operator and constrained motion energy budgets. The visual performance of the human operator, however, is not perfect and is given via a probabilistic modeling from [4]. We started with the case where the sites, which contain the objects to be classified, are preassigned to the robots in order to understand optimum allocation of other resources. We then extended our analysis to further include the optimization of site allocation. Simulation results confirm that considerable resource savings can be achieved using our proposed approach. Overall, our framework allows the robots to collaboratively and optimally decide on when to ask humans for help, when to rely on the initial sensing, and when to gather more information from the field.

Acknowledgements This work was supported in part by NSF NeTS award #1321171 and NSF RI award #1619376.

References

1. Boyd S, Vandenberghe L (2009) Convex Optimization. Cambridge university press, Cambridge
2. Branson S, Wah C, Schroff F, Babenko B, Welinder P, Perona P, Belongie S (2010) Visual recognition with humans in the loop. In: Proceedings of the European conference on computer vision, pp 438–451
3. Burke J, Murphy R, Coovert M, Riddle D (2004) Moonlight in miami: field study of human-robot interaction in the context of an urban search and rescue disaster response training exercise. Hum-Comput Interact 19(1–2):85–116
4. Cai H, Mostofi Y (2015) To ask or not to ask: a foundation for the optimization of human-robot collaborations. In: Proceedings of the American control conference, pp 440–446
5. Cai H, Mostofi Y (2016) Asking for help with the right question by predicting human performance. In: Proceedings of robotics: science and systems
6. Cai H, Mostofi Y (2016) A human-robot collaborative traveling salesman problem: robotic site inspection with human assistance. In: Proceedings of the American control conference

7. Cao M, Stewart A, Leonard N (2008) Integrating human and robot decision-making dynamics with feedback: models and convergence analysis. In: Proceedings of the IEEE conference on decision and control, pp 1127–1132
8. Chipalkatty R (2012) Human-in-the-loop control for cooperative human-robot tasks. Ph.D. thesis, Georgia Institute of Technology
9. Dantzig GB (1957) Discrete-variable extremum problems. Oper Res 5(2):266–288
10. Dias MB, Kannan B, Browning B, Jones E, Argall B, Dias MF, Zinck M, Veloso M, Stentz A (2008) Sliding autonomy for peer-to-peer human-robot teams. In: Proceedings of the international conference on intelligent autonomous systems, pp 332–341
11. Martello S, Toth P (1990) Knapsack problems: algorithms and computer implementations. Wiley, New York
12. Meriçli Ç, Veloso M, Akin H (2011) Task refinement for autonomous robots using complementary corrective human feedback. Int J Adv Robot Syst 8(2):68
13. Srivastava V (2012) Stochastic search and surveillance strategies for mixed human-robot teams. Ph.D. thesis, University of California, Santa Barbara
14. Stewart A, Cao M, Nedic A, Tomlin D, Leonard N (2012) Towards human-robot teams: model-based analysis of human decision making in two-alternative choice tasks with social feedback. Proc IEEE 100(3):751–775
15. Toris R, Suay H, Chernova S (2012) A practical comparison of three robot learning from demonstration algorithms. In: Proceedings of the ACM/IEEE International Conference on Human-Robot Interaction, pp 261–262

Chapter 13
Human-Collaborative Schemes in the Motion Control of Single and Multiple Mobile Robots

Antonio Franchi

13.1 Introduction

A mobile robot is primarily a machine that is able to perceive the surrounding environment and move in it in a safe and effective way for the humans, for itself, and for the given objective. The ability to automatically control its own motion, at both the higher level (guidance, planning) and the lower one (navigation and servoing), constitutes one of the fundamental building blocks of the sought robot autonomy. However, robots, which in general do not exist and operate isolated from humans, can actually improve their effectiveness in achieving the given task by means of a well-designed human-collaborative control schemes. Analyzing and properly designing the way to achieve an optimal human collaboration is therefore a crucial aspect in order to attain a fully mature robotic system that is able to operate in a complex and real world.

In this chapter we show and compare several ways to effectively design the collaborative control. We do so by first introducing a simple yet descriptive model for the robots and the human interfaces. We then propose a list of what we consider fundamental axes for a proper classification of the collaborative scheme, namely the

- *physical domain of the robots*, e.g., whether they are operating and moving on the ground or in the air;
- *degree of autonomy from the human operator*, i.e., at which level and with which frequency the human operator is involved in influencing the robot motion, in particular, in this chapter we mainly focus on the so-called 'shared control' case. Within this case we further specialize our taxonomy depending on the *presence of a time horizon in the reference* that is sent by the operator to the robots.

A. Franchi (✉)
CNRS, LAAS, 7 Avenue du Colonel Roche, 31400 Toulouse, France
e-mail: antonio.franchi@laas.fr

A. Franchi
Univ de Toulouse, LAAS, 31400 Toulouse, France

Y. Wang and F. Zhang (eds.), *Trends in Control and Decision-Making for Human–Robot Collaboration Systems*, DOI 10.1007/978-3-319-40533-9_13

- *kind force interaction with the operator*, i.e., whether the collaboration involves or not mechanical interaction and haptics and in which direction (forward-force, feedback-force (bilateral), unilateral);
- *near-operation versus the teleoperation*;
- *amount of physical interaction with the environment*;
- *use of onboard sensors*;

The proposed taxonomy is then further specialized in the case of the use of multiple robots, for which we introduce additional axes, namely

- *level of centralization*;
- use of a scheme of the kind *master–leader–followers*;
- use of a scheme of the kind *formation-orthogonal*;
- use of a scheme of the kind *global property preservation*;
- *ability to perform cooperative physical maneuvers.*

The common denominator of all the axes presented in this chapter is the presence of a human operator in the control loop. In fact, fully autonomous control algorithms are not in general easy to extend in order to cope with (and to exploit) the presence of a human operator. The main goal of the chapter is therefore to introduce the reader to the several ways to effectively include human operators in the control of both single and multiple aerial mobile robots, and to provide a first-level analysis on these systems. A particular emphasis is set on the aerial robots case, which is at date not yet fully explored and understood, and is among the most challenging cases. Nevertheless, all the presented aerial robotics schemes are easily transferable to many other mobile robots as well.

13.2 Modeling of the Robot and the Interactions

In this section we provide a formal description of the robot and its interactions. The main goal is to introduce a standard model and a nomenclature that is convenient for the description of human-collaborative schemes like the ones that we illustrate in this chapter. For the sake of completeness, the notation introduced goes beyond the examples included in the taxonomy proposed in the next sections.

13.2.1 Mobile Robot

A *mobile robot* can be viewed as a mechanical system possessing some degrees of freedom (DoFs) that is equipped with actuators, sensors, a processing unit, a communication module, and a mathematical model of itself and the surrounding environment. The combination of all these components is used by the robot to control its own DoFs and, as a byproduct, to modify the state of the surrounding environment,

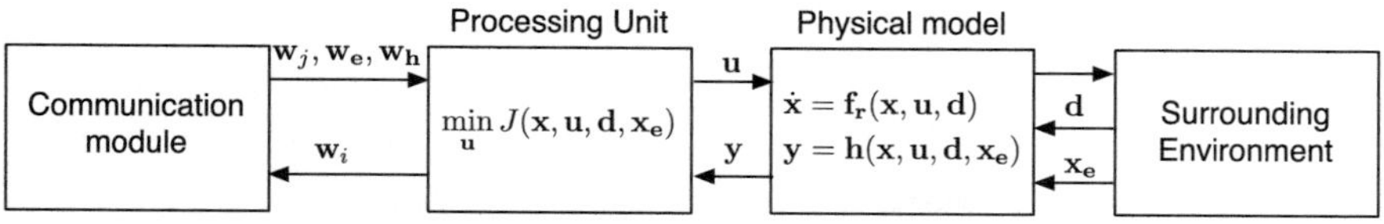

Fig. 13.1 High-level model of the i-th mobile robot

in the presence of external disturbances and uncertainties. We refer to Fig. 13.1 for a high-level model of a mobile robot physically interacting and communicating with the surrounding environment.

The *configuration* of the $n_q \in \mathbb{N}$ DoFs of the robot is represented by a vector $\mathbf{q} \in \mathbb{R}^{n_q}$. The physical model of the robots is implicitly represented by a differential equation

$$\dot{\mathbf{x}} = \mathbf{f_r}(\mathbf{x}, \mathbf{u}, \mathbf{d}), \tag{13.1}$$

where $\mathbf{x} \in \mathbb{R}^{n_x}$ is the *physical state*, $\mathbf{u} \in \mathbb{R}^{n_u}$ is the *control input* (a signal that the processing unit can change arbitrarily and drives the robot actuators), and $\mathbf{d} \in \mathbb{R}^{n_d}$ is the *external input* (a signal that is out of control to the robot and can represent both the interaction with the environment and the inaccuracies of the model). The vector $\mathbf{x}$ includes the configuration $\mathbf{q}$ and may include also its derivative, i.e., $\mathbf{x} = (\mathbf{q}^T \; \dot{\mathbf{q}}^T)^T$.

In presence of more than one robot we assume an index $i \in \{1, \ldots, N\}$ being associated to each robot, where N is the total number of robots. We shall refer to the robot associated to i as the 'i-th robot' or the 'robot i.' The quantities corresponding to the i-th robot are added the subscript i, e.g., $\mathbf{x}$ becomes $\mathbf{x}_i$, $\mathbf{u}$ becomes $\mathbf{u}_i$, and so on so forth.

The sensors of a robot typically provide a partial information about the physical state, the external input, and the surrounding environment. This information is in general represented by the measurement function

$$\mathbf{y} = \mathbf{h}(\mathbf{x}, \mathbf{u}, \mathbf{d}, \mathbf{x_e}), \tag{13.2}$$

where $\mathbf{y} \in \mathbb{R}^{n_y}$ is the *measurement* and $\mathbf{x_e} \in \mathbb{R}^{n_{xe}}$ is the *environment state*, which may include external objects, other robots as seen by the robot i, and a nearby human collaborator.

The *communication module* provides the robots additional signals that can be used by its processing unit to achieve their mission. A signal that comes from another robot j is denoted with $\mathbf{w}_j \in \mathbb{R}^{n_{wj}}$, where $j \in \mathcal{N}_i$, being $\mathcal{N}_i \subseteq \{1, \ldots, N\}$ the set of robots that communicate with the robot i (*communication neighbors*). A signal that comes from the human–robot interface is instead denoted with $\mathbf{w_h} \in \mathbb{R}^{n_{wh}}$. A signal could come also from other sources (e.g., a smart environment) in that case we denote it with $\mathbf{w_e} \in \mathbb{R}^{n_{we}}$. The communication module is also in charge of disseminating to each robot j, with $j \in \mathcal{N}_i$, the signal $\mathbf{w}_i$ coming from the robot processing unit.

The *processing unit* implements a robot control algorithm that chooses $\mathbf{u}$ in order to, typically, minimize an objective function

$$J(\mathbf{x}, \mathbf{u}, \mathbf{d}, \mathbf{x_e}, \{\mathbf{w}_j \mid j \in \mathcal{N}_i\}, \mathbf{w_h}, \mathbf{w_e}). \tag{13.3}$$

In order to do so, the robot control algorithm can rely on the following information:

- a model of the robot (e.g., (13.1)) and a model of the environment
- the measurement $\mathbf{y}$
- the communicated signals $\{\mathbf{w}_j \mid j \in \mathcal{N}_i\}$, $\mathbf{w_h}$, and $\mathbf{w_e}$,

which constitute the input to the control algorithm. The output of the control algorithm is $\mathbf{u}$, i.e., the control input to the actuators.

13.2.2 Communication Infrastructure

A network of robots is a group of robots whose processing units communicate through a communication infrastructure. The simplest way to model the overall communication state is to consider a graph $\mathcal{G} = \{\mathcal{I}, \mathcal{E}\}$, where the set of vertexes $\mathcal{I} = \{1, \ldots, N\}$ represents the robots in the network, and $\mathcal{E} \subset \mathcal{I} \times \mathcal{I}$ is the set of edges for which $(i, j) \in \mathcal{E}$ if and only if $j \in \mathcal{N}_i$, where $\mathcal{N}_i$ has been previously defined as the set of communication neighbors of the robot i.

A graph has an algebraic representation given by the adjacency and the incidence matrixes, denoted with $\mathbf{A}$ and $\mathbf{E}$, respectively (see, e.g., [1, 2] for the formal definitions of these and all the algebraic graph concepts mentioned in this chapter). Another matrix that plays an important role in the control of networks of robots is the Laplacian matrix $\mathbf{L} = \mathbf{E}^T\mathbf{E} \in \mathbb{R}^{N\times N}$, which is always positive semidefinite, i.e., with all real and nonnegative eigenvalues. A fundamental result in algebraic graph theory is that the multiplicity of the eigenvalue $\lambda_1 = 0$ of $\mathbf{L}$ is equal to number of connected components of $\mathcal{G}$ [3]. For this reason, the second smallest eigenvalue of $\mathbf{L}$, typically denoted with λ_2, is called algebraic connectivity, and $\lambda_2 > 0$ if and only if the graph is connected. Those definitions extend seamlessly to the case of weighted graphs, i.e., graphs in which a weight number is associated to each edge, representing the quality of the communication. Algebraic graph theory plays a fundamental role in the control of networked robots, see [4] for an introduction to this vast topic.

13.2.3 Human–Robot Interface

A *human–robot interface* is an input–output device that allows a human operator to retrieve/send information from/to one or more robots at the same time. The interface is constituted by a physical part, which is able to record the human actions and to provide a sensorial feedback to the human, a processing unit, and a communication

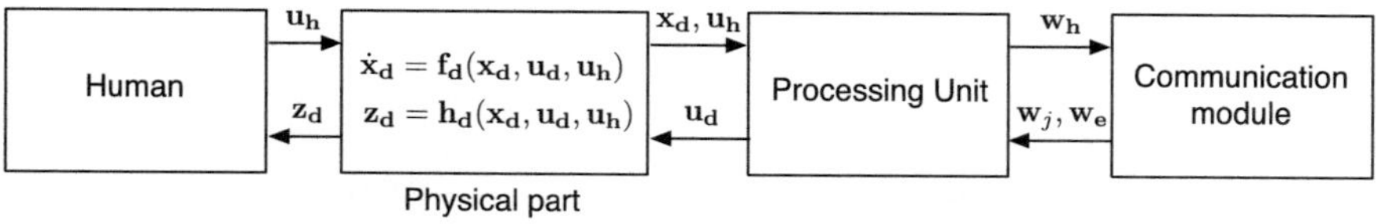

Fig. 13.2 High-level model of the human–robot interface

module. Examples of interfaces are joysticks, control pads, button pads, displays, movement recognition units, and haptic interfaces, just to mention a few. We refer the reader to Fig. 13.2 for a high-level model of a human–robot interface robot interacting with the human and communicating with the robots.

The *physical part* of the interface is implicitly modeled by a differential equation

$$\dot{\mathbf{x}}_\mathbf{d} = \mathbf{f_d}(\mathbf{x_d}, \mathbf{u_d}, \mathbf{u_h}), \tag{13.4}$$

where $\mathbf{x_d} \in \mathbb{R}^{n_{x_d}}$ is the *state of the interface*, $\mathbf{u_d} \in \mathbb{R}^{n_{u_d}}$ is the *control input of the interface* (a signal that can be changed arbitrarily by the processing unit), and $\mathbf{u_h} \in \mathbb{R}^{n_{u_h}}$ is the external *input from the human operator* (a signal that is out of control of the interface and represents the human input to the system). We do not include inaccuracies and disturbances in the interface model since human–robot interfaces are typically well-calibrated instruments that operate in a structured environment. For the same reason we assume that the internal state of the interface $\mathbf{x_d}$ is fully measurable.

The *communication module* feeds the interface with signals coming from a set of robots denoted with $\mathcal{N}_l \subseteq \{1, \ldots, N\}$ and called *interface neighbors*. We denote the signal coming from the robot $l \in \mathcal{N}_d$ with $\mathbf{w_d}_l \in \mathbb{R}^{n_{w_{d_l}}}$. As in the previous case, additional signals coming from other sources (e.g., a smart environment) are denoted with $\mathbf{w_e} \in \mathbb{R}^{n_{w_e}}$. Finally, the communication module is also in charge of sending the signal $\mathbf{w_h}$ to the robots in $\mathcal{N}_d$.

The *processing unit* of the interface has two roles. The first role is to compute $\mathbf{w_h}$ on the basis of $\mathbf{x_d}$ so that the human can influence the motion control of the robots. The second role is to provide a *feedback to the human operator* in the form of an image, a change of color, a beep sound, a force provided back to the operator in a haptic interface, and so on. Generically speaking, any kind of feedback can be modeled by a static function

$$\mathbf{z_d} = \mathbf{h_d}(\mathbf{x_d}, \mathbf{u_d}, \mathbf{u_h}). \tag{13.5}$$

The input $\mathbf{u_d}$ is computed with the purpose of letting $\mathbf{z_d}$ be informative of the communicated signals $\{\mathbf{w}_j \mid j \in \mathcal{N}_i\}$, $\mathbf{w}_h$, and $\mathbf{w}_e$.

13.3 A Taxonomy of Collaborative Human–Robot Control

In this section we present a taxonomy[1] for the situation in which a human collaborates in the motion control of one or multiple mobile robots. We shall refer when needed to the terminology introduced in Sect. 13.2. A particular emphasis will be given to the case of aerial robots, even though many results and considerations hold for any kind of mobile robot.

13.3.1 Physical Domain of the Robots

First of all, the *collaborative human–robot control* can be divided into several categories, corresponding to the physical domain of the robotic system, i.e., terrestrial (ground), aerial, space, marine, and underwater, just to mention the main ones. In the following we provide a quick review on the models used for the first two domains.

The models of ground robots stop typically at the kinematic level. An example is to consider $\mathbf{x} = (\mathbf{p}^T \ \psi)^T \in \mathsf{SE}(2)$ where $\mathbf{p}^T$ and ψ are the position and orientation of the mobile robot, respectively. The simplest model can be represented as $\dot{\mathbf{x}} = \mathbf{u}$ if the robot is omnidirectional and holonomic or as $\dot{\mathbf{x}} = \mathbf{G}(\mathbf{x})\mathbf{u}$ with $\text{rank}(\mathbf{G}) < 3$ if the platform is underactuated and non-holonomic.

Aerial robots cannot be modeled just kinematically because the dynamic effects are nonnegligible, due mainly to (1) the larger speed of aerial robots when compared to ground robots, and (2) the impossibility to aerodynamically generate input forces that are as strong as the one generated using the motor-wheel actuators available in the ground-robot case. Furthermore, in order to increase the energy efficiency, the mechanical simplicity and robustness, and the lightness of the platform, a low number of actuators are used, leading to underactuated robots. The most used aerial robots nowadays are multi-rotor platforms, such as the quadrotor configuration [5]. In this case the robot state is one of the rigid-body floatings in space $\mathbf{x} = ((\mathbf{p}, \mathbf{R}), \mathbf{v}, \boldsymbol{\omega}) \in \mathsf{SE}(3) \times \mathbb{R}^3 \times \mathbb{R}^3$, where $\mathbf{p}$, $\mathbf{R}$, $\mathbf{v}$, and $\boldsymbol{\omega}$ represent the position, attitude, velocity, and angular velocity of the robot, respectively. The dynamical model is then the following:

$$\dot{\mathbf{p}} = \mathbf{v} \tag{13.6}$$

$$\dot{\mathbf{v}} = \mathbf{f}_1(\mathbf{x}) + \mathbf{G}_1(\mathbf{x})\mathbf{u} \tag{13.7}$$

$$\dot{\mathbf{R}} = \mathbf{R}\omega \tag{13.8}$$

$$\dot{\boldsymbol{\omega}} = \mathbf{f}_2(\mathbf{x}) + \mathbf{G}_2(\mathbf{x})\mathbf{u}. \tag{13.9}$$

[1] A taxonomy that is comprehensive of all the possible cases and shows all the most important works in collaborative human–robot control is clearly out of the scope of this chapter, which does not claim to be a complete survey of this huge topic.

Table 13.1 Degrees of autonomy in the human–robot collaborative control

Non-robotic systems	Robotic systems			
Fully human-operated	Collaborative human–robot control			Fully autonomous
	Direct	Shared	Supervisory	

In most of the cases, e.g., for coplanar multirotors, the attitude dynamics (the last two equations) is fully actuated (i.e., rank$(\mathbf{G}_2) = 3$), while the position dynamics (the first two equations) is underactuated (typically rank$(\mathbf{G}_1) = 1$). Furthermore, there is a cascaded structure in which $\mathbf{f}_1$, $\mathbf{G}_1$, $\mathbf{f}_2$, and $\mathbf{G}_2$ depend only on $\mathbf{R}$ and $\boldsymbol{\omega}$.

The increased range applications enabled by the use of aerial robots with respect to the exclusive use of ground robots together with the larger complexity of aerial robots platforms made the design of collaborative human-aerial-robot control, a particularly interest research domain. In this chapter we will therefore focus especially on the aerial robot case. However, the proposed taxonomy and the large majority of presented methods apply for all the other domains as well.

13.3.2 Degree of Autonomy from the Human Operator

The *collaborative human–robot control* can be then roughly divided into three categories, corresponding to the degree of autonomy of the robotic system. This kind of classification, recalled in Table 13.1, is similar to the one used in the teleoperation literature, see [6].

The separation between the three categories cannot be sharp. Typically, real situations show intermediate aspects between the three categories. In this chapter we will present works that belong mainly to shared control category, with some supervisory control aspects as well.

13.3.2.1 Direct Control

Direct control refers to the case in which the robot motion is completely (i.e., in all its DoFs) decided by the human operator. The role of the robot is only to reproduce the human operator motion with as much fidelity as possible. In this case the human–robot interface has the same number of DoFs of the robot ($n_x = n_{x_d}$), $\mathbf{w_h} = \mathbf{x_d}$, and the goal of the robot control $\mathbf{u}$ is to have

$$J = \|\mathbf{x} - \mathbf{w_h}\| < \varepsilon \qquad (13.10)$$

with ε as small as possible.

13.3.2.2 Supervisory Control

Supervisory control refers to the case in which the robot motion is mainly controlled by autonomous decisions of the robot, while the human operator is in charge of providing only high-level directives to be fulfilled. In this case, the human–robot interface has typically a much lower number of DoFs of the robot ($n_x > n_{x_d}$), and the goal of the robot control $\mathbf{u}$ is to minimize

$$J = \|\mathbf{h}(\mathbf{x}(t^\star), \mathbf{x_e}(t^\star)) - \mathbf{w_h}\|, \tag{13.11}$$

where in this case $\mathbf{h}(\mathbf{x}, \mathbf{x_e}(t^\star))$ is a projection map for both the robot and environment state and $t^\star$ are some particular time instants in which the sub-task should be achieved.

13.3.2.3 Shared Control

Shared control refers to the case in which the robot motion is determined by both the human operator and robot decisions in a mostly balanced fashion. In this case typically the human–robot interface has a lower number of DoFs of the robot ($n_x > n_{x_d}$), and the goal of the robot control $\mathbf{u}$ is to minimize

$$J = \|\mathbf{h}(\mathbf{x}) - \mathbf{w_h}\|, \tag{13.12}$$

where $\mathbf{h}(\mathbf{x})$ is a map that projects $\mathbf{x}$ on the lower dimensional space controlled by the human operator.

Time Horizon in the Operator Reference

Another important aspect in the collaborative control schemes that we are analyzing is the length of the time horizon associated to the reference signal (e.g., desired position or velocity) given by the operator to the robot. In the direct control case the length of time horizon is intrinsically zero since the operator reference is instantaneous and refers to the 'now.' On the contrary in supervisory control schemes the time horizon is typically very large and the reference command pertains to some time instant in the relative long-term future, as explained before in this chapter.

In the shared control scenario the presence of intermediate cases is possible. However, in the literature, the majority of the shared control works consider that the human operator is in charge of guiding the aerial robot during the task by specifying the current reference position or velocity, as in [7–9]. This persistent commitment can result rather demanding for the human operator. Furthermore, instantaneous and persistent collaboration may be unnecessary in some parts of the task execution, such as in all the applications in which the robot has to follow a predefined path and the purpose of the operator is only to provide sporadic modification to the planned path in response to unpredicted changes of the external situation.

In the work presented in [10] (see Fig. 13.3 for an illustration of this scheme) this consideration is taken into account and the collaboration is shifted directly at the

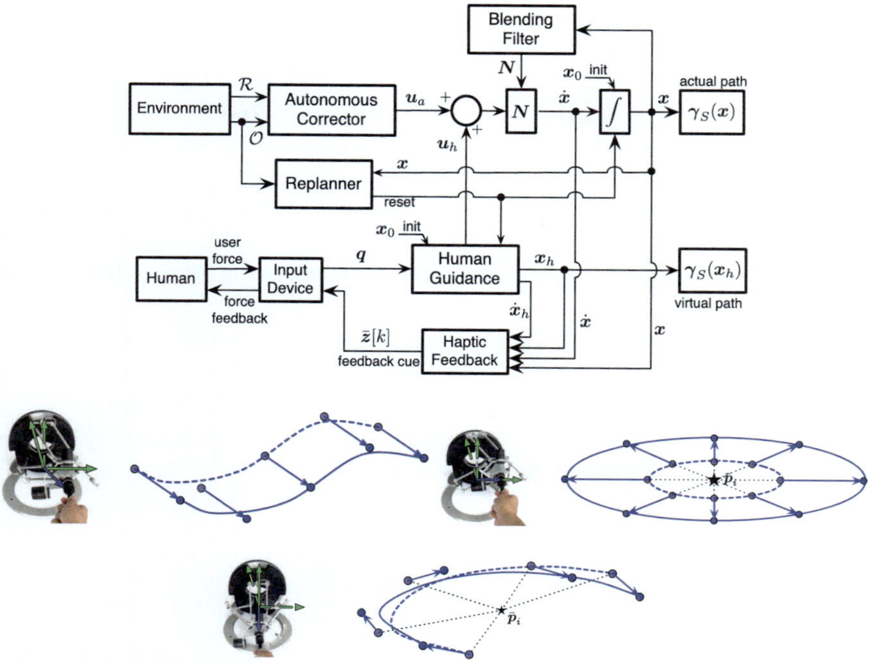

Fig. 13.3 *Top* block diagram of a feedback-force human–robot collaboration scheme with relatively long-time horizon in the operator reference. *Bottom* examples of canonical path transformations. *Green arrows* represent the DoFs. Depending on the transformation the device motion is artificially restricted (*continuous green arrows*): 2 DoFs for translation, 1 DoFs for scaling, 1 DoFs for rotation. *Blue arrows* represent the commands and corresponding motion of the control points

planning level by letting the human operator control the planned path for a given future (and nonnegligible) time horizon. First, the operator is in charge of modifying online the shape of a planned path with a haptic interface (e.g., by changing the location of some control points). Second, the robot autonomously corrects the modified path in order to meet additional requirements such as collision avoidance, dynamic limits, etc. Finally, consistently with the spirit of feedback-force schemes, the haptic cue is computed based on the mismatch between the path modification requests asked by the operator and the autonomous modifications of the robot using an impedance-like algorithm. In our formalism, this kind of paradigm can be summarized by the following equations:

$$J = \|\mathbf{h}(\mathbf{x}_{[t,t+T]}) - \mathbf{x_d}\| \tag{13.13}$$

$$\mathbf{u_d} = \mathbf{K}(\mathbf{h}(\bar{\mathbf{x}}_{[t,t+T]}) - \mathbf{x_d}) + \mathbf{f}(\mathbf{x_d}), \tag{13.14}$$

where $\mathbf{x}_{[t,t+T]}$ represents the future state of the robot in the time horizon $T > 0$, and $\bar{\mathbf{x}}_{[t,t+T]}$ is state re-planned by the robot.

A interesting future direction in this sense would be to study task-dependent automatic adaptation of the length of the time horizon in the operator reference.

13.3.2.4 Non-collaborative Human–Robot Control

Outside the two extremes of collaborative control there are two cases which do not involve collaboration, see Table 13.1.

A robot is *fully autonomous* if the human operator has no influence on it, i.e., the signal $\mathbf{w_h}$ is empty.

The antipodal case is when the human is in charge of steering all the DoFs by means of low-level actions, like while driving a (nonautonomous) car or operating a mechanical tool. In this case the machine is not a robotic system, because it lacks of any relevant autonomous aspect. The machine is said *fully human-operated*. In this case the human acts directly on $\mathbf{u}$.

13.3.3 Force Interaction with the Operator

Another distinction can be made depending on the information value given to the *mechanical exchange* between the human operator and the interface.

A collaboration scheme is *forward-force* if the force exerted by the human operator is informative for the collaborative control. In other words, in a forward-force collaboration, the human input $\mathbf{u_h}$ includes a measure (or an indirect estimation) of the force exerted by the human operator on the human–robot interface and this measure influences $\mathbf{w_h}$ as well.

A collaboration scheme is *feedback-force* (or *bilateral*) if the human–robot interface exerts, towards the human operator, a force that conveys some information related to the collaborative control. In other words, in a feedback-force collaboration the feedback to the human operator $\mathbf{z_d}$ includes a force exerted by the interface toward the human operator. An interface that is not feedback force (bilateral) is said *unilateral*.

Notice that

- in this terminology, 'unilateral' means that there is no force feedback. Other kind of feedback signals (e.g., visual) may be still present in a unilateral scheme;
- a feedback-force scheme does not necessarily imply that the scheme is also forward-force and, vice versa, a forward-force scheme could be unilateral;
- the previous definitions can be extended by replacing force with mechanical quantities related to forces and, more in general, mechanical power such as moments, stiffness/compliance, and friction.

Forward-force schemes include the so-called 'force–position' or admittance-based schemes, in which the force provided by the human operator $\mathbf{u_h}$ is recorded by a force sensor or estimated by an observer on the human interface and it is used to generate online a reference position and velocity that the robot tracks using a local position control loop. In our formalism this kind of scheme can be represented by

$$\dot{\mathbf{x}}_{\text{ref}} = \mathbf{f}(\mathbf{x}_{\text{ref}}, \mathbf{u_h}) \tag{13.15}$$

$$J = \|\mathbf{h}(\mathbf{x}) - \mathbf{x}_{\text{ref}}\|. \tag{13.16}$$

This kind of approach has been used, e.g., in [7] where the authors propose an admittance control modality for controlling an aerial vehicle. The operator applies a force to a haptic interface and the interface responds with a corresponding displacement, and at the same time the force applied by the operator is used to generate a reference for the vehicle controller.

The simplest feedback-force schemes are the so-called 'position-force' or impedance-based schemes. In these schemes the configuration in which a haptic device is set by the operator is used a reference position or velocity for the robot. The force feedback is then proportional to the mismatch between the actual position/velocity and the reference. In our formalism this kind of scheme can be represented by

$$J = \|\mathbf{h}(\mathbf{x}) - \mathbf{x_d}\| \tag{13.17}$$

$$\mathbf{u_d} = \mathbf{K}(\mathbf{h}(\mathbf{x}) - \mathbf{x_d}) + \mathbf{f}(\mathbf{x_d}), \tag{13.18}$$

where $\mathbf{K}$ and $\mathbf{f}(\mathbf{x_d})$ are a gain matrix and a stabilizing term for the device, respectively.

A seminal work in this field is [11], where a force-feedback scheme based on virtual viscoelastic couplings is designed for controlling a ground mobile robot in an obstacle populated environment. Extension to the multiple ground-robot case has been considered, e.g., in [12]. Finally many applications and extensions of this concept to the aerial robot case can be found in the very recent literature, see [8, 13–16].

More advanced feedback-force schemes include in the feedback the measurement of the force exerted by the robot on the environment or vice versa. In [17] the authors propose a scheme that is both forward force and feedback force to control an aerial robot. The forward scheme is similar to an admittance-like one, while the feedback force is used to display obstacles using the concept of dynamic kinesthetic boundary. In [18] a variable force-feedback gain is instead proposed.

Researcher efforts have also been directed toward the critical comparison between different schemes. In [19] a theoretical and experimental comparison between the admittance- and impedance-like schemes is proposed both from the perceptual and performance point of view. In [20] the authors compare two different haptic cueings, namely the aforementioned reference mismatch versus the obstacle force cueing.

The main advantage of forward-force schemes is the availability of an additional intuitive information channel for the human operator in the forward channel. The main disadvantage of forward-force is the necessity to measure the human force, which is usually very noisy and also difficult to give a unique interpretation among the different human operators.

The main advantages of feedback-force schemes are

- the additional channel for providing information to the operator, which hopefully results in a more immersive experience;
- the fact that haptic feedback may result more immediate to interpret than visual feedback due to the longer cognitive pipeline related to the latter;
- the fact that haptic feedback requires less bandwidth than vision (e.g., to feedback the distance from an obstacle), which in turn generates less delay than vision in long-distance internet based communication, see [21].

The main disadvantages of feedback force are

- the fact that the force applied to the interface influences back its the motion in a way that might result involuntarily for the operator;
- the increased difficult to ensure overall stability;
- the sensibility to delays;
- the need to take into account ergonomics issues, which may increase operator fatigue.

Stability issues in the bilateral human-collaborative control of mobile robots have been addressed in several ways, many of which related to passivity-based control such as the passivity observer/passivity controller (PO/PC) approach [22] the passive set-position modulation (PSPM) approach [23], and the two-layer approach [24].

13.3.4 Near-Operation Versus Teleoperation

A human–robot collaborative control is a *near-operation* if the human operator perceives the scene related to the control task by direct sensing, e.g., by line-of-sight, sense of touch, etc. Conversely, the human–robot collaborative control is a *teleoperation* when the perception is made through indirect sensing, e.g., television, haptic interfaces, etc.

A teleoperation is *delayed* when the delays of the communication are nonnegligible (due to, e.g., long distances or long processing times). As it is well known, the presence of delays in the teleoperation may have a dramatic influence on the stability of feedback-force schemes if not properly addressed, see [6] and reference therein.

Large delays (usually greater than 0.15 s) between the operator and the robots are unavoidable in intercontinental collaborative control settings. In addition to stability, those delays generate also practical problems in the quality of video streaming, the safety, and the ability to maneuver the robot. In fact, in long-distance communication, the video transmission has usually a larger delay than the reception of state of the remote robots because of the larger size of the packets of the video streaming.

Examples of intercontinental shared control of aerial robots are still very few at date. In [21] the authors show experiments of collaborative control for aerial robots with German–Korean intercontinental communication using a standard internet channel with average delay of 0.35 s (see Fig. 13.4 and Figs. 2 and 3 of [21]). In

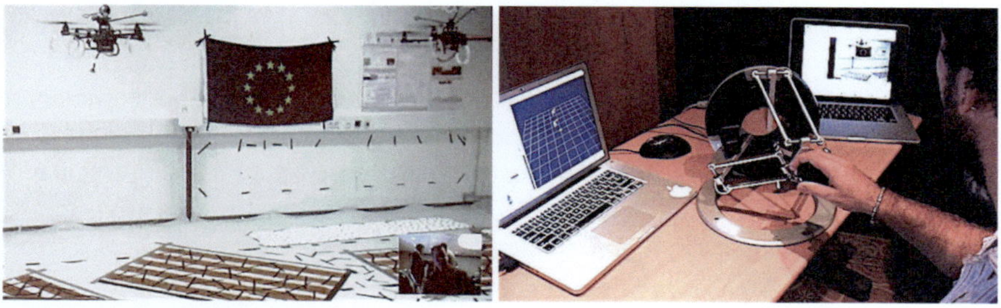

Fig. 13.4 Experiments of intercontinental (Korean-German) feedback-force collaborative human–robot control. Ddemo at the IEEE International Conference on Systems Man and Cybernetics 2012, in Seoul, South Korea

Table 13.2 Amount of physical interaction of the robot(s) with the environment

Amount of physical interaction of the robot with the environment		
With contact	Contact-free	Negligible
The external environment constrains the DoFs of the robot	The forces produced by the external environment do not constrain the DoFs of the robot	The disturbing/constraining external forces are negligible

this work it is shown that the feedback-force scheme can be made stable in the presence of such large delay and packet losses by resorting to the PSPM passivity-based technique [23]. The benefit of using force feedback to detect obstacles in advance with respect to the more delayed visual feedback is also made clear through the experiments. Following the results provided in [21], other intercontinental schemes for human-collaborative control with aerial vehicles have been studied and experimentally validated, such as in [25].

A different, but still passivity based, technique is used in [26] to cope with delays in the collaborative force interaction control of aerial robots. In this case the control scheme resorts to the two-layer architecture, in which a passivity layer (whose goal is to preserve passivity, hence stability, of the interconnection) is built on top of the transparency layer, whose goal is instead to deteriorate the quality of the references and the force feedback as less as possible [24].

13.3.5 *Physical Interaction with the Environment*

Another important distinction concerns the nature of physical interaction of the robot(s) with the environment. This represents a sort of dual aspect with respect to the force interaction on the human operator side. Table 13.2 summarizes the tree cases described in the following.

When the robot interacts with a solid/impenetrable environment which can significantly constrain the robot DoFs, the physical interaction is *with contact*.

The *contact-free physical interaction* case is instead when the robot is subject to external physical disturbances that are not constraining its DoFs, as in the previous case, but are mostly acting as an external disturbance, such as in the presence of wind, magnetic field, etc.

We finally refer to the *negligible physical interaction* case when physical interaction is only part of the actuation mechanism and plays almost no role in constraining or disturbing the robot motion. This is the case of a ground mobile robot moving freely on a flat surface with perfect wheel traction, and of a multi-rotor robot flying in the absence of wind and away from other surfaces (walls, floors, ceilings, etc.) thus not experiencing any undesired aerodynamical effect. Note that a sort of 'physical interaction' is actually present in both the examples and essential for the functioning of the robot (i.e., the wheel traction needed to produce the moving force, and the aerodynamic flow needed to produce the drag and the thrust forces). Nevertheless, these interactions do not constitute an impediment to the robot motion or an external disturbance that has to be compensated by the controller.

The cases of both fixed-base manipulators and grounded mobile manipulators have been deeply studied in the past, see [6, 12, 27] and references therein for an overview of the state of the art.

A mostly open problem is instead the case of collaborative control of aerial robots in contact with the environment. In fact, while the study of control strategies for aerial robots in physical interaction with contact has received a lot of attention in the recent years, a few works exist at date addressing the more challenging problem of having a human in the loop with a feedback-force scheme. An example of these few works is [28] where a teleoperation scheme is presented that allows to convert a contact-free physical interaction controller into a controller for physical interaction with contact, see Fig. 13.5. The approach is based on the aerial physical interaction control algorithm presented in [29], where using a rigid tool attached to the aerial vehicle the robot is able to exert a 3D force on the environment. Following the feedback-force paradigm, the algorithm in [28] provides the user with a force feedback that is proportional to the force measured by a force sensor at the tool-tip,

$$\mathbf{z}_\mathrm{d} = \mathbf{d}. \tag{13.19}$$

During the contact-free approaching phases the algorithm automatically slows down the speed of the robot in order to achieve a smooth passage from free flight to contact constrained motion as shown in the block diagram of Fig. 2 in [28]. This kind of variable/tunable autonomy scheme is a typical example of collaborative shared control with physical interaction extended to the more challenging aerial robotics case.

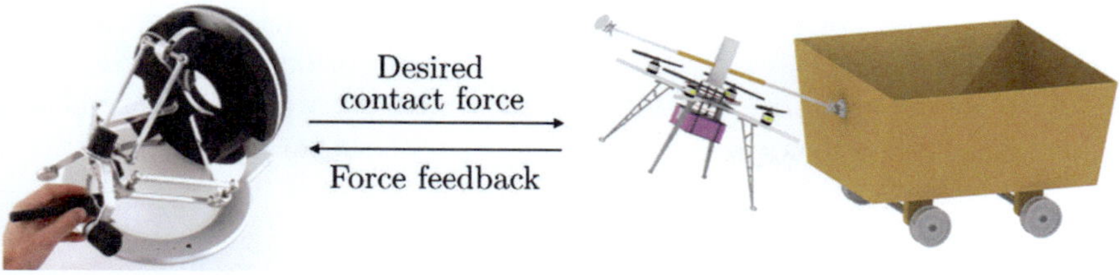

Fig. 13.5 Aerial teleoperation for physical interaction with contact

13.3.6 Use of Onboard Sensors Only

Of fundamental importance in robotics is the ability of the robot to be fully autonomous from the sensing point of view. With respect to this issue we can roughly distinguish between two different classes of approaches: the approaches that use external sensorial infrastructures and the approaches that make use of onboard only sensorial equipment. The large majority of works in collaborative control of aerial robots belong to the first category. Motion capture systems, global positioning systems (GPS), and wireless-based localization are some example of the infrastructures used in these works [8, 14, 30–32].

Some recent efforts have been directed toward the second kind of approach, which presents clearly more challenges, especially in the aerial robotics case. The work in [33], where a bilateral teleoperation scheme for aerial robots is presented, represents a first step in the direction of the second class of approaches. In this work the state of the robot is retrieved using a motion capture system; however, the obstacle detection is performed using a laser range finder which then feeds the obstacle avoidance algorithm with onboard data.

In the experiments presented in [34] a human operator controls a group of aerial robots that use onboard cameras to measure the relative angles between themselves while the velocity is still estimated using an external motion capture system.

One of the first works addressing the problem of full-onboard sensorial equipment in the collaborative feedback-force control of aerial robots is presented in [9]. The human operates with velocity control (and with the feedback-force paradigm) the aerial robot in an unstructured environment, see Fig. 9 in [9]. The state of the robot is estimated using only the onboard IMU and a depth camera (RGB-D). Thanks to the availability of the depth information the obstacle safety is ensured. During the collaboration the aerial robot conducts autonomously some extra tasks that increase the ergonomics of the operation and its safety. The main of these tasks is a continuous pan scanning that temporarily enlarges the field of view of the robot exploiting the yaw DoF that is left free by the human operator collaboration. This movement is compensated in real time using an adaptive filter thus allowing the operator to experience a yaw-rate-free operation. The haptic feedback is based on the impedance paradigm where the actual velocity is estimated using a photometric error approach.

The development of human-collaborative control schemes that can cope with the use of onboard sensors only is crucial for the employment in the real world. However, this aspect has been often overlooked by the recent literature, while it should probably deserve a higher attention.

13.4 A Taxonomy of Collaborative Human–Multi-robot Control

The collaborative scheme may include just one robot (*single-robot* case) or several distinct robots (*multi-robot* case). The use a group of coordinating and cooperating robots represents an enabling technology for a large number of applications. As a matter of fact multi-robot systems can show an improved robustness with respect to single-point system failures and the capacity of adaptation to environments that are both uncertain and dynamic. Multi-robot systems may present also economical advantages compared to monolithic complex robots due to the lower price of small size systems. Range of applications span from data retrieval and collection, distributed sensing on large-scale areas, deep space interferometry, and search and rescue missions. Among multi-robot systems aerial robots play an essential role for their ability to cover large distances due to their speed and large areas, thanks to the presence of the third (altitude) dimension.

Thanks to the appealing features and the great potentials of multiple aerial robot systems, the recent years have seen a sharp increasing in the study of collaborative human–robot schemes for such kind of robotic systems. The goal of this section is to revise and classify some of these approaches from a multi-robot perspective. In [5] the authors provide an overview that summarizes generic collaborative scheme for the bilateral shared control of groups of aerial robots. The proposed approach is able to integrate three main components, i.e., the human operator assistance, the force feedback, and the control of the group topological and metrical properties. Regarding the last component of the work suggests to consider three main cases: the unconstrained-, the connected-, and the constant-topology cases. In this section we aim at generalizing this classification taking into account additional aspects related to the collaborative human/multi-robot interaction.

13.4.1 Level of Centralization

A first classification when dealing with human-collaborative schemes for multi-robot systems is about the level of centralization. Several definitions of decentralization have been given in the literature, which often overlap with the concept of a distributed algorithm. Here we adopt a simple definition that is the following: a scheme is decentralized (or distributed) if the amount of communication (packets exchanged) and computation (complexity) for unit of time, and the size of memory needed by each robot and by the human–robot interface, are constant when the total number of robots is increased. A scheme that is not decentralized is centralized. Centralized schemes are the one that need for example the communication graph $\mathcal{G}$ to be complete, i.e., $(i, j) \in \mathcal{E}\ \forall i \neq j$. Centralized schemes are also the one assuming that the human–robot interface sends the message $\mathbf{w_h}$ to all the robots in the team, i.e., $\mathcal{N}_l = N$.

In general, decentralized algorithms are more scalable, i.e., applicable to higher number of robots, but are less efficient than centralized algorithm in the case of a smaller number of robot and when the communication bandwidth and the computation capabilities of each robot allow centralized implementations.

13.4.2 Master–Leader–Followers Schemes

A possible approach to collaboratively control a group of robots is the *master–leader–followers scheme*. In this scheme all the robots implement any decentralized control strategy resorting to local communication in order to carry on some task while the human operator (the master) communicates with only one robot of the group at time (the leader). All the non-leader robots are called the followers. For example, the robotic team task could be to monitor a given area while keeping a certain formation, and the master, which is supervising the overall task providing velocity references, can be in communication only with one robot at time due to long-distance bandwidth limitations. This scenario is abstractly depicted in Fig. 1 of [35], where an abstraction of master–leader–followers scheme is shown. The human operator (master agent) can communicate only with one robot at the time (the leader) using a low channel bandwidth, while the leader robot and the followers communicate through a time-varying distributed communication topology.

Several challenging problems can arise even from this fairly basic situation when collaborative human–robot control is needed. A first example is the difficulty of maintaining the stability of the feedback-force interconnection between the human operator and such multi-robot system, especially in the case in which the multi-robot team interaction topology can arbitrarily switch even independently of the robot state. In fact, in [30] it is shown that this kind of switching interaction can generate a nonpassive and potentially unstable behaviors. The authors then present a passivity-based approach to guarantee a stable behavior of the leader and the followers regardless of the autonomous motion, the switching topology and the interaction with the environment. At the same time, making the multi-robot system passive turns instrumental for the feedback-force collaborative scheme with the human operator (the master). Experiments using this approach are shown in Fig. 13.5 in [30] where a human operator collaborates with a group of robots using as feedback-force scheme. Passivity of the heterogeneous team of robots is guaranteed despite the switching nature of the communication.

Another challenge arising in this context is called the *online leader selection* problem, i.e., the ability of the system to choose online the best leader that the master has to communicate with. This problem is addressed in [35, 36] for what concerns the maximization of the effectiveness of the velocity reference provided by the human operator. In this work the authors show that this effectiveness is upper-bounded by an index that depends on the current state of the robots and by a graph theoretical index, namely the algebraic connectivity of a special leader-dependent digraph.

13.4.3 Formation-Orthogonal Control Schemes

Another approach to collaboratively control a group of robots is the *formation-orthogonal control* scheme. In this scheme the robots autonomously stabilize the relative geometry of the formation using relative or absolute measurements, while the human operator sends references to steer the remaining DoFs of the multi-robot system.

In [12, 27] the authors propose a method to bilaterally teleoperate a group robots, such as a group of ground mobile manipulators grasping an object. The main idea of the method relies on the concept of passive decomposition, with which it is possible to decompose the multi-robot dynamics in two independent systems while preserving passivity from an energetic point of view. The locked system abstracts the overall team motion and is controlled by the human operator using a feedback-force paradigm. The shape system defines instead the cooperative grasping aspect and is autonomously controlled by the team of robots. This paradigm allows the human to indirectly control the motion of the grasped object without influencing the reliability of the grasping.

Regarding the aerial robot case, in [31] the authors propose a feedback-force collaboration scheme in which the human operator teleoperates the motion of the multiple aerial robots that act as a virtual deformable object. The scheme is based on three layers: the low-level robot control, the online velocity reference generator, and the bilateral teleoperation layer. The low-level robot control is used in order to abstract from the particular kind of dynamics of the robot. The velocity reference generator mixes together the velocity reference provided by the operator, the obstacle avoidance action, and the distance preservation term. The last two terms are based on an artificial potential approach. The haptic cue provided to the operator is proportional to the average of the velocity reference mismatch in a way that is similar to an impedance-like approach, but extended to the whole multi-robot system considered as a virtual large deformable object. In order to implement this formation control scheme the position measurements of each robot (e.g., from GPS) are needed.

An alternative (especially from the sensing point of view) formation-orthogonal control scheme is presented [34], see Figs. 3 and 7 in [34] for a visual illustration. This case differs from [31] for three main aspects. The first is that it uses only measurements of the relative angles between the robots (that can be retrieved from onboard cameras). The second is that the formation control approach used is almost globally convergent and it is not prone to local minima like the artificial potential schemes typically are. The third is that the human operator velocity reference acts on the tangent bundle of the manifold of configurations that keep the relative angles constant, which includes collective rotations, translations, and dilations. In this way the formation control specifications and the human control result always orthogonal and do not interfere with each other thus establishing a clear separation between the fully autonomous and the human-operated domains of intervention.

13.4.4 Group-Property Preservation Schemes

Formation-orthogonal control schemes may excessively limit the relative mobility between the robots in the group by blocking the relative distances and angles to some fixed or overly constrained values. On the other side, for some tasks it is enough to preserve some high-level structural properties of the group rather than determining the exact shape of it. We refer to such approaches as *group-property preservation schemes*.

A first example of high-level structural property of a group of robots is the connectivity of its topology, intended either as the communication network or the sensing graph. Connectivity is a very important property since it ensures that the data flow between the different robots cannot be broken and therefore the information can be successfully distributed.

In [1] the authors present a decentralized method that maintains group the connectivity in a flexible way, i.e., allowing the topology to change over time as long as connectivity is preserved, see Fig. 10 in [1]. Connectivity preservation is ensured by keeping the algebraic connectivity λ_2 (see Sect. 13.2.2) positive during all the robot maneuvers. The algebraic connectivity concept is then extended to take into account complex robot-to-robot interaction going beyond simple max-distance models, which brings to the definition of the generalized connectivity eigenvalue. This is obtained by introducing a state-dependent adjacency matrix $\mathbf{A}(\mathbf{x}_1, \ldots, \mathbf{x}_N)$ and consequently, a state-dependent Laplacian $\mathbf{L}(\mathbf{x}_1, \ldots, \mathbf{x}_N)$. The resulting generalized connectivity is also able to embed mild formation control constrains and obstacle avoidance if needed. Exogenous control actions coming from human operators or from a distributed target visiting algorithm, like the one presented in [37], can be also included. Using a feedback-force collaborative scheme one or more human operators can interact with the group of robots and immersively feel, through haptics, the connectivity force generated by the group.

Connectivity alone might be not enough to achieve certain tasks in which the robots can rely on relative sensing only. In these cases it is more appropriate to consider the concept of rigidity instead, which defines the property of being able to instantaneously reconstruct the group shape using local relative measurements, such as relative distances or relative angles. The rigidity concept has been originally introduced in the structure theory in order to characterize the notion of flexibility and stiffness of rigid-body structures.

The main algebraic object of what is called infinitesimal rigidity [2] is the so-called rigidity matrix $\mathbf{R}(\mathscr{G}, \mathbf{p}_1, \ldots, \mathbf{p}_N)$ which is defined as

$$\mathbf{R}(\mathscr{G}, \mathbf{p}_1, \ldots, \mathbf{p}_N) = [\mathbf{E}_l(\mathscr{G}_1)^T \cdots \mathbf{E}_l(\mathscr{G}_N)^T](\mathbf{I}_N \otimes [\mathbf{p}_1^T \ldots \mathbf{p}_1^T]^T)),$$

where $\mathbf{E}_l(\mathscr{G}_i)$ is the local incidence matrix relative to the i-th robot, $\mathbf{I}_N$ is the $N \times N$ identity matrix, and $\otimes$ denotes the Kronecker product between matrices.

In the context of multi-robot systems, the authors of [2] develop a theoretical machinery to express the degree of rigidity using a single nonnegative number, namely the rigidity eigenvalue, defined as the smallest non-structurally-zero eigenvalue of the symmetric rigidity matrix

$$\mathscr{R} = \big(\mathbf{W}(\mathbf{p}_1, \ldots, \mathbf{p}_N)\mathbf{R}(\mathscr{G}, \mathbf{p}_1, \ldots, \mathbf{p}_N)\big)^T \big(\mathbf{W}(\mathbf{p}_1, \ldots, \mathbf{p}_N)\mathbf{R}(\mathscr{G}, \mathbf{p}_1, \ldots, \mathbf{p}_N)\big),$$

where $\mathbf{W}$ is the diagonal matrix of the state-dependent weights defined on the edges of $\mathscr{E}$. The authors in [2] present a distributed algorithm that ensures the preservation of the rigidity resorting to local relative distance measurements only. Experiments are shown in which multiple human operators can interact with the group of robots providing velocity references while the group of aerial robots preserves the required degree of rigidity in an autonomous way. As in the connectivity case, haptic cues can be used to perceive the group rigidity using the feedback-force scheme.

Another advanced way of establishing a collaboration between human and the robotics team in the group-property preservation schemes is to let the human finely control the index that defines the group properties as the connectivity and the rigidity eigenvalues, using one of the DoFs of the interface. A work that explores this idea in a feedback-force fashion for the connectivity index is presented in [38].

13.4.5 Physical Interaction with Contact

We have already shown examples of coordination of ground robots physical interaction with contact, e.g., [12, 27]. Many other works are present in the literature for the ground mobile robot case, which are not mentioned here. On the other side, human-collaborative control of multiple aerial robots performing physical interaction with contact is a very young research field and not many works have been considering this scenario at date.

In [39] it is proposed a control framework to let a group of aerial robots grasp an object in a way that each robot uses an attached rigid tool to establish a single contact, see Fig. 13.6 and Fig. 1 in [39] where a more detailed block diagram is provided. Each robot acts as a flying finger that collaborates with the others to establish a N-fingered hand. In this way the aerial robots realize a unique hand-like system that can grasp and transport an object. The whole system is operated by a human hand using a two-layer approach. First the operator hand is tracked with an RGB-D camera and some features are extracted and mapped to virtual attraction points. Second, the aerial robots use an impedance force control in order to track the virtual attraction points with some compliance thus enabling grasping and transportation. The approach is

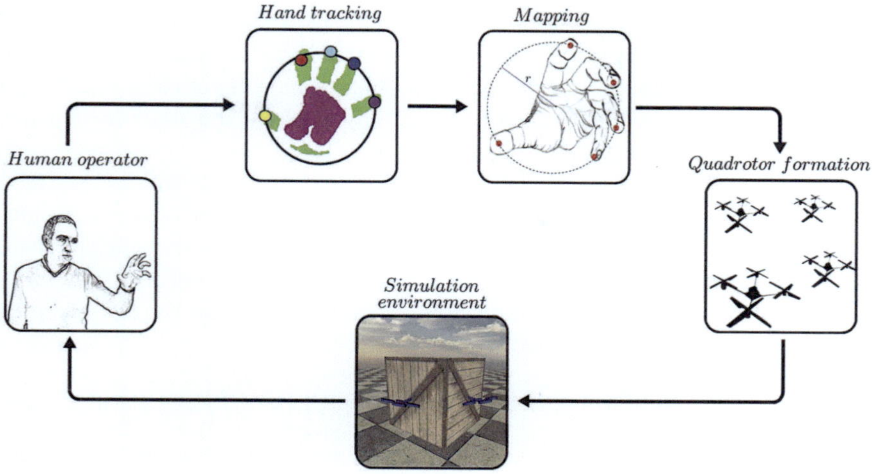

Fig. 13.6 Block diagram of the flying hand collaboration concept. Each aerial robot acts as a single finger to grasp an object and is teleoperated by a human hand through the use of a suitable abstraction layer

validated in simulation only. An extension of this framework is presented in [40], with the use of optimization-based control and haptic feedback.

Another approach that solves a similar problem from the robot cooperative side, but it lacks of the human-collaborative part, is presented in [41]. In this work multiple aerial robots are connected to an object using spherical joints and are used as thrust generators to cooperative manipulate the object in a physical interaction with contact scenario. The method, which is validated with experiments using a team of quadrotors as aerial robots, represents a promising building block for a possible extension to the human-collaborative case.

13.5 Conclusions

In this chapter we have shown and compared several different ways available in the literature to design the human-collaborative control of the motion of mobile robots, with particular emphasis on the aerial robot case. We have introduced a simple yet descriptive model for the robots and the human interfaces and we have proposed a list of fundamental axes for a proper classification of the collaborative scheme, including but not limited to the degree of autonomy, the presence of interaction force with the human or with the environment, the kind multi-robot schemes, and many others.

The common denominator of all the classification axes proposed in this chapter is the presence of a human operator in the control loop. In fact, fully autonomous control algorithms are not in general easily extendable in order to cope with the presence of a human operator and at the same time to take the best out of this collaboration. The main goal of the chapter has been to introduce the reader and provide a first-level analysis on the several ways to effectively include human operators in the control of both single and multiple aerial mobile robots. A particular emphasis has been set on the aerial robots case, which is at date not yet fully explored and understood, and is among the most challenging cases in robotics.

A lot of problems remain still unsolved in this field, among of which is worth to mention the possibility to perform effective bilateral teleoperation of aerial manipulators, the possibility of relying on onboard sensing in any environmental situation, the ability to cope with the large uncertainty in the perception, and control of bilaterally teleoperated aerial robots thus enabling their use in the real world for helping humans in the industry and in their everyday life.

Acknowledgements The author desires to thank all the collaborators that co-authored with him the works cited in this chapter. A special acknowledgement goes to Paolo Robuffo Giordano, with whom many of the algorithms and ideas that are present in those cited works have been conceived and developed.

References

1. Robuffo Giordano P, Franchi A, Secchi C, Bülthoff HH (2013) A passivity-based decentralized strategy for generalized connectivity maintenance. Int J Robot Res 32(3):299–323
2. Zelazo D, Franchi A, Bülthoff HH, Robuffo Giordano P (2014) Decentralized rigidity maintenance control with range measurements for multi-robot systems. Int J Robot Res 34(1):105–128
3. Fiedler M (1973) Algebraic connectivity of graphs. Czechoslov Math J 23(98):298–305
4. Mesbahi M, Egerstedt M (2010) Graph Theoretic methods in multiagent networks, 1st edn. Princeton series in applied mathematics. Princeton University Press, Princeton
5. Franchi A, Secchi C, Ryll M, Bülthoff HH, Robuffo Giordano P (2012) Shared control: Balancing autonomy and human assistance with a group of quadrotor UAVs,. IEEE Robot Autom Mag Special Issue Aerial Robot Quadrotor Platf 19(3):57–68
6. Niemeyer G, Preusche C, Hirzinger G (2008) Telerobotics. In: Siciliano B, Khatib O (eds)Springer handbook of robotics. Springer, Berlin, pp. 741–757
7. Schill F, Hou X, Mahony R (2010) Admittance mode framework for haptic teleoperation of hovering vehicles with unlimited workspace. In: 2010 Australasian conference on robotics and automation, Brisbane, Australia
8. Mersha AY, Stramigioli S, Carloni R (2012) Switching-based mapping and control for haptic teleoperation of aerial robots. In: 2012 IEEE/RSJ international conference on intelligent robots and systems, Vilamoura, Portugal, pp 2629–2634
9. Stegagno P, Basile M, Bülthoff HH, Franchi A (2014) A semi-autonomous UAV platform for indoor remote operation with visual and haptic feedback. In 2014 IEEE international conference on robotics and automation, Hong Kong, China, pp 3862–3869
10. Masone C, Robuffo Giordano P, Bülthoff HH, Franchi A (2014) Semi-autonomous trajectory generation for mobile robots with integral haptic shared control. In: 2014 IEEE international conference on robotics and automation, Hong Kong, China, pp 6468–6475
11. Diolaiti N, Melchiorri C (2002) Teleoperation of a mobile robot through haptic feedback. In: IEEE international workshop on haptic virtual environments and their applications, Ottawa, Canada, pp 62–72
12. Lee D, Martinez-Palafox O, Spong MW (2005) Bilateral teleoperation of multiple cooperative robots over delayed communication network: application. In: 2005 IEEE international conference on robotics and automation, Barcelona, Spain, pp 366–371

13. Stramigioli S, Mahony R, Corke P (2010) A novel approach to haptic tele-operation of aerial robot vehicles. In: 2010 IEEE international conference on robotics and automation, Anchorage, AK, pp 5302–5308
14. Rodríguez-Seda EJ, Troy JJ, Erignac CA, Murray P, Stipanović DM, Spong MW (2010) Bilateral teleoperation of multiple mobile agents: coordinated motion and collision avoidance. IEEE Trans Control Syst Technol 18(4):984–992
15. Franchi A, Robuffo Giordano P, Secchi C, Son HI, Bülthoff HH (2011) A passivity-based decentralized approach for the bilateral teleoperation of a group of UAVs with switching topology. In: 2011 IEEE international conference on robotics and automation, Shanghai, China, pp 898–905
16. Lee D, Franchi A, Robuffo Giordano P, Son HI, Bülthoff HH (2011) Haptic teleoperation of multiple unmanned aerial vehicles over the internet. In: 2011 IEEE international conference on robotics and automation, Shanghai, China, pp 1341–1347
17. Hou X, Mahony R (2013) Dynamic kinesthetic boundary for haptic teleoperation of aerial robotic vehicles. In: 2013 IEEE/RSJ international conference on intelligent robots and systems, Tokyo, Japan, pp 4549–4950
18. Farkhatdinov I, Ryu JH (2010) Improving mobile robot bilateral teleoperation by introducing variable force feedback gain. In: 2010 IEEE/RSJ international conference on intelligent robots and systems, Taipei, Taiwan, pp 5812–5817
19. Hou X, Mahony RE, Schill F (2016) Comparative study of haptic interfaces for bilateral teleoperation of VTOL aerial robot. IEEE Trans Syst Man Cybern: Syst. http://arxiv.org/abs/1305.5719
20. Son HI, Franchi A, Chuang LL, Kim J, Bülthoff HH, Robuffo Giordano P (2013) Human-centered design and evaluation of haptic cueing for teleoperation of multiple mobile robots. IEEE Trans Syst Man Cybern. Part B. Cybernetics 43(2):597–609
21. Riedel M, Franchi A, Bülthoff HH, Robuffo Giordano P, Son HI (2012) Experiments on intercontinental haptic control of multiple UAVs. In: 12th international conference on intelligent autonomous systems, Jeju Island, Korea, pp 227–238
22. Hannaford B, Ryu J-H (2002) Time-domain passivity control of haptic interfaces. IEEE Trans Robot Autom 18(1):1–10
23. Lee DJ, Huang K (2010) Passive-set-position-modulation framework for interactive robotic systems. IEEE Trans Robot 26(2):354–369
24. Franken M, Stramigioli S, Misra S, Secchi C, Macchelli A (2011) Bilateral telemanipulation with time delays: a two-layer approach combining passivity and transparency. IEEE Trans Robot 27(4):741–756
25. Mersha AY, Hou X, Mahony R, Stramigioli S, Corke P, Carloni R (2013) Intercontinental haptic teleoperation of a flying vehicle: a step towards real-time applications. In: 2013 IEEE/RSJ international conference on intelligent robots and systems, Tokyo, Japan, pp 4951–4957
26. Secchi C, Franchi A, Bülthoff HH, Robuffo Giordano P (2012) Bilateral teleoperation of a group of UAVs with communication delays and switching topology. In: 2012 IEEE international conference on robotics and automation, St. Paul, MN, pp 4307–4314
27. Lee D, Spong MW (2005) Bilateral teleoperation of multiple cooperative robots over delayed communication network: theory. In: 2005 IEEE international conference on robotics and automation, Barcelona, Spain, pp 360–365
28. Gioioso G, Mohammadi M, Franchi A, Prattichizzo D (2015) A force-based bilateral teleoperation framework for aerial robots in contact with the environment. In: 2015 IEEE international conference on robotics and automation, Seattle, WA, pp 318–324
29. Gioioso G, Ryll M, Prattichizzo D, Bülthoff HH, Franchi A (2014) Turning a near-hovering controlled quadrotor into a 3D force effector. In: 2014 IEEE international conference on robotics and automation, Hong Kong, China, pp 6278–6284
30. Franchi A, Secchi C, Son HI, Bülthoff HH, Robuffo Giordano P (2012) Bilateral teleoperation of groups of mobile robots with time-varying topology. IEEE Trans Robot 28(5):1019–1033
31. Lee DJ, Franchi A, Son HI, Bülthoff HH, Robuffo Giordano P (2013) Semi-autonomous haptic teleoperation control architecture of multiple unmanned aerial vehicles. IEEE/ASME Trans Mechatron Focus Sect Aerospace. Mechatronics 18(4):1334–1345

32. Grabe V, Riedel M, Bülthoff HH, Robuffo Giordano P, Franchi A (2013) The TeleKyb framework for a modular and extendible ROS-based quadrotor control. In: 6th European Conference on Mobile Robots, Barcelona, Spain, pp 19–25
33. Omari S, Hua MD, Ducard GJJ, Hamel T (2013) Bilateral haptic teleoperation of VTOL UAVs. In: 2013 IEEE international conference on robotics and automation, Karlsruhe, Germany, pp 2385–2391
34. Franchi A, Masone C, Grabe V, Ryll M, Bülthoff HH, Robuffo Giordano P (2012) Modeling and control of UAV bearing-formations with bilateral high-level steering. Int J Robot Res Special Issue 3D Explor Mapp Surveill 31(12):1504–1525
35. Franchi A, Robuffo Giordano P (2016) Online leader selection for improved collective tracking and formation maintenance. IEEE Trans Control Netw Syst. http://dx.doi.org/10.1109/TCNS.2016.2567222
36. Franchi A, Bülthoff HH, Robuffo Giordano P (2011) Distributed online leader selection in the bilateral teleoperation of multiple UAVs. In: 50th IEEE Conference on Decision and Control, Orlando, FL, pp 3559–3565
37. Nestmeyer T, Robuffo Giordano P, Bülthoff HH, Franchi A (2016) Decentralized simultaneous multi-target exploration using a connected network of multiple robots. Auton Robot. http://dx.doi.org/10.1007/s10514-016-9578-9
38. Secchi C, Franchi A, Bülthoff HH, Robuffo Giordano P (2013) Bilateral control of the degree of connectivity in multiple mobile-robot teleoperation. In: IEEE international conference on robotics and automation, Karlsruhe, Germany, May 2013
39. Gioioso G, Franchi A, Salvietti G, Scheggi S, Prattichizzo D (2014) The flying hand: a formation of uavs for cooperative aerial tele-manipulation. In: 2014 IEEE international conference on robotics and automation, Hong Kong, China, pp 4335–4341
40. Mohammadi M, Franchi A, Barcelli D, Prattichizzo D (2016) Cooperative aerial tele-manipulation with haptic feedback. In: 2016 IEEE/RSJ international conference on intelligent robots and systems, Daejeon, South Korea
41. Nguyen H-N, Park S, Lee DJ (2015) Aerial tool operation system using quadrotors as rotating thrust generators. In: 2015 IEEE/RSJ international conference on intelligent robots and systems, Hamburg, Germany

Chapter 14
A Passivity-Based Approach to Human–Swarm Collaboration and Passivity Analysis of Human Operators

T. Hatanaka, N. Chopra, J. Yamauchi and M. Fujita

14.1 Introduction

During the last decade, a number of research works have been devoted to cooperative control of autonomous multi-robot systems, and a variety of automatic control algorithms have been developed. Nevertheless, complex missions over highly uncertain environment still require mediacy of human operators in order to help challenging cognitive tasks and adaptation to environmental changes. Thus, human-operated multi-robot systems have been actively studied not only from technological but also from control theoretic perspectives under the paradigm of human–swarm interactions [8–10] or collaborations [4, 33] (see also other chapters of this book).

The semi-autonomous robotic system design has been in-depth studied in the fields of bilateral teleoperation, which involves a dual robot system where one robot, a slave, tracks the motion of another robot, the master, commanded by a human operator. It is the passivity-based approach [15, 17, 26] that has been recognized as one of the most powerful/elegant design concepts for the control of bilateral teleoperators. The passivity-based teleoperation has also been extended to the operation of

---

This chapter was developed and enhanced from an earlier paper published as [16] © 2015 IEEE.

T. Hatanaka (✉) · J. Yamauchi · M. Fujita
School of Engineering, Tokyo Institute of Technology,
Tokyo, Japan
e-mail: hatanaka@ctrl.titech.ac.jp

J. Yamauchi
e-mail: yamauchi@fl.ctrl.titech.ac.jp

M. Fujita
e-mail: fujita@ctrl.titech.ac.jp

N. Chopra
Department of Mechanical Engineering, University of Maryland,
College Park, MD, USA
e-mail: nchopra@umd.edu

Y. Wang and F. Zhang (eds.), *Trends in Control and Decision-Making for Human–Robot Collaboration Systems*, DOI 10.1007/978-3-319-40533-9_14

multiple slave robots [11, 12, 14, 20, 21, 30, 31]. However, the focus of these results [20, 21, 30] has been primarily on centralized control architectures in the sense that a robot needs access to all of the slave robots. Meanwhile, the papers [11, 31] and their related publications assume fully distributed control architecture, where mobile robots at the slave side implement distributed control to achieve cooperative tasks including formation, connectivity maintenance, and collision avoidance, while a master robot has access only to a member of the slave robots. The architecture in the slave side is closer to the leader-based control architecture presented in [8, 10]. Although extensions to more complex networks with multiple master–operator pairs are presented in [12, 14], each master still interacts with a single slave even within the control system.

In spite of the similarity in control architectures, there are several differences between the various studies on human–swarm interactions/collaborations [4, 8–10, 33] and bilateral teleoperation. First, the former focuses on higher level human–robot interactions and, accordingly, a simpler model for the robot dynamics is utilized as compared to the publications on bilateral teleoperation. For example, [8, 10] employ a single integrator as the robot dynamics. Moreover, [4, 33] do not even explicitly deal with robot motion dynamics. Second, the main objective of bilateral teleoperation is to guarantee good tracking performance, transparency as well ensure closed-loop stability in the presence of factors stemming from remote control, such as communication delays and packet losses. However, these are not always central issues in [4, 8, 10, 33], and, for example, [8, 10] study the relationship between the inter-robot network configuration and human operability, quantified by controllability and manipulability.

Although many inexpensive interfaces including smart phones, tablets, 3-D motion sensor devices have been recently commoditized, control theoretic formulations typically presume traditional interfaces like robot manipulators or joysticks. The paper [32] focused on this aspect, however the authors addressed control of a single robot as compared to the multi-robot case considered in the present chapter. Also, it has been well established in bilateral teleoperation that force feedback improves control performance and transparency. However, force feedback may not be necessarily required for completing high-level tasks. Indeed, many publications on the human–swarm interactions presume only visual feedback (See, e.g., [6, 24, 25, 27].).

In this chapter, we address semi-autonomous synchronization of robots' positions or velocities to a desired reference value, assuming that either the position or velocity control mode is pre-specified by the operator, where the operator is assumed to determine a command signal by utilizing only visual feedback, and to send the signal to the robots through an appropriately defined interface. The present problem is motivated by the navigation of a robotic group to an area of interest, which is required, e.g., in the scenario of environmental monitoring, and search and rescue. In order to achieve the synchronization, we first employ, as a robot controller, the so-called PI consensus estimator [3, 13], and prove passivity of the resulting robotic system. Then, we close a loop between the operator and robots based on passivity, which renders the human–swarm interaction to be bilateral. Differently from [8, 10, 11], the human operator in the presented system interacts with an arbitrary number of robots, thereby enhancing tolerability against hardware and/or communication failures. Although interactions with multiple robots likely increases the human task complexity, it will be demonstrated that the passivity paradigm allows the operator to behave as if he were controlling a single robot. We then prove position/velocity synchronization on the presented system assuming passivity of an appropriately defined decision process of the human operator.

The second goal of this chapter is to analyze passivity of the operator's decision process, using techniques in closed-loop system identification [18]. The human behavior modeling in teleoperation systems is also addressed in [22, 28], but they deal with a single slave robot and the effects of the network connection on the human behavior are not analyzed. In addition, [22, 28] do not focus on passivity of the operator. We first briefly review the results in [16] regarding the position control mode, where the human passivity and the effect of the human learning ability are analyzed and then a human passivation scheme is presented. In this chapter, we demonstrate the human passivity in the velocity control mode, and moreover study the individual variability. To address the latter issue, we build models for another five subjects. Then, we see that some subjects show unstable behavior even after the same number of trials as [16] depending on the network connections, and accordingly undesirable phase lags appear in the low frequency domain, which may violate passivity. Although it is confirmed that additional training passifies all subjects, it is undesirable to require so many trials for the operator. We thus present a scheme to reduce the effort for attaining passivity. It is eventually demonstrated that the operator block for all subjects is passified for three different information structures through limited number of trials by using the presented scheme.

The above results are in part presented in the authors' antecessor [16]. The contributions newly added in the present article are as follows: We first add analytical results for time-varying references both in position and velocity control modes based on $\mathscr{L}_2$ stability. In the human passivity analysis, we address the human passivity analysis in the velocity control mode. We also investigate the individual variability and present novel insights about the human learning ability not presented in [16] together with a new human passivation scheme. The proofs of the technical results are also fully detailed which were either skipped [16] or only a proof sketch was provided.

We finally define passivity, the central concept of this chapter, as a preliminary for the subsequent discussions.

Definition 14.1 ([15, 19]) A system with input $u \in \mathscr{U}$ and output $y \in \mathscr{Y}$ is said to be passive if there exists a constant $\beta \geq 0$ such that

$$\int_0^T y^T(t)u(t)dt \geq -\beta$$

for all input signals $u \in \mathscr{U}$ and for all $T \geq 0$. In addition, the system is said to be very strictly passive if there exist $\beta \geq 0$, $\varepsilon > 0$, and $\gamma > 0$ such that

$$\int_0^T y^T(t)u(t)dt \geq -\beta + \varepsilon \int_0^T \|y(t)\|^2 dt + \gamma \int_0^T \|u(t)\|^2 dt$$

for all input signals $u \in \mathscr{U}$ and for all $T \geq 0$.

14.2 Intended Scenario and Control Goals

In this section, we introduce the intended scenario, illustrated in Fig. 14.1, and control goals addressed in this chapter.

Let us consider a pair of a human operator and n mobile robots $\mathscr{V} = \{1, \dots, n\}$ located on a 2-D plane. Each robot i's dynamics is assumed to be modeled by

$$\dot{q}_i = u_i,\ i \in \mathscr{V}, \tag{14.1}$$

where $q_i \in \mathbb{R}^2$ and $u_i \in \mathbb{R}^2$ are respectively the position and the velocity input of robot i. Every robot can communicate with neighboring robots and the inter-agent information exchanges are modeled by an undirected graph $G = (\mathscr{V}, \mathscr{E})$, $\mathscr{E} \subseteq \mathscr{V} \times \mathscr{V}$. Then, robot i has access to information of robots belonging to the so-called neighbor set $\mathscr{N}_i = \{j \in \mathscr{V} | \ (i, j) \in \mathscr{E}\}$.

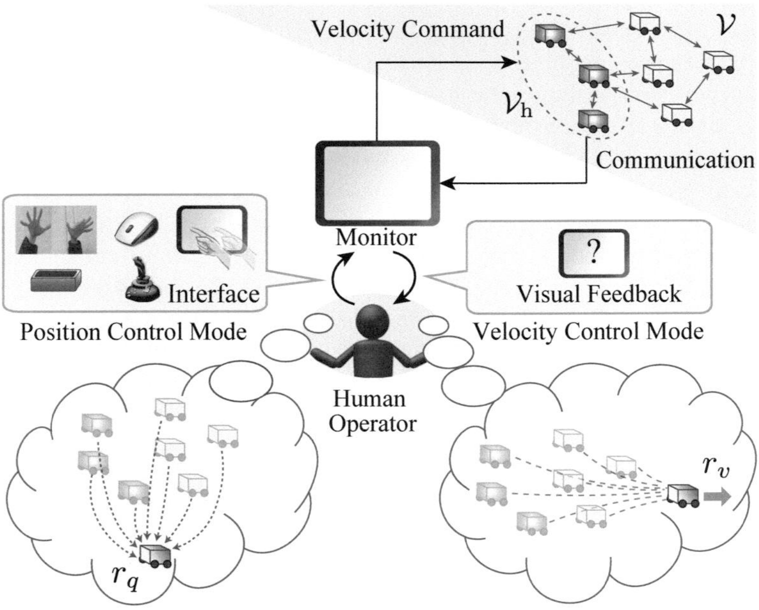

Fig. 14.1 Scenario under consideration in this chapter. A group of mobile robots is in remote environment. Partial information of the robots is fed back to the operator side and it is shown on a display placed in front of the operator. The operator chooses either position or the velocity reference, sends the command signal to a part of the robots so that their motion is synchronized to the reference

The human operator is accessible only to a part of the robotic group, where a robot is said to be accessible from the operator if the operator can send a signal to and receive information from the robot. In the sequel, the set of the accessible robots is denoted by $\mathcal{V}_{\rm h}$. In order to display the information fed back from the robots $\mathcal{V}_{\rm h}$, a monitor is assumed to be placed in front of the operator, and the operator is assumed to visually acquire the information, denoted by $y_{\rm h}$. We also suppose that the operator has a device to manipulate the velocity of the robots, similarly to a video game. Using the device, he/she determines a velocity command signal $u_{\rm h} \in \mathbb{R}^2$ based on $y_{\rm h}$. The device then generates a modified signal $u \in \mathbb{R}^2$ based on $u_{\rm h}$, and sends the common u to all accessible robots, which mitigates the human task complexity.

Once each robot $i \in \mathcal{V}_{\rm h}$ receives the command u, it reflects the signal in the velocity input in the form of

$$u_i = u_{{\rm r},i} + \delta_i u, \tag{14.2}$$

where $\delta_i = 1$ if $i \in \mathcal{V}_{\rm h}$, and $\delta_i = 0$ otherwise. The notation $u_{{\rm r},i}$ represents a signal, designed later, consisting of the feedback information from neighboring robots $\mathcal{N}_i$.

The primary role of the operator is to specify the goal of the robotic group. In this chapter, we address two fundamental goals, namely the position synchronization and velocity synchronization defined below. The operator chooses either of the control modes together with the reference signals. In the sequel, the position and velocity references are denoted by r_q and r_v, respectively. Then, the goal of the position control mode is defined by

$$\lim_{t\to\infty} \|q_i - r_q\| = 0 \ \forall i \in \mathcal{V}, \tag{14.3}$$

while that of the velocity control is by

$$\lim_{t\to\infty} \|\dot{q}_i - r_v\| = 0, \ \lim_{t\to\infty} \|q_i - q_j\| = 0 \ \forall i, j \in \mathcal{V}. \tag{14.4}$$

We assume that the robots do not have access to the selected mode, i.e., the controller determining $u_{{\rm r},i}$ cannot be switched depending on the selected mode. The restriction is imposed in order to ease the implementation by avoiding information broadcasting, required for sharing the mode among all robots. Note also that the references r_q and r_v are only in the operator's brain and the system designer cannot directly utilize/manipulate these signals.

While the primary role is almost static, we also expect, as the secondary role, the human operator to exert flexibility in adapting to environmental changes. For example, as is shown in Sect. 14.5, the dynamics of the robotic group varies when the network G switches due to environmental factors or link failures. Hopefully, the operator is aware of the changes and determines the command to achieve good transient behavior.

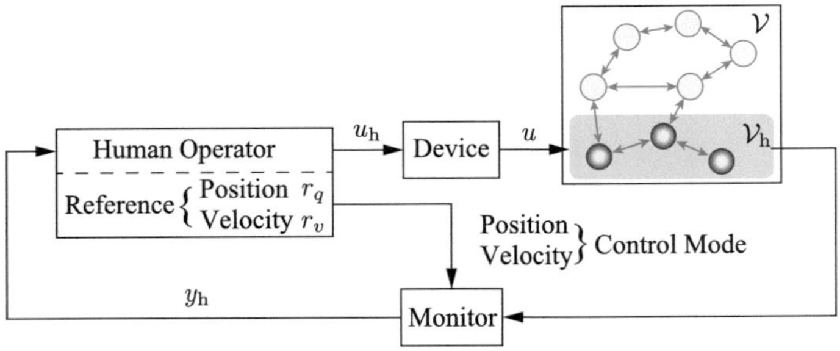

Fig. 14.2 Information flow between human operator and multi-robot system

Third, the operator is expected to seek the command $u_{\rm h}$ such that the control objectives are achieved. This role is usually required only in the velocity control mode since u must be equal to 0 when $q_i = r_q$ $\forall i$ as long as r_q is constant.

The objective here is to then design the robot controller determining $u_{{\rm r},i}$ and the information $y_{\rm h}$ displayed on the monitor so that (14.3) and (14.4) are achieved depending on the selected control mode. Note that, unlike the robot controller, $y_{\rm h}$ can be switched according to the selected control mode, since the switch is executed in the operator side. The block diagram of the complete control system is illustrated in Fig. 14.2.

Remark 14.1 Although (14.3) and the second equation of (14.4) imply collisions among robots, adding biases to the actual robot positions and regarding each q_i as a virtual position with the bias would trivially avoid collisions at least in the final configuration (See, e.g., Chap. 10 of [15]). As the slight modification does not affect the subsequent results, we utilize the above simpler formulation. Collision avoidance in the transient state is left as a future work.

Remark 14.2 Although a commonly used wheeled mobile robot has nonholonomic constraints, it is well known that the dynamics can be reduced to (14.1) using the so-called look ahead control together with input–output feedback linearization [7, 35]. Regarding q_i as the forward reference point [35] instead of the actual position, the subsequent discussions are applied to such wheeled robots.

Remark 14.3 The dynamics (14.1) are much simpler than those utilized in the publications on teleoperation [11, 12, 14, 15, 17, 20, 21, 26, 30, 31]. In other words, our focus is placed on the low frequency domain over which the robot dynamics can be modeled by (14.1). Such simplifications may cause problems in high-speed robot motion control and extensions to more complex dynamics is important, but we take (14.1) for the following reasons. The intended scenario deals with high-level control problems where the performance in a traditional sense is not always an issue. Indeed, (14.1) is often used in cooperative robot control [23, 29] which also addresses high-level missions. Moreover, (14.1) is also commonly employed in the publications on human–swarm-interactions [8, 10].

14.3 Control Architecture and Passivity

In this section, we present a control architecture for the aforementioned human–swarm collaboration problem. To this end, we first design a controller which determines the input $u_{\mathrm{r},i}$. As the robots do not have knowledge of the selected control mode, the controller must be designed so as to achieve both (14.3) and (14.4). To meet this requirement, the following structure presented in [3, 13] is expected to be useful.

$$\dot{\xi}_i = \sum_{j \in \mathcal{N}_i} a_{ij}(q_j - q_i) \tag{14.5a}$$

$$u_{\mathrm{r},i} = \sum_{j \in \mathcal{N}_i} a_{ij}(q_j - q_i) + \sum_{j \in \mathcal{N}_i} a_{ij}(\xi_i - \xi_j) \tag{14.5b}$$

with $a_{ij} = a_{ji}\ \forall i, j \in \mathcal{V}, a_{ij} > 0$ if $(i, j) \in \mathcal{E}$ and $a_{ij} = 0$ otherwise. It is to be noted that the controller (14.5) is itself not new. Our contribution here is in its application to the human–swarm collaboration problem. The following results serve as a bridge between [3, 13] and the present problem.

Substituting (14.5) into (14.1) and (14.2) and collecting the resulting robot dynamics for all $i \in \mathcal{V}$ yields

$$\begin{bmatrix} \dot{\xi} \\ \dot{q} \end{bmatrix} = \begin{bmatrix} 0 & -L \otimes I_2 \\ L \otimes I_2 & -L \otimes I_2 \end{bmatrix} \begin{bmatrix} \xi \\ q \end{bmatrix} + \begin{bmatrix} 0 \\ D \otimes I_2 \end{bmatrix} u, \tag{14.6}$$

where $\xi = [\xi_1^T \ \dots \ \xi_n^T]^T$ and $q = [q_1^T \ \dots \ q_n^T]^T$. The matrix $L \in \mathbb{R}^{n \times n}$ is the so-called Graph Laplacian [29], $D \in \mathbb{R}^{n \times 1}$ is the vector whose ith element is δ_i, I_2 is the 2-dimensional identity matrix and $\otimes$ represents the Kronecker product. In this chapter, we use the following fact without proofs.

Lemma 14.1 ([29]) *The Graph Laplacian L of undirected graphs with n nodes is symmetric and positive semidefinite, and satisfies $L\mathbf{1}_n = 0$ and $\mathbf{1}_n^T L = 0$, where $\mathbf{1}_n$ is the n-D vector whose elements are all* 1.

We next reveal passivity of the system (14.6) when the output is selected as

$$z_q = \frac{1}{m} \sum_{i \in \mathcal{V}_{\mathrm{h}}} q_i = \frac{1}{m} (D \otimes I_2)^T q, \tag{14.7}$$

which describes the average position of the accessible robots. The notation m is the number of elements of $\mathcal{V}_{\mathrm{h}}$. Then, we have the following lemma.

Lemma 14.2 *The system (14.6) describing the robotic group motion is passive from u to z_q with respect to the storage function*

$$S_q = \frac{1}{2m}\|q\|^2 + \frac{1}{2m}\|\xi\|^2 > 0. \tag{14.8}$$

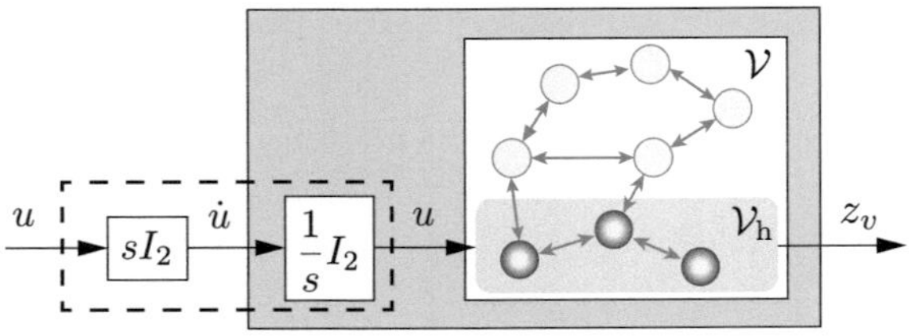

Fig. 14.3 Illustration of the virtual differential/integral operations. The operations are introduced just for analysis and they are not executed in real implementation

Proof The skew-symmetry of the non-diagonal elements of the matrix in the first term of (14.6), as shown in Lemma 14.1, immediately means

$$\dot{S}_q = -\frac{1}{m}q^T(L \otimes I_2)q + \frac{1}{m}q^T(D \otimes I_2)u. \tag{14.9}$$

Using Lemma 14.1 again, (14.9) is rewritten as

$$\dot{S}_q \leq \left(\frac{1}{m}(D \otimes I_2)^T q\right)^T u = z_q^T u \tag{14.10}$$

Integrating (14.10) from time 0 to T completes the proof.

We next focus on the system dynamics in the case of the velocity control mode. Let us assume differentiability of u and virtually rewrite the block diagram of the robots dynamics as in Fig. 14.3. Then, the state equation of the shaded blocks with input $\dot{u}$ and the state ξ, q, and u is formulated by

$$\begin{bmatrix} \dot{u} \\ \dot{\xi} \\ \dot{q} \\ \ddot{\xi} \\ \ddot{q} \end{bmatrix} = \begin{bmatrix} & 0 & 0 \\ & I_n & 0 \\ O & 0 & I_n \\ & 0 & -L \otimes I_2 \\ & L \otimes I_2 & -L \otimes I_2 \end{bmatrix} \begin{bmatrix} u \\ \xi \\ q \\ \dot{\xi} \\ \dot{q} \end{bmatrix} + \begin{bmatrix} I_2 \\ 0 \\ 0 \\ 0 \\ D \otimes I_2 \end{bmatrix} \dot{u}. \tag{14.11}$$

Denote the average velocity of $\mathscr{V}_{\mathrm{h}}$ as

$$z_v = \frac{1}{m}\sum_{i \in \mathscr{V}_{\mathrm{h}}} \dot{q}_i = \frac{1}{m}(D \otimes I_2)^T \dot{q}.$$

Then, we have the following lemma.

Lemma 14.3 *The system (14.11) is passive from $\dot{u}$ to z_v with respect to the storage function*

$$S_v = \frac{1}{2m}\|\dot{q}\|^2 + \frac{1}{2m}\|\dot{\xi}\|^2 \geq 0. \tag{14.12}$$

Proof The time derivative of S_v along with the trajectories of (14.11) is given by

$$\begin{aligned} \dot{S}_v &= -\frac{1}{m}\dot{q}^T(L \otimes I_2)\dot{q} + \frac{1}{m}\dot{q}^T(D \otimes I_2)\dot{u} \\ &\leq \frac{1}{m}\dot{q}^T(D \otimes I_2)\dot{u} = z_v^T\dot{u} \end{aligned} \tag{14.13}$$

from Lemma 14.1 and skew-symmetry of the non-diagonal elements of the matrix in the first term of (14.11). Integrating (14.13) from time 0 to T completes the proof.

14.4 Convergence Analysis

14.4.1 Synchronization in Position Control Mode

Let us first consider the case, where the position control mode is selected by the operator. In this case, we suppose that the device determines the signal u using a map F_q as $u = F_q(u_h)$. Although general descriptions might be possible, it is sufficient, in the context of this chapter, to assume that the map F_q is given by the transfer function

$$F_q(s) = k\frac{T_1 s + 1}{T_2 s + 1}, \quad k > 0,\ T_1 > 0,\ T_2 > 0. \tag{14.14}$$

The filter with $k = 1$ was used in order to passify the human operator in [16], and it is also employed for the same purpose in this chapter. The reason for introducing the gain k will be revealed in Sect. 14.5.

Based on Lemma 14.3, we first determine the information y_h displayed on the monitor as

$$y_h = z_q = \frac{1}{m}\sum_{i \in \mathcal{V}_h} q_i, \tag{14.15}$$

which means that the operator visually feedbacks the average position of the accessible robots. It is additionally assumed that the human operator computes the error $r_q - y_h$, and determines the action u_h according to the error. Then, the block diagram is rewritten as Fig. 14.4, where the operator Φ describes the human decision process to determine the velocity command u_h. Note that each robot $i \in \mathcal{V}_h$ sends its own position q_i to the operator side, and the average (14.15) is computed at the operator side, which ensures distributed information exchanges among robots. In the sequel, the map from the error signal $r_q - y_h$ to u including (14.14) is denoted by Ψ_q.

Let us now prove (14.3) under the following assumptions.

Assumption 14.1 The graph G is fixed and connected.

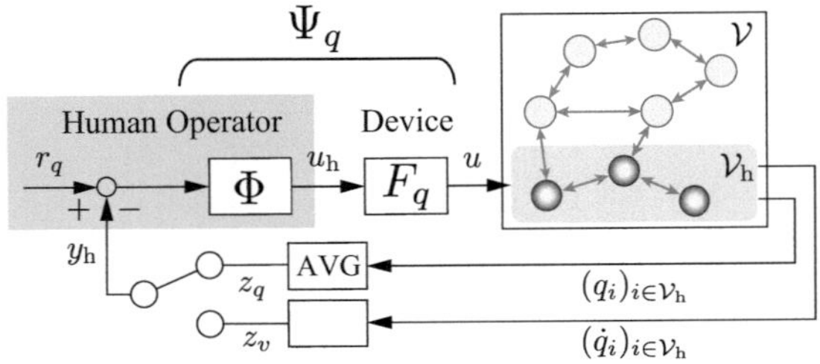

Fig. 14.4 Block diagram of the system in the case of position control mode, where the block AVG describes the computation of the average

Assumption 14.2

- r_q is constant.
- The operator identically sends $u_{\rm h} = 0$ only if $y_{\rm h} = r_q$ identically holds.
- If the input to the human $r_q - y_{\rm h}$ is bounded, the internal states of the decision process Φ is also bounded.
- The map Ψ_q is time invariant (i.e., Φ is time invariant) and passive from $r_q - y_{\rm h} = r_q - z_q$ to u, i.e., $\exists \beta \geq 0$ such that

$$\int_0^T u^T(\zeta)(r_q - y_{\rm h}(\zeta))d\zeta \geq -\beta \ \forall T > 0. \tag{14.16}$$

The first assumption implies the operator changes the goal (r_q and r_v in addition to control mode) infrequently so that the system behavior can be analyzed assuming a fixed mode and constant references. The second assumption is reasonable. Although there is ambiguity on the human internal states in the third item, it is reasonable to assume that variables affected by $r_q - y_{\rm h}$ or affecting $u_{\rm h}$ do not diverge. However, the fourth assumption requires additional explanation and this is addressed in the next section.

Utilizing the above assumption, we have the following result, originally presented in [16].

Theorem 14.1 *Consider the feedback connection (Fig. 14.4) consisting of n robots described by (14.1) and (14.5), connected by a graph G satisfying Assumption 14.1, and a map Ψ_q, with a decision process Φ and the device block F_q in (14.14), satisfying Assumption 14.2. Then, the feedback system achieves the condition (14.3).*

Proof Define $\bar{q} = q - (\mathbf{1}_n \otimes I_2) r_q$. Since r_q is constant, we have

$$\dot{\xi} = -(L \otimes I_2)\bar{q}, \tag{14.17a}$$
$$\dot{\bar{q}} = -(L \otimes I_2)\bar{q} + (L \otimes I_2)\xi + (D \otimes I_2)u, \tag{14.17b}$$

from (14.6). Since the dynamics (14.17a, 14.17b) takes the same form as (14.6), if we define $\bar{S}_q = \frac{1}{2m}\|\bar{q}\|^2 + \frac{1}{2m}\|\xi\|^2 > 0$, we obtain

$$\dot{\bar{S}}_q = -\frac{1}{m}\bar{q}^T(L \otimes I_2)\bar{q} + \frac{1}{m}\bar{q}^T(D \otimes I_2)u \tag{14.18}$$

in the same manner as (14.9) and (14.10). Noticing that $r_q - y_{\mathrm{h}} = -\frac{1}{m}(D \otimes I_2)^T \bar{q}$, (14.16) is rewritten as

$$\frac{1}{m}\int_0^t -u^T(\zeta)((D \otimes I_2)^T \bar{q}(\zeta))d\zeta \geq -\beta. \tag{14.19}$$

Define

$$S_{\mathrm{h}}^q = \frac{1}{m}\int_0^t -u^T(\zeta)((D \otimes I_2)^T \bar{q}(\zeta))d\zeta + \beta. \tag{14.20}$$

The time derivative of the function S_{h}^q is then given by

$$\dot{S}_{\mathrm{h}}^q = -\frac{1}{m}((D \otimes I_2)^T \bar{q})^T u. \tag{14.21}$$

Defining $V_q = \bar{S}_q + S_{\mathrm{h}}^q \geq 0$, then we have

$$\dot{V}_q = -\frac{1}{m}\bar{q}^T(L \otimes I_2)\bar{q} \tag{14.22}$$

from (14.18) and (14.21).

Utilizing the assumption that Ψ_q is time invariant, LaSalle's principle can be applied to the current problem. We see from (14.22) that $\dot{V}_q = 0$ is equivalent to

$$(L \otimes I_2)\bar{q} = 0, \tag{14.23}$$

which implies position synchronization under Assumption 14.1. Consider the trajectories of the system satisfying (14.23). Differentiating (14.23) with respect to time yields

$$0 = (L \otimes I_2)\{(L \otimes I_2)\xi + (D \otimes I_2)u\}. \tag{14.24}$$

Under Assumption 14.1, (14.24) is equivalent to

$$(L \otimes I_2)\xi + (D \otimes I_2)u = (\mathbf{1}_n \otimes I_2)c(t) \tag{14.25}$$

with $c(t) \in \mathbb{R}^2\ \forall t \geq 0$. Then, multiplying $(\mathbf{1} \otimes I_2)^T$ from left to both sides of (14.25) yields $mu(t) = nc(t)$ from Lemma 14.1. Substituting this into (14.25) yields

$$(L \otimes I_2)\xi = (\mathbf{1}_n \otimes I_2 - (n/m)D \otimes I_2)c(t). \tag{14.26}$$

From (14.17a) and (14.23), the left-hand side of (14.26) is constant and hence c and u must be also constant. Suppose that $c \neq 0$. Then, we differentiate $((D \otimes I_2)^T \bar{q})^T u$ as

$$\frac{d}{dt}((D \otimes I_2)^T \bar{q})^T u = u^T(D^T \otimes I_2)\dot{\bar{q}}$$
$$= \frac{n}{m} c^T (D^T \otimes I_2)(\mathbf{1}_n \otimes I_2)c = n\|c\|^2 > 0. \quad (14.27)$$

This implies that $u^T((D \otimes I_2)^T \bar{q}) = -mu^T(r_q - y_{\rm h})$ is monotonically increasing, which contradicts the passivity condition (14.16). Thus, c must equal 0. Then, $mu = nc$ means $u = u_{\rm h} = 0$. From the second item of Assumption 14.2, we have $z_q = r_q$, which implies $q_i = r_q \ \forall i \in \mathscr{V}$ due to position synchronization (14.23). Hence, all system solutions satisfying identically $\dot{V}_q = 0$ must also satisfy (14.3). This completes the proof.

It is easy to confirm from internal model principle that the condition (14.3) is not satisfied in the case where the reference r_q is varying. Even in such a case, we have the following corollary under a stronger assumption on the human characteristics.

Corollary 14.1 *Consider the feedback connection (Fig. 14.4) consisting of n robots described by (14.1) and (14.5) connected by a graph G satisfying Assumption 14.1, and a map Ψ_q with a decision process Φ and the device block F_q in (14.14). Then, if the map Ψ_q is very strictly passive, i.e., $\exists \beta \geq 0, \varepsilon > 0, \gamma > 0$ such that*

$$\int_0^T (r_q(\zeta) - y_{\rm h}(\zeta))^T u(\zeta) d\zeta \geq -\beta + \varepsilon \int_0^T \|u(\zeta)\|^2 d\zeta + \gamma \int_0^T \|r_q(\zeta) - y_{\rm h}(\zeta)\|^2 d\zeta$$

for all $T > 0$, then the system from r_q to $r_q - y_{\rm h}$ is $\mathscr{L}_2$ stable.

Proof Define

$$\tilde{S}_{\rm h}^q = \int_0^t (r_q(\zeta) - y_{\rm h}(\zeta))^T u(\zeta) d\zeta$$
$$- \varepsilon \int_0^t \|u(\zeta)\|^2 d\zeta - \gamma \int_0^t \|r_q(\zeta) - y_{\rm h}(\zeta)\|^2 d\zeta + \beta, \quad (14.28)$$

and $\tilde{V}_q = S_q + \tilde{S}_{\rm h}^q$. Then, from (14.10), the time derivative of $\tilde{V}_q$ along the system trajectories is given as

$$\dot{\tilde{V}}_q \leq y_{\rm h}^T u + (r_q - y_{\rm h})^T u - \varepsilon\|u\|^2 - \gamma\|r_q - y_{\rm h}\|^2$$
$$= r_q^T u - \varepsilon\|u\|^2 - \gamma\|r_q - y_{\rm h}\|^2$$

By completing square, we also have

$$\dot{\tilde{V}}_q + \gamma\|r_q - y_{\rm h}\|^2 \leq \frac{1}{4\varepsilon}\|r_q\|^2.$$

Now, integrating both sides of (14.40) from time 0 to T yields

$$\tilde{V}_q(T) + \gamma \int_0^T \|r_q(\zeta) - y_{\rm h}(\zeta)\|^2 d\zeta \leq \frac{1}{4\varepsilon} \int_0^T \|r_q(\zeta)\|^2 d\zeta + \tilde{V}_q(0). \quad (14.29)$$

Since $V_q(T)$ must be nonnegative, we also obtain

$$\int_0^T \|r_q(\zeta) - y_{\rm h}(\zeta)\|^2 d\zeta \leq \frac{1}{4\varepsilon\gamma} \int_0^T \|r_q(\zeta)\|^2 d\zeta + \frac{1}{\gamma}\tilde{V}_q(0), \quad (14.30)$$

which completes the proof.

14.4.2 Synchronization in Velocity Control Mode

We next consider the velocity control mode. In this case, we determine the signal u by filtering $u_{\rm h}$ using a filter

$$F_v(s) = \frac{1}{T_v s + 1}, \quad T_v > 0 \quad (14.31)$$

in order to ensure differentiability of u. Remark that u is computed in the operator side, and hence, the manipulation of $u_{\rm h}$ can differ between the position and velocity control mode.

We also determine $y_{\rm h}$ fed back to the operator as $y_{\rm h} = z_v$ based on Lemma 14.3. Similarly to Sect. 14.4.1, it is assumed that the human operator computes the error $r_v - y_{\rm h}$ and determines the action $u_{\rm h}$ according to the error. Then, the block diagram of the system is illustrated as in Fig. 14.5. In the sequel, denote the map from $r_v - y_{\rm h}$ to $\dot{u}$ by Ψ_v as shown in Fig. 14.5.

Analogous with Assumption 14.2, the following conditions are assumed.

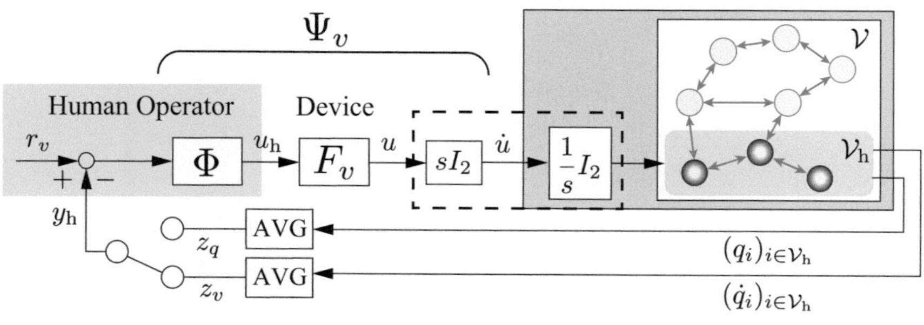

Fig. 14.5 Block diagram of the system in the case of velocity control mode

Assumption 14.3

- r_v is constant.
- $u_{\rm h}$ is identically equal to a constant value if and only if $y_{\rm h} = r_v$ identically holds.
- If the input to the human $r_v - y_{\rm h}$ is bounded, the internal states of Φ and u are also bounded.
- The map Ψ_v is time invariant and passive from $r_v - y_{\rm h} = r_v - z_v$ to $\dot{u}$.

Then, we have the following theorem, originally presented in [16].

Theorem 14.2 *Consider the feedback connection in Fig. 14.5 consisting of n robots described by (14.1) and (14.5), connected by a graph G satisfying Assumption 14.1, and a human operator Φ with the filter F_v satisfying Assumption 14.3. Then, if $m < n$, the feedback system achieves the conditions in (14.4).*

Proof Define $\tilde{q} = q - (\mathbf{1}_n \otimes I_2)(\int_0^t r_v d\tau)$ and $\tilde{\xi} = (L \otimes I_2)\xi$. Then, we have

$$\dot{u} = \dot{u} \tag{14.32a}$$
$$\dot{\tilde{\xi}} = -(L^2 \otimes I_2)\tilde{q}, \tag{14.32b}$$
$$\dot{\tilde{q}} = -(L \otimes I_2)\tilde{q} + \tilde{\xi} + (D \otimes I_2)u - (\mathbf{1}_n \otimes I_2)r_v, \tag{14.32c}$$

from (14.11). We also define $\tilde{S}_v = \frac{1}{2m}\|\dot{\tilde{q}}\|^2 + \frac{1}{2m}\tilde{q}^T(L^2 \otimes I_2)\tilde{q} \geq 0$. Then the time derivative of $\tilde{S}_v$ along (14.32b, 14.32c, 14.32a) satisfies

$$\dot{\tilde{S}}_v = -\frac{1}{m}\dot{\tilde{q}}^T(L \otimes I_2)\dot{\tilde{q}} + \frac{1}{m}\dot{\tilde{q}}^T(D \otimes I_2)\dot{u}, \tag{14.33}$$

which is proved in the same manner as (14.13). Noticing that $r_v - y_{\rm h} = -\frac{1}{m}(D \otimes I_2)^T\dot{\tilde{q}}$, the fourth item of Assumption 14.3 implies

$$S_{\rm h}^v = \frac{1}{m}\int_0^T -((D \otimes I_2)^T\dot{\tilde{q}}(\zeta))^T\dot{u}(\zeta)d\zeta + \beta \geq 0. \tag{14.34}$$

The time derivative of the function $S_{\rm h}^v$ is given by

$$\dot{S}_{\rm h}^v = -\frac{1}{m}((D \otimes I_2)^T\dot{\tilde{q}})^T\dot{u}. \tag{14.35}$$

Defining $V_v = \tilde{S}_v + S_{\rm h}^v \geq 0$, we have

$$\dot{V}_v = -\frac{1}{m}\dot{\tilde{q}}^T(L \otimes I_2)\dot{\tilde{q}} \tag{14.36}$$

From (14.36), $(L \otimes I_2)\tilde{q}$ is bounded. Then, the second item of Assumption 14.3 implies boundedness of the internal states of the operator and u. Equation (14.36) also implies boundedness of $\dot{\tilde{q}}$. Since every term on the right-hand side of (14.32c) is proved to be bounded except for $\tilde{\xi}$ and the left-hand side is also bounded, $\tilde{\xi}$ must be bounded. Namely, there exists a positively invariant set for the system (14.32b, 14.32c, 14.32a) and LaSalle's principle is applicable.

Note from (14.36) that $\dot{V}_v = 0$ is equivalent to $(L \otimes I_2)\dot{\tilde{q}} = 0$ which means velocity synchronization under Assumption 14.1. Consider the trajectories of the system identically satisfying $(L \otimes I_2)\dot{\tilde{q}} = 0$ which implies that the right-hand side of (14.32b) is constant. Since the assumption of $(L \otimes I_2)\tilde{q} \neq 0$ contradicts boundedness of $\tilde{\xi}$, we can conclude position synchronization and the fact that $\tilde{\xi}$ is a constant. Let us now multiply $(\mathbf{1}_n \otimes I_2)^T$ to (14.32c) from the left. Then, from the definition of $\tilde{\xi} = (L \otimes I_2)\xi$ and position synchronization, we have

$$\dot{\tilde{q}}_i = \frac{m}{n}u - r_v. \tag{14.37}$$

Substituting (14.37) into (14.32c) and using the fact of position synchronization, we get

$$\left(\frac{m}{n} - \delta_i\right) u = \tilde{\xi}_i. \tag{14.38}$$

Since $n \neq m$ and $\tilde{\xi}_i$ is constant, u also remains constant. From the second item of Assumption 14.3, we confirm $r_v = z_v$ and hence $r_v = \dot{q}_i \ \forall i \in \mathscr{V}$ from $(L \otimes I_2)\dot{\tilde{q}} = 0$. This completes the proof of (14.4).

We see from (14.37) that u has to converge to $\frac{n}{m}r_v$. Although the ratio $\frac{n}{m}$ may change due to link/robot failures, it is not hard for the operator to find $\frac{n}{m}r_v$ from the information of $r_v - y_{\rm h}$ even without such knowledge.

In the case of $n = m$, we see from the above proof that $z_v = \dot{q}_i = u$ asymptotically holds. Namely, if the human is rational enough to eventually send $u_{\rm h} = r_v$ in such an ideal situation, (14.4) is also achieved in the case of $n = m$.

If r_v is not constant, we can prove the following result similarly to Corollary 14.1.

Corollary 14.2 *Consider the feedback connection (Fig. 14.5) consisting of n robots described by (14.1) and (14.5), connected by a graph G satisfying Assumption 14.1, and a human operator Φ with the filter F_v. If Ψ_v is very strictly passive, i.e., $\exists \beta \geq 0 \varepsilon > 0, \gamma > 0$ such that*

$$\int_0^T (r_v(\zeta) - y_{\rm h}(\zeta))^T \dot{u}(\zeta) d\zeta \geq -\beta + \varepsilon \int_0^T \|\dot{u}(\zeta)\|^2 d\zeta + \gamma \int_0^T \|r_v(\zeta) - y_{\rm h}(\zeta)\|^2 d\zeta$$

for all $T > 0$, then the feedback system from r_v to $(r_v - y_{\rm h})$ is $\mathscr{L}_2$ stable.

Proof Define

$$\tilde{S}_{\rm h}^v = \int_0^t (r_v(\zeta) - y_{\rm h}(\zeta))^T \dot{u}(\zeta) d\zeta - \varepsilon \int_0^t \|\dot{u}(\zeta)\|^2 d\zeta - \gamma \int_0^t \|r_v(\zeta) - y_{\rm h}(\zeta)\|^2 d\zeta + \beta, \tag{14.39}$$

and $\tilde{V}_v = S_v + \tilde{S}_{\rm h}^q$. Then, from (14.13), the time derivative of $\tilde{V}_v$ along the system trajectories is given as

$$\dot{\tilde{V}}_v \leq y_{\rm h}^T u + (r_v - y_{\rm h})^T \dot{u} - \varepsilon \|u\|^2 - \gamma \|r_v - y_{\rm h}\|^2$$
$$= r_v^T \dot{u} - \varepsilon \|u\|^2 - \gamma \|r_v - y_{\rm h}\|^2.$$

By completing square, we also have

$$\dot{\tilde{V}}_v + \gamma \|r_v - y_{\rm h}\|^2 \leq \frac{1}{4\varepsilon} \|r_v\|^2.$$

Now, integrating both sides of (14.40) from time 0 to T yields

$$\tilde{V}_v(T) + \gamma \int_0^T \|r_v - y_{\rm h}\|^2 d\zeta \leq \frac{1}{4\varepsilon} \int_0^T \|r_v(\zeta)\|^2 d\zeta + \tilde{V}_v(0). \tag{14.40}$$

Since $V_q(T)$ must be nonnegative, we also obtain

$$\int_0^T \|r_v - y_{\rm h}\|^2 d\zeta \leq \frac{1}{4\varepsilon\gamma} \int_0^T \|r_v(\zeta)\|^2 d\zeta + \frac{1}{\gamma} \tilde{V}_v(0), \tag{14.41}$$

which completes the proof.

In the next section, we demonstrate the human operator passivity assumed in the aforementioned results similarly to [16], since passivity between force and velocity as input–output pairs has been assumed since early works [1, 5], however it is nontrivial to ascertain whether this assumption also holds for the proposed human operator decision process. Although a human decision process similar to ours is also analyzed in [22, 28], their focus is not placed on human passivity. Also, they are using a traditional interface, joystick, and hence whether the analysis is applied to the novel interfaces is unclear. More importantly, they deal with a single slave robot and the effects of the network connection on the human model are not analyzed. In the next section, we thus examine the human operator passivity through experiments on a human-in-the-loop simulator. It is to be noted that this chapter adds novel analysis on the individual variability and velocity control not presented in [16].

14.5 Passivity of the Human Operator Decision Process

In the previous section, we demonstrated synchronization assuming passivity of the human operator's decision processes. This section examines validity of this assumption by using data acquired on a human-in-the-loop simulator together with system identification techniques. It should be emphasized that we are interested only in the properties of the human working in the presented closed-loop system. Thus, the data

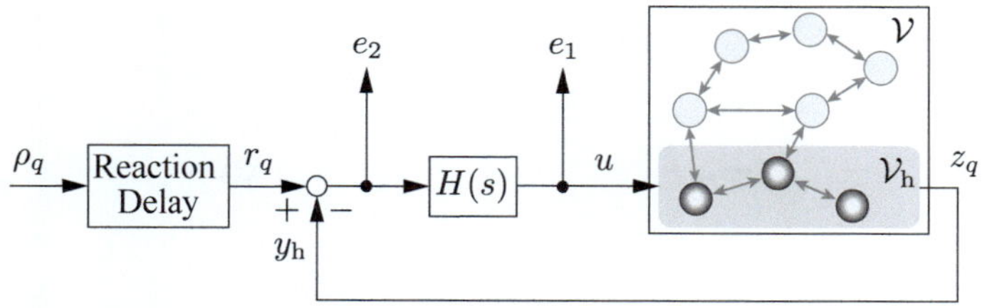

Fig. 14.6 Illustration of the system assumed in the experimental study for the position control mode. The notation ρ_q describes the reference position displayed on the monitor, which are assumed to be recognized with delay as r_q by the operator

must be acquired through simulations of the closed-loop system, and hence we have to employ techniques of closed-loop system identification [18].

Before starting analysis, we list simplifications taken in the section.

- We approximate the human decision process Φ by a stable and proper linear time invariant system.
- We focus only on the 1-D motion assuming that the human decisions for x- and y-coordinates can be decoupled, which seems to be reasonable since the coordinates are completely decoupled at least in the plant (multi-robot system) side but investigations on the coupling effects might be necessary in the future.

Under the above simplifications, the map Ψ_q or Ψ_v including the human decision process is described by a transfer function $H(s)$, and it is passive if the phase of $H(s)$ lies within $\pm 90°$ [2]. The block diagram for the position control mode under these simplifications is depicted in Fig. 14.6, where the computation (14.15) is omitted for simplification.

We also restrict the frequency domain over which we examine if the phase condition holds, since we cannot expect that the human is passive over the whole frequency domain. For example, it is known that the Nyquist frequency of the human visual feedback is equal to about 100 rad/s, and we cannot expect that the phase condition holds over such a high frequency domain. We thus focus only on the domain up to 10 rad/s under the assumption that the human can successfully utilize the information for control at least over the domain. Remark that the selection of the target frequency domain is the same as [22, 28].

14.5.1 Experimental Setup and Approach

We develop a simulator shown in Fig. 14.7. An operator holds the tablet, Apple iPad mini MD533J/A (64 GB), in the figure and touches a point whose x-coordinate specifies the velocity command u_h. The command is depicted by the line in Fig. 14.7. The average position z_q or z_v to be fed back to the operator is shown by a dot in Fig. 14.7. A trial subject is told to lead the dot to the mark $\oplus$ which is regarded as

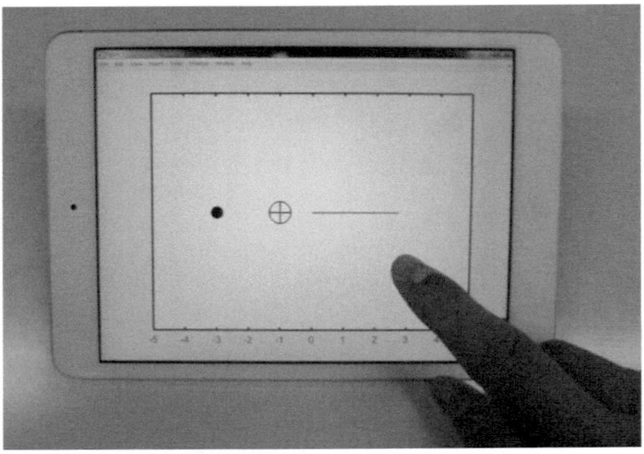

Fig. 14.7 Developed simulator, where the feedback information $y_h = z_q$ or $y_h = z_v$ is displayed by a *dot*. The *mark* $\oplus$ and the *line* describe the externally provided reference ρ_q or ρ_v and human command u_h, respectively

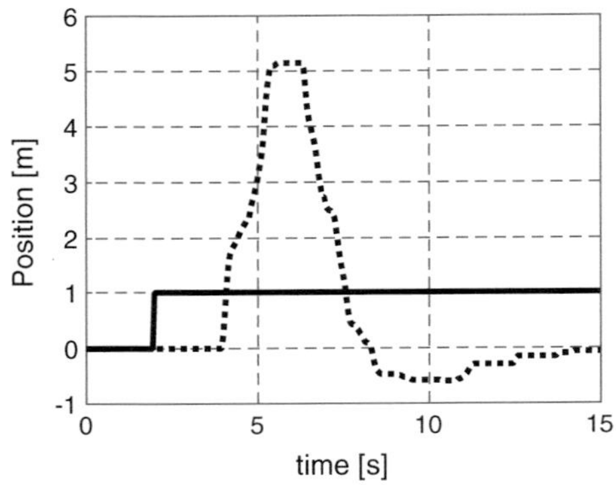

Fig. 14.8 Time responses of ρ_q (*solid*) and u_h (*dashed*), where it is confirmed that the human reacts to the change of ρ_q after unignorable delay

an input for identification and randomly jumps to either of the integers from -3 to 3 at every 20 s. In the sequel, the position of $\oplus$ is denoted by ρ_q in the case of the position control mode, and by ρ_v in the velocity control mode.

A time responses of u_h for a trajectory of ρ_q with a jump is depicted in Fig. 14.8. It is observed from the figure that u_h has an unignorable delay from the rise of ρ_q. The papers [22, 28] take a model, where the delay is included in the feedback loop. However, in the authors' opinion, it is mainly caused by the time for awareness of the new mark positions, which should be placed outside of the loop as shown in Fig. 14.6. If it is true, the phenomenon is unique to the experimental system, since the reference r_q is not externally given but generated in the human operator's brain in practice. In order to eliminate the reaction delay from the identification process, we also add program codes so that the trial subject makes a declaration that the new ρ_q is recognized by tapping the monitor while recording the time. Hereafter, we regard, as r_q, the shifted signal illustrated by the solid line in Fig. 14.9. Remark that there must also be delays in the loop as stated in [22, 28] but they are included in the data, and hence, the resulting model reflects their effects. The same codes to eliminate the reaction delay are also added to the program for the velocity control mode.

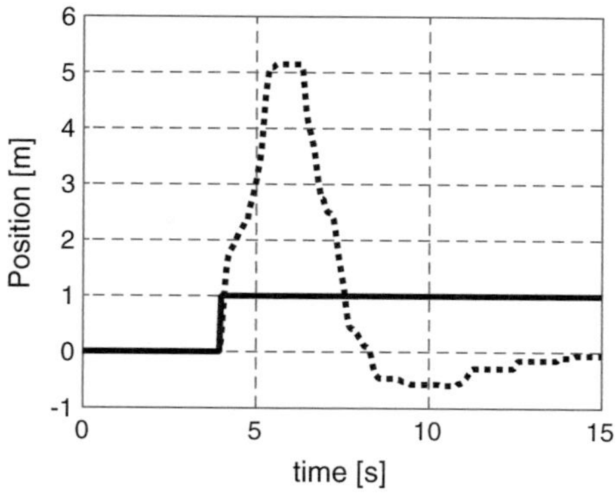

Fig. 14.9 Time responses of the shifted signal r_q (*solid*) and u_h (*dashed*)

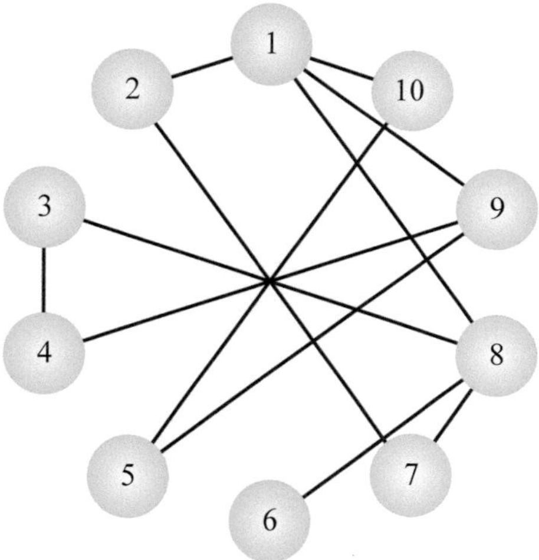

Fig. 14.10 Graph G_1 with sparse edges

Once the experimental data are acquired, we apply the joint input–output method [18] in order to identify $H(s)$. Namely, in the position control mode, we define $e_1 = u$ and $e_2 = r_q - y_h$ as in Fig. 14.6 and conduct open-loop identification of the transfer function from r_q to e_1, denoted by $H_1(s)$, and the function from r_q to e_2, denoted by $H_2(s)$. Then, $H(s)$ can be identified by the computation $H(s) = \frac{H_1(s)}{H_2(s)}$. However, since we are interested only in the phase diagram of H, it is sufficient to acquire the frequency response characteristics. In this chapter, we thus employ a nonparametric system identification method to identify H_1 and H_2 in order to avoid the issue regarding the order selection. A similar procedure is also taken in the velocity control mode.

Throughout this section, we consider ten robots and two different connected graphs G_1 and G_2 in Figs. 14.10 and 14.11 consisting of sparse and dense edges respectively. We focus on three types of human–robotic networks: Type 1 with $\mathcal{V}_h = \mathcal{V}$ and G_1, Type 2 with $\mathcal{V}_h = \{5\}$ and G_2 and Type 3 with $\mathcal{V}_h = \{7\}$ and G_1. Here, we let $a_{ij} = 1 \, \forall (i, j) \in \mathcal{E}$. Then, the Bode diagrams of these systems from the

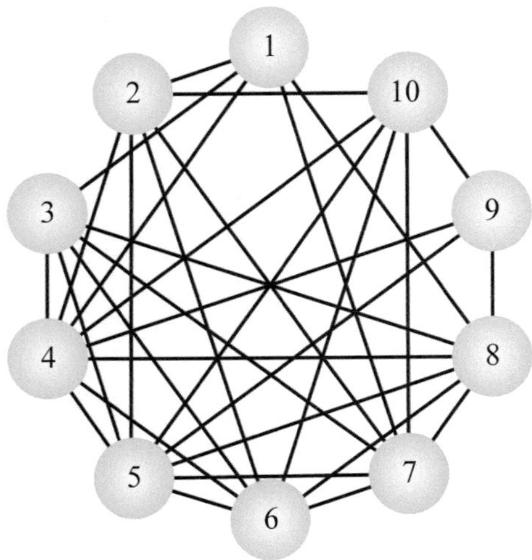

Fig. 14.11 Graph G_2 with dense edges

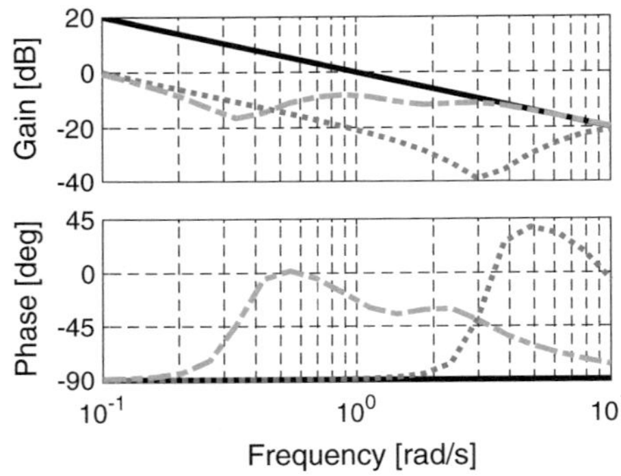

Fig. 14.12 Bode diagram of plants (multi-robot system) for the position control mode, where the *solid*, *dotted* and *dashed curves* respectively describe Type 1, 2 and 3

input u to $y_h = z_q$ are illustrated in Fig. 14.12, where the solid, dotted, and dashed curves respectively describe the systems of Type 1, 2, and 3. The line colors are also consistent with the subsequent figures. We see that the diagram for Type 1 with $\mathscr{V}_h = \mathscr{V}$ is equal to the integrator while the other (gain) diagrams include notches at different frequency domains. Note that the frequency of the notch for the dense graph is higher than that for the sparse graph. In the case of the velocity control mode, the dynamics from u to $y_h = z_v$ is given by simply adding a differentiator to the output and the Bode diagrams are illustrated in Fig. 14.13.

14.5.2 Analysis on Human Passivity in Position Control Mode

This subsection treats the position control mode and reviews the results of [16]. We first set $F_q(s) = 1$ and have a trial subject repeat 2 min trials ten times. Then, we

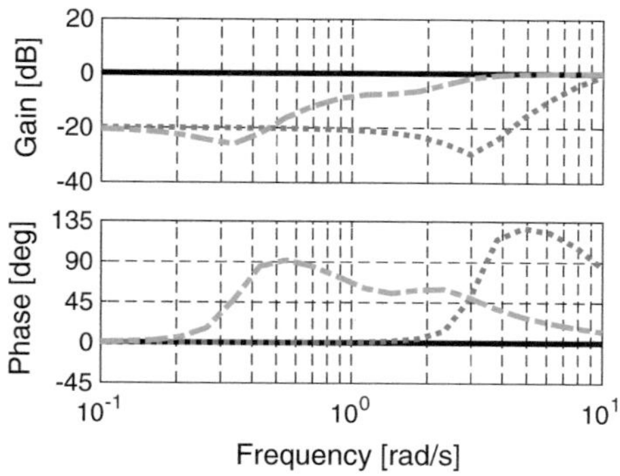

Fig. 14.13 Bode diagram of plants (multi-robot system) for the velocity control mode, where the *solid*, *dotted* and *dashed curves* respectively describe Type 1, 2 and 3

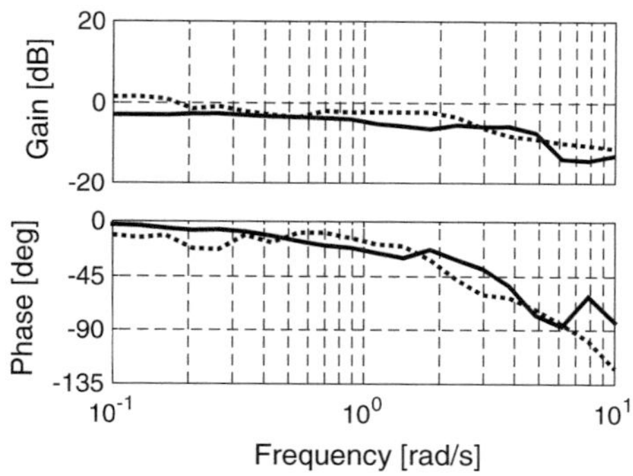

Fig. 14.14 Bode diagram of the identified human decision process Φ for Type 1

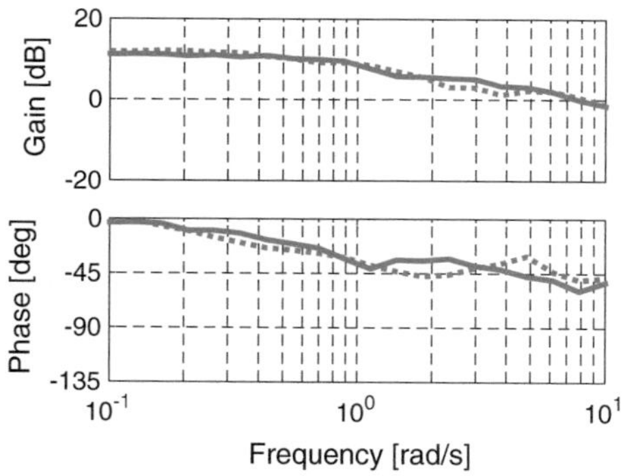

Fig. 14.15 Bode diagram of the identified human decision process Φ for Type 2

build two human models using the former five trial data and the latter respectively in order to analyze the effects of the human learning. The Bode diagrams of the identified models are depicted in Figs. 14.14, 14.15, and 14.16, where the dashed curve and solid curve respectively show the model for the former data and that for the latter.

Regarding the models, [16] pointed out that:

- The human decision process is dependent on the multi-robot dynamics.
- Good operability does not always imply human passivity, which is confirmed from the fact that the phase lag of Fig. 14.14 is larger than that of Fig. 14.15 although

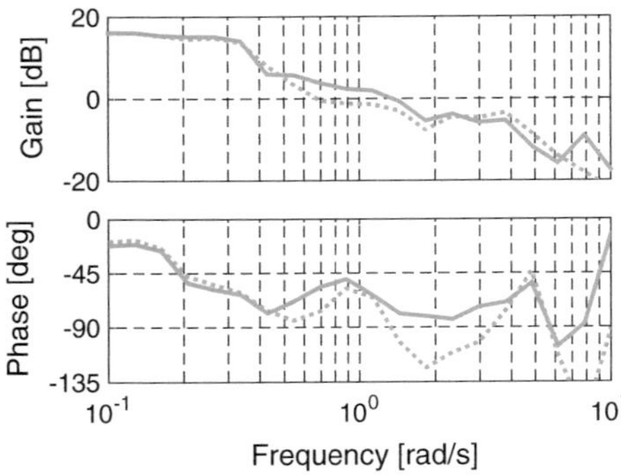

Fig. 14.16 Bode diagram of the identified human decision process Φ for Type 3

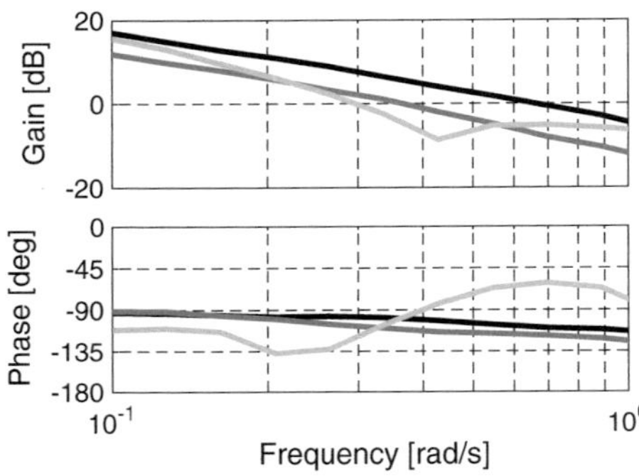

Fig. 14.17 Bode diagram of open-loop transfer function

Type 1 has the best operability in the sense that u_{h} almost directly affect the velocity of z_p as seen from Fig. 14.12.

- Human learning ability works so that the phase lags over the high frequency domain at around 1–10 rad/s are reduced, namely the decision process approaches a passive system through trials.
- In the low frequency domain at around 0.1–1 rad/s, the human learns the inverse model of the plant and accordingly the open-loop transfer function is shaped to have a sufficient DC gain and a smooth form. They can be confirmed from the differences in the DC gains between Fig. 14.14 and Figs. 14.15 and 14.16, and the phase lag at around 0.3–0.5 rad/s in Fig. 14.16. This investigation is at least partially consistent with the crossover model in [22, 28]. The resulting open-loop transfer functions over the domain are illustrated in Fig. 14.17.

Please refer to [16] for more detailed analysis.

In [16], we also focused on the fact that the phase in Fig. 14.16 steps over -90° in the domain greater than 5 rad/s. Then, based on the hypothesis that the human operator was not able to manage well the high frequency elements, we added the high-pass filter

$$F_q(s) = \frac{0.24s + 1}{0.04s + 1} \tag{14.42}$$

(Fig. 14.18) in order to automatize the management of the high frequency signals. The resulting Ψ_q including both of Φ and F_q for Type 1, 2, and 3 are shown by the solid curves in Figs. 14.19, 14.20 and 14.21, respectively. The dashed curves in the figures describe the models shown in Figs. 14.14, 14.15 and 14.16 obtained without the filter. We can observe that the phases lie within $\pm 90°$ and hence Ψ_q get passive for all network types. However, all of the above results are obtained only for a specific trial subject.

Remark 14.4 Here, we use the data sampled at every 0.01s. Since identification is known to be sensitive to the selection of the sampling period, we also conducted the identification by decimating the data so that the sampling period is 0.1s but we could not confirm any noticeable changes.

14.5.3 Analysis on Human Passivity in Velocity Control Mode

Let us next consider the velocity control mode. Here, we select the filter F_v as

$$F_v(s) = \frac{1}{0.083s + 1} \tag{14.43}$$

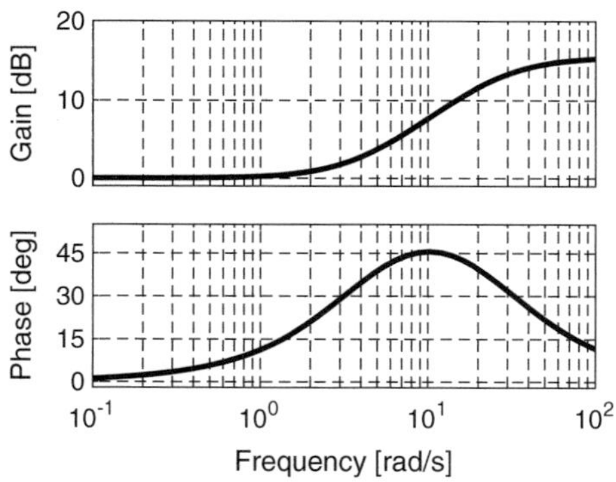

Fig. 14.18 Bode diagram of the high-pass filter in (14.42) added in order to passify the human operator

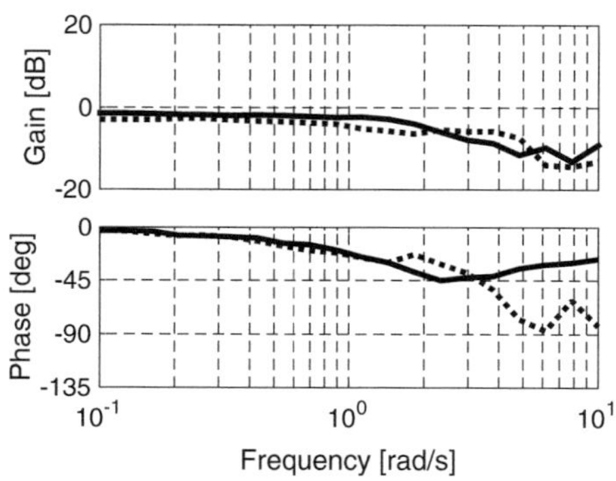

Fig. 14.19 Bode diagram of the identified Ψ_q including the decision process Φ and the filter $F_q(s)$ in (14.42) for Type 1

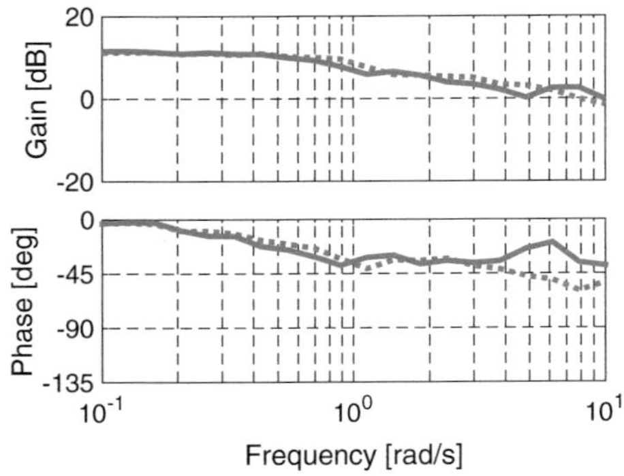

Fig. 14.20 Bode diagram of the identified Ψ_q including the decision process Φ and the filter $F_q(s)$ in (14.42) for Type 2

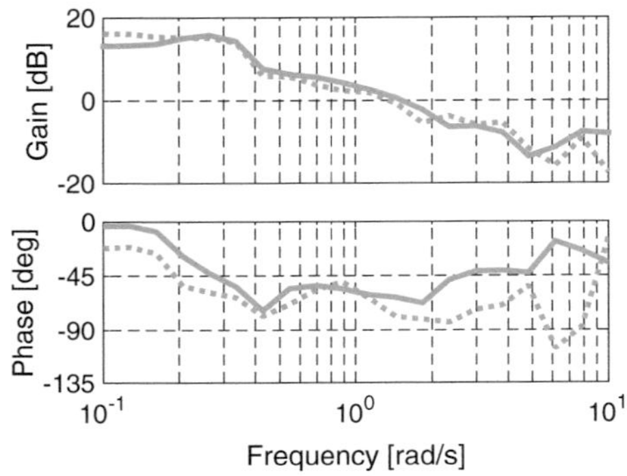

Fig. 14.21 Bode diagram of the identified Ψ_q including the decision process Φ and the filter $F_q(s)$ in (14.42) for Type 3

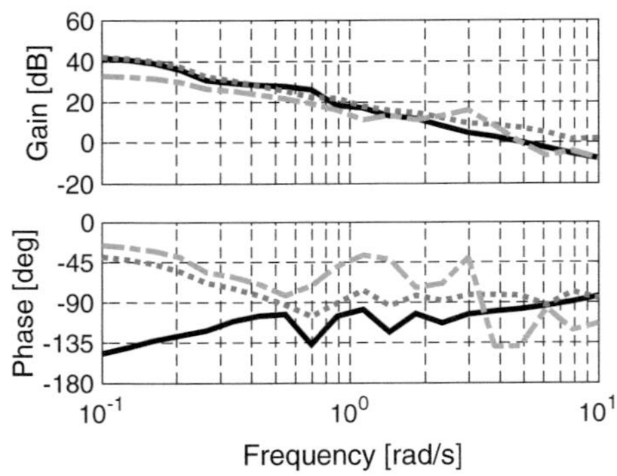

Fig. 14.22 Bode diagram of the identified system from $r_v - y_h$ to u for the former five trial data. The *line colors* are consistent with those in Figs. 14.12 and 14.13

so that it does not have effects on the target frequency domain. Remark that, in the sequel, we let the reference signals jump at every 10 s and reduce the period for each trial to 1 min in order to save time, which brings about slight changes to the shape over the low frequency domain but they are not significant enough to change the subsequent discussions.

We apply the same identification procedure as the position control mode. Then, the Bode diagrams of the identified systems from $r_v - y_h$ to u are shown in Figs. 14.22 and 14.23, where Fig. 14.22 illustrates the model for the former five trial data and

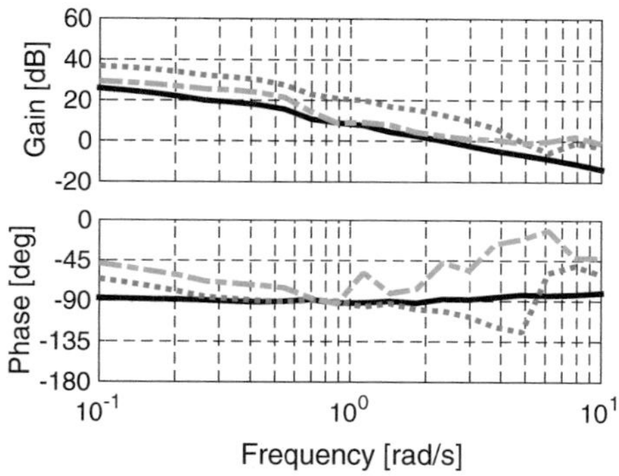

Fig. 14.23 Bode diagram of the identified system from $r_v - y_h$ to u for the latter five trial data

Fig. 14.23 for the latter. We observe that the shapes of the phase diagrams get smooth as repeating the trials especially over the low frequency domain.

Assumption 14.3 requires passivity for the system with output not u but $\dot{u}$, which is satisfied if the phases in Figs. 14.22 and 14.23 lie from -180 to 0. We see from the figures that the condition is satisfied for every network type and for both figures. Thus, we conclude that, in the case of the velocity control mode, the human attains passivity without any remedy like the position control mode and without many trials.

14.5.4 Analysis on Individual Variability

In this subsection, we again consider the position control mode and analyze the individual variability. For this purpose, we have another five trial subjects conduct the same experiments with the filter in (14.42) integrated. The resulting models of Ψ_q built using the latter five trial data for Type 1, 2, and 3 are illustrated in Figs. 14.24, 14.25 and 14.26, respectively. We see from the figures that one subject shows behavior similar to the above but others do not. In the phase diagram of Fig. 14.26, a subject slightly steps over $-90°$ and hence do not exactly attain passivity even with the help of the filter. More importantly, we observe large phase lags for four subjects over the low frequency domain from 0.1 to 0.4 rad/s in Fig. 14.24, and three of them violate the passivity condition. It was confirmed from questionnaires to the subjects that it is not easy to correctly react to the fast robot motion for Type 1. As a result, the subjects fail to learn the inverse model and sometimes cause over/undershoots even after five trials, which might supply unstable zeros to the model.

In summary, we conclude from the above results that the number of trials required for correctly learning the model depends on the human characteristics. If it is true, adding more trainings would passify all the operator. To confirm it, we have the subjects conduct more than 50 trials. Then, Figs. 14.24, 14.25 and 14.26 are changed to Figs. 14.27, 14.28 and 14.29, respectively. We see from Fig. 14.27 that the phase lags in the low frequency domain disappear. In addition, the lags in Fig. 14.29 are

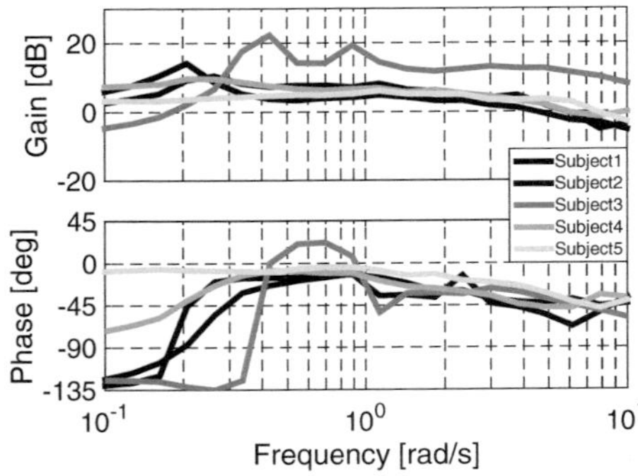

Fig. 14.24 Bode diagram of the identified Ψ_q with the filter in (14.42) for another five subjects for Type 1

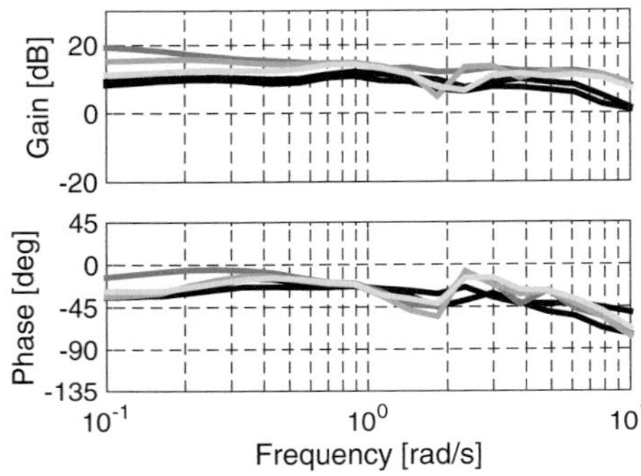

Fig. 14.25 Bode diagram of the identified Ψ_q with the filter in (14.42) for another five subjects for Type 2

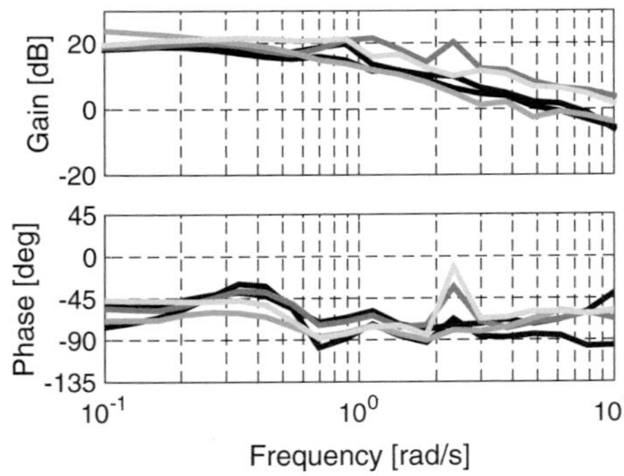

Fig. 14.26 Bode diagram of the identified Ψ_q with the filter in (14.42) for another five subjects for Type 3

reduced over the entire domain. Consequently, all the subjects attain passivity for all network types. More trials did not provide any significant changes in the human property. Namely, the time invariance in Assumption 14.2 means that he has already gotten skillful through sufficient trials.

Despite the eventual recovery of passivity, repeating such many trials might be demanding in the actual system operation. We thus present a method to reduce efforts for attaining passivity. Remark that the subsequent trials are conducted just after those corresponding to Figs. 14.24, 14.25 and 14.26, namely before the sufficient number of trials to obtain Figs. 14.27, 14.28 and 14.29.

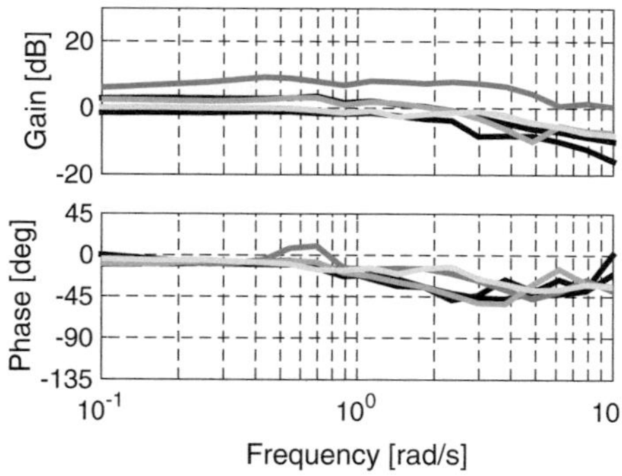

Fig. 14.27 Bode diagram of the identified Ψ_q with the filter in (14.42) after over 50 trials for Type 1

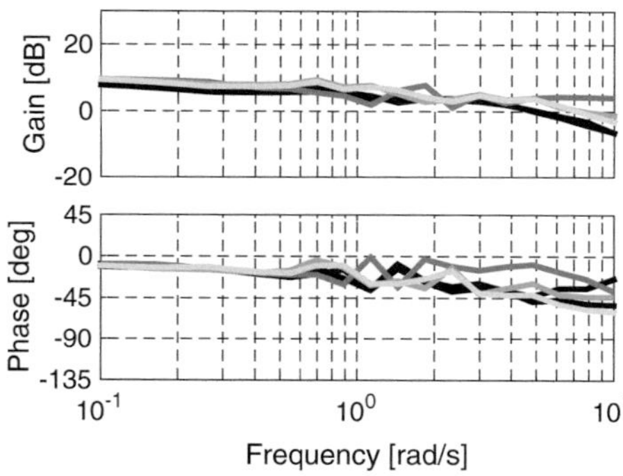

Fig. 14.28 Bode diagram of the identified Ψ_q with the filter in (14.42) after over 50 trials for Type 2

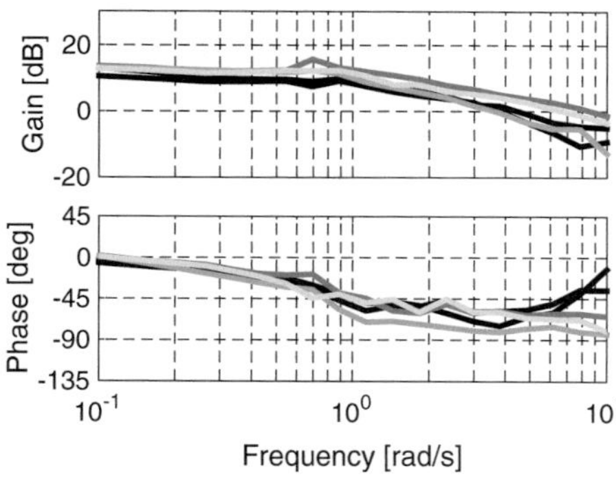

Fig. 14.29 Bode diagram of the identified Ψ_q with the filter in (14.42) after over 50 trials for Type 3

We again focus on the lag in the low frequency domain in Fig. 14.24. From the aforementioned questionnaires and the time series data, too much operability and the resulting fast reactions of the robots cause the lags over the low frequency domain. Following the insight, we replace the filter (14.42) by

$$F_q(s) = 0.5\frac{0.24s + 1}{0.04s + 1} \tag{14.44}$$

by multiplying 0.5 in order to reduce operability. This clearly makes the reaction of the robots lumpish, and it is fully expected that learning such slow motion is much easier.

Then, we let the subjects conduct ten more trials and build their models using the data of the latter five. As we should not change F_q depending on the network connection, we repeated the same process not only for Type 1 but also for the other network types. The identified Ψ_q with the filter in (14.44) are depicted in Figs. 14.30, 14.31 and 14.32. We see from Fig. 14.30 that the phase lags at the low frequency domain are drastically reduced. We also see from Fig. 14.32 that the phase lags over the entire frequency domain are reduced and every subject attains passivity. The authors are neutral on whether the reduction of the phase lags over the high frequency domain is caused by the new filter or the training through additional trials. In any case, we can at least conclude that all of the subjects get passive for all network types after only few trials by using (14.44).

From the above results together with a common sense, we can conclude that there is a trade-off between operability and learning speed. The relation of these two factors would strongly depend on the interface and even the size of the display. Testing other devices is left as a future work.

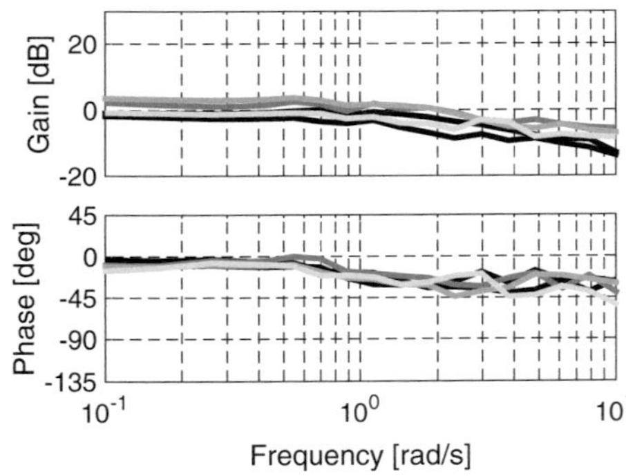

Fig. 14.30 Bode diagram of the identified Ψ_q including the decision process Φ and the filter $F_q(s)$ in (14.44) for Type 1

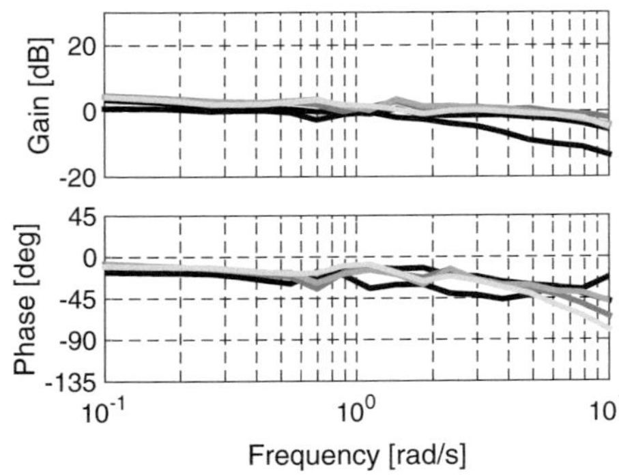

Fig. 14.31 Bode diagram of the identified Ψ_q including the decision process Φ and the filter $F_q(s)$ in (14.44) for Type 2

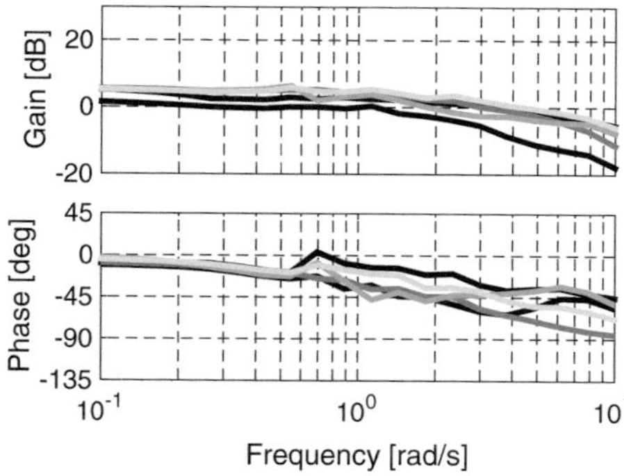

Fig. 14.32 Bode diagram of the identified Ψ_q including the decision process Φ and the filter $F_q(s)$ in (14.44) for Type 3

14.6 Summary

In this chapter, we studied passivity-based distributed operation of multi-robot systems so that the mobile robots synchronize their positions/velocities to reference values provided by a human operator. We first applied the controller presented in [13] in order to guarantee not only position but also velocity synchronization. The resulting system then has been proved to be passive from an externally added human command to average positions or velocities of the accessible robots. From the passivity property, we closed the loop between the robots and the operator by having the operator feedback the averages depending on the selected (position or velocity) control modes. Then, synchronization has been proved assuming passivity of the human operator. We then examined validity of the human passivity through experiments on a human-in-the-loop simulator. We first reviewed the results in [16] regarding the position control mode. Then, passivity in the velocity control mode has been confirmed. We then analyzed individual variability. To this end, additional five trial subjects conducted the same experiments and it was discovered subjects could not attain passivity even after some trials. In order to enhance the passivity recovery, we reduced the human operability by multiplying a gain to the velocity command. The experimental results demonstrated that the presented method allows all the subjects to meet the passivity condition.

A variety of problems still remain open. For example, we have not investigated the effects of robot-to-robot or robot-to-operator communication delays, which is one of the central topic in bilateral teleoperation. In addition, we need to investigate more complex robot dynamics than the single integrator. Demonstrations of the other interfaces are also required. The extension to 3-D coordination is also interesting. The theoretical results are easily extended to 3D, but the projection from 3D to 2D inherent in the human visual feedback process would affect the human passivity. Although this chapter passifies the human operator based on heuristics, a more systematic solution might be provided using the techniques in [34].

References

1. Anderson RJ, Spong MW (1989) Bilateral control of teleoperators with time delay. IEEE Trans Autom Control 34(5):494–501
2. Bai H, Arcak M, Wen J (2011) Cooperative control design: a systematic, passivity-based approach. Communications and control engineering series. Springer, New York
3. Bai H, Shafi YS (2015) Synchronization under space and time-dependent heterogeneities. Automatica 62(12):274–283
4. Cai H, Mostofi Y (2015) To ask or not to ask: A foundation for the optimization of human-robot collaborations. In: Proceedings of the 2015 american control conference, pp 440–446
5. Colgate JE, Hogan N (1988) Robust control of dynamically interacting systems. Int J Control 48(1):65–88
6. Cummings ML (2004) Human supervisory control of swarming networks. In: 2nd annual swarming: autonomous intelligent networked systems conference
7. d'Andrea-Novel B, Bastin G, Campion G (1991) Modelling and control of non holonomic wheeled mobile robots. In: Proceedings of the 1991 international conference on robotics and automation, pp 1130–1135
8. de la Croix JP, Egerstedt M (2012) Controllability characterizations of leader-based swarm interactions. In: AAAI symposium on human control of bio-inspired swarms, Arlington, DC, November 2012
9. Egerstedt M (2014) Human interactions with complex networks. In: Samad T, Annaswamy AM (eds) The Impact of Control Technology, 2nd edn. IEEE Control Systems Society. http://www.ieeecss.org
10. Egerstedt M, de la Croix J-P, Kawashima H, Kingston P (2014) Interacting with networks of mobile agents. Large-scale networks in engineering and life sciences. Springer, Heidelberg, pp 199–224
11. Franchi A, Secchi C, Son HI, Bulthoff HH, Giordano PR (2012) Bilateral teleoperation of groups of mobile robots with time-varying topology. IEEE Trans Robot 28(5):1019–1033
12. Franchi A, Secchi C, Ryll M, Bulthoff HH, Giordano PR (2012) Shared control: balancing autonomy and human assistance with a group of quadrotor UAVs. IEEE Robotics Autom Mag 19(3):57–68
13. Freeman RA, Yang P, Lynch KM (2006) Stability and convergence properties of dynamic average consensus estimators. In: Proceedings of the 45th ieee conference on decision and control, pp 398–403
14. Giordano PR, Franchi A, Secchi C, Son HI, Bulthoff HH (2013) A passivity-based decentralized strategy for generalized connectivity maintenance. Int J Robot Res 32(3):299–323
15. Hatanaka T, Chopra N, Fujita M, Spong MW (2015) Passivity based control and estimation in networked robotics., Communications and control engineering series. Springer International Publishing
16. Hatanaka T, Chopra N, Fujita M (2015) Passivity-based bilateral human-swarm-interactions for cooperative robotic networks and human passivity analysis. In: Proceedings of the 54th IEEE conference on decision and control, pp 1033–1039
17. Hokayem PF, Spong MW (2006) Bilateral teleoperation: an historical survey. Automatica 42(12):2035–2057
18. Katayama T (2005) Subspace methods for system identification., Communications and control engineering series. Springer, London
19. Kottenstette N, McCourt MJ, Xia M, Gupta V, Antsaklis PJ (2014) On relationships among passivity, positive realness, and dissipativity in linear systems. Automatica 50(4):1003–1016
20. Lee D, Spong MW (2005) Bilateral teleoperation of multiple cooperative robots over delayed communication network: theory. In: Proceedings of the IEEE international conference on robotics and automation, pp 360–365
21. Liu Y, Chopra N (2012) Controlled synchronization of heterogeneous robotic manipulators in task space. IEEE Trans Robot 28(1):268–275

22. McRuer D (1980) Human dynamics in man-machine systems. Automatica 16(3):237–253
23. Martinez S, Cortes J, Bullo F (2007) Motion coordination with distributed information. IEEE Control Syst Mag 27(4):75–88
24. McLurkin J, Smith J, Frankel J, Sotkowitz D, Blau D, Schmidt B (2006) Speaking swarmish: human-robot interface design for large swarms of autonomous mobile robots. In: Proceedings of the AAAI spring symposium
25. Mekdeci B, Cummings ML (2009) Modeling multiple human operators in the supervisory control of heterogenous unmanned vehicles. In: Proceedings of the 9th conference on performance metrics for intelligent systems
26. Nuno E, Basanezb L, Ortega R (2011) Passivity-based control for bilateral teleoperation: a tutorial. Automatica 47(3):485–495
27. Olsen DR, Wood B (2004) Fan-out: measuring human control of multiple robots. In: Proceedings of the SIGCHI conference on human factors in computing systems
28. Penin LF, Caballero A, Aracil R, Barrientos A (1998) Human behavior modeling in master-slave teleoperation with kinesthetic feedback. In: Proceedings of the 1998 IEEE international conference on robotics and automation, vol 3, pp 2244–2249
29. Ren W, Beard RW (2008) Distributed consensus in multi-vehicle cooperative control., Communications and control engineering seriesSpringer, London
30. Rodriguez-Seda EJ, Troy JJ, Erignac CA, Murray P, Stipanovic DM, Spong MW (2010) Bilateral teleoperation of multiple mobile agents: coordinated motion and collision avoidance. IEEE Trans Control Syst Technol 18(4):984–992
31. Secchi C, Franchi A, Bulthoff HH, Giordano PR (2012) Bilateral teleoperation of a group of UAVs with communication delays and switching topology. In: Proceedings of the 2012 IEEE international conference on robotics and automation, pp 4307–4314
32. Varnell P, Zhang F (2013) Dissipativity-based teleoperation with time-varying communication delays. In: Proceedings of the 4th workshop on distributed estimation and control in networked systems, pp 369–376
33. Wang X, Shi Z, Zhang F, Wang Y (2015) Mutual trust based scheduling for (semi) autonomous multi-agent systems. In: Proceedings of the 2015 american control conference, pp 459–464
34. Xia M, Antsaklis PJ, Gupta V (2014) Passivity analysis of human as a controller. ISIS Technical report, ISIS-2014-002
35. Yun X, Yamamoto Y (1993) Internal dynamics of a wheeled mobile robot. In: Proceedings of the 1993 IEEE/RSJ international conference on intelligent robots and systems, pp 1288–1294

Chapter 15
Human–Swarm Interactions via Coverage of Time-Varying Densities

Yancy Diaz-Mercado, Sung G. Lee and Magnus Egerstedt

15.1 Introduction

Multi-agent and swarm robotics has received increased research interest in the last few decades. There are a number of benefits for considering the use of multi-robot systems: incorporating redundancy into a system for increased robustness [4, 23], task and resource allocation [25], and complementing heterogeneous capabilities [29, 67, 74]. On the other hand, increasing the number of agents can lead to problems with the scalability of algorithms [24, 70], emergence of unaccounted artifacts in local rules at a global level (e.g., deadlocks [10], local optima, and local traps [61]), and the requirement for considerations on coordination and collision avoidance [47, 61].

A number of applications have been approached from a multi-robot system perspective. In the multi-robot foraging paradigm, agents wander around an environment searching for items of interest [3, 48, 76]. These items are then collected and returned to specified locations. Foraging is of interest from a biology standpoint, as many species forage for food (e.g., ants [79], birds [52]). However, the foraging task can also relate to many real-world problems, such as waste or specimen collection in hazardous environments, explosive detection and disposal, and search and rescue operations after a crisis such as a flooding or an earthquake.

This chapter was developed and enhanced from two earlier papers published as [17] and [41]

Y. Diaz-Mercado (✉) · S.G. Lee · M. Egerstedt
Georgia Institute of Technology, Atlanta, GA 30332, USA
e-mail: yancy.diaz@gatech.edu

S.G. Lee
e-mail: slee656@gatech.edu

M. Egerstedt
e-mail: magnus@gatech.edu

Y. Wang and F. Zhang (eds.), *Trends in Control and Decision-Making for Human–Robot Collaboration Systems*, DOI 10.1007/978-3-319-40533-9_15

Another application of interest in the research community and in industry is cooperative assembly [6, 56, 73, 77] and self-assembly [30, 38, 83]. In the cooperative assembly scenario, multiple robots need to coordinate their motion in order to cooperatively assemble a structure using a possibly heterogeneous set of tools, end effectors, skills, or parts. Similarly, in the self-assembly scenario, multiple robots need to coordinate their motion in order to collectively form a structure or achieve a geometric configuration. In either scenario, kinematic and other constraints, such as skill or tool compatibility, play an important role in the multi-agent coordination. Real-world applications include the automotive assembly industry, space structure assembly [22], and collective construction [42, 65].

There has been a recent push for robotic farming and precision agriculture [2, 57, 63], where a fleet of robots is sent to gather data on the status of crops, tend to and harvest them. Mobile sensor networks take advantage of sensor node mobility to relocate sensors in order to enhance sensing and maximize coverage [34, 60, 69], provide robustness to sensor node failure [8, 82], and reduce overall energy consumption by the network [32]. In some communication architectures, mobile agents called data MULEs or message ferries [36, 74, 81, 85] are used to transport data between sensors, access points, or base stations, in situations where it is impractical to establish physical communication. Multi-robot simultaneous localization and mapping (SLAM) takes advantage of the robot team's size to attain a more complete (or more detailed) map faster than with a single robot by combining information from multiple sources and coordinating motion [12, 33, 62, 78, 86]. Other applications include transportation systems (e.g., intelligent highways, air traffic control) [53], and the convoy protection scenario (e.g., surveillance, coverage) [18].

A factor in this increased research effort is how miniaturization of technology has allowed for a rise in cost-effective but powerful microprocessors and microcontrollers, facilitating the study and future viability of multi-robot systems and their applications. In [71], a low-cost and fully scalable robot called the Kilobot was introduced that moves via the use of two vibration motors, sliding along a surface. This led to the first 1000-robot swarm to operate without human intervention [72]. A swarm of 20 micro-quadrotors was developed and programmed in [40] to achieve tight formations and trajectory following. The experiments demonstrate agility and coordination among the robots in real time. In [66], a low-cost and scalable differential-drive micro-robot called the GRITSBot was presented as a part of a robotic ecosystem. The modularity of sensing and actuation capabilities allows for high adaptability to environmental constraints, as well as to the functional requirements of the application setting. Other swarm robots include the SwarmBot [49], the r-one robot [50], the Jasmine robot [37], and the Alice robot [11].

What all these applications and technological trends have in common is the need for some form of coordination among the agents in the multi-robot team. Further, in many of these situations there is a need for adapting to changes in the environment or the mission parameters. Although robots can be very good at low-level control objectives (e.g., stabilization, formation maintaining, tracking), they lack the capacity human beings possess of making high-level decisions on the go. The *human–swarm interactions* (HSI) paradigm allows for a human operator to influence a swarm of robots at a high-level while the swarm continues their low-level control objectives. This allows for a combination of high-level and low-level decision-making. Human operators can, for example, switch swarm behavior, change overall location, or introduce new objectives for the robot swarm.

Here, we present a novel approach for influencing a team of robots using coverage algorithms for general time-varying density functions. This has potential implications for how human operators can interact with large teams of mobile robots, where one of the main challenges is the construction of suitable abstractions that make the entire team amenable to human control, e.g., [39]. For such abstractions to be useful, they need to scale gracefully as the number of robots increases. Density functions are promising abstractions, for they are independent of the robot team's size and instead directly manipulate the environment where the robot swarm operates. We propose two methods for HSI with which to influence teams of robots: one is tailored for obtaining desired geometric configurations, the other for quick manipulation of the swarm as a whole. These methods are tested through simulation and robotic implementation.

The organization of this chapter is as follows: in Sect. 15.2 we begin discussion of the coverage problem as a mechanism to influence the swarm; in Sect. 15.2.2 we propose update laws for optimal coverage of time-varying densities; in Sect. 15.2.3 we find approximations to the centralized schemes presented in previous sections; in Sect. 15.3 we propose two methods of generating density functions for the sake of HSI; and finally, in Sect. 15.4 we validate these HSI methods by implementing them in two different robot platforms.

15.2 Human–Swarm Interactions via Coverage

The *human–swarm interactions* (HSI) paradigm allows for a human operator to introduce high-level decision making into a robot swarm low-level control objectives. The subject has been approached from different aspects in the literature. One approach allows human operators to guide the swarm in the environment by injecting information through interfaces. In [5], a set of trials was performed on a swarm that was autonomously performing radiation search and localization in which human input injected various levels of guidance into the swarm. These trials suggested that injecting appropriate levels of guidance information allowed the swarm to achieve its objective more efficiently. In [80], an architecture was proposed to allow the human operator to guide the swarm by introducing attraction and repulsion points for the sake of obstacle avoidance. In [59], they studied the effect of providing additional information, such as haptic feedback, to the human operator in order to help navigate the swarm around obstacles.

Other approaches focus on the converse – the swarm aids the humans achieve their tasks. For example, in [54], assistive swarm robotics was used to help navigate firefighters and support them with search and rescue operations. In [28, 55], they used the swarm of robots to obtain multiple views of human gestures, which provided multiple samples for machine learning algorithms on gesture recognition. Practical considerations for HSI were explored in [58], where the foraging paradigm was explored under bandwidth restrictions, such as one might experience in underwater missions or on other remote locations.

There have also been approaches which focus on finding a medium that affords the control and manipulation of a swarm of robots. In [16], they used clay as a deformable medium to specify the desired geometric configuration of the swarm. The desired configuration was then processed using image recognition algorithms and distributed formation controllers were used to achieve desired configurations. In our work, we take this latter approach – enabling human–swarm interactions by providing the human operator a mechanism that affords the manipulation of the swarm and that scales with the number of agents in the robot swarm. In particular, we will impose the desired concentration, *or density*, of agents directly on the environment. This way, the robots' task is to spread or concentrate in the different parts of the environment in order to satisfy the desired density of agents imposed by the user. We will achieve this by performing *coverage control* over a human-provided density function.

Coverage control is one area of multi-agent control that has received significant attention lately, e.g., [27, 46], and it is concerned with how to position agents in such a way that "surveillance" of a domain of interest is maximized. In this context, an idea that has been widely adopted to describe how interesting a "domain of interest", is to associate a density function to the domain, as was done in [13–15, 31, 46, 68]. However, the focus of previous coverage algorithms has largely been on static density functions, which does not provide enough flexibility when human operators are to adaptively interact with a team through a dynamic re-shaping of the density functions.

To enable this line of inquiry, we require an algorithm that can guarantee multi-robot optimal coverage given general time-varying density functions. Applications to this beyond the means for multi-robot influence can be found in a number of domains. For example, in search and rescue scenarios, the density function could represent the probability of a lost person being at a certain point in an area, e.g., [45]. Additionally, optimal coverage of density functions for multi-robot surveillance and exploration was used in [31], where the density function was modeled to be a function of the explored "frontier" (for other examples, see [26] and references therein).

To date, relatively little work has been done on coverage with time-varying density functions. In [14], the time-varying case was investigated under a set of simplifying assumptions on the density functions. Another stab at the problem was pursued in [68], where time-varying density functions where used as a means to tracking moving targets. While simulations and experiments verified that coverage was indeed achieved, formal guarantees were absent.

In contrast to [14] and [68], in this paper we derive an algorithm that guarantees optimal coverage with quite general, time-varying density functions [41]. As with the algorithms for time-invariant density functions, centroidal Voronoi tessellations (CVT) will play a key role. A CVT is a configuration where the positions of each robot coincide with the centroids of their Voronoi cells, given a so-called Voronoi tessellation of the domain of interest for which the robots are the Voronoi generators themselves. The CVT may result in the optimality of a commonly used metric on the coverage of a domain by the robot swarm. To that end, the control laws presented here seek to drive the agents into a CVT given a user-provided density describing the areas of importance.

We further find decentralized approximations to the control laws presented here that allow the individual agents in the swarm to compute their update law with only local information. The fact that only adjacency information is required in their update law means that a single operator could potentially influence an arbitrarily large number of agents. Due to the scalability of the algorithm, and the level of abstraction that the generation of density functions offers (in that it does not care for the number of agents performing the coverage), we consider these decentralized update laws as tools for HSI [17]. The idea presented here is to allow a human operator to influence teams of robots by generating density functions, letting robot swarm perform coverage over this density.

In subsequent sections, we offer two different means of generating density functions for HSI. The first method allows the user to easily specify the geometric configuration of the swarm and "move" them around, but at an added computational cost. The second method reduces computational cost by using predefined Gaussian functions to allow the human operator to quickly assign importance to the area of interest. It is fast and allows the user to "move" the team of robot easily, but is not as amenable to "shaping" the swarm as the first method. Figure 15.1 illustrates how the proposed approaches allow a user to manipulate a team of robots to drive to certain areas of the environment. Toward the end of the chapter, we present more discussion on the robotic implementation of these HSI methods.

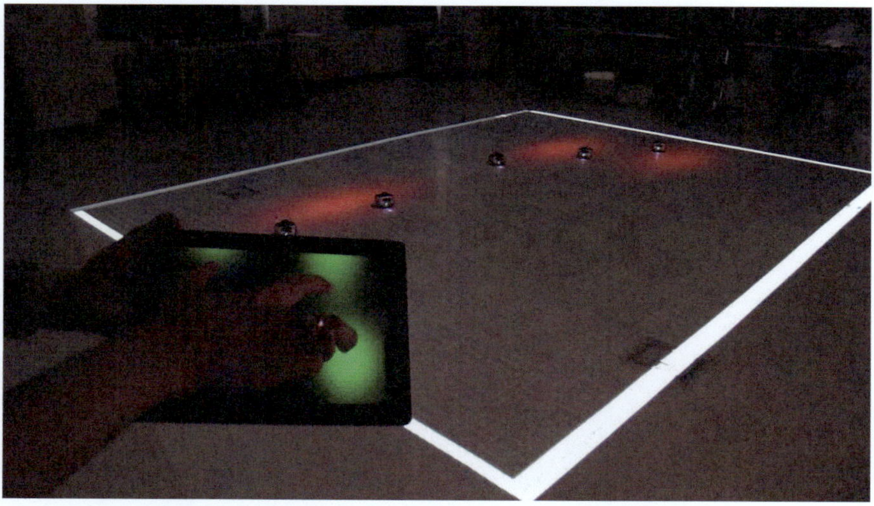

Fig. 15.1 A team of five Khepera III robots are influenced by a human operator to drive to certain parts of the environment. The tablet is representative of the robot environment. An overhead projector allows us to overlay the broadcasted density on the environment in real time

15.2.1 The Coverage Problem

We are on the hunt for decentralized update laws that allow a human operator to influence the swarm of agents by providing a density function characterizing the areas of importance within the domain. The agents in the robot swarm will seek to optimize their coverage over the domain under the influence of the user-provided density. In order to discuss optimal coverage, one first has to associate a cost to a robot configuration that describes how well a given area is being covered. For this, we will follow the construction of the so-called locational cost, as was done, for example, in [14].

Let $D \subset \mathbb{R}^2$ be the two-dimensional[1] convex domain representing the robot workspace. Moreover, let $\phi : D \times [0, \infty) \to (0, \infty)$ be the associated density function, which we will assume is bounded and continuously differentiable in both arguments, and where $\phi(q, t)$ captures the relative importance of a point $q \in D$ at time t. For the sake of HSI, this density function will be a user-provided input to the robot swarm.

The coverage problem involves placing n robots in D, and we let $p_i \in D$, $i = 1, \ldots, n$ be the position of the ith robot. Moreover, the domain itself will be divided into regions of dominance, e.g., [15], $P_1, \ldots, P_n$ (forming a proper partition of D), where the idea is to let robot i be in charge of covering region P_i. One can then ask how good the choice of p and P is, where $p = [p_1^T, \ldots, p_n^T]^T$, and $P = \{P_1, \ldots, P_n\}$. The final piece needed to answer this question is a measure of how well a given point $q \in D$ is covered by robot i at position $p_i \in D$ (see [20] and references therein). In the most general setting, one can consider the locational cost given by

$$\mathcal{H}(p, P, t) = \sum_{i=1}^{n} \int_{P_i} f(\|q - p_i\|)\phi(q, t)dq.$$

for $f : \mathbb{R} \to [0, \infty)$. As the performance of a large class of sensors deteriorate with a rate proportional to the square of the distance, [1, 51], here we will instead consider the locational cost given by

$$\mathcal{H}(p, P, t) = \sum_{i=1}^{n} \int_{P_i} \|q - p_i\|^2 \phi(q, t)dq. \tag{15.1}$$

At a given time t, when a configuration of robots p together with the partition P minimize (15.1), the domain is said to be optimally covered with respect to ϕ. However, it is possible to view the minimization problem as a function of p alone, [14], by observing that given p, the choices of P_i that minimize (15.1) is

$$V_i(p) = \{q \in D \mid \|q - p_i\| \leq \left\|q - p_j\right\|, i \neq j\}.$$

[1] Here we focus on the two-dimensional case, but note that the math also extends to higher dimensional spaces as well.

This partition of D is a Voronoi tessellation — hence the use of V_i to denote the region. With this choice of region, we can remove the partition as a decision variable and instead focus on the locational cost

$$\mathcal{H}(p,t) = \sum_{i=1}^{n} \int_{V_i(p)} \|q - p_i\|^2 \phi(q,t)dq \tag{15.2}$$

where $\phi(q,t)$ is the user-provided input to the system.

In [20, 35] it was shown that

$$\frac{\partial \mathcal{H}}{\partial p_i} = \int_{V_i} -2(q - p_i)^T \phi(q,t)dq, \tag{15.3}$$

and since $\phi > 0$, one can define the mass m_i and center of mass c_i of the ith Voronoi cell, V_i, as

$$m_i(p,t) = \int_{V_i(p)} \phi(q,t)dq, \tag{15.4}$$

$$c_i(p,t) = \frac{\int_{V_i(p)} q\phi(q,t)dq}{m_i}. \tag{15.5}$$

For convenience, we will stack the masses into a single diagonal (positive-definite) matrix denoted by

$$M = \begin{bmatrix} m_1 & & & \\ & m_2 & & \\ & & \ddots & \\ & & & m_n \end{bmatrix} \otimes I_d, \tag{15.6}$$

where "$\otimes$" is the Kronecker product and I_d is the $d \times d$ identity matrix where d corresponds to the dimension of the domain ($d = 2$ is considered here). We also group the centers of mass as a single vector $c = \left[c_1^T, \dots, c_n^T\right]^T$. As was shown in [20, 35], using these quantities allows us to rewrite the partial derivative in (15.3) as

$$\frac{\partial \mathcal{H}}{\partial p_i} = 2m_i(p_i - c_i)^T. \tag{15.7}$$

From this expression, we can see that a critical point of (15.2) is

$$p_i(t) = c_i(p,t), \quad i = 1, \dots, n, \tag{15.8}$$

and a minimizer to (15.2) is necessarily of this form, [21]. Moreover, when (15.8) is satisfied, p is a so-called centroidal Voronoi tessellation (CVT).

The robots being in a CVT configuration does not, however, imply that the global minimum of (15.2) is attained, e.g., [20]. In fact, the CVT is in general not unique given a density function ϕ. Finding the globally minimizing configuration is a difficult problem due to the nonlinearity and nonconvexity of (15.2), as discussed in [43]. As such, in this chapter, we are interested in designing algorithms that guarantee convergence to local minima with respect to time-varying density functions (as our goal is to merely be able to influence the swarm), and we make no claims about finding the global minimum.

In light of (15.7), the gradient direction (with respect to p_i) is given by $\frac{(p_i - c_i)}{\|p_i - c_i\|}$. As such, a (scaled) gradient descent motion for the individual robots to execute would be

Lloyd:

$$\dot{p}_i(t) = -\kappa(p_i(t) - c_i(p, t)) \tag{15.9}$$

where κ is a positive gain. This is a continuous-time version of Lloyd's algorithm [44] for obtaining CVTs as long as ϕ does not depend on t. The way to see this, as was done in [15], is to take $\mathcal{H}(p)$ in (15.2) (note that we assume that $\mathcal{H}$ only depends on p and not on t for the purpose of this argument) as the Lyapunov function,

$$\frac{d}{dt}\mathcal{H}(p) = \sum_{i=1}^{n} \frac{\partial}{\partial p_i}\mathcal{H}(p)\dot{p}_i = -2\kappa \sum_{i=1}^{n} m_i \|p_i - c_i(p)\|^2 .$$

By LaSalle's invariance principle, the multi-robot system asymptotically converges to a configuration $\{\|p_i - c_i(p)\|^2 = 0, \ i = 1, \ldots, n\}$, i.e., to a CVT ([15]).

However, if ϕ is time varying, the same control law does not stabilize the multi-robot system to a CVT. This point can be hinted at by investigating the evolution of a time-dependent $\mathcal{H}(p, t)$,

$$\begin{aligned}\frac{d}{dt}\mathcal{H}(p(t), t) &= \sum_{i=1}^{n} \frac{\partial}{\partial p_i}\mathcal{H}(p(t), t)\dot{p}_i + \frac{\partial}{\partial t}\mathcal{H}(p(t), t)\\ &= \sum_{i=1}^{n}\int_{V_i} \|q - p_i(t)\|^2 \frac{\partial \phi}{\partial t}(q, t)dq - 2\kappa \sum_{i=1}^{n} m_i \|p_i(t) - c_i(p(t), t)\|^2 .\end{aligned}$$

There is no reason, in general, to assume that this expression is negative since we do not want to impose assumptions on slowly varying, or even quasi-static, density functions. Instead, what is needed is a new set of algorithms for handling the time-varying case.

To get around the problem associated with nonslowly varying density functions, timing information must be included in the motion of the robots. In [14], this was done through the assumption that $\phi(q, t)$ is such that

$$\frac{d}{dt}\left(\sum_{i=1}^{n}\int_{V_i}\|q-c_i\|^2\phi(q,t)dq\right)=0. \tag{15.10}$$

Letting

$$\dot{m}_i=\int_{V_i}\dot{\phi}(q,t)dq, \qquad \dot{c}_i=\frac{1}{m_i}\left(\int_{V_i}q\dot{\phi}(q,t)dq-\dot{m}_ic_i\right), \tag{15.11}$$

the algorithm in [14] for time-varying density functions is given by

Cortes:

$$\dot{p}_i=\dot{c}_i-\left(\kappa+\frac{\dot{m}_i}{m_i}\right)(p_i-c_i). \tag{15.12}$$

Under the previously mentioned assumption on ϕ, $\mathcal{H}(p,t)$ again becomes a Lyapunov function when the agents move according to (15.12), and convergence to a time-varying CVT is established.

Unfortunately, the assumption required to make (15.12) work is rather restrictive for the sake of HSI as it basically states that the density function must be designed in such a way that the locational cost achieves a minimum that remains constant as the agents move. Thus, for the remainder of the paper, we will develop new methods for handling time-varying density functions that do not impose major assumptions on $\phi(q,t)$. In fact, if the density function is to be thought of as an external, human-generated input to the system, there are no a priori reasons why the human operator would restrict the interactions to satisfy particular regularity assumptions on ϕ.

15.2.2 Centralized Coverage of Time-Varying Densities

We are now in pursuit of a control laws that allow us achieve a CVT, even if the density function is time varying. In order to move forward with our analysis, we need to understand some of the properties associated with the center of mass of a given Voronoi tessellation. We first note that the partial derivatives of the center of mass with respect to the agents positions has eigenvalues bounded by one, as long as the agents are close enough to a CVT, and under the assumption that at the CVT the locational cost is locally strongly convex.

Lemma 15.1 *If the locational cost is locally strongly convex at the CVT, then there exists* $\delta > 0$ *such that if* $\|p(t) - c(p(t), t)\| < \delta$ *then* $\mathrm{Re}(\lambda) < 1$ *for all* $\lambda \in \mathrm{eig}(\frac{\partial c}{\partial p})$.

Proof As it was shown in [19], under the assumption that the locational cost is locally strongly convex around the CVT, we have that

$$\begin{aligned}
\left.\frac{\partial^2 \mathcal{H}}{\partial p^2}\right|_{p=c} &= 2M\left(I - \frac{\partial c}{\partial p}\right) \succ 0 \\
\Rightarrow M^{1/2}\left(I - \frac{\partial c}{\partial p}\right)M^{-1/2} &\succ 0 \\
\Rightarrow \left(I - M^{1/2}\frac{\partial c}{\partial p}M^{-1/2}\right) &\succ 0
\end{aligned}$$

where M is the positive-definite diagonal matrix of Voronoi cell masses defined in (15.6), which implies that for all $\lambda \in \mathrm{eig}(\frac{\partial c}{\partial p})$ we have that $\mathrm{Re}(\lambda) < 1$.

Let $f(\lambda) = \det\left(\lambda I - \frac{\partial c}{\partial p}\right) = \sum_{i=1}^{2N} a_i \lambda^i$ be the characteristic polynomial for $\frac{\partial c}{\partial p}$, whose roots ($\lambda$ such that $f(\lambda) = 0$) are the eigenvalues of $\frac{\partial c}{\partial p}$. The coefficients a_i, $i = 1, \ldots, 2N - 1$, are themselves polynomials of the entries of $\partial c/\partial p$, with $a_{2N} = 1$.

Since $m_i > 0$ for all i, the different entries of $\partial c/\partial p$ are continuous (as we will see below). Further, since the roots of any polynomial vary continuously with all but its highest order coefficient [84], we have that the eigenvalues of $\partial c/\partial p$ are continuous.

One can apply a chain of continuity arguments to show that there exists $\delta > 0$ so $\mathrm{Re}(\lambda) < 1$ whenever $\|p(t) - c(p(t), t)\| < \delta$. □

Corollary 15.1 *As long as the agents are close enough to a CVT configuration for which the locational cost is locally strongly convex, the matrix inverse* $(I - \frac{\partial c}{\partial p})^{-1}$ *exists.*

We use this corollary to obtain our first result on maintaining a CVT for all time, as originally stated in [41].

Theorem 15.1 (Maintaining CVT [41]) *Let* $p(t_0) = c(p(t_0), t_0)$. *If*

$$\dot{p} = \left(I - \frac{\partial c}{\partial p}\right)^{-1} \frac{\partial c}{\partial t}, \qquad t \geq t_0$$

then

$$\|p(t) - c(p(t), t)\| = 0, \qquad t \geq t_0$$

as long as the inverse $\left(I - \partial c / \partial p\right)^{-1}$ *is well defined.*

Proof Assume the agents begin from a CVT configuration, i.e., $p(t_0) = c(p(t_0), t_0)$. We need to ensure that

$$\frac{\mathrm{d}}{\mathrm{d}t}\left(p(t) - c(p(t), t)\right) = 0, \qquad \forall t \geq t_0.$$

But this implies that $\dot{p} = \dot{c} = \frac{\partial c}{\partial p}\dot{p} + \frac{\partial c}{\partial t}$, which can be rearranged into the form

$$\dot{p} = \left(I - \frac{\partial c}{\partial p}\right)^{-1} \frac{\partial c}{\partial t}$$

as long as the inverse is well defined. From Lemma 15.1, a sufficient condition for this is that the initial CVT makes the locational cost locally strongly convex. □

In order for this theorem to be meaningful, we require that the agents first achieve a CVT configuration which is then maintained. This control law also assumes the agents will be able to instantaneously accelerate to the required velocities, which may not be possible in practice. In order to compensate for saturation, modeling errors, and deviations from the CVT, in [41] a proportional term is introduced that forces the agents into a CVT. This update law was called the *TVD-C* which stands for time-varying densities, centralized case.

$$\dot{p}(t) = \left(I - \frac{\partial c}{\partial p}\right)^{-1} \left(-\kappa\left(p(t) - c(p(t), t)\right) + \frac{\partial c}{\partial t}\right) \tag{15.13}$$

where $\kappa > 0$ is a proportional gain. It is noteworthy that this proportional term influences the team of robots to move toward a (scaled) gradient descent direction to achieve a CVT configuration, and that once a CVT is achieved the proportional term does not contribute to the update law and Theorem 15.1 holds.

As long as the inverse is well defined, this update law will drive the agents into a CVT. This is encapsulated in the following theorem.

Theorem 15.2 (Convergence of TVD-C) *If* $1 \notin \text{eig}\left(\frac{\partial c}{\partial p}\right)$ *and we let*

$$\dot{p} = \left(I - \frac{\partial c}{\partial p}\right)^{-1} \left(-\kappa\,(p - c) + \frac{\partial c}{\partial t}\right), \qquad t \geq t_0$$

then $\|p(t) - c(p(t), t)\| \to 0$ *as* $t \to \infty$ *with rate of decrease* $\exp\left(-\kappa(t - t_0)\right)$.

Proof Consider the total derivative for $\|p(t) - c(p(t), t)\|^2$ below

$$\begin{aligned}
\frac{\mathrm{d}}{\mathrm{d}t}\left(\|p - c\|^2\right) &= 2\,(p - c)^T\,(\dot{p} - \dot{c}) = 2\,(p - c)^T\left(\dot{p} - \frac{\partial c}{\partial p}\dot{p} - \frac{\partial c}{\partial t}\right) \\
&= 2\,(p - c)^T\left(\left(I - \frac{\partial c}{\partial p}\right)\dot{p} - \frac{\partial c}{\partial t}\right) \\
&= 2\,(p - c)^T\left(\left(I - \frac{\partial c}{\partial p}\right)\left(I - \frac{\partial c}{\partial p}\right)^{-1}\left(-\kappa\,(p - c) + \frac{\partial c}{\partial t}\right) - \frac{\partial c}{\partial t}\right) \\
&= -2\kappa\,\|p - c\|^2
\end{aligned}$$

which tells us that

$$\frac{\mathrm{d}}{\mathrm{d}t}\left(\|p - c\|^2\right) = -2\kappa\,\|p - c\|^2.$$

This is a homogeneous linear differential equation which for initial condition $\|p(t_0) - c(p(t_0), t_0)\|$ has solution given by

$$\begin{aligned}
&\|p(t) - c(p(t), t)\|^2 = \exp(-2\kappa(t - t_0))\,\|p(t_0) - c(p(t_0), t_0)\|^2 \\
\Rightarrow\quad &\|p(t) - c(p(t), t)\| \;= \exp(-\kappa(t - t_0))\,\|p(t_0) - c(p(t_0), t_0)\|.
\end{aligned}$$

Hence $\|p(t) - c(p(t), t)\| \to 0$ as $t \to \infty$. Further, we see that

$$\begin{aligned}
&\|p(t) - c(p(t), t)\| = \exp(-\kappa(t - t_0))\,\|p(t_0) - c(p(t_0), t_0)\| \\
\Rightarrow\quad &\frac{\|p(t) - c(p(t), t)\|}{\|p(t_0) - c(p(t_0), t_0)\|} = \exp(-\kappa(t - t_0))
\end{aligned}$$

such that the rate of decrease is given by $\exp\left(-\kappa(t - t_0)\right)$. □

Recalling Lemma 15.1, as long as the agents are close enough to a CVT and the locational cost is locally strongly convex around the CVT, the inverse in control law TVD-C will drive the agents to the CVT. However, note that it is not necessary for the locational cost to be locally strongly convex at a CVT. Indeed, in [19] they present several CVT configurations that result in saddle points instead. What has been observed in robotic implementations of these algorithms (Sect. 15.4) is that the addition of the gradient descent term seeks to reduce the overall locational cost. It has also been observed that all the disturbances seem to drive the agents into stable CVT configurations. This in turn suggests the achieved CVT are in fact local minimizers to the locational cost, as opposed to saddle points.

It is also noteworthy that Theorem 15.2 only requires that $1 \notin \text{eig}(\partial c/\partial p)$, and as such the agents are driven into a CVT even if $\text{Re}(\lambda) > 1$ for any $\lambda \in \text{eig}(\partial c/\partial p)$. If this is the case, then the problem lies in that due to the continuity of these eigenvalues, we will inevitably obtain $1 \in \text{eig}(\partial c/\partial p)$. In practice, however, these ill-conditioned matrices are hardly a problem as the set $\{t | 1 \in \text{eig}(\frac{\partial c}{\partial p}(p(t), t))\}$ seems to often be a zero-measure set.

We would still like to avoid the case of ill-conditioned matrices all together, and perhaps more importantly, this matrix inverse has a dense structure, which makes this a centralized approach — this will not scale well, and will hinder our effort in manipulating large teams of robots. We will address these issues in the following section where we proceed to obtain distributed approximations to our centralized update laws.

15.2.3 Distributed Coverage of Time-Varying Densities

In order to implement update law (15.13), we need to deal with two subtle difficulties associated with it. The first is the computation of the deceivingly innocent looking terms $\partial c/\partial p$ and $\partial c/\partial t$, the second is addressing the computation of the matrix inverse $\left(I - \partial c/\partial p\right)^{-1}$. We will address the matrix inverse difficulty in Sect. 15.2.3.2. In the next section, we address the issue of the computation of the partial derivatives.

15.2.3.1 Computing the Partial Derivatives

Recall that by combining Eqs. (15.4) and (15.5), we have that

$$c_i\,(p,t) = \frac{\int_{V_i(p)} q\,\phi(q,t)\;\mathrm{d}q}{\int_{V_i(p)} \phi(q,t)\;\mathrm{d}q},$$

which depends on p in the boundary of the area over which the two integrals are taken.

In order to compute these partials, we first need to make use of Leibniz rule, e.g., [19].

Lemma 15.2 *Let $\Omega(p)$ be a region that is a smooth function of p such that the unit outward normal vector n is uniquely defined almost everywhere on $\partial\Omega$, which is the boundary of Ω. Let*

$$F = \int_{\Omega(p)} f(q)dq.$$

Then

$$\frac{\partial F}{\partial p} = \int_{\partial\Omega(p)} f(q)\frac{\partial q}{\partial p} \cdot n(q)dq$$

where $\frac{\partial q}{\partial p}$ *is the derivative of the points on* $\partial\Omega$ *with respect to* p.

In [19], it was investigated how Voronoi cells changed as functions of p_i. In fact, it was shown in [19] that for any point $q \in \partial V_{ij}$ (the boundary between adjacent cells V_i and V_j),

$$\begin{aligned}
\frac{\partial q}{\partial p_j^{(b)}} \cdot (p_j - p_i) &= \frac{1}{2} e_b \cdot (p_j - p_i) - e_b \cdot \left(q - \frac{p_i + p_j}{2}\right), \\
\frac{\partial q}{\partial p_i^{(b)}} \cdot (p_j - p_i) &= \frac{1}{2} e_b \cdot (p_j - p_i) + e_b \cdot \left(q - \frac{p_i + p_j}{2}\right),
\end{aligned}$$

where $p_j^{(b)}$ denotes the bth component of the vector p_j and e_b is the bth elementary unit vector. Note that in this chapter, $b = 1, 2$ since we are considering the case $D \subset \mathbb{R}^2$ only.

Substituting this into Leibniz rule, we obtain

$$\begin{aligned}
\frac{\partial c_i^{(a)}}{\partial p_j^{(b)}} = &-\left(\int_{\partial V_{ij}} \phi q^{(a)} \frac{q^{(b)} - p_j^{(b)}}{\|p_j - p_i\|} dq\right) \Big/ m_i \\
&+ \left(\int_{V_i(P)} \phi q^{(a)} dq\right)\left(\int_{\partial V_{ij}} \phi \frac{q^{(b)} - p_j^{(b)}}{\|p_j - p_i\|} dq\right) \Big/ m_i^2
\end{aligned} \tag{15.14}$$

where $a = 1, 2$, $b = 1, 2$ and where $i \neq j$. When $i = j$ we must consider the contribution from all neighbors

$$\begin{aligned}
\frac{\partial c_i^{(a)}}{\partial p_i^{(b)}} = \sum_{k \in N_{V_i}} \Bigg[&\left(\int_{\partial V_{ik}} \phi q^{(a)} \frac{q^{(b)} - p_i^{(b)}}{\|p_k - p_i\|} dq\right) \Big/ m_i \\
&- \left(\int_{V_i(P)} \phi q^{(a)} dq\right)\left(\int_{\partial V_{ik}} \phi \frac{q^{(b)} - p_i^{(b)}}{\|p_k - p_i\|} dq\right) \Big/ m_i^2 \Bigg].
\end{aligned} \tag{15.15}$$

We can rewrite these equations more compactly using block matrix notation

$$\left[\frac{\partial c}{\partial p}\right]_{ij} = \frac{\partial c_i}{\partial p_j} = -\frac{\int_{\partial V_{ij}} (q - c_i)(q - p_j)^T \phi \, \mathrm{d}q}{m_i \| p_j - p_i \|}, \tag{15.16}$$

$$\left[\frac{\partial c}{\partial p}\right]_{ii} = \frac{\partial c_i}{\partial p_i} = \sum_{k \in N_{V_i}} \frac{\int_{\partial V_{ik}} (q - c_i)(q - p_i)^T \phi \, \mathrm{d}q}{m_i \| p_j - p_k \|}. \tag{15.17}$$

where $[\cdot]_{ij}$ corresponds to the ith, j^{th} $d \times d$ block matrix.

It is noteworthy that given a continuously differentiable density function ϕ, computing $\partial c/\partial p$ at any given time t becomes an exercise in finding line and area integrals. In implementation, it suffices to use numerical approximations to compute these integrals (e.g., Riemann sums, Gaussian quadrature rule).

One more partial derivative is required for update law (15.13), namely $\partial c/\partial t$. Another application of Leibniz rule results in

$$\frac{\partial c_i}{\partial t} = \frac{m_i \int_{V_i} q \frac{\partial \phi}{\partial t}(q, t) \, \mathrm{d}q - \int_{V_i} q \phi(q, t) \, \mathrm{d}q \int_{V_i(p)} \frac{\partial \phi}{\partial t}(q, t) \, \mathrm{d}q}{m_i^2} \tag{15.18}$$

or more compactly

$$\frac{\partial c_i}{\partial t} = \frac{\int_{V_i} (q - c_i) \frac{\partial \phi}{\partial t} \, \mathrm{d}q}{m_i}. \tag{15.19}$$

with $\frac{\partial c}{\partial t}^T = [\frac{\partial c_1}{\partial t}^T, \ldots, \frac{\partial c_n}{\partial t}^T]$.

When implementing the update law, it suffices to numerically compute the integrals in $\partial c/\partial t$. However, unlike with the computation of the integrals in $\partial c/\partial p$, we require knowledge of $\partial \phi/\partial t$. If ϕ is not provided analytically in t, then one could:

1. Utilize a finite difference scheme to approximate $\partial \phi/\partial t$. This could however give rise to difficulties with noisy measurements.
2. Alternatively, the user could provide the time evolution of the density function directly via a continuous function $\partial \phi/\partial t$ such that $\phi(q, t) = \int_{t_0}^{t} \frac{\partial \phi}{\partial t}(q, \tau) \, \mathrm{d}\tau$. In the implementations found in Sect. 15.3, the "shape" of the density is defined as a continuously differentiable function $\phi(q, t_0)$ defined over D, and the user-provided input corresponds to $\partial \phi/\partial t$.

As a final implementation note, finding the integrals over V_i may be computationally intensive. However, the expanded versions of these partial derivatives reveal that only a few integrals actually need to be computed per agent. To compute all expressions it suffices to compute the integrals of ϕ, $d\phi/dt$, $q\phi$, and $qd\phi/dt$ over V_i, which may be computed once for each agent (and may be computed in a distributed fashion).

15.2.3.2 Circumventing the Matrix Inverse Computation

The second subtle difficulty with update law (15.13) is ensuring that the inverse $(I - \partial c/\partial p)^{-1}$ is well defined, which is generally difficult. Lemma 15.1 provides us with a sufficient condition for the existence of this matrix. In [19] it was also shown that in the time-invariant case, the inverse is well defined as long as $\phi(p)$ is a log-concave function of p. We would also need ϕ to be continuously differentiable in both arguments, so these two conditions are enough to ensure that the inverse exists. Since the motivation for this work is to have a human operator influence the team of robots by generating these density functions, the former constraint is quite an unsatisfying one, for it would greatly reduce the types of density functions allowable. Moreover, the computation of this $2n \times 2n$ matrix inversion does not scale well with increase in number of agents, and requires information on every agent which makes it a centralized scheme. Fortunately, it is possible to alleviate these concerns by performing a distributed approximation of update law (15.13). The desired approximation can be found by using the Neumann series, e.g., [75].

Lemma 15.3 (Neumann series) *Let A be a square matrix. If $\lim_{k\to\infty} A^k = 0$, then $I - A$ is invertible and*

$$(I - A)^{-1} = I + A + A^2 + A^3 + \cdots .$$

Moreover, for an $m \times m$ square matrix A, $\lim_{k\to\infty} A^k = 0$ if and only if $|\lambda_i| < 1$ for all $i = 1, 2, \ldots, m$, where λ_i are the eigenvalues of A. As such, let λ_{max} denote the eigenvalue with the largest magnitude of the matrix $\partial c/\partial p$. Using the Neumann series, we can express $(I - \partial c/\partial p)^{-1}$ as

$$\left(I - \frac{\partial c}{\partial p}\right)^{-1} = I + \frac{\partial c}{\partial p} + \left(\frac{\partial c}{\partial p}\right)^2 + \cdots \tag{15.20}$$

as long as $|\lambda_{max}| < 1$.

Our goal will be to truncate this series to obtain the well-defined approximation to the matrix inverse, but then the question arises: how many terms should be kept? The answer lies in the sparsity structure of $\partial c / \partial p$.

Given a Voronoi partition of the area of interest, we denote the boundary between the two cells V_i and V_j by ∂V_{ij}. Since we are only considering the planar case, there are three possibilities for ∂V_{ij}:

1. ∂V_{ij} is empty, meaning that cells V_i and V_j do not intersect.
2. ∂V_{ij} consist of a single point, meaning that cells V_i and V_j share a single vertex.
3. ∂V_{ij} is a line, meaning that cells V_i and V_j share a face.

We will denote N_{V_i} to be set of indexes pertaining to the agents whose Voronoi cells V_j share a face with agent i's Voronoi cell V_i.

Lemma 15.4 $j \notin N_{V_i} \implies \dfrac{\partial c_i}{\partial p_j} = 0.$

Proof For the first two cases, i.e., ∂V_{ij} is either empty or consists of a singleton, from (15.14) and (15.15) we see that the integrals over ∂V_{ij} would be zero. Note that for these two cases, this will be true for all four elements in $\partial c_i / \partial p_j$. Since these two cases correspond to agents i and j not sharing a face, we conclude that $j \notin N_{V_i}$ implies that $\partial c_i / \partial p_j = 0$. □

This lemma tells us that $\partial c / \partial p$ actually encodes adjacency information of the graph induced by the Voronoi tessellation. This induced graph is known as the *Delaunay graph.*

To obtain a distributed update law, we must insist that the update for $\dot{p}_i$ depends only on information from itself (p_i and $\phi(q, t)$ for $q \in V_i$) and information on neighboring agents (p_j and $\phi(q, t)$ for $q \in V_j$, for all $j \in N_{V_i}$). To this end, we truncate the Neumann series in (15.20) after just two entries, i.e., $\left(I - \partial c / \partial p\right)^{-1} \approx I + \partial c / \partial p$. By modifying update law (15.13) with this approximation, we obtain the update law

$$\dot{p} = \left(I + \frac{\partial c}{\partial p}\right)\left(-\kappa\,(p - c) + \frac{\partial c}{\partial t}\right),$$

which at the individual robot level results in the update law called *TVD-D*$_1$ for Time-Varying Densities, Distributed case with 1-hop adjacency information:

$$\dot{p}_i = \frac{\partial c_i}{\partial t} - \kappa\,(p_i - c_i) + \sum_{j \in \overline{N}_{V_i}} \frac{\partial c_i}{\partial p_j}\left(\frac{\partial c_j}{\partial t} - \kappa\left(p_j - c_j\right)\right) \tag{15.21}$$

where $\overline{N}_{V_i}$ is the closed neighborhood set to Voronoi cell i in the Delaunay graph.

It should be noted that (15.21) is always well defined (as long as ϕ is continuously differentiable). In other words, even if the Neumann series is not convergent or if the inverse does not exist, the entries in (15.21) are well defined. In fact, it turns out that during the robotic experiment, even in cases where $|\lambda_{max}| > 1$, the robots consistently evolve in a manner that achieves coverage. In [41], a comparison is made between both the centralized and decentralized approaches shown here to other available techniques for coverage of time-varying densities — we refer the readers to it for a discussion on how they compare.

One can now investigate what happens when higher order terms are kept in the Neumann series. For this, we let $\text{dist}(i, j)$ denote the edge distance, or number of edges in the shortest path, between i and j in the Delaunay graph induced by the Voronoi tessellation. And, as $\partial c / \partial p$ is a (block) adjacency matrix, we have that

$$\left[(\partial c / \partial p)^k\right]_{ij} \neq 0 \quad \Longrightarrow \quad \text{dist}(i, j) = k,\ k = 0, 1, 2, \ldots$$

where $[\ \cdot\ ]_{ij}$ denotes the block corresponding to cell c_i and robot position p_j.

The k-hop version of *TVD-D*$_1$ thus becomes

$$\dot{p} = \sum_{\ell=0}^{k} \left(\frac{\partial c}{\partial p}\right)^{\ell} \left(-\kappa(p - c) + \frac{\partial c}{\partial t}\right). \tag{15.22}$$

15.3 Designing Density Functions

At the heart of our HSI implementation lies update law (15.21) which allows a human operator to influence the position of the swarm by broadcasting the time-varying density ϕ defined over the domain D. In this section, we describe two different design methods that allow a human operator to generate the density function based on simple inputs such as the general shape of the density or its position in the domain.

15.3.1 Diffusion of Drawn Geometric Configurations

We now present an approach that allows the user to specify the geometric configuration of the swarm. The process consists of allowing the human operator to draw the desired shape in the tablet-like interface. This drawing, taken as a binary image over the area of interest $\Psi : D \to \{0, 1\}$, can be made smooth and turned into a continuously differentiable function in the spatial argument by evolving it as a diffusion process [64]. The result is a smooth, nonparametric density function with the pixels determining the density intensities. One could then simply pass the image to the numerical integrating tools to obtain update law for the agents, interpolating between pixel values as needed. Figure 15.2a illustrates a potential graphical user interface to provide the input density to the swarm by allowing a user to draw the desired configuration. Figure 15.2c illustrates an example of the smoothed drawing that represents the desired density.

However, in order to reduce the amount of information needed to be communicated to the agents, it is possible to use the level sets of the resulting image to come up with a Gaussian Mixture Model (GMM), which would collapse the dimensionality from the pixel count (which could be significant) to k centroids and covariances, k being the amount of Gaussian functions desired to approximate the nonparametric density, thus reducing significantly the amount of information required to represent the density. A GMM is given in the form

$$\phi_{shape}(q) = \sum_{i=1}^{k} f_i(q) = \sum_{i=1}^{k} \frac{\alpha_i}{2\pi\sqrt{|\Sigma_i|}} \exp\left(-\frac{1}{2}(q - \mu_i)^T \Sigma_i^{-1} (q - \mu_i)\right)$$

(a) GUI allows the human operator to draw the desired density shape.

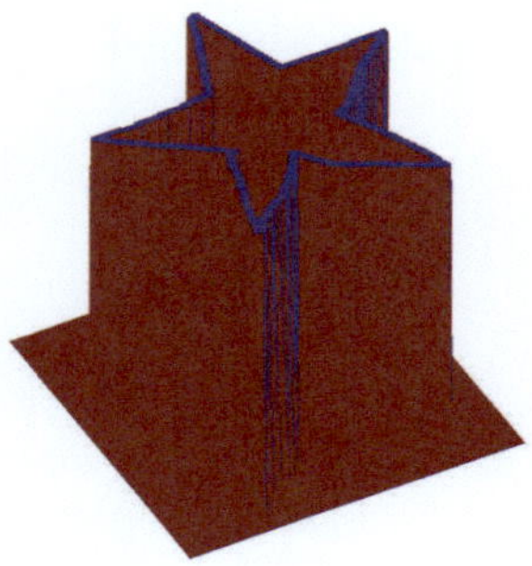

(b) Drawn shape as binary image over domain.

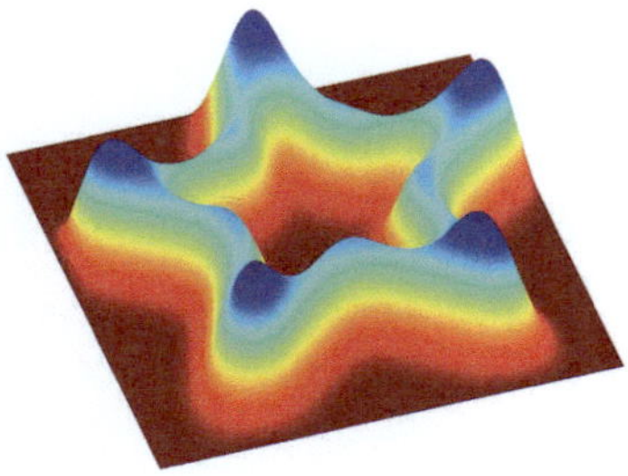

(c) Image turned smooth by diffusion process.

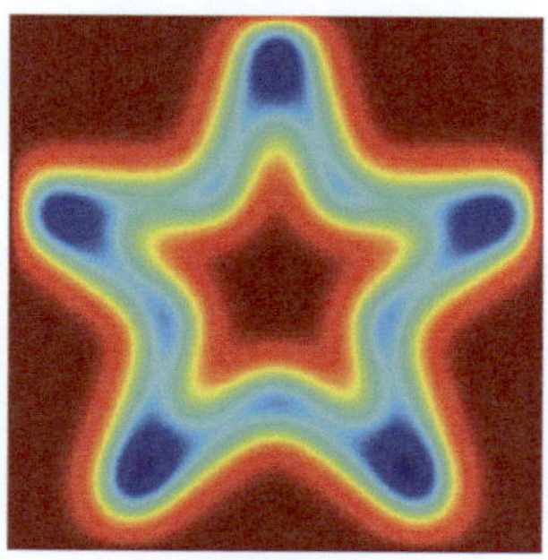

(d) Level sets of desired density.

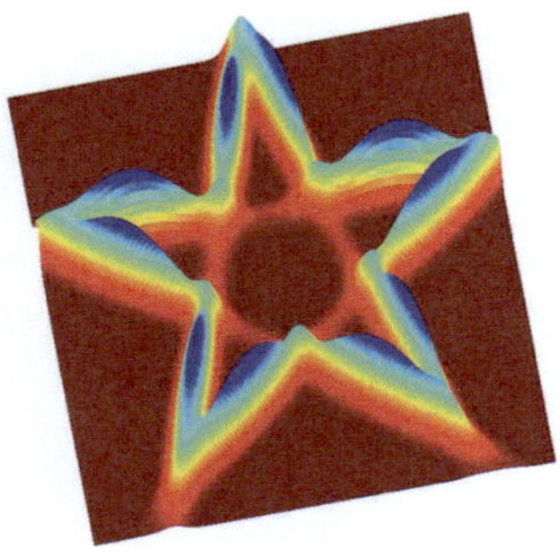

(e) GMM approximation of desired shape.

Fig. 15.2 Graphical user interface (GUI) allows the user to generate density shapes where μ_i, Σ_i and α_i are the mean, covariance, and weight of the ith mixture, with $\sum \alpha_i = 1$ and $\sum \mu_i = 0$.

In [9], a process to obtain a GMM from a data set with redundancies over the data is presented. This process can be utilized to obtain a parametric approximation of the desired density. In order to generate the required redundant data sets, sample points can be selected from the contour level sets of the desired density (Fig. 15.2d). The points in this data set are then grouped into k clusters, where the design parameter k is the number of Gaussian models used in the GMM. A larger k can be used for a finer approximation of the desired density whereas a smaller k can be used for coarser approximation. The data from each cluster is used to determine the parameters for each of the Gaussian models. Figure 15.2e illustrates the GMM found to approximate the density with the use of ten Gaussian functions.

In order to move the density around, one could evolve the mixture means to track the desired user-provided position, e.g.,

$$\phi(q,t) = \sum_{i=1}^{k} f_i\left(q - v(t)\right), \quad \frac{\partial \phi}{\partial t} = \sum_{i=1}^{k} f_i\left(q - v(t)\right)\left(q - v(t) - \mu_i\right)^T \Sigma_i^{-1} \dot{v}(t)$$

where $v(t)$ is the solution to $\dot{v}(t) = -\kappa\left(v(t) - r(t)\right)$ given $v(t_0) \in D$ for some $\kappa > 0$, and $r(t) \in D$ is the human-operator provided desired location of the density at t.

15.3.2 Control of Gaussian Functions

The previous approach allows a user to generate a density function that captures the important areas to be covered by simply drawing over the domain. In order to reduce the amount of information required to describe the generated nonparametric density, this was approximated by a Gaussian Mixture Model (GMM) with k Gaussian functions. If the geometric shape the swarm takes is not as important as actively manipulating the swarm, then the GMM concept from the previous method can be modified by fixing the a priori weights and "shape" of the Gaussian functions.

In this method the human operator taps the desired location of the swarm on a tablet representative of the domain D. Influence on the swarm is achieved by performing coverage of the density function made of Gaussian functions in the form

$$\phi(q,t) = \varepsilon_0 + \sum_{i=1}^{R} \ell_i(t) e^{-\frac{1}{2}(q-\mu_i(t))^T \Sigma^{-1}(q-\mu_i(t))}$$

$$\frac{\partial \phi}{\partial t}(q,t) = \sum_{i=1}^{R} \left[\ell_i(t)\left(q-\mu_i(t)\right)^T \Sigma^{-1} \dot{\mu}_i(t) + \dot{\ell}_i(t)\right] e^{-\frac{1}{2}(q-\mu_i(t))^T \Sigma^{-1}(q-\mu_i(t))}$$

where Σ is a positive-definite two-dimensional matrix that determines the "shape" for the Gaussian functions (e.g., $\Sigma = \sigma^2 I_2$, $\sigma > 0$, provides a circular shape), R is the number of fingers the user is tapping with, ε_0 is a small positive constant so that $\phi > 0$ even if there are no fingers presents, and $\mu_i(t)$ is the solution to

$$\dot{\mu}_i = -\kappa(\mu_i - r_i(t))$$

for some $\kappa > 0$ and where $r_i(t) \in D$ is the position of the ith desired location in D (provided by the number of "taps" on the tablet-interface) which directly influences the centroid of the individual Gaussian functions, $i = 1, \ldots, R$. The Gaussian terms are weighted by transition functions in order to smoothly introduce and remove them from the domain of interest. These are of the form

$$\ell_i(t) = \frac{1}{1 + \exp\left(-\alpha((t - \tau_{i,1}) - \frac{1}{2})\right)} - \frac{1}{1 + \exp\left(-\alpha((t - \tau_{i,2}) - \frac{1}{2})\right)}$$

where $\tau_{i,1}$ corresponds to the time at which the ith Gaussian term is added, $\tau_{i,2}$ corresponds to the time at which the ith Gaussian term is removed ($\tau_{i,1} < \tau_{i,2}$), and $\alpha > 0$ (e.g., $\alpha = 10$) determines the rate of transition.

15.4 Robotic Experiments

Both density design methods described in Sects. 15.3.1 and 15.3.2 were implemented on robotic platforms. In particular, two different multi-robot systems were used to compare the effectiveness of the methodology across platforms. The first robot platform used for these experiments were a smaller and lower cost multi-robot system consisting of GRITSBots[2] [66]. The GRITSBots possess a main processor that runs at 80 MHz and handles user code and WiFi data processing, and a co-processor for handling motor control running at 8 MHz, stepper motors for differential drive, and WiFi-based communication that uses UDP-based sockets. For these experiments, a computer handles tracking of the robots based on standard webcams (1280×720 resolution) that detect ArUco tags for tracking, and that runs at approximately 25 Hz. The tracking system was used to provide position and orientation data to the agents and was also used to compute the Voronoi tessellation. Matlab is used for visualization of the user-provided density function and the Voronoi tessellation.

The density design method described in Sect. 15.3.2 was also implemented on a different robotic platform. The robotic platforms used for the experiments were Khepera III robots from K-team. The Khepera III differential-drive wheeled robots each have a 600 MHz ARM processor with 128 Mb RAM, embedded Linux, and are equipped with a wireless card for communication over a wireless router. An Ubuntu

[2]For more information on the GRITSBots and their testbed, see www.robotarium.org.

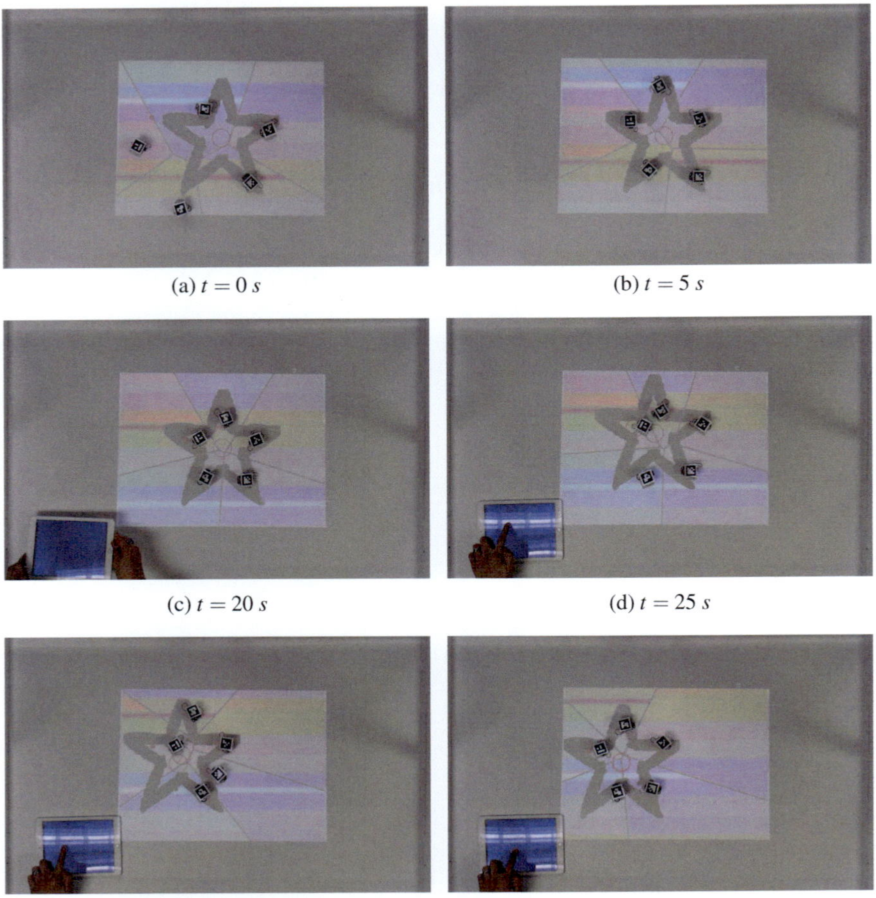

(a) $t = 0\ s$ (b) $t = 5\ s$

(c) $t = 20\ s$ (d) $t = 25\ s$

(e) $t = 30\ s$ (f) $t = 35\ s$

Fig. 15.3 Robotic implementation of a drawn star-shaped density on a team of GRITSBots. A tablet is later used (**c–f**) to directly influence the location of the swarm formation. An overhead projector is used to visualize in real time: the broadcasted density function, Voronoi tessellation, and current desired location of the swarm provided by the user through the tablet

computer with an Intel dual core CPU 2.13 GHz and 4 GB of memory, running ROS (Robot Operating System, version Diamondback) received state information from ten OptiTrack S250e motion capture cameras, including position and orientation data for the agents. This data was also used to compute the Voronoi tessellation. The rviz ROS package was used for visualization of the user-provided density function.

In both experiments, an overhead projector is used to project the visualization onto the real physical environment of the robots to give a real-time visual representation of the user-provided density function. The human operator used a tablet device to input the density information. These were transmitted over WiFi and UDP to a central

(a) User observes position of swarm and redirects it with the use of a tablet.

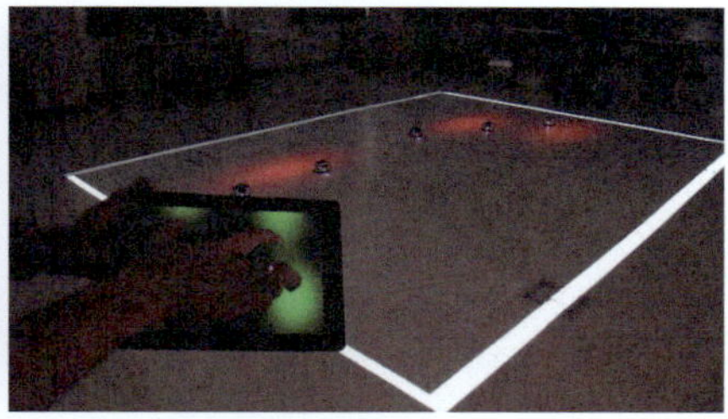

(b) User splits the swarm by introducing multiple locations of interest.

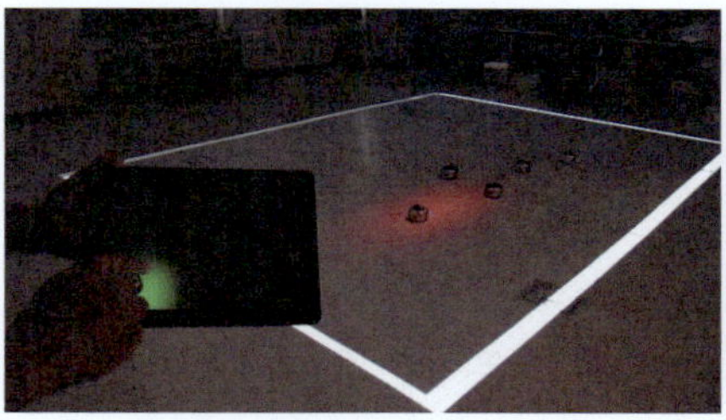

(c) User influencing swarm to center of domain.

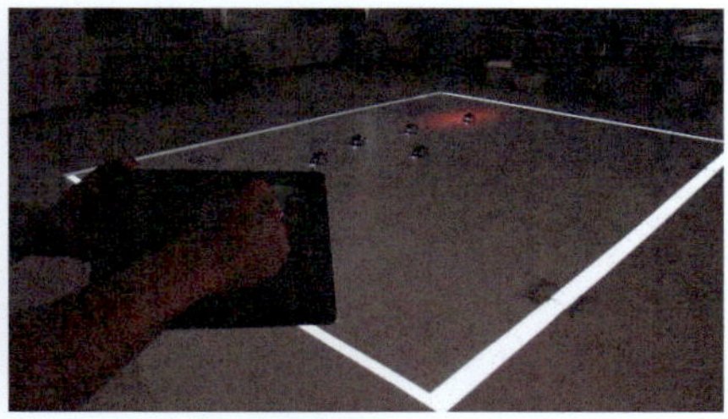

(d) User influencing swarm to corner of domain.

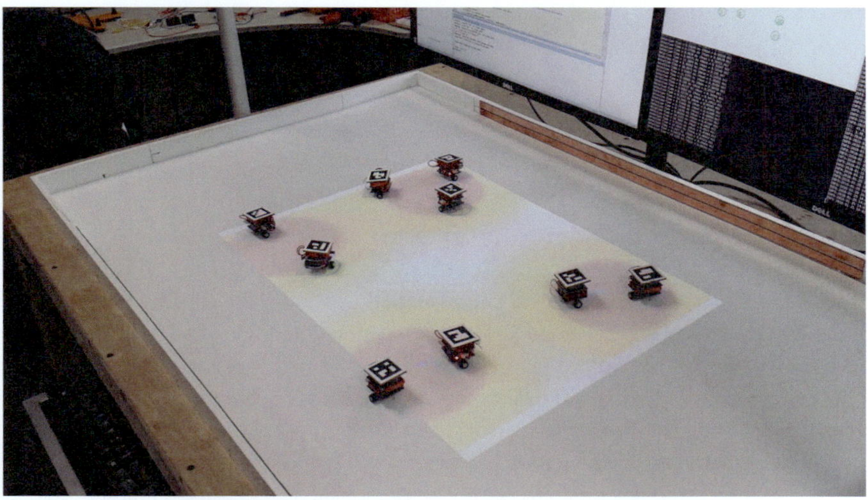

(e) Nine *GRITSBots* influenced by the *control of Gaussian functions* HSI scheme.

Fig. 15.4 Robotic implementation of HSI via coverage of Gaussian functions. In **a** the user observes the position of swarm and redirects it with the use of a tablet. A tablet is used (**b–d**) to directly influence the location of the Khepera III swarm. In **e** four fingers are used on a swarm of GRITSBots. An overhead projector is used to visualize the user-provided density function in real time over the robot workspace. See https://youtu.be/gsNHJwA7V-U for a video of the implementation computer which calculated the update law for each agent. Even though a central computer is used for the computation, the control is computed for every agent only using adjacency information, and then transmitted to the pertinent robot via WiFi and UDP. Both robotic platforms may be modeled with unicycle dynamics

$$\dot{x} = \left[v_i \cos(\theta_i),\ v_i \sin(\theta_i)\right]^T, \qquad \dot{\theta}_i = \omega_i$$

where $x_i \in \mathbb{R}^2$ is agent i's position in D, θ_i its heading, and (v_i, ω_i) are its linear and angular velocity inputs. A transformation similar to the one found in [15] is used to map the update law to the velocity inputs, i.e.,

$$v_i = \left[\cos(\theta_i),\ \sin(\theta_i)\right] \cdot \dot{p}_i, \qquad \omega_i = \kappa \arctan\left(\frac{\left[-\sin(\theta_i),\ \cos(\theta_i)\right] \cdot \dot{p}_i}{\left[\cos(\theta_i),\ \sin(\theta_i)\right] \cdot \dot{p}_i}\right)$$

for some $\kappa > 0$. In Fig. 15.3, we see the GRITSBots achieving the desired geometric configuration while the human operator changes the location of the desired shape, and in Fig. 15.4e, we see the GRITSBots clustering over the four reference locations provided by the user. Figure 15.4 illustrates the HSI strategy from Sect. 15.3.2 on the Khepera III platforms as well. Measurements on the coverage performance on these platforms, as well as comparison to other coverage algorithms, can be found in [41].

15.5 Conclusions

In this chapter, we developed a framework for human–swarm interactions in which a human operator broadcasts a density function, encoding geometric configurations or areas of interest within the domain, which is then optimally covered by the swarm of robots. By executing the distributed update laws presented here, individual robots in the swarm only need to coordinate their motion with nearby agents. As such, these update laws scale as the size of the robot swarm increases. Further, the human input to the system (i.e., the time-varying density function) manipulates the environment directly, which makes it independent of the size of the robot swarm. In this way, the human operator can manipulate the entire multi-robot system without having to take into account the individual agents. We also proposed two methods for allowing a human operator to generate density functions: by drawing shapes on an input device (e.g., tablet), and dynamically changing the location of the density in the environment (e.g., tapping on the areas of interest). These HSI strategies were validated on two different robot platforms.

In order to pursue these HSI strategies, contributions were made to the problem of coverage of time-varying density functions. In particular, important spectral properties are provided which are used to provide convergence guarantees for the presented update laws. In addition, a family of distributed approximations is presented which allow the agents to provide optimal coverage of the human-generated density functions while only relying on local information.

Although collision avoidance is not explicitly considered in this chapter, collisions are rarely observed during the robotic experiments. This is in part due to the nature of the coverage problem, in which the goal (to an extent) is for the agents to spread out. In fact, if agents were point-particles, then no collisions would be observed given that agents remain in the interior of their Voronoi cells, which are nonoverlapping. Practically, since agents are not point-particles, and depending on how concentrated the density function is, it is possible for the agents to get close enough to the boundary of their Voronoi cells to collide. A common way to handle collisions is to introduce an artificial repulsive force term in the agents' update law. Although this can be implemented with simple modifications to the update law, it does not allow us to provide formal guarantees on collision avoidance or convergence to a CVT. There has been recent work on using *barrier certificates* to modify swarm controllers in a minimally invasive way that results in provably collision-free controllers [7]. It may be possible to use the sets described by these barrier certificates to reach conclusions on convergence to a CVT.

Acknowledgements This work is sponsored by grant number FA9550-13-1-0029 from the US Air Force Office for Scientific Research.

References

1. Adlakha S, Srivastava MB (2003) Critical density thresholds for coverage in wireless sensor networks. In: Proceedings of the wireless communications and networking conference (WCNC), IEEE, pp 1615–1620
2. Åstrand B, Baerveldt AJ (2002) An agricultural mobile robot with vision-based perception for mechanical weed control. Auton. Robots 13(1):21–35
3. Balch T (1999) The impact of diversity on performance in multi-robot foraging. In: Proceedings of the third annual conference on autonomous agents. ACM, pp 92–99
4. Balch T, Arkin RC (1998) Behavior-based formation control for multirobot teams. IEEE Trans Robot Autom 14(6):926–939
5. Bashyal S, Venayagamoorthy G (2008) Human swarm interaction for radiation source search and localization. In: Proceedings of the swarm intelligence symposium, SIS 2008. IEEE, pp 1–8. doi:10.1109/SIS.2008.4668287
6. Bonert M, Shu L, Benhabib B (2000) Motion planning for multi-robot assembly systems. Int J Comput Integr Manuf 13(4):301–310
7. Borrmann U, Wang L, Ames AD, Egerstedt M (2015) Control barrier certificates for safe swarm behavior. IFAC-PapersOnLine 48(27):68–73
8. Butler Z, Rus D (2003) Event-based motion control for mobile-sensor networks. IEEE Pervasive Comput 2(4):34–42
9. Calinon S, Guenter F, Billard A (2007) On learning, representing and generalizing a task in a humanoid robot. IEEE Trans Syst Man Cybern Part B 37(2):286–298
10. Cao YU, Fukunaga AS, Kahng A (1997) Cooperative mobile robotics: antecedents and directions. Auton Robots 4(1):7–27
11. Caprari G, Balmer P, Piguet R, Siegwart R (1998) The autonomous micro robot "alice": a platform for scientific and commercial applications. In: Proceedings of the 1998 international symposium on micromechatronics and human science, MHS'98. IEEE, pp 231–235

12. Chang HJ, Lee CG, Hu YC, Lu YH (2007) Multi-robot slam with topological/metric maps. In: IEEE/RSJ international conference on intelligent robots and systems, IROS 2007. IEEE, pp 1467–1472
13. Cortés J, Bullo F (2005) Coordination and geometric optimization via distributed dynamical systems. SIAM J Control Optim 44(5):1543–1574. doi:10.1137/S0363012903428652
14. Cortes J, Martinez S, Karatas T, Bullo F (2002) Coverage control for mobile sensing networks: variations on a theme. In: Proceedings of the mediterranean conference on control and automation, Lisbon, Portugal, Electronic Proceedings
15. Cortes J, Martinez S, Karatas T, Bullo F (2004) Coverage control for mobile sensing networks. IEEE Trans Robot Autom 20(2):243–255
16. Diana M, de la Croix JP, Egerstedt M (2013) Deformable-medium affordances for interacting with multi-robot systems. In: Proceedings of the 2013 IEEE/RSJ international conference on intelligent robots and systems (IROS), pp 5252–5257. doi:10.1109/IROS.2013.6697116
17. Diaz-Mercado Y, Lee SG, Egerstedt M (2015) Distributed dynamic density coverage for human-swarm interactions. In: Proceedings of the american control conference (ACC), pp 353–358. doi:10.1109/ACC.2015.7170761
18. Ding XC, Rahmani AR, Egerstedt M (2010) Multi-uav convoy protection: an optimal approach to path planning and coordination. IEEE Trans Robot 26(2):256–268
19. Du Q, Emelianenko M (2006) Acceleration schemes for computing centroidal Voronoi tessellations. Numer Linear Algebra Appl 13(2–3):173–192. doi:10.1002/nla.476
20. Du Q, Faber V, Gunzburger M (1999) Centroidal voronoi tessellations: applications and algorithms. SIAM Rev 41(4):637–676. doi:10.1137/S0036144599352836
21. Du Q, Emelianenko M, Ju L (2006) Convergence of the Lloyd algorithm for computing centroidal voronoi tessellations. SIAM J Numer Anal 44(1):102–119. doi:10.1137/040617364
22. Dubowsky S, Boning P (2007) The coordinated control of space robot teams for the on-orbit construction of large flexible space structures. In: Proceedings of the 2007 IEEE international conference robotics and automation
23. Egerstedt M, Hu X (2001) Formation constrained multi-agent control. IEEE Trans Robot Autom 17(6):947–951. doi:10.1109/70.976029
24. Georgeff M (1988) Communication and interaction in multi-agent planning. Read Distrib Artif Intell 313:125–129
25. Gerkey BP, Matarić MJ (2004) A formal analysis and taxonomy of task allocation in multi-robot systems. Int J Robot Res 23(9):939–954
26. Ghaffarkhah A, Yan Y, Mostofi Y (2011) Dynamic coverage of time-varying environments using a mobile robot – A communication-aware perspective. In: GLOBECOM Workshops (GC Wkshps). IEEE, pp 1297–1302. doi:10.1109/GLOCOMW.2011.6162394
27. Ghosh A, Das SK (2008) Coverage and connectivity issues in wireless sensor networks: a survey. Pervasive Mobile Comput 4(3):303–334
28. Giusti A, Nagi J, Gambardella L, Bonardi S, Di Caro G (2012) Human-swarm interaction through distributed cooperative gesture recognition. In: Proceedings of the 7th ACM/IEEE international conference on human-robot interaction (HRI). pp 401–401
29. Grocholsky B, Keller J, Kumar V, Pappas G (2006) Cooperative air and ground surveillance. IEEE Robot Autom Mag 13(3):16–25
30. Groß R, Bonani M, Mondada F, Dorigo M (2006) Autonomous self-assembly in swarm-bots. IEEE Trans Robot 22(6):1115–1130
31. Haumann D, Willert V, Listmann KD (2013) DisCoverage: from coverage to distributed multi-robot exploration. In: Proceedings of the 4th IFAC workshop on distributed estimation and control in networked systems, Koblenz, Germany, vol 4, pp 328–335, doi:10.3182/20130925-2-DE-4044.00009
32. Heo N, Varshney PK (2005) Energy-efficient deployment of intelligent mobile sensor networks. IEEE Trans Syst Man Cybern Part A: Syst Hum 35(1):78–92
33. Howard A (2006) Multi-robot simultaneous localization and mapping using particle filters. Int J Robot Res 25(12):1243–1256

34. Howard A, Matarić MJ, Sukhatme GS (2002) An incremental self-deployment algorithm for mobile sensor networks. Auton Robots 13(2):113–126
35. Iri M, Murota K, Ohya T (1984) A fast Voronoi-diagram algorithm with applications to geographical optimization problems. In: Thoft-Christensen P (ed) System modelling and optimization, vol 59., Lecture notes in control and information sciences. Springer, Berlin, pp 273–288
36. Jea D, Somasundara A, Srivastava M (2005) Multiple controlled mobile elements (data mules) for data collection in sensor networks. Distributed computing in sensor systems. Springer, Berlin, pp 244–257
37. Kernbach S, Thenius R, Kernbach O, Schmickl T (2009) Re-embodiment of honeybee aggregation behavior in an artificial micro-robotic system. Adapt Behav 17(3):237–259
38. Klavins E (2007) Programmable self-assembly. IEEE Control Syst 27(4):43–56
39. Kolling A, Sycara K, Nunnally S, Lewis M (2013) Human swarm interaction: an experimental study of two types of interaction with foraging swarms. J Hum-Robot Interact 2(2):103–128
40. Kushleyev A, Mellinger D, Powers C, Kumar V (2013) Towards a swarm of agile micro quadrotors. Auton Robots 35(4):287–300
41. Lee S, Diaz-Mercado Y, Egerstedt M (2015) Multirobot control using time-varying density functions. IEEE Trans Robot 31(2):489–493. doi:10.1109/TRO.2015.2397771
42. Lindsey Q, Mellinger D, Kumar V (2012) Construction with quadrotor teams. Auton Robots 33(3):323–336
43. Liu Y, Wang W, Lévy B, Sun F, Yan DM, Lu L, Yang C (2009) On centroidal voronoi tessellation - energy smoothness and fast computation. ACM Trans Graph 28(4):1–17. doi:10.1145/1559755.1559758
44. Lloyd S (1982) Least squares quantization in PCM. IEEE Trans Inform Theory 28(2):129–137. doi:10.1109/TIT.1982.1056489
45. Macwan A, Nejat G, Benhabib B (2011) Target-motion prediction for robotic search and rescue in wilderness environments. IEEE Trans Syst Man Cybern Part B 41(5):1287–1298
46. Martinez S, Cortes J, Bullo F (2007) Motion Coordination with Distributed Information. Control Systems Magazine, IEEE 27(4):75–88
47. Mastellone S, Stipanović DM, Graunke CR, Intlekofer KA, Spong MW (2008) Formation control and collision avoidance for multi-agent non-holonomic systems: Theory and experiments. Int J Robot Res 27(1):107–126
48. Matarić MJ (1997) Reinforcement learning in the multi-robot domain. Robot colonies. Springer, Heidelberg, pp 73–83
49. McLurkin J, Smith J, Frankel J, Sotkowitz D, Blau D, Schmidt B (2006) Speaking swarmish: Human-robot interface design for large swarms of autonomous mobile robots. In: Proceedings of the AAAI spring symposium: to boldly go where no human-robot team has gone before, pp 72–75
50. McLurkin J, Lynch AJ, Rixner S, Barr TW, Chou A, Foster K, Bilstein S (2013) A low-cost multi-robot system for research, teaching, and outreach. Distributed autonomous robotic systems. Springer, Heidelberg, pp 597–609
51. Meguerdichian S, Koushanfar F, Qu G, Potkonjak M (2001) Exposure in wireless Ad-Hoc sensor networks. In: Proceedings of the 7th annual international conference on mobile computing and networking, MobiCom '01, ACM, New York, NY, USA, pp 139–150. doi:10.1145/381677.381691
52. Morse DH (1970) Ecological aspects of some mixed-species foraging flocks of birds. Ecol Monogr 40(1):119–168
53. Murray RM (2007) Recent research in cooperative control of multivehicle systems. J Dyn Syst, Meas Control 129(5):571–583
54. Naghsh A, Gancet J, Tanoto A, Roast C (2008) Analysis and design of human-robot swarm interaction in firefighting. In: Proceedings of the 17th IEEE international symposium on robot and human interactive communication, RO-MAN 2008, pp 255–260. doi:10.1109/ROMAN.2008.4600675
55. Nagi J, Ngo H, Giusti A, Gambardella L, Schmidhuber J, Di Caro G (2012) Incremental learning using partial feedback for gesture-based human-swarm interaction. In: Proceedings of the RO-MAN. IEEE, pp 898–905. doi:10.1109/ROMAN.2012.6343865

56. Nguyen W, Mills JK (1996) Multi-robot control for flexible fixtureless assembly of flexible sheet metal auto body parts. In: Proceedings of the 1996 IEEE international conference on robotics and automation, vol 3. IEEE, pp 2340–2345
57. Noguchi N, Will J, Reid J, Zhang Q (2004) Development of a master-slave robot system for farm operations. Comput Electron Agric 44(1):1–19
58. Nunnally S, Walker P, Kolling A, Chakraborty N, Lewis M, Sycara K, Goodrich M (2012) Human influence of robotic swarms with bandwidth and localization issues. In: Proceedings of the 2012 IEEE international conference on systems, man, and cybernetics (SMC), pp 333–338. doi:10.1109/ICSMC.2012.6377723
59. Nunnally S, Walker P, Chakraborty N, Lewis M, Sycara K (2013) Using coverage for measuring the effect of haptic feedback in human robotic swarm interaction. In: Proceedings of the 2013 IEEE international conference on systems, man, and cybernetics (SMC), pp 516–521. doi:10.1109/SMC.2013.94
60. Ögren P, Fiorelli E, Leonard NE (2004) Cooperative control of mobile sensor networks: adaptive gradient climbing in a distributed environment. IEEE Trans Autom Control 49(8):1292–1302
61. Olfati-Saber R (2006) Flocking for multi-agent dynamic systems: algorithms and theory. IEEE Trans Autom Control 51(3):401–420
62. Özkucur NE, Akın HL (2010) Cooperative multi-robot map merging using fast-slam. RoboCup 2009: robot soccer world cup XIII. Springer, Heidelberg, pp 449–460
63. Pedersen SM, Fountas S, Have H, Blackmore B (2006) Agricultural robots-system analysis and economic feasibility. Precis Agric 7(4):295–308
64. Perona P, Malik J (1990) Scale-space and edge detection using anisotropic diffusion. IEEE Trans Pattern Anal Mach Intell 12(7):629–639. doi:10.1109/34.56205
65. Petersen KH, Nagpal R, Werfel JK (2011) Termes: An autonomous robotic system for three-dimensional collective construction. In: Proceedings of robotics: science and systems conference (RSS)
66. Pickem D, Lee M, Egerstedt M (2015) The GRITSBot in its natural habitat - A multi-robot testbed. In: Proceedings of the 2015 IEEE international conference on robotics and automation (ICRA), pp 4062–4067. doi:10.1109/ICRA.2015.7139767
67. Pimenta LC, Kumar V, Mesquita RC, Pereira GA (2008) Sensing and coverage for a network of heterogeneous robots. In: Proceedings of the 47th IEEE conference on decision and control, CDC 2008. IEEE, pp 3947–3952
68. Pimenta LCA, Schwager M, Lindsey Q, Kumar V, Rus D, Mesquita RC, Pereira GAS (2010) Simultaneous coverage and tracking (SCAT) of moving targets with robot networks. In: Chirikjian GS, Choset H, Morales M, Murphey T (eds) Algorithmic foundation of robotics VIII, Springer tracts in advanced robotics, vol 57. Springer, Berlin, pp 85–99. doi:10.1007/978-3-642-00312-7_6
69. Poduri S, Sukhatme GS (2004) Constrained coverage for mobile sensor networks. In: Proceedings of the 2004 IEEE international conference on robotics and automation, ICRA'04, vol 1. IEEE, pp 165–171
70. Rana OF, Stout K (2000) What is scalability in multi-agent systems? In: Proceedings of the fourth international conference on autonomous agents. ACM, pp 56–63
71. Rubenstein M, Ahler C, Nagpal R (2012) Kilobot: A low cost scalable robot system for collective behaviors. In: Proceedings of the 2012 IEEE international conference on robotics and automation (ICRA). IEEE, pp 3293–3298
72. Rubenstein M, Cornejo A, Nagpal R (2014) Programmable self-assembly in a thousand-robot swarm. Science 345(6198):795–799
73. Sánchez G, Latombe JC (2002) On delaying collision checking in prm planning: application to multi-robot coordination. Int J Robot Res 21(1):5–26
74. Shah RC, Roy S, Jain S, Brunette W (2003) Data mules: modeling and analysis of a three-tier architecture for sparse sensor networks. Ad Hoc Netw 1(2):215–233
75. Stewart G (1998) Matrix algorithms: basic decompositions, vol 1. Society for Industrial and Applied Mathematics, Philadelphia

76. Sugawara K, Kazama T, Watanabe T (2004) Foraging behavior of interacting robots with virtual pheromone. In: Proceedings of the 2004 IEEE/RSJ international conference on intelligent robots and systems (IROS 2004), vol 3, pp 3074–3079. doi:10.1109/IROS.2004.1389878
77. Sun D, Mills J (2002) Adaptive synchronized control for coordination of multirobot assembly tasks. IEEE Trans Robot Autom 18(4):498–510. doi:10.1109/TRA.2002.802229
78. Thrun S, Liu Y (2005) Multi-robot slam with sparse extended information filers. Robotics research. The eleventh international symposium. Springer, Berlin, pp 254–266
79. Traniello JF (1989) Foraging strategies of ants. Ann Rev Entomol 34(1):191–210
80. Vasile C, Pavel A, Buiu C (2011) Integrating human swarm interaction in a distributed robotic control system. In: Proceedings of the 2011 IEEE conference on automation science and engineering (CASE), pp 743–748. doi:10.1109/CASE.2011.6042493
81. Vasilescu I, Kotay K, Rus D, Dunbabin M, Corke P (2005) Data collection, storage, and retrieval with an underwater sensor network. In: Proceedings of the 3rd international conference on Embedded networked sensor systems. ACM, pp 154–165
82. Wang G, Cao G, Porta TL, Zhang W (2005) Sensor relocation in mobile sensor networks. In: Proceedings of the 24th annual joint conference of the IEEE computer and communications societies, INFOCOM 2005, vol 4. IEEE, pp 2302–2312
83. Yim M, Shen WM, Salemi B, Rus D, Moll M, Lipson H, Klavins E, Chirikjian GS (2007) Modular self-reconfigurable robot systems [grand challenges of robotics]. IEEE Robot Autom Mag 14(1):43–52
84. Zedek M (1965) Continuity and location of zeros of linear combinations of polynomials. Proc Am Math Soc 16(1):78–84
85. Zhao W, Ammar M, Zegura E (2005) Controlling the mobility of multiple data transport ferries in a delay-tolerant network. In: Proceedings of the 24th annual joint conference of the IEEE computer and communications societies, INFOCOM 2005, vol 2. IEEE, pp 1407–1418
86. Zhou XS, Roumeliotis S, et al (2006) Multi-robot slam with unknown initial correspondence: the robot rendezvous case. In: Proceedings of the 2006 IEEE/RSJ international conference on intelligent robots and systems. IEEE, pp 1785–1792

Chapter 16
Co-design of Control and Scheduling for Human–Swarm Collaboration Systems Based on Mutual Trust

Xiaotian Wang and Yue Wang

16.1 Introduction

Swarm agent systems/robotics seek to coordinate large numbers of basic agents, which take inspiration from social insects [33]. A swarm system typically consists of a large number of relatively simple agents that act independently and in parallel on various tasks in order to achieve its overall goal [3]. In [21], Keller and Segel present a coupled reaction-diffusion partial derivative dynamic equation to model the motions of swarms. In [5], Bonabeau et al. provide a response threshold model to divide the labor of insects swarms. Each swarm responds to task-related stimuli and corresponds to its response threshold. In [32], Romanczuk and Schimansky-Geier propose a model of collective motion for swarm agents based on escape and pursuit responses. In [46], Yu et al. investigate swarming behaviors in multi-agent systems with nonlinear dynamics. In [12], Gazi and Passino propose N-member "individual-based" continuous-time swarm models by two fundamentally different approaches, i.e., the spatial and nonspatial approaches, to analyze the swarm dynamics. Comparing these aforementioned models, we adopt the swarm dynamic model set up by Gazi and Passino [12] to consider both cohesiveness and collision avoidance properties among agents. Furthermore, we consider a swarm network model inspired from biological systems [30]. Wild geese in biology usually arrange in different formations when they migrate seasonally. There is generally a leading wild goose in a flock. Similarly, for land animals, such as zebra and horses, when they move to other territories, there always exists one leader in the herd [45]. Besides, fish is observed always dividing into numerous schools based on different destinations during mass migration [6]. Inspired by these phenomena in biology, we imagine a leading agent in

X. Wang · Y. Wang (✉)
Department of Mechanical Engineering, Clemson University, Clemson, SC, USA
e-mail: yue6@g.clemson.edu

Y. Wang
e-mail: xiaotiw@g.clemson.edu

Y. Wang and F. Zhang (eds.), *Trends in Control and Decision-Making for Human–Robot Collaboration Systems*, DOI 10.1007/978-3-319-40533-9_16

a swarm cluster as a leading animal in a flock and clusters of swarms that coordinate with each other in a network [18]. That is to say, each leading agent is responsible for its belonging swarm cluster and coordinate with the neighboring clusters in the network while other agents in the swarm cluster will follow the motion of this leading agent.

Swarm systems are distributed in nature and well-suited for tasks that are concerned with the space [28, 33]. Environmental monitoring and freightage are examples of swarms applications [25]. Take freightage as an example, when heavy objects need to be moved in constrained terrains, a swarm of agents can be deployed to save labor cost and satisfy space requirements. Based on the object's size and shape, several clusters of swarms can be deployed. Hence, swarm coordination and agent control should be addressed to satisfy these requirements and the field of swarm control has received considerable attention. In [22], Kim et al. present a framework for decentralized control of self-organizing swarm systems based on artificial potential functions. Agents in a swarm self-organize to flock and achieve formation through attractive and repulsive forces among themselves using the artificial potential functions. Gazi et al., in [13], design continuous-time control schemes for swarms via constructive analysis based on artificial potential functions and sliding mode control techniques. Sayama, in [36], presents new methods of decentralized control with which swarms of agents can spontaneously organize themselves and collectively form the desired shape. The swarm formation control method proposed in [1] based on potential fields satisfies scalability (to varying numbers of swarm members) and supports multiple formations. In [23], Kolling et al. investigate two basic principles for controlling a robot swarm, i.e., selection and beacon control. Both are based on setting the modes of robots, which range from trivial behaviors, such as stop, to complex autonomous behaviors, such as optimal deployment, in order to control their behavior. Here in this chapter, we focus on planar motion of mobile swarm agents with space requirements and will construct agent control laws according to desired geometric relationship, which will be introduced in Sect. 16.2.

However, in a network consisted of several swarm clusters, communication among leading agents is not easily available due to limitation of bandwidth and channel capability. The leading agent will always need to keep in touch with its followers to direct them, which occupies its limited communication resources. These agents have strict communication constraints requiring neighbors to be in close physical proximity in order to communicate [2]. As the size of a swarm cluster and the number of clusters in the network increase, the available bandwidth to the leader further decreases [15]. This is also often observed in biology swarms [39]. To address this issue, a centralized control system may be deployed to distribute information [10], which nevertheless is very expensive for a large-scale network composed of a large number of swarm clusters and vulnerable to uncertainties. On the other hand, it is necessary to include humans at some level of decision-making within these swarm networks both as a safety check and to ensure that the automation is truly supporting overall mission goals [9]. It is crucial to enable the control of swarms by a human operator who could interact with the autonomy and adapt to specific challenges in a variety of conditions [23]. Furthermore, it is envisioned that a small number of humans can collaborate with a swarm network to reduce manpower. Therefore, in this work, we consider a human–swarm collaboration systems where a network of human operators join the swarms and work together to achieve the common task goal. In [8], Coppin and Legras analyze the human factor issues raised by the deployment of swarms of autonomous robots under the supervision of a human operator. Specifically, they study how representational gaps between humans and agents impact global collaboration performances. Most previous works investigating the possibility of humans controlling swarms [8, 9, 23] have human on the swarms' loop as supervisors. In our case, we instead consider human–swarm collaboration, where humans and swarms collaborate with each other and build mutual trust toward each other.

Furthermore, we consider the modeling of human networks in the human–swarm collaboration. In [27], Lieberman et al. study evolutionary dynamics for homogeneous or spatially extended populations. They generalize the population structure by arranging individuals on a graph. In [7], Champagnat et al. unify basic evolutionary models to account for mutation bias and random drift between multiple evolutionary attractors and describe a population in which the adaptive traits of individuals influence their birth rate, the mutation process, their death rate, and how they interact with each other and their external environment. Replicator-mutator (RM) dynamics are used to describe the dynamics of complex adaptive systems in population genetics, biochemistry and models of language learning [17, 24]. In [19], Hussein uses the RM dynamical equations to model the process of building individual behavioral inclinations. In [42], Wang et al. extend the RM dynamics into a distributed version for multi-agent networks with local interactions. In this chapter, we will use the RM dynamics in [42] to model the individual human operator preference inclinations that depend on his/her subjective choices, individual mutation characteristics, and interactions with other individual humans in the network.

Mutual trust is the basis of collaboration and hence building trust between human operators and basic agents is important for effective human–robot collaboration [44]. The meta-analysis [16] studies trust factors involved in human–robot interaction (HRI), which includes robot-related, human-related, and environmental-related factors. In [11], Freedy et al. study the critical performance attributes of trust in HRI and develop a collaborative performance model. Itoh and Tanaka, in [20], propose a mathematical model of trust in automation based on the expectation of humans from the automation, dependability of the automation, and predictability of automation behavior. Similar to the human–agent trust and agent–human trust models proposed in [44], we will develop computational models for human–swarm trust and swarm–human trust in Sect. 16.3.

Furthermore, we propose a novel measurement, called "fitness," to evaluate the effectiveness of the collaboration between human and swarm networks. Based on this novel measurement, we propose a dynamic scheduling algorithm to allocate the human attention to each swarm cluster and guarantee successful human–swarm collaboration, as Fig. 16.1 shows. The algorithm is extended from the authors' previous work on the scheduling of one-human and multi-agent systems [44]. Besides, we assume that one human can only collaborate with the leading agent in one swarm

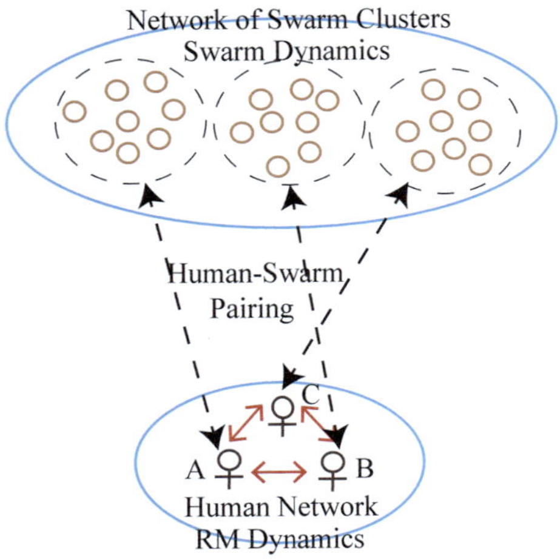

Fig. 16.1 Schematic diagram of the proposed multi-human multi-swarm collaboration network

cluster at a time, hence the allocation of human resources to multiple swarm clusters may lead to time delay of provided information. Hence, the agent control laws will be developed to incorporate the dynamic scheduling states. A similar method has been used in [38] where the time delay between sampling and actuation, represented by the states of scheduling, exists in the continuous-time control signal. Model predictive control (MPC) is designed to optimize future system behaviors, depending on prediction of the delay time. In our work, time delay specifically refers to the communication delay between the current swarm cluster and its neighbors due to human attention allocation. The co-design of time delay in control, scheduling states, as well as human attention allocation is the novelty of this work.

The organization of the rest of the chapter is as follows. In Sect. 16.2, we set up the swarms system dynamics and control. We design motion control laws for the leading agent in each swarm cluster together with scheduling states. We also develop motion control for following agents for both collision avoidance and cohesiveness. In Sect. 16.3, we introduce a collaboration framework to pair humans and swarm clusters. A dynamic scheduling algorithm is proposed in Sect. 16.4. Simulation results and discussion are provided in Sect. 16.5. Finally, we concluded this chapter in Sect. 16.6.

16.2 Swarm Setup

16.2.1 Dynamic Timing Model and Collaboration Delay

We focus on the planar motion of a network of mobile swarm clusters, where each swarm cluster has its own coordinates. Assume a human–swarm system with a network of humans $h_j,\ j = 1, 2, \dots, N$ and a network of swarm clusters s_m, $m = 1, 2, \ldots, M$. Here we consider a formation control problem and the formation shape is constructed by the leading agents in each swarm cluster. Two modes of motion control are considered for the leading agent in each swarm cluster, i.e., manual mode and autonomous mode. The manual mode means that a human operator provides neighboring swarm cluster's information, such as position and velocity, to the leader of the current collaborating swarm cluster. If the operator can provide such information at every time instant, the swarm is under the exclusively manual mode and hence has access to all necessary information to achieve its task. However, since a human needs to collaborate with multiple swarm clusters and share his/her attention among them, continuous communication with all swarm clusters is impossible. On the other hand, under the autonomous mode, the leading agent can only calculate its motion control law based on neighbors' information upon last time when the human collaborated with it. That is, the neighbors' information is not updated. Hence, under the manual mode, the more information a leading agent knows, the better the swarm performance. On the contrary, under the autonomous mode, the longer the time delay, the less information a leader knows, the worse the performance. Before we formally define this time delay, let us introduce the dynamic timing model $\mathbf{Z}^j(t) = [\mathbf{Q}^j(t), \mathbf{S}^j(t), \mathbf{O}^j(t)]$ that describes the current status of each swarm cluster under collaboration with human operator h_j at time step t [37].

The state $\mathbf{Q}^j(t) = [q_1^j(t), \ldots, q_M^j(t)]$ indicates how long after time instant, t, the next collaboration request from swarm cluster s_m, $m = 1, 2, \ldots, M$ will launch, and it satisfies the following update equations:

$$\begin{cases} q_m^j(t+1) = q_m^j(t) - 1, & \text{if } q_m^j(t) > 0 \qquad (1a) \\ q_m^j(t+1) = L_m^j, & \text{if } q_m^j(t) = 0 \qquad (1b) \end{cases}$$

where M is the number of swarm clusters in the network, and L_m^j is the period within which the human operator h_j interacts with the leading agent in swarm cluster s_m.

The state $\mathbf{S}^j(t) = [s_1^j(t), \ldots, s_M^j(t)]$ indicates the remaining collaboration time required after time instant, t, by swarm cluster s_m, and it satisfies the following update equations:

$$\begin{cases} s_m^j(t+1) = s_m^j(t) - 1, & \text{if } q_m^j(t) > 0 \text{ and } ctr_m^j(t) = 1 \qquad (2a) \\ s_m^j(t+1) = s_m^j(t), & \text{if } q_m^j(t) > 0 \text{ and } ctr_m^j(t) = 0 \qquad (2b) \\ s_m^j(t+1) = I_m^j, & \text{if } q_m^j(t) = 0 \qquad (2c) \end{cases}$$

where $ctr_m^j(t)$ represents the scheduling decision. If a human operator h_j is scheduled to collaborate with swarm cluster s_m at time instant t, $ctr_m^j(t) = 1$ and 0 otherwise. I_m^j is the length of time the human operator h_j collaborates with swarm cluster s_m within each period L_m^j.

The state $\mathbf{O}^j(t) = [o_1^j(t), \ldots, o_M^j(k)]$ indicates the dynamic response time and has two situations: (1) When a human h_j still collaborates with the swarm cluster s_m,

$o_m^j(t)$ denotes the length of time from the initial request to the current time instant, t; (2) When the collaboration between the human h_j and cluster s_m is completed, $o_m^j(t)$ denotes the length of time from the initial request to the collaboration completion time, and it satisfies the following update equations:

$$\begin{cases} o_m^j(t+1) = o_m^j(t) + 1, & \text{if } q_m^j(t) > 0 \text{ and } s_m^j(t) > 0 \quad \text{(3a)} \\ o_m^j(t+1) = o_m^j(t), & \text{if } q_m^j(t) > 0 \text{ and } s_m^j(t) = 0 \quad \text{(3b)} \\ o_m^j(t+1) = 0, & \text{if } q_m^j(t) = 0 \quad \text{(3c)} \end{cases}$$

We now define the time delay as δ_m^j for human h_j to collaborate with swarm cluster s_m at time t, which satisfies the following update equation:

$$\delta_m^j(t) = o_m^j(t) - \max\{0, I_m^j - s_m^j(t)\}, \text{ if } s_m^j(t) = s_m^j(t-1) \text{ and } o_m^j(t) > 0, \quad (16.4)$$

where $o_m^j(t)$ is the dynamic response time for the human h_j and swarm s_m pair, and $s_m^j(t)$ is the remaining collaboration time. $I_m^j - s_m^j(t)$ denotes the collaborated time. $s_m^j(t) = s_m^j(t-1)$ means that the human h_j stops collaboration with the swam cluster s_m. When the human stops collaborating with the swarm, the propagation of dynamic response time minus the collaborated time represents the delay time.

16.2.2 Cooperative Control for Swarm Agents

16.2.2.1 Leading Agent Dynamics and Formation Control Law

Consider the following dynamics of the leading agent in swarm cluster s_m [14, 31]:

$$\dot{\mathbf{x}}_m(t) = \begin{cases} \Phi\, \mathbf{x}_m(t) + \Gamma\, \mathbf{u}_m(t), & \text{(manual mode)} \\ \Phi\, \mathbf{x}_m(t) + \Gamma\, \mathbf{u}_m(t - \delta_m(t)), & \text{(autonomous mode)} \end{cases} \quad (16.5)$$

$$\Phi = \begin{bmatrix} 0 & 1 & 0 & 0 \\ 0 & -1 & 0 & 0 \\ 0 & 0 & 0 & 1 \\ 0 & 0 & 0 & -1 \end{bmatrix}, \Gamma = \begin{bmatrix} 0 & 0 \\ 1 & 0 \\ 0 & 0 \\ 0 & 1 \end{bmatrix}$$

$$\mathbf{x}_m(t) = [x_m(t), \dot{x}_m(t), y_m(t), \dot{y}_m(t)]^T, \ \mathbf{u}_m(t) = [u_{mx}(t), u_{my}(t)]^T$$

where $m = 1, 2, \ldots, M$, $[x_m, \dot{x}_m(t), y_m(t), \dot{y}_m(t)]^T$ indicates the position and velocity of the leading agent of swarm cluster s_m in the Cartesian coordinate and $\delta_m = \min_j \delta_m^j$, $j = 1, 2, \ldots, N$ gives the shortest time delay when a swarm s_m is not collaborating with any human.

The term $\mathbf{u}_m$ is the formation control input to the leading agent in swarm cluster s_m, aiming to shape a regular M-polygon for a network of M swarm clusters:

$$\mathbf{u}_m(t) = [u_{mx}(t)\;\; u_{my}(t)]^T, \tag{16.6}$$

$$u_{mx} = -\sum_{n=1,n\neq m}^{M} k_f \frac{(x_m(t)-x_n(t))[(x_m(t)-x_n(t))^2 - Rdx^2_{m,n}]}{\sqrt{[(x_m(t)-x_n(t))^2 - Rdx^2_{m,n}]^2 + [(y_m(t)-y_n(t))^2 - Rdy^2_{m,n}]^2}}$$

$$u_{my} = -\sum_{n=1,n\neq m}^{M} k_f \frac{(y_m(t)-y_n(t))[(y_m(t)-y_n(t))^2 - Rdy^2_{m,n}]}{\sqrt{[(x_m(t)-x_n(t))^2 - Rdy^2_{m,n}]^2 + [(y_m(t)-y_n(t))^2 - Rdy^2_{m,n}]^2}} \tag{16.7}$$

where $Rdx_{m,n}$ and $Rdy_{m,n}$ represent the desired distance between two leaders in the m and n swarm clusters in the x and y direction, respectively, k_f is a positive feedback gain, and $(x_n, y_n)^T$ is for position of the leading agent in neighboring swarm cluster n. Under the autonomous mode, the information from neighboring swarms undergoes time delay and hence the control input to the leader in swarm m becomes:

$$\mathbf{u}_m(t-\delta_m) = [u_{mx}(t-\delta_m)\;\; u_{my}(t-\delta_m)]^T, \tag{16.8}$$

$$u_{mx}(t-\delta_m) = -\sum_{n=1,n\neq m}^{M} k_f \frac{(x_m(t)-x_n(t-\delta_m))[(x_m(t)-x_n(t-\delta_m))^2 - Rdx^2_{m,n}]}{\sqrt{[(x_m(t)-x_n(t-\delta_m))^2 - Rdx^2_{m,n}]^2 + [(y_m(t)-y_n(t-\delta_m))^2 - Rdy^2_{m,n}]^2}}$$

$$u_{my}(t-\delta_m) = -\sum_{n=1,n\neq m}^{M} k_f \frac{(y_m(t)-y_n(t-\delta_m))[(y_m(t)-y_n(t-\delta_m))^2 - Rdy^2_{m,n}]}{\sqrt{[(x_m(t)-x_n(t-\delta_m))^2 - Rdy^2_{m,n}]^2 + [(y_m(t)-y_n(t-\delta_m))^2 - Rdy^2_{m,n}]^2}}. \tag{16.9}$$

Theorem 16.1 *If a leading agent in the swarm cluster m is under exclusively manual mode, the control law* $\mathbf{u}_m(t)$ *given by (16.6)* locally *drives* $\dot{x}_m(t) \to 0$, $\dot{y}_m(t) \to 0$, $x_m(t) - x_n(t) \to Rdx_{m,n}$ *, and* $y_m(t) - y_n(t) \to Rdy_{m,n}$ $(n \neq m)$.

Proof Based on the agent dynamics shown in Eq. (16.5), we first define a Lyapunov function as [14],

$$V(t) = \frac{1}{2}\sum_{m=1}^{M}\Bigg\{\dot{x}_m(t)^2 + \dot{y}_m(t)^2 + \sum_{n=1,n\neq m}^{M} k_f \sqrt{[(x_m(t)-x_n(t))^2 - Rdx^2_{m,n}]^2 + [(y_m(t)-y_n(t))^2 - Rdy^2_{m,n}]^2}\Bigg\},$$

further computing $\dot{V} = \nabla V \cdot \dot{\mathbf{Q}}$ gives

$$\dot{V} = \sum_{m=1}^{M} \begin{bmatrix} \sum_{n=1,n\neq m}^{M} k_f \frac{(x_m-x_n)[(x_m-x_n)^2-Rdx^2_{m,n}]}{\sqrt{[(x_m-x_n)^2-Rdx^2_{m,n}]^2+[(y_m-y_n)^2-Rdy^2_{m,n}]^2}} \\ \dot{x}_m \\ \sum_{n=1,n\neq m}^{M} k_f \frac{(y_m-y_n)[(y_m-y_n)^2-Rdy^2_{m,n}]}{\sqrt{[(x_m-x_n)^2-Rdx_{m,n})^2]^2+[(y_m-y_n)^2-Rdy^2_{m,n}]^2}} \\ \dot{y}_m \end{bmatrix}^T \cdot \begin{bmatrix} \dot{x}_m \\ -\dot{x}_m + u_{mx} \\ \dot{y}_m \\ -\dot{y}_m + u_{my} \end{bmatrix}$$

$$= \sum_{m=1}^{M}\{-(\dot{x}_m^2 + \dot{y}_m^2)\} \leq 0. \tag{16.10}$$

Therefore, we can conclude that the swarm network under the exclusively manual mode is stable holding for any M. □

Remark 16.1 Note that since there does not exist a strict Lyapunov function, based on LaSalle's invariance principle, we can only derive a local stability result, which means the convergency of states in the above dynamic system can only be achieved within a compact set for these states (Theorem 1 in [31]).

Remark 16.2 When a leading agent in the swarm m is under the exclusively autonomous mode, the control law $\mathbf{u}_m(t-\delta_m)$ is given by (16.8). We cannot guarantee the formation stability because the term $\mathbf{u}_m(t-\delta_m)$ in the derivative of Lyapunov function (16.10) cannot be cancelled out and further $\dot{V} \leq 0$ cannot be confirmed.

16.2.2.2 Following Agent Dynamics and Motion Control Law

Inspired by the swarm model in [12], we propose the equation of motion for each individual agent i, $i = 1, 2, \ldots, C_m$, in a swarm cluster m as follows

$$\dot{\mathbf{x}}^i_m(t) = -A^i_m\,(\mathbf{x}^i_m(t) - \mathbf{x}_m(t)) + \Psi\;\mathbf{u}^i_m(t), \tag{16.11}$$

$$\Psi = \begin{bmatrix} 1 & 0 \\ 0 & 0 \\ 0 & 1 \\ 0 & 0 \end{bmatrix},\;\; \mathbf{x}^i_m(t) = [x^i_m(t), \dot{x}^i_m(t), y^i_m(t), \dot{y}^i_m(t)]^T,\;\; \mathbf{u}^i_m(t) = [u^i_{mx}(t), u^i_{my}(t)]^T$$

where A^i_m is a positive coefficient for agent i in the swarm m.

When each agent in a swarm is following the leader, collision avoidance and cohesion among all agents should be considered. Hence, the formula of $\mathbf{u}^i_m$ is given by

$$\mathbf{u}^i_m = \begin{bmatrix} \sum\limits_{j=1, j\neq i}^{C_m} g(x^i_m - x^j_m) + g(x^i_m - x_m) \\ \sum\limits_{j=1, j\neq i}^{C_m} g(y^i_m - y^j_m) + g(y^i_m - y_m) \end{bmatrix}.$$

The function $g(\cdot)$ represents the function of mutual repulsion between the individual agents. The explicit formulae with respect to $x^i_m - x^j_m$ are given by

$$g(x^i_m - x^j_m) = \begin{cases} b\,(x^i_m - x^j_m - d_{mx})\;exp\left(-\frac{(x^i_m - x^j_m - d_{mx})^2}{c}\right), & x^i_m - x^j_m \geq d_{mx} \\ -b\,(x^i_m - x^j_m - d_{mx})\;exp\left(\frac{c'}{x^i_m - x^j_m}\right), & 0 \leq x^i_m - x^j_m < d_{mx} \\ -b\,(x^i_m - x^j_m + d_{mx})\;exp\left(-\frac{c'}{x^i_m - x^j_m}\right), & -d_{mx} \leq x^i_m - x^j_m < 0 \\ b\,(x^i_m - x^j_m + d_{mx})\;exp\left(-\frac{(x^i_m - x^j_m + d_{mx})^2}{c}\right), & x^i_m - x^j_m < -d_{mx} \end{cases} \tag{16.12}$$

$$g(y_m^i - y_m^j) = \begin{cases} b\,(y_m^i - y_m^j - d_{my})\; exp\left(-\frac{(y_m^i - y_m^j - d_{my})^2}{c}\right), & y_m^i - y_m^j \geq d_{my} \\ -b\,(y_m^i - y_m^j - d_{my})\; exp\left(\frac{c'}{y_m^i - y_m^j}\right), & 0 \leq y_m^i - y_m^j < d_{my} \\ -b\,(y_m^i - y_m^j + d_{my})\; exp\left(-\frac{c'}{y_m^i - y_m^j}\right), & -d_{my} \leq y_m^i - y_m^j < 0 \\ b\,(y_m^i - y_m^j + d_{my})\; exp\left(-\frac{(y_m^i - y_m^j + d_{my})^2}{c}\right), & y_m^i - y_m^j \leq -d_{my} \end{cases} \tag{16.13}$$

where b, c, c', d_{mx}, and d_{my} are positive constants.

Lemma 16.1 *There exists a local maximum value for the $g(\cdot)$ function, i.e., $g(z) \leq b^* = b\sqrt{\frac{c}{2}}\, exp(-\frac{1}{2})$ if $z \geq d$ or $z < -d$.*

Proof According to the explicit formulae of the function $g(\cdot)$ shown in Eqs. (16.12) and (16.13), we can summarize the $g(\cdot)$ function in one formula, as

$$g(z) = \begin{cases} b\,(z-d)\; exp\left(-\frac{(z-d)^2}{c}\right), & z \geq d \\ -b\,(z-d)\; exp\left(\frac{c'}{z}\right), & 0 \leq z < d \\ -b\,(z+d)\; exp\left(-\frac{c'}{z}\right), & -d \leq z < 0 \\ b\,(z+d)\; exp\left(-\frac{(z+d)^2}{c}\right), & z < -d \end{cases}$$

where b, c, c', and d are positive constants. Note that $g(z)$ is a continuous function in the piecewise value range.

I. When $z \geq d$, we have $\frac{dg}{dz} = b\; exp\left(-\frac{(z-d)^2}{c}\right)\left[1 - \frac{2(z-d)^2}{c}\right]$.

Let $\frac{dg}{dz} = 0$. We obtain

$$\begin{aligned} b\; exp\left(-\frac{(z-d)^2}{c}\right)\left[1 - \frac{2(z-d)^2}{c}\right] &= 0 \\ \Rightarrow (z-d)^2 &= \frac{c}{2} \\ \Rightarrow z &= \sqrt{\frac{c}{2}} + d \end{aligned}$$

When $(z-d)^2 < \frac{c}{2} \Rightarrow d < z < \sqrt{\frac{c}{2}} + d$, we have $\frac{dg}{dz} > 0$.

When $(z-d)^2 > \frac{c}{2} \Rightarrow z > \sqrt{\frac{c}{2}} + d$, we have $\frac{dg}{dz} < 0$.

When $(z-d)^2 = 0 \Rightarrow z = d$, we have $g(d) = 0$.

Hence, we can conclude that when $z \geq d$, the maximum value of function $g(z)$ is $b\sqrt{\frac{c}{2}}\; exp(-\frac{1}{2})$ at $z = \sqrt{\frac{c}{2}} + d$.

II. When $-d \leq z < d$, we have $g(z)$ is unbounded. This guarantees that agents will not collide with each other.

III. When $z < -d$, we have $\frac{dg}{dz} = b\; exp\left(-\frac{(z+d)^2}{c}\right)\left[1 - \frac{2(z+d)^2}{c}\right]$.

Let $\frac{dg}{dz} = 0$. We obtain

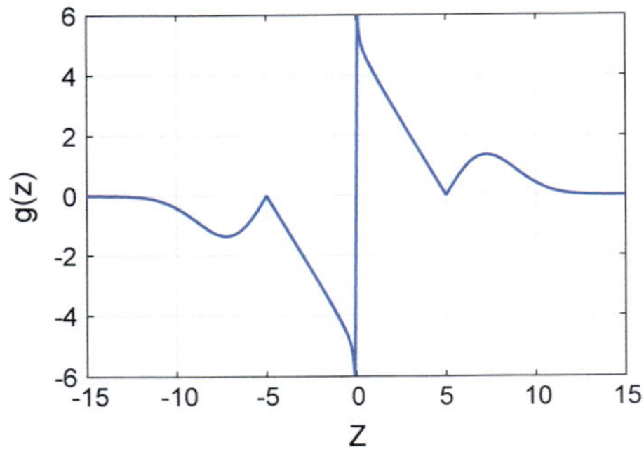

Fig. 16.2 $g(z)$ function

$$\begin{aligned} b\, exp\left(-\tfrac{(z+d)^2}{c}\right)\left[1-\tfrac{2(z+d)^2}{c}\right] &= 0 \\ \Rightarrow (z+d)^2 &= \tfrac{c}{2} \\ \Rightarrow z &= -\sqrt{\tfrac{c}{2}}-d \end{aligned}$$

When $(z+d)^2 < \frac{c}{2} \Rightarrow -\sqrt{\frac{c}{2}} - d < z < -d$, we have $\frac{dg}{dz} > 0$.
When $(z+d)^2 > \frac{c}{2} \Rightarrow z < -\sqrt{\frac{c}{2}} - d$, we have $\frac{dg}{dz} < 0$.
When $(z+d)^2 = 0 \Rightarrow z = -d$, we have $g(-d) = 0$.
Hence, we can conclude that when $z < -d$, the maximum value of function $g(z) = 0$.

Therefore, we can make a conclusion that there exists no global maximum value for $g(z)$ function. However, when $z \geq d$ or $z < -d$, we have $g(z) \leq b\sqrt{\frac{c}{2}}\, exp(-\frac{1}{2})$ for $z \in \mathbb{R}^1$. □

For the $z \in \mathbb{R}^1$ case with $b = 1$, $c = 10$, $c' = 0.02$, and $d = 5$, this function is shown in Fig. 16.2.

We now analyze the cohesiveness and collision avoidance properties for agents in a swarm cluster. We first define the differences between a following agent i and the leading agent in swarm m, as $\mathbf{E}^i = \mathbf{x}_m^i - \mathbf{x}_m$. Note that $\dot{\mathbf{E}}^i = \dot{\mathbf{x}}_m^i - \dot{\mathbf{x}}_m$, which is

$$\dot{\mathbf{E}}^i = -A_m^i \mathbf{E}^i + \Psi \mathbf{u}_m^i - \Phi \mathbf{x}_m - \Gamma \mathbf{u}_m.$$

Theorem 16.2 *Consider the follower agents in a swarm cluster described by the model in Eq. (16.11). Then as $t \to \infty$, we have $\mathbf{q}_m^i \to B_\varepsilon(\mathbf{q}_m)$, where*

$$B_\varepsilon(\mathbf{q}_m) = \{(x_m^i, y_m^i) : \sqrt{(x_m^i - x_m)^2 + (y_m^i - y_m)^2} \leq \varepsilon_m\}$$

$$\varepsilon_m = \sqrt{\left\{\frac{C_m b^* + \zeta_x}{\min\{A_m^i\}}\right\}^2 + \left\{\frac{C_m b^* + \zeta_y}{\min\{A_m^i\}}\right\}^2}$$

In addition, we have

$$-\frac{\zeta_x + k_f(M-1)\varepsilon_x}{A_m^i} \leq (\dot{x}_m^i - \dot{x}_m) \leq \frac{\zeta_x + k_f(M-1)\varepsilon_x}{A_m^i}$$

$$-\frac{\zeta_y + k_f(M-1)\varepsilon_y}{A_m^i} \leq (\dot{y}_m^i - \dot{y}_m) \leq \frac{\zeta_y + k_f(M-1)\varepsilon_y}{A_m^i}$$

Proof Defining the Lyapunov function as $\bar{V}_i = \frac{1}{2}||\mathbf{E}^i||^2 = \frac{1}{2}\mathbf{E}^{iT}\mathbf{E}^i$. $\dot{\bar{V}}_i$ is further given by $\dot{\bar{V}}_i = \dot{\mathbf{E}}_i^T\mathbf{E}_i = -A_m^i(\mathbf{E}^i)^T\mathbf{E}^i + (\Psi\mathbf{u}_m^i)^T\mathbf{E}^i - (\Phi\mathbf{x}_m)^T\mathbf{E}^i - (\Gamma\mathbf{u}_m)^T\mathbf{E}^i$. Expanding $\dot{\bar{V}}_i$, we can obtain

$$\begin{aligned}
\dot{\bar{V}}_i &= -A_m^i||\mathbf{E}^i||^2 + u_{mx}^i(x_m^i - x_m) + u_{my}^i(y_m^i - y_m) \\
&\quad -\dot{x}_m(x_m^i - x_m) + \dot{x}_m(\dot{x}_m^i - \dot{x}_m) - \dot{y}_m(y_m^i - y_m) + \dot{y}_m(\dot{y}_m^i - \dot{y}_m) \\
&\quad -u_{mx}(\dot{x}_m^i - \dot{x}_m) - u_{my}(\dot{y}_m^i - \dot{y}_m) \\
&= -A_m^i||x_m^i - x_m||^2 + \left\{ \sum_{j=1, j\neq i}^{C_m} g(x_m^i - x_m^j) + g(x_m^i - x_m) \right\} (x_m^i - x_m) - \dot{x}_m(x_m^i - x_m) \\
&\quad -A_m^i||\dot{x}_m^i - \dot{x}_m||^2 + \dot{x}_m(\dot{x}_m^i - \dot{x}_m) - u_{mx}(\dot{x}_m^i - \dot{x}_m) \\
&\quad -A_m^i||y_m^i - y_m||^2 + \left\{ \sum_{j=1, j\neq i}^{C_m} g(y_m^i - y_m^j) + g(y_m^i - y_m) \right\} (y_m^i - y_m) - \dot{y}_m(y_m^i - y_m) \\
&\quad -A_m^i||\dot{y}_m^i - \dot{y}_m||^2 + \dot{y}_m(\dot{y}_m^i - \dot{y}_m) - u_{my}(\dot{y}_m^i - \dot{y}_m).
\end{aligned}$$

From the swarm systems in Eq. (16.5) and the Lyapunov function shown in Eq. (16.10), we can see the velocities of the leading agent in two directions will converge to a neighborhood of zero and the distances between two leading agents will locally converge to the desired relative distance. Hence, we set $\dot{x}_m < \zeta_x$ and $\dot{y}_m < \zeta_y$. Furthermore, we have $x_m - x_n < \varepsilon_x$ and $y_m - y_n < \varepsilon_y$ with $\varepsilon_x > 0,\ \varepsilon_y > 0$. $|u_{mx}|$ and $|u_{my}|$ are then less than or equal to $k_f(M-1)\varepsilon_x$ and $k_f(M-1)\varepsilon_y$, respectively, from the formula of the formation control $\mathbf{u}_m$. Combining with $g(z) < b^*$ from **Lemma 1**, we have

$$\begin{aligned}
\dot{\bar{V}}_i \leq\ & -A_m^i||x_m^i - x_m||^2 + C_m b^*\ ||x_m^i - x_m|| + \zeta_x\ ||x_m^i - x_m|| \\
& -A_m^i||\dot{x}_m^i - \dot{x}_m||^2 + \zeta_x\ ||\dot{x}_m^i - \dot{x}_m|| + k_f(M-1)\varepsilon_x\ ||\dot{x}_m^i - \dot{x}_m|| \\
& -A_m^i||y_m^i - y_m||^2 + C_m b^*\ ||y_m^i - y_m|| + \zeta_y\ ||y_m^i - y_m|| \\
& -A_m^i||\dot{y}_m^i - \dot{y}_m||^2 + \zeta_y\ ||\dot{y}_m^i - \dot{y}_m|| + k_f(M-1)\varepsilon_y\ ||\dot{y}_m^i - \dot{y}_m|| \\
\leq\ & -A_m^i||x_m^i - x_m|| \left\{ ||x_m^i - x_m|| - \frac{C_m b^* + \zeta_x}{A_m^i} \right\} \\
& -A_m^i||\dot{x}_m^i - \dot{x}_m|| \left\{ ||\dot{x}_m^i - \dot{x}_m|| - \frac{\zeta_x + k_f(M-1)\varepsilon_x}{A_m^i} \right\} \\
& -A_m^i||y_m^i - y_m|| \left\{ ||y_m^i - y_m|| - \frac{C_m b^* + \zeta_y}{A_m^i} \right\} \\
& -A_m^i||\dot{y}_m^i - \dot{y}_m|| \left\{ ||\dot{y}_m^i - \dot{y}_m|| - \frac{\zeta_y + k_f(M-1)\varepsilon_y}{A_m^i} \right\}.
\end{aligned}$$

Therefore, we can conclude that as long as $||x_m^i - x_m|| \geq \frac{C_m b^* + \zeta_x}{A_m^i}$, $||y_m^i - y_m|| \geq \frac{C_m b^* + \zeta_y}{A_m^i}$, $||\dot{x}_m^i - \dot{x}_m|| \geq \frac{\zeta_x + k_f(M-1)\varepsilon_x}{A_m^i}$, and $||\dot{y}_m^i - \dot{y}_m|| \geq \frac{\zeta_y + k_f(M-1)\varepsilon_y}{A_m^i}$, we have $\dot{\bar{V}}_i \leq 0$.

Hence, we will guarantee that as these differences decrease, eventually the bounds for position and velocity will be achieved. □

Remark 16.3 Notice that no matter whether a leading agent in the swarm m is under the autonomous mode or under the manual mode, both of these two formation control laws are bounded with $k_f(M-1)\varepsilon_x$ and $k_f(M-1)\varepsilon_y$. It indicates that the control modes of the leading agent in a swarm will not affect the cohesiveness of the following agents in this swarm.

16.3 Collaboration Framework

For large-scale human–swarm collaboration systems, the critical problem is to pair human operators with swarm clusters such that human–swarm mutual trust is kept within the desired range for effective HRI and the overall task can be achieved. In this section, we propose a fitness-based measurement to pair human operators and swarm clusters and guarantee successful human–swarm collaborations.

16.3.1 Trust Model

We consider human–swarm mutual trust model, i.e., human-to-swarm trust $T_{h_j \to s_m}(t)$ and swarm-to-human trust $T_{s_m \to h_j}(t)$. Combining the qualitative trust model [34] and time-series trust model proposed in [26], a similar human-to-swarm trust model is proposed as follows:

$$T_{h_j \to s_m}(t) = A_1 T_{h_j \to s_m}(t-1) + B_1 P_{s_m}(t) - B_2 P_{s_m}(t-1) + D_1 F_{s_m}(t) - D_2 F_{s_m}(t-1), \tag{16.14}$$

where $P_{s_m}(t)$ denotes the swarm performance and $F_{s_m}(t)$ denotes the swarm fault rate. A_1, B_1, B_2, D_1, and D_2 are constant coefficients whose values depend on the human operator, the swarm cluster, and the collaborative task. As the above equation shows, the current trust level $T_{h_j \to s_m}(t)$ is determined by the prior trust level $T_{h_j \to s_m}(t-1)$, change of swarm performance, and change of swarm fault rate.

Next, we consider the swarm-to-human trust model $T_{s_m \to h_j}(t)$. Because a swarm cluster is composed of a leading agent and its followers, the swarm-to-human trust should consider every agent's trust in human within the swarm cluster. We propose the following swarm-to-human trust model:

$$T_{s_m \to h_j}(t) = \frac{1}{C_m} \sum_{i=1}^{C_m} T_{a_m^i \to h_j}(t), \tag{16.15}$$

where a_m^i denotes agent i in the swarm cluster s_m, and $T_{a_m^i \rightarrow h_j}(t)$ indicates agent-to-human trust.

The agent-to-human trust model, $T_{a_m^i \rightarrow h_j}(t)$, will depend on the change of performance of the human collaborator, P_{h_j}, and the human fault rate, F_{h_j}. We propose the following agent-to-human trust model:

$$\begin{aligned} T_{a_m^i \rightarrow h_j}(t) = A_2 T_{a_m^i \rightarrow h_j}(t-1) + C_1 P_{h_j}(t) - C_2 P_{h_j}(t-1) \\ + E_1 F_{h_j}(t) - E_2 F_{h_j}(t-1), \end{aligned} \tag{16.16}$$

where A_2, C_i, E_i, $i = 1, 2$ are constant coefficients.

16.3.2 Human Performance Model

The Yerkes–Dodson law [47] describes human performance as an empirical model with respect to human arousal and task difficulty. Here, we refer human performance as the capability and efficiency of the human operator collaborating with a swarm cluster. We represent the following performance model, as [4]

$$P_{h_j}(t) = \left(P_{h_j,\max} - P_{h_j,\min}\right) \left(\frac{r_j(t)}{\beta_j}\right)^{\beta_j} \left(\frac{1-r_j(t)}{1-\beta_j}\right)^{1-\beta_j} + P_{h_j,\min}, \tag{16.17}$$

where $\beta_j \in (0, 1)$ represents the difficulty of a task for a human (a smaller value of β_j represents a more difficult task [29]), $r_j(t)$ represents the human utilization, and $P_{h_j,\max}$ and $P_{h_j,\min}$ represent the maximum and minimum human performance value, respectively. Note that the human performance model (16.17) guarantees that P_{h_j} is bounded between $[P_{h_j,\min}, P_{h_j,\max}]$.

Inspired by the single-server queue model [35, 43], we use the following utilization model for the case when one-human operator collaborates with multiple leading agents

$$\dot{r}_j(t) = -\frac{1}{\tau_{h_j}} r_j(t) + \frac{1}{\tau_{h_j}} \sum_{m=1}^{M} ctr_m^j(t), \tag{16.18}$$

$$ctr_m^j(t) = \begin{cases} 1 & \text{manual mode} \\ 0 & \text{autonomous mode} \end{cases},$$

where $ctr_m^j(t)$ denotes the control mode of a leading agent as introduced in Equation (2) and $\tau_{h_j} > 0$ is a time constant that determines the extent to which past utilization affects the current state. The time constant τ_{h_j} here represents the inverse of the sensitivity of the operator h_j to his/her recent utilization history, which means that a larger τ_{h_j} corresponds to lower sensitivity and vice versa. As shown in Eq. (16.18), the utilization ratio $r_j(t)$ is determined by the control modes of all leading agents in the network. $r_j(t)$ increases in the manual mode ($ctr_m^j(t) = 1$) and decreases in the autonomous mode ($ctr_m^j(t) = 0$). The utilization ratio $r_j(t)$ is bounded between 0 and 1.

16.3.3 Swarm Performance Model

We propose the swarm performance model based on the cohesiveness of the cluster and swarm formation performance. The cohesiveness is measured by the distances between the leader and each following agent in the cluster. The formation performance is measured by the actual relative distance between the current swarm leader and its neighboring leaders compared to the desired distances. The performance model is updated according to the following equation

$$P_{s_m}(t) = (1 - k_m)\left\{1 - \frac{\sum\limits_{i \in C_m} e_m^i(t)}{\sum\limits_{i \in C_m} \varepsilon_m}\right\} + k_m \frac{\sum\limits_{n=1, n \neq m}^{M} ARd_{m,n}(t)}{\sum\limits_{n=1, n \neq m}^{M} Rd_{m,n}} \tag{16.19}$$

where $e_m^i(t) = \sqrt{(x_m^i(t) - x_m(t))^2 + (y_m^i(t) - y_m(t))^2}$ is the distance between the position of an individual following agent i and the leader in the swarm cluster m at current time instant t, ε_m is the bounded size of the swarm cluster m shown in **Theorem** 16.2, $ARd_{m,n}(t) = \sqrt{(x_m(t) - x_n(t))^2 + (y_m(t) - y_n(t))^2}$ denotes the instantaneous relative distance between the swarm cluster m and its neighbor n, and $Rd_{m,n} = \sqrt{Rdx_{m,n}^2 + Rdy_{m,n}^2}$ represents the desired relative distance between m and n. $k_m \in [0, 1]$ is a constant weighting factor to normalize the swarm performance.

16.3.4 Human Attention Preference

In the individual-based replicator-mutator (RM) dynamics, each individual in a network decides how to allocate its resources to different tasks based on its own tasking priorities and the local communications with its neighbors as well [40]. In a similar fashion, we model the multi-human network based on the individual-based RM dynamics.

Consider a multi-human network that forms a weighted graph $G_h = (V_h, E_h)$ with a set of vertices V_h of human h_j, $j = 1, 2, \ldots, N$ and edges connecting h_j with h_k. For the swarm network, consider a swarm cluster relational graph $G_s^j = (V_s^j, E_s^j)$ associated with human h_j with a set of nodes representing cluster s_m, $m = 1, 2, \ldots, M$ (see Fig. 16.1 for an illustration for the human–swarm collaboration system). Let $\mathbf{A}^j = [a_{mn}^j]$, $a_{mm}^j = 1$, denote the *rewards matrix* of G_s^j, where a_{mn}^j describes the prioritization of the cluster s_m over s_n evaluated by human h_j under only local interactions with its neighbors. We set $a^j_{mn} < a^j_{mm} = 1$, which indicates the cluster s_m itself has the largest priority. Let $\mathbf{p}^j(t) = (p^j_1(t), p^j_2(t), \ldots, p^j_M(t))^T \in \mathbb{R}^M$ be a vector of allocated attention of human h_j, $j = 1, 2, \ldots, N$ to the leading agent in swarm cluster s_m, $m = 1, 2, \ldots, M$ at time t. That is, p^j_m represents the percentage of attention that human h_j allocates to the leading agent of cluster s_m. Let the total attention of human h_j be 1, we have $c^T\mathbf{p}^j(t) = 1$, where $c = (1, 1, \ldots, 1)^T \in \mathbb{R}^M$. Here, a local-interaction version of the individual-based evolutionary dynamics with social interactions model developed in [19] is adopted to model the evolution of $\mathbf{p}^j(t)$.

For each human h_j, define the expected rewards gained by preferring to swarm cluster s_m at time t as $f^j_m(t) = \sum_{n=1}^{M} a^j_{mn}(t)p^j_n(t)$, $m = 1, 2, \ldots, M$, $j = 1, 2, \ldots, N$. The expected rewards vector associated with human h_j at time t can be given in matrix form, as $\mathbf{f}^j(t) = \mathbf{A}^j(t)\mathbf{p}^j(t) \in \mathbb{R}^M$. Let $\mathbf{F}^j(t) = diag(\mathbf{f}^j(t))$. Define $\phi^j(t) = \mathbf{f}^j(t)^T\mathbf{p}^j(t)$ as the total rewards gained by human h_j via the attention allocation of all the swarm clusters. Let $\mathbf{W}^j(t) = [w^j_{mn}]$ be the mutation matrix associated with human h_j, which is a row stochastic matrix satisfying $\sum_{n=1}^{M} w^j_{mn}(t) = 1$. The component w^j_{mn} is the likelihood that human h_j reallocates his/her attention from cluster s_m to s_n ($s_m \neq s_n$) at time t.

Now consider the effect of local interactions between human operators within the network. Let us first define two parameters. $f^j_h > 0$ describes the capability differences of human operators. $\mathbf{A}^h = [a^h_{jk}]$, $j, k = 1, 2, \ldots, N$ describes the strength of the link or connectedness of human h_j to his/her neighbor h_k. Let the parameter ϕ_{h_j} be the total capability of human h_j's neighborhood and it is defined as

$$\phi_{h_j} = \sum_{k=1}^{N} f^k_h a^h_{jk},$$

where N is the number of human operators.

Human h_j's individual evolutionary RM equation can then be written in the matrix form, as

$$\dot{\mathbf{p}}^j(t) = [\mathbf{W}^j(t)]^T\mathbf{F}^j(t)\mathbf{p}^j(t) - \phi^j(t)\mathbf{p}^j(t) + \sum_{k=1}^{N} f^k_h a^h_{jk}\mathbf{p}^k(t) - \phi_{h_j}(t)\mathbf{p}^j(t) \quad (16.20)$$

where $\mathbf{p}^k(t)$ represents the neighboring human h_k's attention allocation. According to this model, human h_j's attention allocation vector $\mathbf{p}^j(t)$ updates according to his/her own valuation of different swarm clusters (given by the first two terms) as well as the effects of the neighboring human's allocation $\mathbf{p}^k(t)$ (given by the latter two terms).

16.3.5 Fitness

We now define a "fitness" relation for a human–swarm pair. Let us first consider the following *fitness*(i, j) model representing relationship between a task i and a computing node j in Task Manager [41]:

$$fitness(i, j) = \frac{100000}{1 + |W_i/S_j - E_i|}$$

where W_i is the workload of the ith task in the Task Manager. S_j is the CPU speed of the jth node according to the Resource Information Service. Hence, W_i/S_j represents the actual execution time of the ith task with respect to the jth node. E_i is the expected execution time of the ith task. Users can set E_i of tasks, or the Task Manager will set it automatically [41]. This method uses difference of the actual and expected execution time of the ith task, i.e., $W_i/S_j - E_i$, to estimate how "fit" a node is for a task. The fitness value shows how suitable the task is to the node. The fitness value ranges from 0 to 100000. Larger fitness values indicate greater suitability between a given task and a node.

Inspired by the above relationship, we define a new "fitness" between an individual h_j and a swarm cluster s_m. This fitness value is proposed as follows:

$$fitness(m, j, t) = \frac{1}{1 + |p_m^j(t)P_{s_m}(t) - P_{s_m}^*|} \tag{16.21}$$

where p_m^j represents the allocated attention of human h_j to the leading agent in the swarm m, $P_{s_m}(t)$ represents the performance of the swarm m given by Eq. (16.19), and $P_{s_m}^*$ represents the desired performance of the swarm m. The multiplication $p_m^j(t)P_{s_m}(t)$ presents the actual performance of swarm m collaborating with human h_j. The desired swarm performance $P_{s_m}^*$ denotes the performance when a cluster leader can maintain desired relative distances with its neighbors by collaborating with human h_j and each follower agent in this swarm cluster approaches the leading agent as closely as possible but still keeps safety distances with each other to avoid collisions. This formula analogously uses difference between the actual and desired performance of the swarm cluster s_m under human h_j's collaboration, i.e., $p_m^j(t)P_{s_m}(t) - P_{s_m}^*$, to estimate how "fit" a human is to collaborate with a swarm. Note that $p_m^j(t)$, $P_{s_m}(t)$, and $P_{s_m}^*$ are guaranteed to be within the open interval (0, 1) based on Eqs. (16.20) and (16.19). To normalize the fitness, we scale the numerator from 10000 to 1. Larger fitness values indicate greater suitability between a given swarm and a human. The swarm clusters that have the largest fitness values will be paired with the human collaborator h_j.

16.4 Real-Time Scheduling

Based on the "fitness" measurement, we can simplify a large-scale multi-human and swarms collaboration network into several one-human and swarms collaboration systems. Consider a gap vector $E_{h_j \to s}(t) = [e_{h_j \to s_1}(t), \ldots, e_{h_j \to s_m}(t), \ldots, e_{h_j \to s_M}(t)]$ where any $e_{h_j \to s_m}(t)$ represents the difference between the current human-to-swarm trust level $T_{h_j \to s_m}(t)$ with respect to the swarm s_m and its desired trust lower limit $T^l_{m,j}$, i.e., $e_{h_j \to s_m}(t) = T_{h_j \to s_m}(t) - T^l_{m,j}$. In a similar fashion, we define another gap vector as $E_{s_m \to h}(t) = [e_{s_m \to h_1}(t), \ldots, e_{s_m \to h_j}(t), \ldots, e_{s_m \to h_N}(t)]$ where any $e_{s_m \to h_j}(t)$ represents the difference between the current swarm-to-human trust level $T_{s_m \to h_j}(t)$ and its desired trust upper limit $T^u_{m,j}$, i.e., $e_{s_m \to h_j}(t) = T^u_{m,j} - T_{s_m \to h_j}(t)$. In this scheduling method, the swarm with the smallest gap gets the highest priority from a human operator. That is, when $e_{h_j \to s_m}(t)$ is smallest with lowest human-to-swarm trust, or when $e_{s_m \to h_j}(t)$ is largest with highest swarm-to-human trust, this swarm s_m will be selected to collaborate with a human operator h_j. We call this as the "Minimum-Gap-First" scheduling. The swarm with the minimum gap will be chosen to collaborate with. When the human-to-swarm trust level approaches to the lower limit, this implies that the swarm performance degrades (i.e., the formation deviates from the desired situation). Hence, in order to improve the swarm performance, the human operator should interact with this swarm and provide more accurate neighboring information to the swarm cluster. When the agent-to-human trust level approaches to the upper limit, this indicates that the swarm cluster desires to collaborate with a human operator. Algorithms 16.1 and 16.2 provide the detailed implementation of the proposed trust-based codesign of control and scheduling.

Description of Algorithm 16.1: We first update the state variables $[\mathbf{Q}^j(t), \mathbf{S}^j(t), \mathbf{O}^j(t)]$ based on the period $\{L^j_m\}^M_{m=1}$, collaboration time $\{I^j_m\}^M_{m=1}$, and control modes $\{ctr^j_m(t)\}^M_{m=1}$. Furthermore, we update the following system states: $\{P_{h_j}(t+1)\}^N_{j=1}$, $\{\{fitness(m, j, t), p^j_m(t+1), T_{h_j \to s_m}(t+1), T_{s_m \to h_j}(t+1)\}^M_{m=1}\}^N_{j=1}$ and $\{\mathbf{x}_m(t+1), P_{s_m}(t+1)\}^M_{m=1}$.

Description of Algorithm 16.2: We first calculate the gap vectors for the next time instant, $t+1$, based on the difference between the current trust level and the lower limit in the human-to-swarm case and the difference between the upper limit and the current trust level in the swarm-to-human case. Next, based on the fitness values for each human operator h_j, we select the clusters that are most suitable (highest fitness values) to collaborate with h_j and categorize these clusters into set $FITSET(j, t)$ (Line 6). Each human operator will further be scheduled to collaborate with swarm clusters based on the Minimum-Gap-First scheduling, as Lines 7–16 show.

Algorithm 16.1 Co-design of Control and Scheduling

Data: $\{r_j(t)\}_{j=1}^N$, $\{\{ctr_m^j(t)\}_{m=1}^M\}_{j=1}^N$, $\{\mathbf{x}_m(t), \mathbf{u}_m(t)\}_{m=1}^M$, $\{\{C_m^j, I_m^j\}_{m=1}^M\}_{j=1}^N$, $\{P_{h_j}(t)\}_{j=1}^N$, $\{P_{s_m}(t)\}_{m=1}^M$, $\{\{p_m^j(t)\}_{m=1}^M\}_{j=1}^N$, $\{\{\mathbf{x}_m^i, \mathbf{u}_m^i\}_{i=1}^{C_m}\}_{m=1}^M$, $\{\{e_m^i(t)\}_{i=1}^{C_m}\}_{m=1}^M$, $\{\sum_{n=1}^M ARd_{m,n}(t)\}_{m=1}^M$, $\{\{T_{h_j\to s_m}(t), T_{s_m\to h_j}(t), q_m^j(t), s_m^j(t), o_m^j(t)\}_{m=1}^M\}_{j=1}^N$

for *each human j* **do**

- $P_{h_j}(t+1) \xleftarrow{\text{Eq.(16.17)}} \{r_j(t), \beta_j, P_{h_j,\min}, P_{h_j,\max}\}$;
- **for** *each swarm cluster* s_m **do**
  - $q_m^j(t+1) \xleftarrow{\text{Eq.(1)}} \{q_m^j(t), L_m^j\}$;
  - $s_m^j(t+1) \xleftarrow{\text{Eq.(2)}} \{s_m^j(t), ctr_m^j(t), q_m^j(t), I_m^j\}$;
  - $o_m^j(t+1) \xleftarrow{\text{Eq.(3)}} \{s_m^j(t), q_m^j(t)\}$;
  - $\delta_m^j(t) \xleftarrow{\text{Eq.(4)}} \{o_m^j(t), s_m^j(t)\}$;
  - $r_j(t+1) \xleftarrow{\text{Eq.(16.18)}} \{r_j(t), ctr_m^j(t)\}$;
  - $fitness(m, j, t) \xleftarrow{\text{Eq.(16.21)}} \{P_{s_m}(t), p_m^j(t)\}$;
  - $\sum_{n=1}^M ARd_{m,n}(t+1) \leftarrow \{\mathbf{x}_m(t), \mathbf{x}_n(t)\}$;
  - $\mathbf{x}_m(t+1) \xleftarrow{\text{Eq.(16.5)}} \{\mathbf{x}_m(t), \mathbf{x}_n(t), ctr_m^j(t), \delta_m(t)\}$;
  - $p_m^j(t+1) \xleftarrow{\text{Eq.(16.20)}} \{p_m^j(t), p_m^k(t)\}$;
  - **for** *each agent i in* s_m **do**
    - $e_m^i(t+1) \leftarrow \{\mathbf{x}_m^i(t+1)\} \xleftarrow{\text{Eq.(16.11)}} \{\mathbf{x}_m^i(t), \mathbf{u}_m^i(t)\}$;
    - $P_{s_m}(t+1) \xleftarrow{\text{Eq.(16.19)}} \{e_m^i(t+1), \sum_{n=1}^M ARd_{m,n}(t+1)\}$;
    - $T_{a_m^i\to h_j}(t+1) \xleftarrow{\text{Eq.(16.16)}} \{T_{a_m^i\to h_j}(t), P_{h_j}(t+1), P_{h_j}(t), F_{h_j}(t+1), F_{h_j}(t)\}$;
  - $T_{h_j\to s_m}(t+1) \xleftarrow{\text{Eq.(16.14)}} \{T_{h_j\to s_m}(t), P_{s_m}(t+1), P_{s_m}(t), F_{s_m}(t+1), F_{s_m}(t)\}$;
  - $T_{s_m\to h_j}(t+1) \xleftarrow{\text{Eq.(16.15)}} \sum T_{a_m^i\to h_j}(t+1)$;

Algorithm 16.2

return $\{r_j(t+1)\}_{j=1}^N$, $\{\{ctr_m^j(t+1)\}_{m=1}^M\}_{j=1}^N$, $\{\{fitness(m, j, t)\}_{m=1}^M\}_{j=1}^N$, $\{P_{h_j}(t+1)\}_{j=1}^N$, $\{P_{s_m}(t+1)\}_{m=1}^M$, $\{\{T_{h_j\to s_m}(t+1), T_{s_m\to h_j}(t+1)\}_{m=1}^M\}_{j=1}^N$, $\{\{q_m^j(t+1), s_m^j(t+1), o_m^j(t+1)\}_{m=1}^M\}_{j=1}^N$

16.5 Simulation Results

16.5.1 Parameter Setup

We simulate the scenario such that two human operators collaborate with four swarm clusters led by four heterogeneous agents. Each leading agent is followed by 10 agents. The states of each leading agent can be updated by Eq. (16.5). The choice

Algorithm 16.2 Minimum-Gap-First Scheduling

Data: $\{\{ctr_m^j(t)\}_{m=1}^M\}_{j=1}^N, \{P_{s_m}(t)\}_{m=1}^M, \{\{fitness(m, j, t)\}_{m=1}^M\}_{j=1}^N,$
$\{\{T_{h_j \to s_m}(t), T_{s_m \to h_j}(t)\}_{m=1}^M\}_{j=1}^N, \{s_m(t+1)\}_{m=1}^M$

for *each* h_j **do**
  /* G_j is a set of swarm clusters with the nonzero remaining collaboration time*/
  $G_j = [\,]$;
  for *each* s_m **do**
    $e_{h_j \to s_m}(t+1) = T_{h_j \to s_m}(t) - T_{m,j}^l$;
    $e_{s_m \to h_j}(t+1) = T_{m,j}^u - T_{s_m \to h_j}(t)$;
  $FITSET(j,t) \leftarrow \{\text{argmax}_m fitness(m, j, t)\}$;
  if $s_m^j(t+1) > 0$ **then**
    $G_j = [G_j, FITSET(j,t)]$;

for *each* h_j **do**
  if G_j *is not empty* **then**
    $m_j^* = \min_{m_j \in G_j}([e_{h_j \to s_m}(t+1), e_{s_m \to h_j}(t+1)])$;
  for *each* s_m **do**
    if $m == m_j^*$ **then**
      $ctr_m^j(t+1) = 1$;
    else if $m \neq m_j^*$ **then**
      $ctr_m^j(t+1) = 0$;

return $\{\{ctr_m^j(t+1)\}_{j=1}^N\}_{m=1}^M$;

Table 16.1 Initial conditions for the leading agent in each swarm cluster

	$x(0)$ (m)	$y(0)$ (m)	$\dot{x}(0)$ (m/s)	$\dot{y}(0)$ (m/s)
s_1	100	105	5	8
s_2	101	106	6	7
s_3	108	103	7	6
s_4	102	107	8	5

of initial states and desired distances between each leading agent and its neighbors are listed in Tables 16.1 and 16.2. The following agent states can be updated by Eq. (16.11). The choice of parameters with respect to control inputs The initial conditions for each following agent in a swarm cluster are set randomly. Each swarm has its initial performance as $[P_{s_1}(0), P_{s_2}(0), P_{s_3}(0), P_{s_4}(0)] = [0.23, 0.27, 0.25, 0.29]$.

The human-to-swarm trust level follows the dynamic model discussed in Eq. (16.14). The constant coefficients in this equation are chosen as $A_1 = 1$, $B_1 = 0.5$, $B_2 = 0.5$, $D_1 = 0.005$, $D_2 = 0.005$, and the fault rates follow the standard normal distribution $N(0, 1)$. The initial trust values between the human operator and four swarms are assumed to be $T_{h_j \to s_m}(0) = 1.6$ and $T_{s_m \to h_j}(0) = 1.6$. Note that the swarm-to-

Table 16.2 Desired relative distances (x and y coordinates) between each swarm cluster

Rd_x/Rd_y	s_1 [m]	s_2 [m]	s_3 [m]	s_4 [m]
s_1	0/0	100/0	100/50	0/50
s_2	100/0	0/0	0/50	100/50
s_3	100/50	0/50	0/0	100/0
s_4	0/50	100/50	100/0	0/0

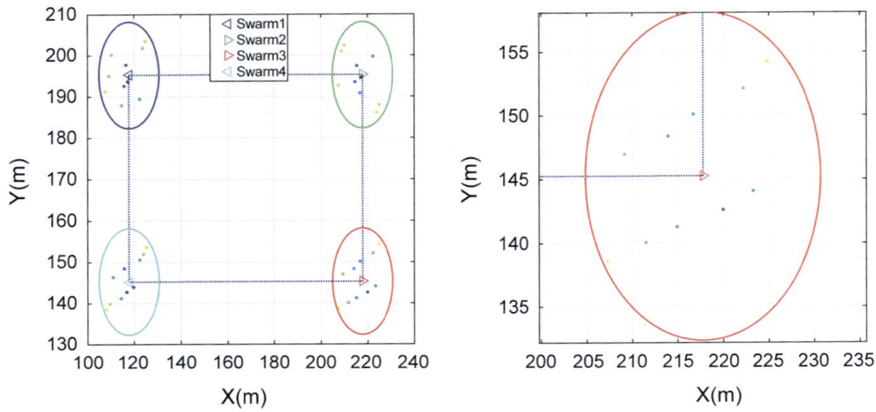

(a) Global graph (b) Local graph for Swarm 1

Fig. 16.3 Formation under the exclusively manual mode

human trust levels are combination of all agent-to-human trust levels. The constant coefficients for agent-to-human trust are chosen as $A_2 = 1$, $C_1 = -0.5$, $C_2 = 0.5$, $E_1 = 0.005$, $E_2 = 0.005$, and the fault rates follow the standard normal distribution $N(0, 1)$. The initial mutual trust value for each agent-to-human in each swarm is assumed to be $T_{a_m^i \to h_j} = 1.6$. In this simulation, we choose the trust regions with the lower bounds $T_{m,j}^l = 1.2$ and the upper bounds $T_{m,j}^u = 3.2$.

16.5.2 Results and Discussions

Figure 16.3 shows the formation shape in the coordinate plane of the human–swarm collaboration system under exclusively manual mode at time t = 100. Because each leading agent under manual mode can obtain complete neighbor information, from Fig. 16.3a, we can observe that the desired formation is achieved based on the defined relative distances. Figure 16.3b depicts the layout of Swarm 1 and shows that each following agent can avoid collision from each other while staying within the bound.

Figure 16.4 shows the formation shape in the coordinate plane of the human–swarm collaboration system under the exclusively autonomous mode at time instant

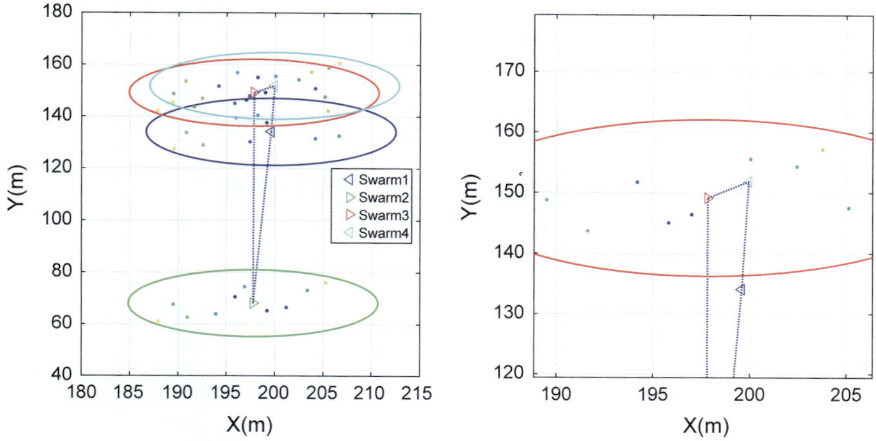

(a) Global graph (b) Local graph for Swarm 1

Fig. 16.4 Formation under the exclusively autonomous mode

$t = 100$. Each leading agent under the autonomous mode can only obtain neighboring information at the initial time. Hence, we can observe from Fig. 16.4a that the formation shape deviates from the desired shape. Figure 16.4b depicts the layout of Swarm 1 as well.

In addition, we observe from Figs. 16.3 and 16.4 that the circle areas defined in Lemma 16.1 can cover all agents in each swarm indicating that the cohesiveness properties are satisfied. As discussed, the cohesiveness of each swarm cluster will not be affected by the mode of the leading agents. Note that we use same parameters in these two modes to obtain the same bounded area for each swarm. Here, we have $\varepsilon_x = \varepsilon_y = 5$. The radius of the area is $\varepsilon_m = \sqrt{\left\{\frac{10b^*+\varepsilon_x}{\min\{A_m^i\}}\right\}^2 + \left\{\frac{10b^*+\varepsilon_y}{\min\{A_m^i\}}\right\}^2}$ and $\min\{A_m^i\}$ is assigned as 9.7 in the simulation.

Figure 16.5 shows the formation shape in the coordinate plane of the human–swarm collaboration system scheduled by the Minimum-Gap-First algorithm under mixed manual and autonomous mode. Furthermore, by comparing the final achieved formation shapes under these three cases, i.e., exclusively manual, exclusively autonomous, and mixed manual and autonomous, from Figs. 16.3, 16.4, and 16.5, we can clearly see that the formation shape scheduled by the Minimum-Gap-First algorithm is close to the desired formation shape. Figure 16.6 shows the dynamic evolution of states of the leading agents in each swarm cluster. Figure 16.7 shows the allocation of attention from two human operators.

Finally, we discuss the tradeoffs between human workload and the acceptable formation shape. According to the Minimum-Gap-First scheduling algorithm, the control mode depends on the trust gaps between trust levels and trust bounds, which in turn decides the formation shape. That is to say, we need to balance these two trust levels in order to have an acceptable shape. Since the workload can be reflected

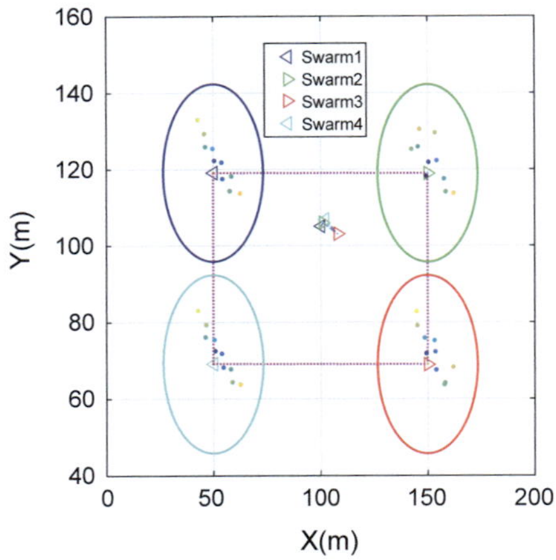

(a) Global graph

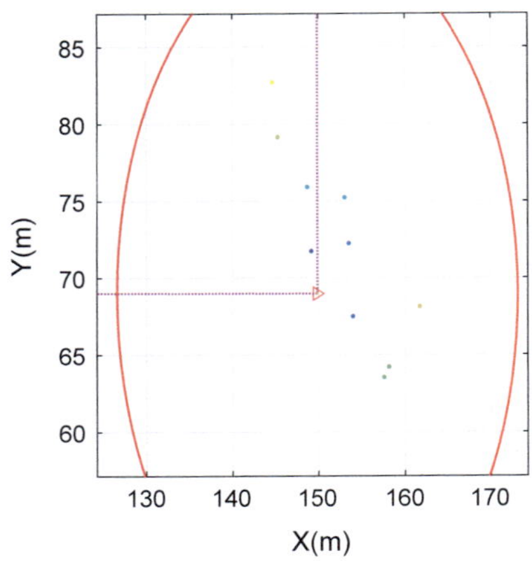

(b) Local graph for Swarm 1

Fig. 16.5 Formation scheduled by the minimum-gap-first algorithm

by the time delay, we directly analyze the statistics of the time delay. Because the ranges of these two unilateral trust levels are known, the upper and lower trust bounds should be set to make these two trust gaps measurable. Tables 16.3 and 16.4 show the formation shapes under different trust bounds. Note that we run the simulation under each trust bound 20 times and take the average. From these two tables, we can observe that the deviation from the desired shape is small if the standard deviation of time delay, i.e., Std(δ_m), is relatively small. The standard deviation of delay time reflects the distribution of human workload to each swarm. The smaller the standard deviation of time delay is, the more equal the distribution of human workload to each swarm is. This indicates if each swarm can obtain relatively more information from the human operators compared with the case under the autonomous mode, the formation constructed by the swarm network is better. Besides, we can observe as the bound increases, the standard deviation of time delay has the trend of increasing. Combining with the range of trust levels shown Figs. 16.6 and 16.7, we can conclude when the human operator makes a decision, if he or she cares both swarm-to-human trust levels and human-to-swarm trust levels, that is, the human operator cares both his/her own opinions and swarm's willingness, the decision for control mode is more reasonable to get a better formation shape. However, even if the distribution of human workload is not equal, the formation shape only deviates slightly from the desired formation shape and is still acceptable.

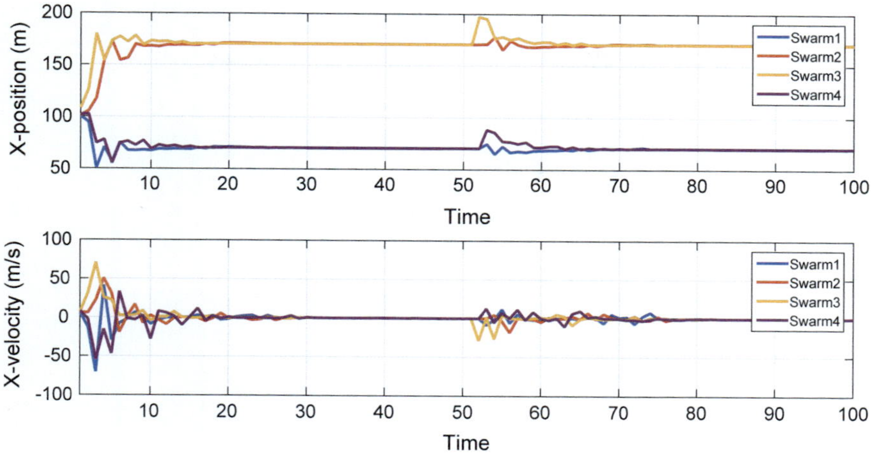

(a) X-direction

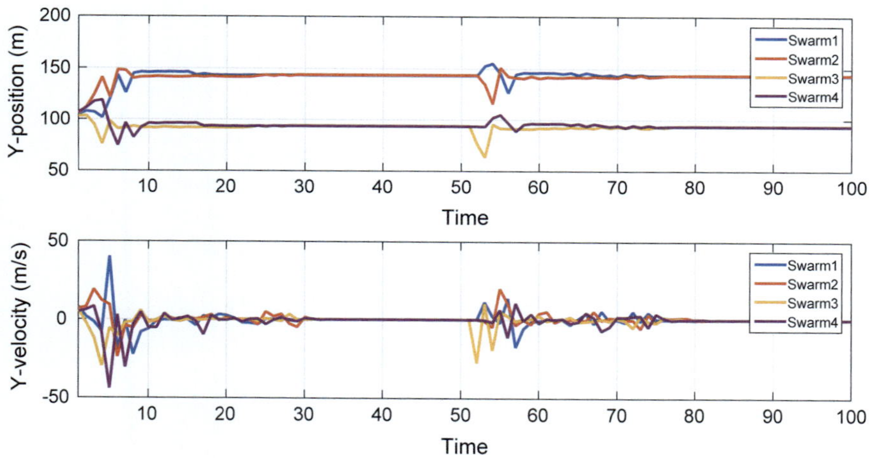

(b) Y-direction

Fig. 16.6 Leading agents dynamics

Table 16.3 Relation between upper trust bounds and formation shape

Upper bound	δ_1	δ_2	δ_3	δ_4	Mean(δ)	Std(δ)	Scheduled shape	Desired shape	Worst shape
6.2	55	61	80	4	50	32.5	303.5	300	N/A
5.2	56	81	61	2	50	33.5	304.5	300	N/A
4.2	52	10	97	41	50	36	308.5	300	N/A
3.2	71	28	52	49	50	17.6	300	300	N/A

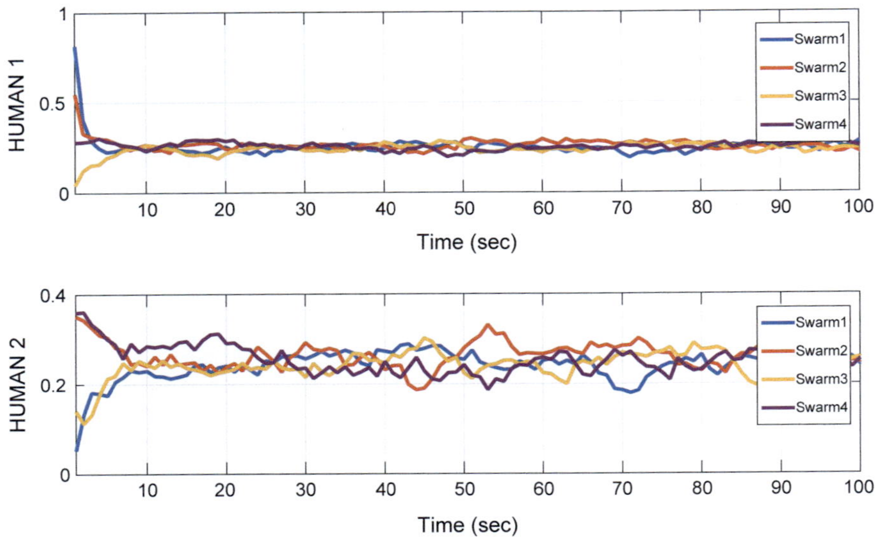

Fig. 16.7 Evolution of human attention allocation

Table 16.4 Relation between lower trust bounds and formation shape

Lower bound	δ_1	δ_2	δ_3	δ_4	Mean(δ)	Std(δ)	Scheduled shape	Desired shape	Worst shape
2.2	50	66	80	4	50	33.1	303.5	300	N/A
1.2	81	18	49	52	50	25.7	304.5	300	N/A
0.2	52	20	87	41	50	28.1	308.5	300	N/A

16.6 Conclusions

In this paper, we investigate the collaboration between multi-human network and swarm-based agents network. Swarm systems are set up based on the Gazi's model and biological phenomenon. Cooperative controls are designed for each leading agent in a swarm cluster so that the desired formation shape constituted by several clusters can be met. Cohesion and collision avoidance controls are designed for each following agent to maintain bounded swarm and avoid collisions. To evaluate the collaboration between the human–swarm networks, we set up bilateral trust models for humans and swarms. We also propose a novel measurement, called "fitness" to pair human-robot collaboration in such a large-scaled network. Besides, we propose a dynamic scheduling algorithm, called "Minimum-Gap-First" to guarantee successful allocation of human resources to different swarms and further codesign the scheduling and control algorithms. Our simulation results show that the proposed algorithms can guarantee effective real-time scheduling of the collaboration systems while ensuring an acceptable formation shape of the swarms.

Acknowledgements This work was supported in part by the National Science Foundation under the grant no. CMMI-1454139.

References

1. Barnes L, Alvis W, Fields M, Valavanis K, Moreno E (2006) Swarm formation control with potential fields formed by bivariate normal functions. In: 14th mediterranean conference on control and automation, 2006. MED'06. IEEE, pp 1–7
2. Beard RW, McLain TW, Nelson DB, Kingston D, Johanson D (2006) Decentralized cooperative aerial surveillance using fixed-wing miniature UAVs. Proc IEEE 94(7):1306–1324
3. Beni G (2005) From swarm intelligence to swarm robotics. Swarm robotics. Springer, Berlin, pp 1–9
4. Bertrand JWM, Van Ooijen HPG (2002) Workload based order release and productivity: a missing link. Prod Plan Control 13(7):665–678
5. Bonabeau E, Theraulaz G, Deneubourg J-L (1996) Quantitative study of the fixed threshold model for the regulation of division of labour in insect societies. Proc R Soc Lond. Ser B: Biol Sci 263(1376):1565–1569
6. Breder CM (1954) Equations descriptive of fish schools and other animal aggregations. Ecology 35:361–370
7. Champagnat N, Ferrière R, Méléard S (2006) Unifying evolutionary dynamics: from individual stochastic processes to macroscopic models. Theor Popul Biol 69(3):297–321
8. Coppin G, Legras F (2012) Autonomy spectrum and performance perception issues in swarm supervisory control. Proc IEEE 100(3):590–603
9. Cummings ML (2004) Human supervisory control of swarming networks. In: 2nd annual swarming: autonomous intelligent networked systems conference, pp 1–9
10. Dimeas AL, Hatziargyriou ND (2005) Operation of a multiagent system for microgrid control. IEEE Trans Power Syst, 20(3):1447–1455
11. Freedy A, DeVisser E, Weltman G, Coeyman N (2007) Measurement of trust in human-robot collaboration. In: International symposium on collaborative technologies and systems, 2007. CTS 2007. IEEE, pp 106–114
12. Gazi V, Passino KM (2004) Stability analysis of social foraging swarms. IEEE Trans Syst Man Cybern Part B: Cybern 34(1):539–557
13. Gazi V, Fidan B, Hanay YS, Ilter Köksal M (2007) Aggregation, foraging, and formation control of swarms with non-holonomic agents using potential functions and sliding mode techniques. Turk J Electr Eng 15(2):149–168
14. Goodwine B, Antsaklis P (2010) Multiagent coordination exploiting system symmetries. In: American Control Conference (ACC), 2010. IEEE, pp 830–835
15. Gu DL, Pei G, Ly H, Gerla M, Zhang B, Hong X (2000) UAV aided intelligent routing for ad-hoc wireless network in single-area theater. In: 2000 IEEE wireless communications and networking conference, 2000. WCNC, vol 3. IEEE, pp 1220–1225
16. Hancock PA, Billings DR, Schaefer KE, Chen JYC, De Visser EJ, Parasuraman R (2011) A meta-analysis of factors affecting trust in human-robot interaction. Hum Factors: J Hum Factors Ergon Soc 53(5):517–527
17. Hofbauer J, Sigmund K (2003) Evolutionary game dynamics. Bull Am Math Soc 40(4):479–519
18. Hong Y, Chen G, Bushnell L (2008) Distributed observers design for leader-following control of multi-agent networks. Automatica 44(3):846–850
19. Hussein I (2009) An individual-based evolutionary dynamics model for networked social behaviors. In: American control conference, 2009. ACC'09. IEEE, pp 5789–5796
20. Itoh M, Tanaka K (2000) Mathematical modeling of trust in automation: trust, distrust, and mistrust. In: Proceedings of the human factors and ergonomics society annual meeting, vol 44. SAGE Publications, pp 9–12

21. Keller EF, Segel LA (1971) Model for chemotaxis. J Theor Biol 30(2):225–234
22. Kim DH, Wang H, Shin S (2006) Decentralized control of autonomous swarm systems using artificial potential functions: analytical design guidelines. J Intell Robot Syst 45(4):369–394
23. Kolling A, Nunnally S, Lewis M (2012) Towards human control of robot swarms. In: Proceedings of the seventh annual ACM/IEEE international conference on human-robot interaction. ACM, pp 89–96
24. Komarova NL (2004) Replicator-mutator equation, universality property and population dynamics of learning. J Theor Biol 230(2):227–239
25. Kumar V, Sahin F (2003) Cognitive maps in swarm robots for the mine detection application. In: IEEE international conference on systems, man and cybernetics, 2003, vol 4. IEEE, pp 3364–3369
26. Lee J, Moray N (1992) Trust, control strategies and allocation of function in human-machine systems. Ergonomics 35(10):1243–1270
27. Lieberman F, Hauert C, Nowak MA (2005) Evolutionary dynamics on graphs. Nature 433(7023):312–316
28. Liu Y, Passino KM (2000) Swarm intelligence: literature overview. Department of electrical engineering, the Ohio State University
29. Mendl M (1999) Performing under pressure: stress and cognitive function. Appl Anim Behav Sci 65(3):221–244
30. Okubo A (1986) Dynamical aspects of animal grouping: swarms, schools, flocks, and herds. Adv Biophys 22:1–94
31. Olfati-Saber R, Murray RM (2002) Distributed cooperative control of multiple vehicle formations using structural potential functions. In: IFAC world congress. Citeseer, pp 346–352
32. Romanczuk P, Schimansky-Geier L (2012) Swarming and pattern formation due to selective attraction and repulsion. Interface Focus 2(6):746–756
33. Şahin E (2005) Swarm robotics: from sources of inspiration to domains of application. Swarm robotics. Springer, Berlin, pp 10–20
34. Sanders T, Oleson KE, Billings DR, Chen JYC, Hancock PA (2011) A model of human-robot trust theoretical model development. In: Proceedings of the human factors and ergonomics society annual meeting, vol 55. SAGE Publications, pp 1432–1436
35. Savla K, Frazzoli E (2012) A dynamical queue approach to intelligent task management for human operators. Proc IEEE 100(3):672–686
36. Sayama H (2007) Decentralized control and interactive design methods for large-scale heterogeneous self-organizing swarms. Advances in artificial life. Springer, Berlin, pp 675–684
37. Shi Z (2014) Non-worst-case response time analysis for real-time systems design. Ph.D. thesis, Georgia Institute of Technology, 2014
38. Shi Z, Zhang F (2013) Predicting time-delays under real-time scheduling for linear model predictive control. In: 2013 international conference on computing, networking and communications (ICNC). IEEE, pp 205–209
39. Sumpter DJT (2010) Collective animal behavior. Princeton University Press, Princeton
40. Wang Q (2014) Modeling and design strategy of online advertising ecosystem. Master's thesis, Clemson University, 2014
41. Wang S-D, Hsu I-T, Huang Z-Y (2005) Dynamic scheduling methods for computational grid environments. In: 11th international conference on parallel and distributed systems, 2005. Proceedings, vol 1. IEEE, pp 22–28
42. Wang Y, Hussein II, Hera A (2010) Evolutionary task assignment in distributed multi-agent networks with local interactions. In: 2010 IEEE/RSJ international conference on intelligent robots and systems (IROS). IEEE, pp 4749–4754
43. Wang Y, Shi Z, Wang C, Zhang F (2014) Human-robot mutual trust in (semi) autonomous underwater robots. In: Cooperative robots and sensor networks 2014. Springer, Berlin, pp 115–137
44. Wang X, Shi Z, Zhang F, Wang Y (2015) Dynamic real-time scheduling for human-agent collaboration systems based on mutual trust. Cyber-Phys Syst (ahead-of-print):1–15

45. Warburton K, Lazarus J (1991) Tendency-distance models of social cohesion in animal groups. J Theor Biol 150(4):473–488
46. Wenwu Y, Chen G, Cao M, Lü J, Zhang H-T (2013) Swarming behaviors in multi-agent systems with nonlinear dynamics. Chaos: an interdisciplinary. J Nonlinear Sci 23(4):043118
47. Yerkes RM, Dodson JD (1908) The relation of strength of stimulus to rapidity of habit-formation. J Comp Neurol Psychol 18(5):459–482

Index

Y. Wang and F. Zhang (eds.), *Trends in Control and Decision-Making for Human–Robot Collaboration Systems*, DOI 10.1007/978-3-319-40533-9

MIX
Papier aus verantwortungsvollen Quellen
Paper from responsible sources
FSC® C105338
FSC
www.fsc.org

If you have any concerns about our products,
you can contact us on
ProductSafety@springernature.com

In case Publisher is established outside the EU,
the EU authorized representative is:
Springer Nature Customer Service Center GmbH
Europaplatz 3, 69115 Heidelberg, Germany

Printed by Libri Plureos GmbH
in Hamburg, Germany